Systems Science and Nonlinear Intelligence Dynamics

Volume 2

Systems Science and Nonlinear Intelligence Dynamics provides a fenestra concerned with the understanding of diverse systems, from life at its simplest level to other corresponding reciprocal interactions getting complex gradually across incredible spectra with a profound view of a system-of-systems interdisciplinary theory, applications, techniques, methods and tools directed towards the distinctive applicable solutions, mathematical systems and modern intelligent methods to highly complex problems. This sort of a diverse spectrum ranges from mathematics to medicine, engineering, technology and science, spanning across formal, natural, social, mathematical, applied and computational sciences with the aim of developing transformative interdisciplinary foundations and merging thereof. The related applications include biological systems, complex systems, systems science, computer science, data science, computational systems, chaotic systems, chaos, systems design, dynamic-dynamical systems, fractional systems, system dynamics, physical laws, intelligence systems, Artificial Intelligence (AI) systems, control systems, robotic systems, systems ecology, medical systems, complexity theory, nonlinear systems, systems theory and real-world related systems, among many others. Furthermore, rigorous mathematical procedures and applications, as the impetus behind discovery and innovation, involve an extensive array of disciplines to implement the multidisciplinary applications with an emphasis of the universal character of mathematical science and modeling. The key aspects commonly foregrounded in systems sciences need to encompass a holistic and integrative view, considering the importance of interaction between a system and its embedded environment as well as complex or subtle trajectories of dynamic behavior.

Nonlinear intelligence dynamics, as a research field, serves to understand and realize the multiple layers of intelligence with its dynamics viewed as a thought-provoking subject matter owing to the requirement of system-of-systems considerations in order to peruse research and practice at different levels with varying goals to fulfill the compelling needs of both academic, practical and industrial areas. Along with these points, the capabilities of predictability, estimation, interpretation, relying on the scope from mathematical sciences, computer sciences to applied sciences as well as engineering mathematics with nonlinear intelligence dynamics lie at the pedestal and intersection with different fields characterized by complex, chaotic, nonlinear, dynamic and transient components to reveal the significance of optimized approaches in real systems and related realms. Significant developments based on these critical points and perspectives, with sound mathematical modeling grounded in the integration of systems in science and nonlinear intelligence dynamics, can pave the way for the future research towards the formulation, design, deployment, analysis and multi-criteria optimization, and extension of the predictability horizons of the applicable solutions to problems arising as formidable challenges in the ever-evolving and rapidly-changing global landscape. Across this line of thought, the integrated solutions with the dynamic approach can support critical multi-stage decision-making processes through the lenses of realistic, actionable and timely information oriented towards the solution of problems in a customized, tailored and to-the-point manner in systems of sciences.

Accordingly, our book series, from an interdisciplinary and multidisciplinary approach, fundamentally deals with the solution and answers of complex problems characterized by nonlinearity, dynamic aspects, ambiguity, uncertainty and transiency, to name some. Each system and its interactional units entail optimal or near-optimal solutions peculiar to their own parameters, so the book series will focus particularly on the most up-to-date developments, findings, advances in theories, novel models, diverse applications along with related analyses and simulations. By catering the needs of different aspects emblematic of systems in science characteristics with regards to each domain, the books within the series are expected to provide methodology, tools, means, applications, techniques and theories relying on theoretical and applied dimensions so that current pressing challenges, which at times may exceed the possible limits of human comprehension, can be inquired and found through these aforementioned means. This integrative sophisticated approach is prominent in dynamic systems, facilitating to achieve enabling solutions, optimization processes, numerical simulations besides technical analyses and related applications. With the presentation of these elements, it will also be possible for individuals including researchers, academics, clinicians and practitioners to be well equipped so that they can come up with applicable and timely conceptualization, design, generation and implementation of the robust and optimized actual solutions, along with the functions through enhanced reliability, precision and realism with respect to solid mathematical, intelligent and computational foundations, towards complex problems under uncertain, transient and dynamic complex elements of science with intricate dissipative structures.

The overarching framework of our book series concerning systems in science and nonlinear intelligent dynamics includes exact sciences, namely mathematics, physics, biology and chemistry, besides medicine, neurology, neuroscience, epidemiology, engineering, biomedical engineering, bioinformatics, life sciences, mathematical biology, computational modeling, clinical/medical research applications, applied sciences, mathematical science, applied mathematics, advanced AI applications, image / signal processing, data science, computer science, computational science, economics, finance and social sciences, and so forth. By elaborating on different, yet, potentially interrelated disciplines, the essential phenomena underlying the entire behaviors of systems in science and nonlinear intelligent dynamics will be covered. We intend to provide a theoretical-based and practical way towards original inter-, intra as well as multidisciplinary research relying upon advanced mathematical modeling and computational foundations. By addressing the theoretical, computational and applied scopes, converging mathematical analyses, methods, demonstrations, simulations, experimental designs, analyses, observations, interpretations, case studies, computer-assisted translations, computing technologies, and so forth, it is intended to validate the significance of up-to-date developments, approaches, findings, applications and schemes in real systems as well as in the related domains.

For information on all publications under the SSNID book series,
visit our website at https://www.worldscientific.com/series/ssnid

Published by

World Scientific Publishing Co. Pte. Ltd.
5 Toh Tuck Link, Singapore 596224
USA office: 27 Warren Street, Suite 401-402, Hackensack, NJ 07601
UK office: 57 Shelton Street, Covent Garden, London WC2H 9HE

Library of Congress Cataloging-in-Publication Data
Names: Baleanu, Dumitru editor | Özdemir, Necati editor | Karaca, Yeliz editor |
Janardhanan, Mukund N. editor | Evirgen, Fırat editor | Küçükkoç, İbrahim editor
Title: Nonlinear Dynamical Control, Computer Simulation and Optimization Systems:
Theory and Applications / [edited by] Dumitru Baleanu, Necati Özdemir, Yeliz Karaca,
Mukund N. Janardhanan, Fırat Evirgen, İbrahim Küçükkoç.
Description: New Jersey : World Scientific, [2026] | Series: Systems Science and
Nonlinear Intelligence Dynamics, 2972-4988 ; vol. 2 |
Includes bibliographical references and index.
Identifiers: LCCN 2025026614 | ISBN 9789819815425 hardcover |
ISBN 9789819815432 ebook for institutions | ISBN 9789819815449 ebook for individuals
Subjects: LCSH: Nonlinear control theory | Engineering mathematics
Classification: LCC QA402.35 .N675 2026
LC record available at https://lccn.loc.gov/2025026614

British Library Cataloguing-in-Publication Data
A catalogue record for this book is available from the British Library.

Cover Designer: Yeliz Karaca

For any available supplementary material, please visit
https://www.worldscientific.com/worldscibooks/10.1142/14373#t=suppl

Desk Editors: Eshak Nabi Akbar Ali/Joy Quek

Typeset by Stallion Press
Email: enquiries@stallionpress.com

Systems Science and Nonlinear Intelligence Dynamics

Series Editors: Yeliz Karaca, Dumitru Baleanu, Yu-Dong Zhang, Osvaldo Gervasi

Nonlinear Dynamical Control, Computer Simulation and Optimization Systems: Theory and Applications

Editors

Dumitru Baleanu, Lebanese American University, Lebanon

Necati Özdemir, Balıkesir University, Turkey

Yeliz Karaca, University of Massachusetts Chan Medical School, USA

Mukund N. Janardhanan, University of Warwick, UK

Fırat Evirgen, Balıkesir University, Turkey

İbrahim Küçükkoç, Balıkesir University, Turkey

World Scientific

NEW JERSEY · LONDON · SINGAPORE · BEIJING · SHANGHAI · HONG KONG · TAIPEI · CHENNAI · TOKYO

Preface

The current technological landscape that has been evolving at a rapid pace has made it mandatory to keep abreast with the latest advances. The recent advances have entailed the integration of science, technology, engineering and mathematics. Thus, the interplay between applied mathematics and engineering has become more foregrounded, gaining further critical position for addressing complex, real-world challenges. This book, comprising 36 chapters, is the culmination of insights and research presented at the Third International Conference on Applied Mathematics in Engineering (ICAME'24). In the light of these contributions, the book explores a wide array of topics, including optimization, control, artificial intelligence (AI), machine learning and operations research, which demonstrate the essential role of mathematics in solving complex engineering problems through sophisticated means.

This volume can be regarded as distinctive owing to its placing dual emphasis on theoretical advancements and their practical implementations. By blending rigorous methodologies with real-world applications, the book aims to offer a comprehensive view of how mathematical tools can address the pressing demands of fields such as robotics, energy systems, finance, mechanics and supply chain management by considering the relevant processes therein as well as directions for future research and applications. Furthermore, detailed case studies and practical examples illustrate the direct impact of these approaches, bridging the gap between abstract theories and the needs of industry.

The book also highlights the importance of interdisciplinary collaboration, showcasing how innovative solutions often emerge at the intersection of academic research and industrial practice. From this perspective, it, in a way, seeks to inspire researchers and practitioners to leverage mathematical techniques for effectively addressing the dynamic challenges faced across the various domains of engineering.

We would like to express our heartfelt gratitude to the speakers and participants of ICAME'24 from numerous countries who contributed with their expertise and through their diverse perspectives. Their invaluable insights and dedication have enriched this volume, making it a robust resource for researchers, students, practitioners and professionals. We would like to convey our sincere thanks to the Series Editors of the *Systems Science and Nonlinear Intelligence Dynamics* book series and to Editor Joy Quek as well as to everyone at World Scientific publishing who have put in efforts in the publication process of the book.

As readers resort to the chapters, it is anticipated that they will find not only a wealth of knowledge but also inspiration to explore new frontiers in applied mathematics and its transformative role as the foundational solution spring in engineering, applied sciences along with other related disciplines and domains.

January 2025

Dumitru Baleanu
Lebanese American University, Lebanon

Necati Özdemir
Balıkesir University, Türkiye

Yeliz Karaca
University of Massachusetts, USA

Mukund N. Janardhanan
University of Warwick, UK

Fırat Evirgen
Balıkesir University, Türkiye

İbrahim Küçükkoç
Balıkesir University, Türkiye

About the Editors

Dumitru Baleanu is a Professor at the Department of Computer Science and Mathematics, Lebanese American University, Beirut, Lebanon. Dumitru Baleanu got his PhD from the Institute of Atomic Physics in 1996. His fields of interest include the fractional dynamics and its applications, fractional differential equations and their applications, discrete mathematics, image processing, bio-informatics, mathematical biology, mathematical physics, soliton theory, Lie symmetry, dynamic systems on time scales, computational complexity, the wavelet method and its applications, quantization of systems with constraints, the Hamilton–Jacobi formalism, geometries admitting generic and non-generic symmetries. Dumitru is a pioneer of the fractional variational principles and their applications in control theory. He is one of the co-authors of the seminal paper entitled "Anomalous diffusion expressed through fractional order differential operators in the Bloch-Torrey equation", published in *Journal of Magnetic Resonance* (2008), which plays now a fundamental role within diffusion weighted MRI. Dumitru had an important role in developing the non-singular operators with Mittag-Leffler kernels and their applications in real world phenomena.

Necati Özdemir is a Professor in the Mathematics Department at Balıkesir University where he has been a faculty member since 2001. Professor Özdemir completed his PhD at University of Exeter in 2000, his MSc at University of Warwick in 1997 and his undergraduate studies at Marmara University in 1992. His research interests lie in the area of control theory and applications, optimization theory, fractional calculus and applications, financial mathematics, biological model and numerical methods, ranging from theory to design to implementation. He has collaborated actively with researchers in several other disciplines.

He has served on several conference and workshop program committees and served as the co-chair for the ICAME conference series in 2018, 2021 and 2024. He also served as a guest editor for reputed international journals and publishers including *Journal of Computational and Applied Mathematics* (Elsevier), Springer and *International Journal of Nonlinear Sciences and Numerical Simulation*. He currently serves on the editorial board of IJOCTA.

Yeliz Karaca is a Professor of Applied Mathematics at the Department of Mathematics and Senior Researcher Professor at the Department of Neurology at University of Massachusetts (UMass), USA. She holds a medical card as the only mathematician entitled to it, as a result of the long-term medical and clinical-related certified trainings she received. She received Outstanding Young Scientist Award, Best Paper Awards and Outstanding Reviewer Award in her specialized discipline, among other national and international awards in different categories as well as grants. She has undertaken editorial tasks in SCI indexed journals, edited books and proceeding books besides having authored books. Her research interests mainly focus on complex systems with applications in various terrains, applied mathematics, computational methods, AI applications, computational complexity, fractional calculus, multifractals, multifractional methods, stochastic processes, wavelet, entropy, solutions of advanced mathematical challenges, mathematical neuroscience and biology as well as advanced data analysis in medicine other related theoretical, computational and applied domains.

Mukund N. Janardhanan is currently working in the prestigious Warwick Manufacturing Group at University of Warwick as Course Director for Intelligent Manufacturing Systems. He previously worked as Director of Education in School of Engineering at University of Leicester, UK. Dr. Mukund has over 10 years of experience including 8+ years in academia and 2 years in an engineering consultancy company.

Dr. Mukund graduated with a PhD in Manufacturing Optimization from Monash University. He also worked as a postdoctoral fellow at Aalborg University, Denmark and has also previously worked in product development in an Engineering Design company. He is an ambitious researcher with special interests in Industry 4.0, manufacturing system optimization and simulation.

Till date, to his credit he has published more than 60 referred research papers in leading manufacturing and operations journals. He is involved in funded research projects. He has also edited special issues in reputed international journals and has served as programme committee members of reported international conferences. He is also a Chartered Engineer with Institution of Mechanical Engineers and Fellow of Higher Education academy, UK.

Fırat Evirgen is a Professor in the Department of Mathematics at Balıkesir University, Turkey. He has been a part of the university's faculty since completing his PhD in Mathematics at Balıkesir University in 2009. Prof. Evirgen's research interests are diverse and include nonlinear optimization, continuous optimization, fractional calculus, mathematical modeling, and optimal control theory. His work in these areas has made significant contributions to the field of mathematics, particularly in the application and development of advanced mathematical techniques and theories. Throughout his career, Prof. Evirgen has published numerous articles in prestigious journals, demonstrating the impact and relevance of his research. His publications can be found in esteemed journals such as *AIMS Bioengineering, Fractals, Journal of Computational and Applied Mathematics, Discrete & Continuous Dynamical Systems-S, Fractal and Fractional, and the Bulletin of the Malaysian Mathematical Sciences Society.* Prof. Evirgen continues to contribute to the advancement of mathematical sciences through his research, publications, and teaching, making a lasting impact on both his students and the academic community at large.

İbrahim Küçükkoç is a Professor in the Industrial Engineering Department at Balıkesir University (Türkiye), where he has been a faculty member since December 2009. He completed his Ph.D. at the University of Exeter (UK) in 2015 upon completing his thesis on modelling and optimization of complex assembly lines (funded by a 4-year scholarship from Turkish Council of Higher Education).

His research interests lie in the area of modelling and optimization of modern manufacturing systems (assembly line balancing in particular), production planning of additive manufacturing machines and application of modern algorithms on sophisticated combinatorial optimization problems. He has collaborated actively with researchers in several other disciplines, such as computer science and mechanical engineering. His publications appeared in various esteemed journals including *European Journal of Operational Research, Computers & Operations Research, International Journal of Production Economics, International Journal of Production Research, Computers & Industrial Engineering and Production Planning & Control.* His research has been cited more than 1900 times (based on Scopus metrics) by national and international scholars.

Contents

Chapter 1

On Better Approximation Order for the Nonlinear Truncated Favard-Szász-Mirakjan Operator

Sezin Çit* and Ogün Doğru
Department of Mathematics, Gazi University, Ankara, Turkey
**sezincit@gazi.edu.tr*

Bede *et al.* presented the nonlinear max-product Truncated Favard-Szász-Mirakjan operator [1]. In [1], the approximation order of this operator is obtained as $1/\sqrt{r}$ with the help of the modulus of continuity. And it is noted that this approximation order cannot be improved except for some subclasses of functions, such as concave functions. The main topic of the current paper is to obtain a better approximation order for the nonlinear max-product truncated Favard-Szász-Mirakjan operator than in [1], without reducing the function classes. Our motivation is that the Cauchy Schwarz inequality is not used when finding the approximation of these operators (with modulus of continuity). We can obtain the approximation order for this operator as $1/r^{1-\frac{1}{\alpha}}$ for $\alpha = 2, 3, \ldots$. In fact, these results for $\alpha = 2$ return to [1]. Since $1-\frac{1}{\alpha}$ tends to 1 for enough big α, we improve this degree of approximation.

1. Introduction

Nonlinear operators are generated by replacing probability by maximum set functions, starting from the idea of obtaining linear and positive operators by probabilistic approximation. Nonlinear operators are presented by Bede *et al.* by the open problem 5.5.4, pp. 324–326 in [2]. The construction of nonlinear operators takes into account the possibility of transforming the algebraic structure from the linear one to another more general one. Thus, the results obtained in classical approximation theory are generalized.

To construct our nonlinear operators that approximate a continuous function $\gamma : [a, b] \to \mathbf{R}$, we analyze the algebraic structure of an ordered semiring, namely the max-product semiring of positive real numbers. The concept of constructing max-product approximation operators involves changing the algebraic structure from the field of real numbers to the ordered, max-product semiring of positive real numbers.

A semi-ring structure: $(\mathbf{R}_+, \vee, \cdot)$ is formed by $\vee$ (maximum) and $\cdot$ (product) operations over the positive real numbers $\mathbf{R}_+$.This structure is referred to as a max-product algebra.

In general, error estimates in classical approximation theory are given in relation to the known modulus of continuity. In relation to our newly introduced algebraic structure, we have the following:

Definition 1. [3] Consider the compact metric space (T, d) and also the metric space $([0,\infty), |.|)$, where $|.|$ is the usual Euclidean distance. $\gamma : T \to [0,\infty)$ is a bounded function and the function ω is defined as

$$\omega(\gamma,\delta) = \bigvee \{|\gamma(t) - \gamma(q)|, t, q \in T, d(t,q) \leq \delta\}$$

and is referred the modulus of continuity of γ.

Let us take a bounded or unbounded subinterval J of real numbers, and

$$CB_+(J) = \{\gamma : J \to \mathbf{R}_+ : \gamma \text{ continuous, bounded on } J\}.$$

Now, let's recall the well-known definition of the classical modulus of continuity.

$$\omega(\gamma,\delta) = \max\{|\gamma(t) - \gamma(q)|; t, q \in J, |t-q| \leq \delta\}. \tag{1}$$

Starting from Shepard operators [4], the topic of max-product operators was investigated for important classes of linear and positive operators as Bernstein operators [5, 6], Favard-Szász-Mirakyan operators [6, 7], Baskakov operators [8], Meyer-König-Zeller operators [9], and Bleimann-Butzer-Hann operators [10]. These studies are also collected in [11]. In these papers, various max-product type operators are presented, and also their convergence rates are analyzed.

There are many studies on max-product operators in the literature. In [13], the Kantorovich types of these operators are defined and their approximation degrees are found. In [14], statistical approximations of max-product operators are presented. On the other hand, in [15], the q-Bernstein max-product operators based on q-integers, first studied by Duman, are defined and their approximation properties are examined. Then, in [16], the approximation properties of the max-product q-Favard-Szász-Mirakjan operator are studied.

For $\gamma \in C[0,\infty)$, the classical Favard-Szász-Mirakjan operators defined as

$$F_r(\gamma;t) = e^{-rt}\sum_{s=0}^{\infty}\gamma\left(\frac{s}{r}\right)\frac{(rt)^s}{s!},$$

were introduced in [12].

The nonlinear Favard-Szász-Mirakjan operator is defined as

$$F_r^{(M)}(\gamma)(t) = \frac{\bigvee_{s=0}^{\infty}\frac{(rt)^s}{s!}\gamma\left(\frac{s}{r}\right)}{\bigvee_{s=0}^{\infty}\frac{(rt)^s}{s!}},$$

where $\gamma \in C[0,\infty)$, $t \in [0,\infty)$, $r \in \mathbf{N}$. In [7], the approximation degree of the $F_r^{(M)}(\gamma)(t)$ operator was found to be $\omega\left(\gamma; \sqrt{t}/\sqrt{r}\right)$ and it was claimed that it cannot be improved except for some subclasses of functions. On the other hand, in [17] we were able to improve the approximation degree for the $F_r^{(M)}(\gamma)(t)$ operator without narrowing the function classes. (See also [18, 19].)

In this study, we will obtain a better approximation degree for the truncated Favard-Szász-Mirakjan operator than the approximation degree found in [1].

2. The Concept of Nonlinear Max-Product Operators

We will review the fundamental definitions and theorems of nonlinear operators from [3, 6, 11] before the main results are presented.

Let J be a bounded or unbounded subinterval of the real numbers and let us show the general form of nonlinear operator sequences as follows
$L_r : CB_+(J) \to CB_+(J)$,

$$L_r(\gamma)(t) = \bigvee_{i=0}^{r} K_r(t, t_i)\,\gamma(t_i)$$

or

$$L_r(\gamma)(t) = \bigvee_{i=0}^{\infty} K_r(t, t_i)\,\gamma(t_i),$$

where $r \in \mathbf{N}$, $\gamma \in CB_+(J)$, $K_r(., t_i) \in CB_+(J)$ and $t_i \in J$, for all i. $L_r(\gamma)(t)$ operators are positive and nonlinear, in addition, they fulfil the following pseudo-linearity condition of the form

$$L_r(\tau\gamma \vee \beta\zeta)(t) = \tau\, L_r(\gamma)(t) \vee \beta\, L_r(\zeta)(t), \forall \tau, \beta \in \mathbf{R}_+,\ \gamma, \zeta : J \to \mathbf{R}_+.$$

We offer some general findings on these types of operators in this part that are used in the following sections.

Lemma 1. *[6] Let $L_r : CB_+(J) \to CB_+(J)$ and $r \in \mathbf{N}$ be a sequence of operators, for $\gamma, \zeta \in CB_+(J)$ fulfilling the following properties:*
(i) $\gamma \le \zeta$ then $L_r(\gamma) \le L_r(\zeta)$ for all $r \in \mathbf{N}$.
(ii) $L_r(\gamma + \zeta) \le L_r(\gamma) + L_r(\zeta)$.
Then for all $t \in J$ we obtain

$$|L_r(\gamma)(t) - L_r(\zeta)(t)| \le L_r(|\gamma - \zeta|)(t).$$

Corollary 1. *[6] Let the operator sequence $L_r : CB_+(J) \to CB_+(J)$ and $r \in \mathbf{N}$ fulfil the positive homogenous property in addition to the conditions in Lemma 1. Let $L_r(e_0) = e_0$, for each $r \in \mathbf{N}$. Thus for each $\gamma \in CB_+(J)$, $r \in \mathbf{N}$ and $t \in J$ we obtain*

$$|L_r(\gamma)(t) - \gamma(t)| \le \left[1 + \frac{1}{\delta} L_r(\varphi_t)(t)\right] \omega(\gamma, \delta)$$

where $\omega(\gamma, \delta)$ represents the classical modulus of continuity as defined by (1) and $\delta > 0$, $\varphi_t(w) = |w - t|$ for each $w \in J$.

3. Nonlinear Truncated Favard-Szász-Mirakjan Operator of Max-Product Kind

Max-product type truncated Favard-Szász-Mirakjan operator is defined as

$$T_r^{(M)}(\gamma)(t) = \frac{\bigvee_{s=0}^{r} \frac{(rt)^s}{s!}\,\gamma\left(\frac{s}{r}\right)}{\bigvee_{s=0}^{r} \frac{(rt)^s}{s!}}, \tag{2}$$

where $\gamma \in C[0,1]$, $t \in [0,1]$, $r \in \mathbf{N}$, and its approximation and shape preserving properties are examined in [1].

Remark 1. [1] (1) The operator $T_r^{(M)}(\gamma)(t)$ provides each condition of Lemma 1 and Corollary 1 for $J = [0,1]$.

(2) It also provides the following property, which is stronger than (i) of Lemma 1, which is the pseudo-linearity property:

$$T_r^{(M)}(\gamma \vee \zeta)(t) = T_r^{(M)}(\gamma)(t) \vee T_r^{(M)}(\zeta)(t), \gamma, \zeta \in CB_+([0,1]).$$

It is also positive homogenous, i.e., $T_r^{(M)}(\varsigma\gamma) = \varsigma T_r^{(M)}(\gamma)$ for all $\varsigma \geq 0$.

(3) The nonlinear $T_r^{(M)}(\gamma)(t)$ operator is positive, bounded, and continuous on $[0,1]$ for arbitrary bounded functions $\gamma : [0,1] \to \mathbf{R}_+$. Additionally, it fulfils $T_r^{(M)}(\gamma)(0) = \gamma(0)$.

4. Auxiliary Results

In this section, we will also give the necessary lemmas for the approximation theorem without proofs. We will also give Remark 2, which generalizes Lemma 3 and allows us to improve the degree of approximation.

The fact that $T_r^{(M)}(\gamma)(0) - \gamma(0) = 0$ for all r, we are going to concentrate on $t > 0$ in the notations and proofs of all estimate findings.

The expressions $M_{s,r,m}(t)$ and $m_{s,r,m}(t)$ for each $s \in \{0,1,2,...,r\}$, $m \in \{0,1,2,...,r-1\}$, $t \in \left[\frac{m}{r}, \frac{m+1}{r}\right]$ are presented in [1]. Similarly, we give these expressions below:

$$M_{s,r,m}(t) = \frac{p_{r,s}(t)}{p_{r,m}(t)}\left|\frac{s}{r} - t\right|, \quad m_{s,r,m}(t) = \frac{p_{r,s}(t)}{p_{r,m}(t)}.$$

It is obvious that if $s \geq m+1$

$$M_{s,r,m}(t) = \frac{p_{r,s}(t)}{p_{r,m}(t)}\left(\frac{s}{r} - t\right)$$

and if $s \leq m-1$

$$M_{s,r,m}(t) = \frac{p_{r,s}(t)}{p_{r,m}(t)}\left(t - \frac{s}{r}\right)$$

where $p_{r,s}(t) = \frac{(rt)^s}{s!}$.

Lemma 2. *[1] For each $s \in \{0,1,2,...,r\}$, $m \in \{0,1,2,...,r-1\}$ and $t \in \left[\frac{m}{r}, \frac{m+1}{r}\right]$ we get*

$$m_{s,r,m}(t) \leq 1.$$

Lemma 3. *[1] Let us conisder $t \in \left[\frac{m}{r}, \frac{m+1}{r}\right]$.*

(i) If $s \in \{m+1, m+2, ..., r-1\}$ is such that $s - \sqrt{s+1} \geq m$, then we obtain

$$M_{s,r,m}(t) \geq M_{s+1,r,m}(t). \tag{3}$$

(ii) If $s \in \{1,2,...,m-1\}$ is such that $s + \sqrt{s} \leq m$, then we obtain

$$M_{s,r,m}(t) \geq M_{s-1,r,m}(t). \tag{4}$$

Remark 2. (*i*) Let us write the inequality (3), hypothesis (*i*) of Lemma 3, as follows.

$$s+1 \leq (s-m)^2$$
$$(s+1)(s-m)^{\alpha-2} \leq (s-m)^{\alpha},$$

where $\alpha = 2, 3, ...$ In this case since $s \geq m+1$ i.e. $s-m \geq 1$ we find $1 \leq (s-m)^{\alpha-2}$. If we write this above, we get

$$(s+1) \leq (s-m)^{\alpha}$$

or

$$m \leq s - (s+1)^{\frac{1}{\alpha}}. \tag{5}$$

Therefore, $s \in \{m+1, m+2, ..., r-1\}$ is such that inequality (5) and for every $\alpha = 2, 3, ...$ we obtain $M_{s,r,m}(t) \geq M_{s+1,r,m}(t)$.

(*ii*) Similarly, if we revise inequality (4) in hypothesis (*ii*) of Lemma 3

$$(m-s)^2 \geq s$$
$$(m-s)^{\alpha} \geq s(m-s)^{\alpha-2},$$

where $\alpha = 2, 3, ...$ Since $s \leq m-1$ i.e. $1 \leq m-s$ so, we find $1 \leq (m-s)^{\alpha-2}$. If we write this above, we get

$$(m-s)^{\alpha} \geq s$$

or

$$m \geq s + s^{\frac{1}{\alpha}}. \tag{6}$$

Therefore, $s \in \{1, ..., m-1\}$ is such that inequality (6) and for every $\alpha = 2, 3, ...$ we obtain $M_{s,r,m}(t) \geq M_{s-1,r,m}(t)$.

Lemma 4. *[1] One has*

$$\bigvee_{s=0}^{r} p_{r,s}(t) = p_{r,m}(t), \text{ for each } t \in \left[\frac{m}{r}, \frac{m+1}{r}\right], \quad m = 0, 1, ..., r-1.$$

where $p_{r,s}(t) = \frac{(rt)^s}{s!}$.

5. Approximation Results

Our main aim in this section is to prove that the degree of approximation can be improved for $T_r^{(M)}(\gamma)(t)$ operators. As seen in Theorem 1, $1 - \frac{1}{\alpha} \geq \frac{1}{2}$ for $\alpha = 2, 3, ...$ so the degree of approximation is improved. In fact, the results found for $\alpha = 2$ become the results in paper [1].

Theorem 1.

Let $\gamma : [0,1] \to \mathbf{R}_+$ *be continuous function on* $[0,1]$. *For the operators in (2) we have*

$$\left|T_r^{(M)}(\gamma)(t) - \gamma(t)\right| \leq 6\,\omega\left(\gamma; \frac{1}{r^{1-\frac{1}{\alpha}}}\right), \quad \text{for all } r \in \mathbf{N},$$

where $\alpha = 2, 3, ...$ *and* $\omega(\gamma, \delta)$ *is the classical modulus of continuity defined in (1).*

Proof. For the nonlinear truncated max-product Favard-Szász-Mirakjan operator by utilising the properties of $\omega(\gamma;\delta)$ and considering the conditions given in Corollary 1, demonstrably

$$\left|T_r^{(M)}(\gamma)(t)-\gamma(t)\right| \leq \left[1+\frac{1}{\delta_r}T_r^{(M)}(\varphi_t)(t)\right]\omega(\gamma,\delta), \tag{7}$$

where $\varphi_t(w)=|w-t|$. Now let us define the expression E_r as follows

$$E_r(t):=T_r^{(M)}(\varphi_t)(t)=\frac{\bigvee_{s=0}^{r}\frac{(rt)^s}{s!}\left|\frac{s}{r}-t\right|}{\bigvee_{s=0}^{r}\frac{(rt)^s}{s!}}, \quad t\in[0,1].$$

For an arbitrary constant $m\in\{0,1,...,r-1\}$, let $t\in\left[\frac{m}{r},\frac{m+1}{r}\right]$. Let us observe from Lemma 4 that clearly

$$E_r(t)=\max_{s=0,1,\ldots,r}\{M_{s,r,m}(t)\},\ t\in\left[\frac{m}{r},\frac{m+1}{r}\right].$$

At first, let us consider the case $m=0$, where $t\in\left[0,\frac{1}{r}\right]$ and $\alpha=2,3,\ldots$. Since $p_{r,0}(t)=1$, then we obtain

$$M_{s,r,0}(t)=\frac{(rt)^s}{s!}\left|\frac{s}{r}-t\right|.$$

For $s=0$, we have

$$M_{0,r,0}(t)=t\leq\frac{1}{r}\leq\frac{1}{r^{1-\frac{1}{\alpha}}}$$

and for every $s\geq 1$, we have

$$M_{s,r,0}(t)\leq\frac{(rt)^s}{s!}\frac{s}{r}\leq\frac{\left(r\frac{1}{r}\right)^s}{(s-1)!r}\leq\frac{1}{r}\leq\frac{1}{r^{1-\frac{1}{\alpha}}}.$$

As a consequence, we have found an upper estimate for the expression $E_r(t)\leq\frac{1}{r^{1-\frac{1}{\alpha}}}$ for each $s=0,1,...,r$ when $m=0$. It is still necessary to get an upper estimate for $M_{s,r,m}(t)$ for any $m=1,2,...,r-1$, when $t\in\left[\frac{m}{r},\frac{m+1}{r}\right]$, $s\in\{0,1,...,r\}$ and $\alpha=2,3,\ldots$.

Indeed, if we can show that

$$M_{s,r,m}(t)\leq\frac{3}{r^{1-\frac{1}{\alpha}}} \tag{8}$$

for each $t\in\left[\frac{m}{r},\frac{m+1}{r}\right]$ and $s=0,1,...,r$ it is obviously

$$E_r(t)\leq\frac{3}{r^{1-\frac{1}{\alpha}}},\ for\ all\ t\in[0,1],\ r\in\mathbf{N}$$

and if we take $\delta_r=\frac{1}{r^{1-\frac{1}{\alpha}}}$ in (7), we may quickly get the expression in the estimate.

In order to show (8), let us examine the below cases for $t \in \left[\frac{m}{r}, \frac{m+1}{r}\right]$
(1) $s = m$;
(2) $s \geq m+1$;
(3) $s \leq m-1$.

Case (1). For $s = m$, we find

$$M_{m,r,m}(t) = \left|\frac{m}{r} - t\right| \leq \left|\frac{m}{r} - \frac{m+1}{r}\right| = \frac{1}{r} \leq \frac{1}{r^{1-\frac{1}{\alpha}}}.$$

Case (2). Subcase (a). Consider $s \geq m+1$ and firstly suppose that $s - (s+1)^{\frac{1}{\alpha}} < m$. From Lemma 2, we find

$$\begin{aligned} M_{s,r,m}(t) &= m_{s,r,m}(t)\left(\frac{s}{r} - t\right) \leq \frac{s}{r} - t \\ &\leq \frac{s}{r} - \frac{m}{r} \leq \frac{s}{r} - \frac{s-(s+1)^{\frac{1}{\alpha}}}{r} \\ &= \frac{(s+1)^{\frac{1}{\alpha}}}{r} \leq \frac{(r+1)^{\frac{1}{\alpha}}}{r} \\ &\leq \frac{(2r)^{\frac{1}{\alpha}}}{r} = \frac{2^{\frac{1}{\alpha}}}{r^{1-\frac{1}{\alpha}}}. \end{aligned}$$

Subcase (b). Consider $s \geq m+1$ and suppose that $s - (s+1)^{\frac{1}{\alpha}} \geq m$. For the function $\psi(t) :– t - (t+1)^{\frac{1}{\alpha}}$, we have

$$\psi'(t) = 1 - \frac{1}{\alpha(t+1)^{1-\frac{1}{\alpha}}} > 0.$$

Therefore, we can conclude that the function $\psi(t)$ is nondecreasing on $[0,\infty)$. Consequently, there exists a maximum value $\bar{s} \in \{1,2,...,r\}$ which fulfils the condition $\bar{s} - (\bar{s}+1)^{\frac{1}{\alpha}} < m$. Now, for $s_1 = \bar{s}+1$, we get $s_1 - (s_1+1)^{\frac{1}{\alpha}} \geq m$ and

$$\begin{aligned} M_{\bar{s}+1,r,m}(t) &= m_{\bar{s}+1,r,m}(t)\left(\frac{\bar{s}+1}{r} - t\right) \\ &\leq \frac{\bar{s}+1}{r} - t \leq \frac{\bar{s}+1}{r} - \frac{m}{r} \\ &\leq \frac{\bar{s}+1}{r} - \frac{\bar{s}-(\bar{s}+1)^{\frac{1}{\alpha}}}{r} = \frac{(\bar{s}+1)^{\frac{1}{\alpha}}+1}{r} \\ &\leq \frac{(2r)^{\frac{1}{\alpha}}+1}{r} \leq \frac{3r^{\frac{1}{\alpha}}}{r} = \frac{3}{r^{1-\frac{1}{\alpha}}}. \end{aligned}$$

In the light of the inequality above, we get $\bar{s} - (\bar{s}+1)^{\frac{1}{\alpha}} < m$. Furthermore, it is correct that $s_1 \geq m+1$. In fact, this is a result of the fact that ψ is nondecreasing and it is evident that $\psi(m) = m - (m+1)^{\frac{1}{\alpha}} < m$. From Remark 2, (i) that's why

$$M_{\bar{s}+1,r,m}(t) \geq M_{\bar{s}+2,r,m}(t) \geq ....$$

Thereby, we obtain $M_{s,r,m}(t) \leq \frac{3}{r^{1-\frac{1}{\alpha}}}$ for any $s \in \{\bar{s}+1, \bar{s}+2, ..., r\}$.

Case (3). Subcase (a). Consider $s \le m-1$ and assume first that $s+s^{\frac{1}{\alpha}} > m$. Then we find

$$\begin{aligned} M_{s,r,m}(t) &= m_{s,r,m}(t)\left(t-\frac{s}{r}\right) \le \frac{m+1}{r}-\frac{s}{r} \\ &\le \frac{s+s^{\frac{1}{\alpha}}+1}{r}-\frac{s}{r}=\frac{s^{\frac{1}{\alpha}}+1}{r} \\ &\le \frac{r^{\frac{1}{\alpha}}+1}{r} \le \frac{2r^{\frac{1}{\alpha}}}{r}=\frac{2}{r^{1-\frac{1}{\alpha}}}. \end{aligned}$$

Subcase (b). Consider $s \le m-1$ and assume now that $s+s^{\frac{1}{\alpha}} \le m$. Let $\tilde{s} \in \{0,1,...,r\}$ represent the minimum value such that $\tilde{s}+\tilde{s}^{\frac{1}{\alpha}} > m$. So, $s_2=\tilde{s}-1$ provides $s_2+s_2^{\frac{1}{\alpha}} \le m$ and

$$\begin{aligned} M_{\tilde{s}-1,r,m}(t) &= m_{\tilde{s}-1,r,m}(t)\left(t-\frac{\tilde{s}-1}{r}\right) \\ &\le \frac{m+1}{r}-\frac{\tilde{s}-1}{r} \le \frac{\tilde{s}+\tilde{s}^{\frac{1}{\alpha}}+1}{r}-\frac{\tilde{s}-1}{r} \\ &= \frac{\tilde{s}^{\frac{1}{\alpha}}+2}{r} \le \frac{r^{\frac{1}{\alpha}}+2}{r} \le \frac{3}{r^{1-\frac{1}{\alpha}}}. \end{aligned}$$

Furthermore, since $m \ge 1$ it follows that $s_2 \le m-1$.

From Remark 2, (ii) then $M_{\tilde{s}-1,r,m}(t) \ge M_{\tilde{s}-2,r,m}(t) \ge ... \ge M_{0,r,m}(t)$. Thus, we obtain $M_{s,r,m}(t) \le \frac{3}{r^{1-\frac{1}{\alpha}}}$ for any $s \le m-1$ and $t \in \left[\frac{m}{r},\frac{m+1}{r}\right]$.

If we use

$$\max\left\{\frac{1}{r^{1-\frac{1}{\alpha}}},\frac{2}{r^{1-\frac{1}{\alpha}}},\frac{3}{r^{1-\frac{1}{\alpha}}}\right\}=\frac{3}{r^{1-\frac{1}{\alpha}}}$$

in (8), we have the desired result. □

Corollary 2. *In [1], the authors found that the order of approximation for max-product truncated Favard-Szász-Mirakjan operators is* $1/\sqrt{r}$*, using the modulus of continuity. They assert that this order of approximation cannot be improved, with the exception of special subclasses of functions, for instance, the concave functions. However, Theorem 1 indicates that we can improve the degree of approximation for* $T_r^{(M)}(\gamma)(t)$ *(without the need to reduce the class of functions). For* $\alpha = 2,3,...$ *the approximation degree is* $1/r^{1-\frac{1}{\alpha}}$*. For each value of* α*, the order of approximation improves since* $1-\frac{1}{\alpha} \ge \frac{1}{2}$*. Also for sufficiently big* α*, the expression* $1-\frac{1}{\alpha}$ *approaches* 1*. In addition, the results for* $\alpha = 2$ *return to [1].*

Declarations

Conflict of interest: The authors declare that they have no conflict of interest.
Data Availability Statement: Data sharing not applicable to this article as no datasets were generated or analysed during the current study.

Acknowledgements

The authors are extremely thankful to the referee(s) for their valuable opinions.

References

[1] B. Bede, L. Coroianu and S. G. Gal, Approximation by truncated Favard-Szász-Mirakjan operator of max-product kind, *Demonstratio Mathematica,* **44**(1), 105–122, (2011).

[2] S. G. Gal, *Shape-Preserving Approximation by Real and Complex Polynomials,* (Birkhäuser, Boston-Basel-Berlin, 2008).

[3] B. Bede, H. Nobuhara, M. Daňková and A. Di Nola, Approximation by pseudo-linear operators, *Fuzzy Sets and Systems,* **159**, 804–820, (2008).

[4] B. Bede, H. Nobuhara, J. Fodor and K. Hirota, Max-product Shepard approximation operators, *Journal of Advanced Computational Intelligence and Intelligent Informatics,* **10,** 494–497, (2006).

[5] B. Bede, L. Coroianu and S. G. Gal, Approximation and shape preserving properties of the Bernstein operator of max-product kind, *International Journal of Mathematics and Mathematical Sciences,* Article ID 590589, 26 pages, (2009) `https://doi.org/10.1155/2009/590589`.

[6] B. Bede, L. Coroianu and S. G. Gal, Approximation by Nonlinear Bernstein and Favard-Szász-Mirakjan operators of max-product kind, *Journal of Concrete and Applicable Mathematics,* **8**(2), 193–207, (2010).

[7] B. Bede, L. Coroianu and S. G. Gal, Approximation and shape preserving properties of the nonlinear Favard-Szász-Mirakjan operator of max-product kind, *Filomat,* **24**(3), 55–72, (2010).

[8] B. Bede, L. Coroianu and S. G. Gal, Approximation and shape preserving properties of the nonlinear Baskakov operator of max-product kind, *Studia Universitatis Babeș-Bolyai Mathematica,* **55**(4), 193–218, (2010).

[9] B. Bede, L. Coroianu and S. G. Gal, Approximation and shape preserving properties of the nonlinear Meyer–König and Zeller operator of max-product kind, *Numerical Functional Analysis and Optimization,* **31**(3), 232–253, (2010).

[10] B. Bede, L. Coroianu and S. G. Gal, Approximation and shape preserving properties of the nonlinear Bleimann-Butzer-Hahn operators of max-product kind, *Commentationes Mathematicae Universitatis Carolinae,* **51**(3), 397–415, (2010).

[11] B. Bede, L. Coroianu and S. G. Gal, *Approximation by Max-Product Type Operators,* (Springer International Publishing, Switzerland 2016).

[12] O. Szász, Generalization of S. Bernstein polynomials to the infinite interval, *Journal of Research of the National Bureau of Standards,* **45** 239–245, (1950).

[13] B. Bede, L. Coroianu and S. G. Gal, Approximation by max-product operators of Kantorovich type. *Studia Universitatis Babeş-Bolyai Mathematica,* **64**(2), 207–223. (2019). `https://doi10.24193/subbmath.2019.2.07`.

[14] O. Duman, Statistical convergence of max-product approximating operators. *Turkish Journal of Mathematics,* 34, 501–514 (2010).

[15] O. Duman, Nonlinear Approximation: q-Bernstein operators of max-product kind, *Intelligent Mathematics II: Applied Mathematics and Approximation Theory,* **441**, Springer (2016).

[16] D. Karahan and E. Acar, Nonlinear Approximation by q-Favard-Szász-Mirakjan Operators of Max-Product Kind. *Communications in Advanced Mathematical Sciences,* **6**(2), 104–114. (2023).

[17] S. Çit and O. Dogru, On better approximation order for the nonlinear Favard-Szász-Mirakjan operator of max-product kind, *Matematički Vesnik*, `https://doi.org/10.57016/MV-nkMX9624`.

[18] S. Çit and O. Dogru, On better approximation order for the nonlinear Bernstein operator of max-product kind, *Filomat*, **38**(13), 4767–4774, (2024). `https://doi.org/10.2298/FIL2413767C`.

[19] S. Çit and O. Dogru, On better approximation order for the nonlinear Bleimann-Butzer-Hahn operator of max-product kind, *Khayyam Journal of Mathematics*, **9**(2), 225–245, (2023). `https://doi.org/10.22034/KJM.2023.357068.2641` .

Chapter 2

Comparison of Modified Karnik-Mendel Algorithm Based Interval Type-2 ANFIS and Type-1 ANFIS Controllers

Muhammet Öztürk*,† and İbrahim Özkol‡,§

*Department of Aeronautical Engineering, Necmettin Erbakan University, Konya, 42090, Turkey

‡Department of Aeronautical Engineering, Istanbul Technical University, Istanbul, 34467, Turkey

†mozturk@erbakan.edu.tr

§ozkol@itu.edu.tr

This study aims to test Interval Type-2 Adaptive Neuro-Fuzzy Inference System (IT2-ANFIS) based Neuro-Fuzzy Control (NFC) by using the Modified Karnik Mendel Algorithm (M-KMA). Determining the FLS antecedent and consequent parameters is the main problem in engineering issues. The Adaptive Neuro-Fuzzy Inference System (ANFIS) is the most used training method in the literature. The ANFIS model works very well for Type-1 FLS (T1-FLS) but is inconvenient for Interval Type-2 FLS (IT2-FLS) due to uncertainties. In 2021, a new IT2-ANFIS method is proposed to tune the IT2-FLS parameters by designing a reduction method named M-KMA. The M-KMA based IT2-ANFIS is superior to T1-ANFIS, so it is used in this paper. The M-KMA based IT2-NFC that consists of IT2-ANFIS structure is implemented to a DC motor control problem and the results are compared to Type-1 NFC. The results show that IT2-NFC performs better compared to T1-NFC.

1. Introduction

The increase in disturbances and the existence of uncertainties create the need to develop advanced control approaches to handle unpredictable conditions. The model-based approaches such as PID [1], and LQR [2] are effectively used in literature. These model-based controllers can deliver the best performance without uncertainties in the system. However, real systems have disturbances and uncertainties, especially in aerial vehicles, such as sensor noise and environmental conditions. The robustness of model-based controllers depends on the system dynamics that any change in the model will lead to control signal corruption [3]. However, intelligent controllers such as Neural Network (NN) [4], Genetic Algorithms (GA), and Fuzzy Logic (FL) [5] based controllers can work regardless of the actual mathematical model.

Conventional Fuzzy Logic Controllers (FLC) parameters are defined by expert opinions or by using neural network methods such as the ANFIS [6] regarding the data. These approaches define the controller parameters considering a previously fully specified model. So, the mentioned controllers only work with some desired value limits and assumptions. To overcome these limits, neural network-based structures are used online to adjust FIS antecedent and consequent parameters while the system is running. Implementing Type-1 Neuro-Fuzzy Control (T1-NFC) to an induction motor with two NFC inputs is compared to the PI controller in 2007 [7]. A three-input T1-NFC structure was tested on a DC motor and compared with PI in 2010 [8]. In 2014, the T1-NFC structure with two inputs was tested on a 2 DOF helicopter system and compared to PID [9]. In 2021, three different T1-NFC structures were tested and compared to each other for a 2 DOF system to achieve the best combination of NFC inputs [10]. All of the mentioned NFCs use the Type-1 ANFIS (T1-ANFIS) structure, but the T1-FLS membership functions are precise, so they cannot handle uncertainties.

Type-2 Fuzzy Logic System (T2-FLS) is a three-dimensional system whose membership functions include uncertainties [11]. Due to the calculation problem of T2-FLS, the interval type-2 FLS (IT2-FLS) is proposed [12]. IT2-FLS is a reduced form of generalized T2-FLS from a three-dimensional to a two-dimensional form that can be easily calculated by using some reduction methods [13]. The IT2-FLS membership functions still have uncertainties. So, reduction methods are proposed in the literature such as Karnik–Mendel Algorithm [14], Enhanced Karnik–Mendel Algorithm (EKMA) [15], Iterative Algorithm with Stop Condition (IASC) [16], Enhanced Iterative Algorithm with Stop Condition (EIASC) [17] and Direct Approach [18].

IT2-FLS parameters must be well-defined for use in control problems. However, because of the uncertainties of membership functions, it is impossible to tune antecedent and consequent parameters. An extended ANFIS structure is proposed for IT2-FLS and compared to T1-ANFIS in 2016 [19]. The results show that the IT2-FLS based ANFIS has bigger error values compared to T1-ANFIS due to uncertainties in the membership functions. In 2017, another IT2-FLS-based ANFIS structure was proposed and compared with T1-FLS for previously defined offline data. The results show that it is not possible to achieve better results compared to T1-ANFIS. The IT2-ANFIS has more computational load compared to T1-ANFIS [20]. In 2021, a new IT2-ANFIS model is proposed by using M-KMA [21]. The M-KMA is an adapted KMA reduction method that provides crisp values for IT2-ANFIS. The results showed that the proposed M-KMA-based IT2-ANFIS is superior compared to T1-ANFIS. In addition, the superiority changes linearly with uncertainties, that is, when the uncertainty increases, the superiority of the IT2- ANFIS increases and vice versa.

This study first proposes using the IT2-ANFIS structure given in the open literature in 2021 [21]. So, the effect of the IT2-ANFIS based NFC was compared to T1-ANFIS based NFC. This will provide a perspective for real-time future studies.

2. Interval Type-2 FIS

Although T1-FLS, proposed by Lotfi A. Zadeh, is very effective in mathematically unsolvable problems and in determining complex systems, it is inadequate in terms of uncertainty [22]. So, T2-FLS was proposed to overcome uncertainties in exceptional circumstances in 1975 [11]. However, it was impossible to get a crisp output from T2-FLS due to the uncertain membership functions. Because of that, IT2-FLS is proposed that still had uncertainties in the membership functions as seen in Figure 1 [13]. In Figure 1, w is the membership values of the inputs, LMF is the Lower Membership Function (MF) and UMF is the Upper Membership Function. As seen in Figure 1, there are two boundaries for every input. The MF values are calculated for both MFs, and a switch point is calculated using reduction methods. The switch point will be used to determine which membership values (LMF, UMF) will be used.

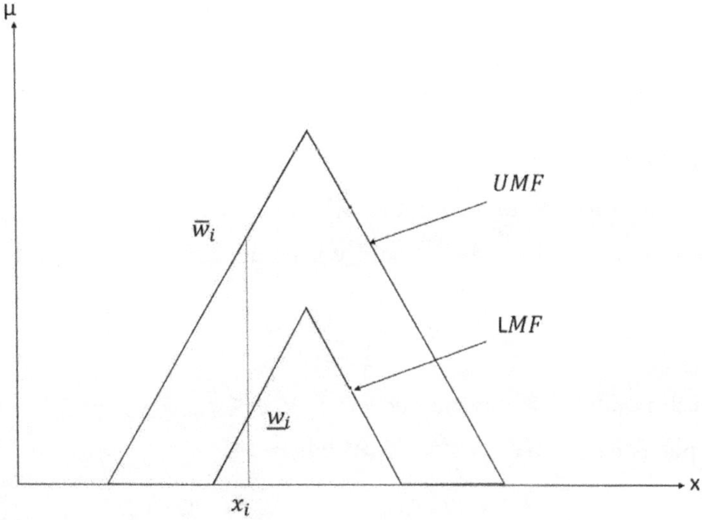

Fig. 1. Interval type-2 membership function.

This paper considers KMA, one of the type-reduction methods, because it is used in the IT2-ANFIS structure.

2.1. *Modified Karnik-Mendel Algorithm*

Karnik-Mendel Algorithm (KMA) is the first algorithm proposed to determine switch points [23]. As stated in the literature, the KMA is inconvenient for ANFIS due to uncertainties in the parameters [19, 20]. So, M-KMA is proposed [21].

The M-KMA calculates an "X" matrix to use in ANFIS that involves the calculated output weights, so it is possible to use them in IT2-ANFIS. The M-KMA is given below:

Computing y_l:

(1) $\underline{f}_i$ are sorted in ascending order and MF weights are coupled with $\underline{f}_i$.

(2) w_i is initialized with $w_i = \frac{\underline{w}_i + \overline{w}_i}{2}$ and y is computed

$$y = \frac{\sum_{i=1}^{N} \underline{f}_i w_i}{\sum_{i=1}^{N} w_i} \tag{1}$$

(3) The switch point "l" is determined for $\underline{f}_l < y \leq \underline{f}_{l+1}$

(4) The output is calculated as $y' = X_l W$ where

$$X_l = \frac{\begin{bmatrix} \overline{w}_1 x_1 & \overline{w}_1 x_2 & \overline{w}_1 & \ldots & \overline{w}_l x_1 & \overline{w}_l x_2 & \overline{w}_l \ldots \\ \underline{w}_{l+1} x_1 & \underline{w}_{l+1} x_2 & \underline{w}_{l+1} & \ldots & \underline{w}_N x_1 & \underline{w}_N x_2 & \underline{w}_N \end{bmatrix}}{\sum_{i=1}^{l} \overline{w}_i + \sum_{i=l+1}^{N} \underline{w}_i} \tag{2}$$

(5) If the output is equal to previously computed y value ($y' = y$), then the iteration is over ($y_l = y$); if not, assume as ($y = y'$) and return to Step 3.

Computing y_r:

(1) $\overline{f}_i$ are sorted in ascending order and MF weights are coupled with $\overline{f}_i$.

(2) w_i is initialized with $w_i = \frac{\underline{w}_i + \overline{w}_i}{2}$ and y is computed

$$y = \frac{\sum_{i=1}^{N} \overline{f}_i w_i}{\sum_{i=1}^{N} w_i} \tag{3}$$

(3) The switch point "r" is determined for $\overline{f}_l < y \leq \overline{f}_{l+1}$

(4) The output is calculated as $y' = X_r W$ where

$$X_r = \frac{\begin{bmatrix} \underline{w}_1 x_1 & \underline{w}_1 x_2 & \underline{w}_1 & \ldots & \underline{w}_l x_1 & \underline{w}_l x_2 & \underline{w}_l \ldots \\ \overline{w}_{l+1} x_1 & \overline{w}_{l+1} x_2 & \overline{w}_{l+1} & \ldots & \overline{w}_N x_1 & \overline{w}_N x_2 & \overline{w}_N \end{bmatrix}}{\sum_{i=1}^{l} \overline{w}_i + \sum_{i=l+1}^{N} \underline{w}_i} \tag{4}$$

(5) If the output is equal to previously computed y value ($y' = y$), then the iteration is over ($y_r = y$); if not, assume as ($y = y'$) and return to Step 3.

Result y:

$$y = \frac{y_r + y_l}{2} = \frac{X_l W + X_r W}{2} = \frac{X_l + X_r}{2} W = XW \tag{5}$$

The defined "X" matrix is a necessary matrix to reach the same result as the Karnik–Mendel algorithm.

3. Interval Type-2 ANFIS (M-KMA)

The IT2-ANFIS structure is given below for two inputs and one output [21].

Layer 1: The MF results are calculated. For every crisp input, two MF results are obtained. These MF results are named UMF and LMF values.

$$\begin{aligned} \underline{O}_i^1 &= \underline{\mu}_{A_i}(x) \qquad \underline{O}_i^1 = \underline{\mu}_{B_i}(y) \\ \overline{O}_i^1 &= \overline{\mu}_{A_i}(x) \qquad \overline{O}_i^1 = \overline{\mu}_{B_i}(y) \end{aligned} \tag{6}$$

Layer 2: All LMF and UMF weights are multiplied separately. LMF results are multiplied with each other and UMF results are multiplied with each other.

$$\begin{aligned} \underline{O}_i^2 &= \underline{\mu}_{A_i}(x) * \underline{\mu}_{B_i}(y) = \underline{w}_i \\ \overline{O}_i^2 &= \overline{\mu}_{A_i}(x) * \overline{\mu}_{B_i}(y) = \overline{w}_i \end{aligned} \tag{7}$$

Layer 3: The M-KMA previously mentioned is used to find switch points. To find the switch points, the weights are sorted regarding the f values, where $f_i = p_i x_1 + q_i x_2 + r_i$.

$$\begin{aligned} \underline{O}_i^3 &= sort(\underline{w}_i \;\; \overline{w}_i \;\; \underline{f}_i) \\ \overline{O}_i^3 &= sort(\underline{w}_i \;\; \overline{w}_i \;\; \overline{f}_i) \end{aligned} \tag{8}$$

Layer 4: The previously mentioned M-KMA is used to find X_r, X_l matrices.

$$\underline{O}_i^4 = X_l = \frac{\begin{bmatrix} \overline{w}_1 x_1 & \overline{w}_1 x_2 & \overline{w}_1 & \ldots & \overline{w}_l x_1 & \overline{w}_l x_2 & \overline{w}_l \ldots \\ \underline{w}_{l+1} x_1 & \underline{w}_{l+1} x_2 & \underline{w}_{l+1} & \ldots & \underline{w}_n x_1 & \underline{w}_n x_2 & \underline{w}_n \end{bmatrix}}{\sum_{n=1}^{l} \overline{w}_n + \sum_{n=l+1}^{N} \underline{w}_n} \tag{9}$$

$$\overline{O}_i^4 = X_r = \frac{\begin{bmatrix} \underline{w}_1 x_1 & \underline{w}_1 x_2 & \underline{w}_1 & \ldots & \underline{w}_r x_1 & \underline{w}_r x_2 & \underline{w}_r \ldots \\ \overline{w}_{r+1} x_1 & \overline{w}_{r+1} x_2 & \overline{w}_{r+1} & \ldots & \overline{w}_n x_1 & \overline{w}_n x_2 & \overline{w}_n \end{bmatrix}}{\sum_{n=1}^{r} \underline{w}_n + \sum_{n=r+1}^{N} \overline{w}_n} \tag{10}$$

Layer 5: The M-KMA is used to calculate a crisp X matrix. So, a clearly defined X matrix can be used in ANFIS.

$$O_i^5 = X_{sorted} = \frac{\underline{O}_i^4 + \overline{O}_i^4}{2} = \frac{X_l + X_r}{2} \tag{11}$$

Layer 6: In layer 5, the X matrix was calculated by sorted weights. So, the X matrix has been unsorted.

$$O_i^6 = X = unsort(X_{sorted}) \tag{12}$$

Layer 7: The output of the IT2-ANFIS is calculated by using the X and W matrices.

$$O_i^7 = y = XW \tag{13}$$

The output (y) can be calculated by using the X values. So, training approaches such as Least Square Estimation (LSE) and Gradient Descent (GD) can be used for IT2-ANFIS [24]. In the control problem, the GD method will be used and so the following equations are given below where "y" is the output value, "y^t" is the target value, and "lr" is the learning rate.

$$E = \frac{1}{2}(y - y^t)^2 \qquad y = XW \tag{14}$$

$$W(t+1) = W(t) - lr\frac{\partial E}{\partial W_i} \tag{15}$$

$$\frac{\partial E}{\partial W_i} = (y - y^t)X \tag{16}$$

$$W(t+1) = W(t) - lr(y - y^t)X \tag{17}$$

The error "$y - y^t$" and the X matrix are used to adjust the consequent parameters "W" as given in Eq. (17).

4. Neuro-Fuzzy Control

The NFC consists of online ANFIS and Fuzzy Logic Controller (FLC) steps. Both blocks run simultaneously. The FLC controls the system, but the ANFIS tunes the consequent parameters of the FLS as seen in Figure 2. FLC has three inputs: errors, output derivative, and the sum of errors [10]. These three inputs are used for the control of the system by FLC and used to tune the subsequent parameters of FLS by ANFIS. In addition, a target function must be defined for ANFIS optimization. In this study, the target function is assumed as "error", so the ANFIS tried to minimize error.

The NFC is tested on a DC motor simulation, where the system input is voltage and the output is angular velocity. The DC motor physical parameters are given in Table 1.

The DC motor equations are:

$$J\frac{\partial^2\theta}{\partial t^2} + b\frac{\partial\theta}{\partial t} = K_t i \tag{18}$$

$$L\frac{\partial i}{\partial t} + Ri = V - K_e\frac{\partial\theta}{\partial t} \tag{19}$$

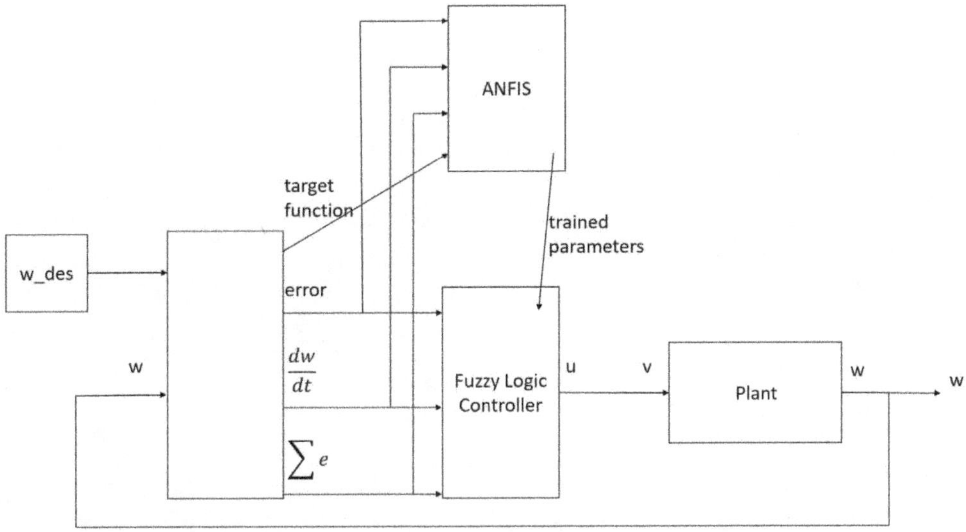

Fig. 2. Three input NFC structure.

Table 1. DC motor physical parameters.

Year	Parameters	Values
J	Moment of inertia of the rotor	0.01 kgm^2
b	Motor viscous friction constant	0.1 Nms
Ke	Electromotive force constant	0.01 $V/rad/sec$
Kt	Motor torque constant	0.01 Nm/Amp
R	Electric Resistance	1 Ohm
L	Electric Inductance	0.5 H

5. Results

The T1-NFC and IT2-NFC are implemented in the DC motor. IT2-FLS antecedent membership functions are chosen as IT2 membership functions and consequent parameters are chosen as T1 membership functions. So, the uncertainties of inputs can be represented by the FLS.

The same learning rate ($lr = 0.07$) and sample time ($dt = 10^{-4}$) is used for a fair comparison. The controllers have three inputs and every input are represented by three triangular membership functions. This means that there are 27 rules. The Sugeno Fuzzy Inference System (FIS) is used: $f_i = p_i * x_1 + q_i * x_2 + s_i * x_3 + r_i$ where $x_{1,2,3}$ are inputs and p, q, s, r are the related coefficients. So, the W matrix must be defined as a 108x1 vector.

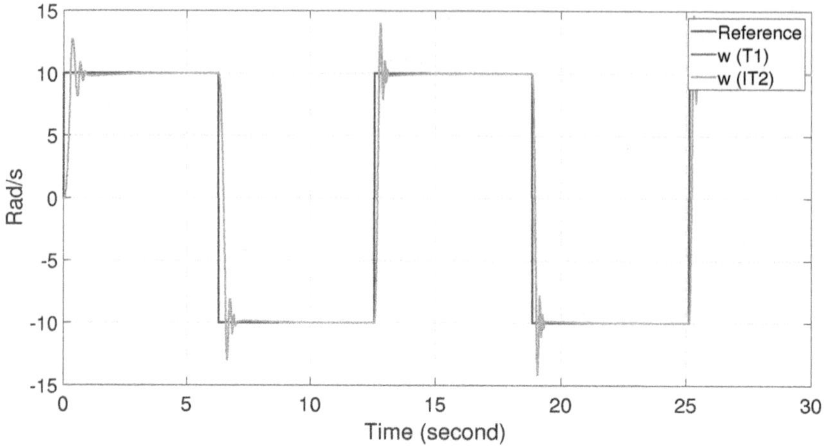

Fig. 3. NFC results for inputs (error, output derivative, sum of error) where w=10.

Table 2. Comparison of T1-NFC and IT2-NFC with three inputs (error, output derivative, sum of error) for $w = 10$.

Method	% OS	T_s	RMSE	Computation Time (second)
Sugeno NFC	27.1	1.39	0.2815	47.3358
IT2 NFC	27.5	0.83	0.2801	74.3662

IT2-FLS membership function type is chosen as A2-C1 (Antecedent MFs type is IT2-MF, consequent MFs type is T1-MF). The IT2-FLS antecedent membership functions LMF tip point is chosen as 0.5. So, the IT2-MFs include uncertainties. The results are given in Figure 3 that both of the NFCs have similar performance.

The DC motor is a Single Input Single Output System (SISO), so the difference is not precious but as seen in Table 2 the IT2-NFC is superior to T1-NFC at settling time, Root Mean Square Error (RMSE). It is clear that the IT2-NFC needs more computational load, but it can handle uncertainties and overcome unexpected situations.

6. Conclusion

The Modified Karnik-Mendel Algorithm and IT2-ANFIS algorithm are newly developed methods and it is seen that the IT2-ANFIS superior to T1-ANFIS in simulation studies. There is no IT2-ANFIS based real-time implementation in the open literature. So, as future works, it is assumed to test in real-time studies by overcoming calculation load and control complexity.

References

[1] B. Han, Y. Zhou, K. K. Deveerasetty, and C. Hu, A review of control algorithms for quadrotor, in *IEEE International Conference on Information and Automation (ICIA)*, (2018), pp. 951–956.
[2] A. Tayebi and S. McGilvray, Attitude stabilization of a VTOL quadrotor aircraft, *IEEE Transactions on Control Systems Technology*, vol. 14, no. 3, pp. 562–571, (2006).
[3] M. Öztürk, Antropomorphic Robot's Dynamics and Adaptive Control Applications: Matlab/Simulink Modeling, Istanbul Technical University, 2014.
[4] T. Shmelova, Y. Shmelov, and S. Vladov, Concept of building intelligent control systems for aircraft, unmanned aerial vehicles and aircraft engines, in *IEEE 6th International Conference on Methods and Systems of Navigation and Motion Control (MSNMC)*, (2020), pp. 14–19.
[5] S. K. Jha, P. Gaur, and A. K. Yadav, Various intelligent control techniques for attitude control of an aircraft system, in *3rd International Conference on Computing for Sustainable Global Development (INDIACom)*, (2016), pp. 2476–2480.
[6] J.-S. Jang, ANFIS: adaptive-network-based fuzzy inference system, *IEEE Transactions on Systems*, Man, and Cybernetics, vol. 23, no. 3, pp. 665–685, (1993).
[7] M. N. Uddin and H. Wen, Development of a self-tuned neuro-fuzzy controller for induction motor drives, *IEEE Transactions on Industry Applications*, vol. 43, no. 4, pp. 1108–1116, (2007).
[8] A. C. Aras, E. Kayacan, Y. Oniz, O. Kaynak, and R. Abiyev, An adaptive neuro-fuzzy architecture for intelligent control of a servo system and its experimental evaluation, in *IEEE International Symposium on Industrial Electronics*, (2010), pp. 68–73.
[9] A. C. Aras and O. Kaynak, Trajectory tracking of a 2-DOF helicopter system using neuro-fuzzy system with parameterized conjunctors, in *IEEE/ASME International Conference on Advanced Intelligent Mechatronics*, (2014), pp. 322–326.
[10] M. Öztürk and İ. Özkol, Comparison of self-tuned Neuro-Fuzzy controllers on 2 DOF helicopter: An application, *SN Applied Sciences*, vol. 3, no. 1, p. 124, (2021).
[11] L. A. Zadeh, The concept of a linguistic variable and its application to approximate reasoning—I, *Information Sciences*, vol. 8, no. 3, pp. 199–249, (1975).
[12] A. Taşkın, A Matlab/simulink Toolbox For Interval Type-2 Fuzzy Logic Systems, Istanbul Technical University, Institute of Science and Technology, M.Sc. Thesis, (2015).
[13] Q. Liang and J. M. Mendel, Interval type-2 fuzzy logic systems: theory and design, *IEEE Transactions on Fuzzy Systems*, vol. 8, no. 5, pp. 535–550, (2000).
[14] N. N. Karnik, J. M. Mendel, and Q. Liang, Type-2 fuzzy logic systems, *IEEE Transactions on Fuzzy Systems*, vol. 7, no. 6, pp. 643–658, (1999).
[15] D. Wu and J. M. Mendel, Enhanced karnik–mendel algorithms, *IEEE Transactions on Fuzzy Systems*, vol. 17, no. 4, pp. 923–934, (2008).
[16] K. Duran, H. Bernal, and M. Melgarejo, Improved iterative algorithm for computing the generalized centroid of an interval type-2 fuzzy set, in *NAFIPS 2008-2008 Annual Meeting of the North American Fuzzy Information Processing Society*, (2008), pp. 1–5.
[17] D. Wu and M. Nie, Comparison and practical implementation of type-reduction algorithms for type-2 fuzzy sets and systems, in *IEEE International Conference on Fuzzy Systems (FUZZ-IEEE 2011)*, (2011), pp. 2131–2138.
[18] C. Chen, R. John, J. Twycross, and J. M. Garibaldi, A direct approach for determining the switch points in the Karnik–Mendel algorithm, *IEEE Transactions on Fuzzy Systems*, vol. 26, no. 2, pp. 1079–1085, (2017).

[19] C. Chen, R. John, J. Twycross, and J. M. Garibaldi, An extended ANFIS architecture and its learning properties for type-1 and interval type-2 models, in *IEEE International Conference on Fuzzy Systems (FUZZ-IEEE)*, (2016), pp. 602–609.

[20] C. Chen, R. John, J. Twycross, and J. M. Garibaldi, Type-1 and interval type-2 ANFIS: A comparison, in *IEEE International Conference on Fuzzy Systems (FUZZ-IEEE)*, (2017), pp. 1–6.

[21] M. Öztürk and İ. Özkol, Comparison of modified Karnik-Mendel algorithm-based interval type-2 ANFIS and type-1 ANFIS, *Aircraft Engineering and Aerospace Technology*, vol. 93, no. 10, pp. 1526–1532, (2021).

[22] L. A. Zadeh, Fuzzy sets, *Information and Control*, vol. 8, no. 3, pp. 338–353, (1965).

[23] N. N. Karnik and J. M. Mendel, Centroid of a type-2 fuzzy set, *Information Sciences*, vol. 132, no. 1–4, pp. 195–220, (2001).

[24] M. Öztürk, *A Modified Anfis System for Aerial Vehicles Control*, Istanbul Technical University, (2022).

Chapter 3

Artificial Neural Network Solution of Two-Dimensional Convection-Diffusion Equation

Furkan Beyazlı* and Harun Selvitopi†

Department of Mathematics, Faculty of Sciences, Erzurum Technical University, Erzurum, 25050, Turkey

**furkan.beyazli07@erzurum.edu.tr*

†*harun.selvitopi@erzurum.edu.tr*

In this study, we have analyzed the two-dimensional steady state convection-diffusion equation with Dirichlet boundary conditions using feed−forward artificial neural networks. Convection-diffusion equation takes an important role in engineering, physics, environmental sciences and many disciplines. It is possible to solve this type of equations with conventional numerical methods however precision issues may appear in the problems are that have sharp boundary conditions. Therefore, this study searches the potential of solving this type of problems more effectively using a feed−forward artificial neural network (ANN) model. Obtained results shows that feed-forward ANNs can produce precise and accurate solutions of convection-diffusion equations with dirichlet boundary condition. It emphasizes ability to handle problems with this type of difficult boundary conditions of ANN. Feed-Forward ANN offers new opportunities to science, engineering and industrial applications in solving numerical differential equations. The increasing role of artificial neural networks in this field represents the capacity to solve more complex and large-scale problems more effectively.

1. Introduction

An artificial neural network (ANN) is a type of neural network inspired by the functioning of the human brain. It finds applications across various scientific fields, including data analysis and solving numerical problems associated with partial differential equations. The use of ANNs in numerical analysis is increasingly expanding. In [1], the ANN has been applied to obtain the solution of the ordinary and partial differential equations. The architecture of finite element neural networks has been developed to address the solution of differential equations in [2]. In [3], the solution of ANN of ordinary differential has been compared. The unified deep artificial neural network approach has been proposed to solve the partial differential equations in [4]. In [5], the feed-forward artificial neural network has been applied to solve differential equations. The classical Galerkin methods has been used to obtain the artificial neural network to acquire the solution of the partial differential equations in [6].

Convection-diffusion equations are used in the transfer of movement or processes, in the circulation of flows from one region, in the transfer of a gas or liquid, in fluid mechanics, and in the transfer of heat or other atmospheric properties in the atmosphere. Therefore, the numerical solution of this type of equations has been acquired various traditional numerical methods such, finite element method, finite difference method, boundary element method etc. In [7], the finite element method has been applied convection–diffusion equations to obtain the numerical solution and the error analysis has also been given. In [8], convection–diffusion equation with the discrete maximum principle has been solved numerically using finite element method. The numerical solution of the two-dimensional convection–diffusion equation has been obtained using fourth-order finite difference method in [9]. In [10], the numerical solution of the two-dimensional convection–diffusion equation has been acquired using the boundary element method. The nonlinear convection–diffusion equation has been solved numerically proposed multiple–relaxation-time lattice Boltzmann method in [11], and the extended version of this method called rectangular multiple–relaxation-time lattice Boltzmann method has also been proposed to solve the nonlinear convection–diffusion equation in [12].

In this study, we have analyzed the two-dimensional convection–diffusion equation numerically using the artificial neural network. The paper is organized as follows: In Section 2, mathematical formulation of the two-dimensional convection–diffusion equation has been given and the problem parameters has defined. The artificial neural network application for considered problem has been presented in Section 3. Numerical results for the different optimization techniques has been presented with graphics, in Section 4. Our work is summarized in Section 5.

2. Mathematical Formulation of the Two-Dimensional Convection–Diffusion Equation

The mathematical formulation of the two-dimensional convection–diffusion equation as:

$$\begin{aligned} -\epsilon\nabla^2\psi(x,y) + \mathbf{k}\nabla\psi(x,y) &= f(x,y), \\ \psi(x,y) = f_0(x,y) \qquad (x,y) &\in \partial\Omega \subseteq \mathbb{R}^2, \end{aligned} \tag{1}$$

where Ω is the solution domain and the $\partial\Omega$ is the boundary of the solution domain.

3. Artificial Neural Network for Two-Dimensional Convection–Diffusion Equation

The artificial neural network process consist of Prediction Function:

$$\psi_t = \mathbf{A}(x,y) + \mathbf{B}(x,y)\mathrm{Net}(x,y,\vec{p}) \tag{2}$$

where $A(x, y)$ as:

$$\begin{aligned}\mathbf{A}(x,y) &= (1-x)f_0(y) + xf_1(y) + (1-y)(g_0(x) - ((1-x)g_0(0) \\ &\quad + xg_0(1))) + y(g_1(x) - ((1-x)g_1(0) + xg_1(1)))\end{aligned} \tag{3}$$

$B(x, y)$ as:

$$\mathbf{B}(x,y) = x(1-x)y(1-y) \tag{4}$$

Then the derivatives of the prediction function as:

$$\begin{aligned}\psi_{tx} &= \mathbf{A}_x(x,y) + \mathbf{B}_x(x,y)\mathrm{Net}(x,y,\vec{p}) + \mathbf{B}(x,y)\mathrm{Net}_x(x,y,\vec{p}) \\ \psi_{txx} &= \mathbf{A}_{xx}(x,y) + \mathbf{B}_{xx}(x,y)\mathrm{Net}(x,y,\vec{p}) + 2\mathbf{B}_x(x,y)\mathrm{Net}_x(x,y,\vec{p}) \\ &\quad + \mathbf{B}(x,y)\mathrm{Net}_{xx}(x,y,\vec{p}) \\ \psi_{ty} &= \mathbf{A}_y(x,y) + \mathbf{B}_y(x,y)\mathrm{Net}(x,y,\vec{p}) + \mathbf{B}(x,y)\mathrm{Net}_y(x,y,\vec{p}) \\ \psi_{tyy} &= \mathbf{A}_{yy}(x,y) + \mathbf{B}_{yy}(x,y)\mathrm{Net}(x,y,\vec{p}) + 2\mathbf{B}_y(x,y)\mathrm{Net}_y(x,y,\vec{p}) \\ &\quad + \mathbf{B}(x,y)\mathrm{Net}_{yy}(x,y,\vec{p})\end{aligned} \tag{5}$$

The Net function is as:

$$\mathrm{Net}(x_j, y_j, \vec{p}) = \sum_i \{a_i \cdot \rho(z_i)\} \tag{6}$$

where

$$\begin{aligned} z_i &= -w_1 * x_j - w_2 * y_j - b \\ \rho(z_i) &= \frac{1}{1+e^{-w_{xi}*x - w_{yi}*y - b_i}}\end{aligned} \tag{7}$$

We use the derivatives of the Net function (6) with respect to x and y to calculate the derivative of the Cost function,

$$\begin{aligned}\frac{\partial Net(x_j,y_j,\vec{p})}{\partial x_j} &= \textstyle\sum_i \{a_i \cdot w_{xi}(1-\rho(z_i)) \cdot \rho(z_i)\} \\ \frac{\partial^2 Net(x_j,y_j,\vec{p})}{\partial x_j^2} &= \textstyle\sum_i \{a_i \cdot w_{xi}^2(1-\rho(z_i)) \cdot \rho(z_i) \cdot (1 - 2\cdot\rho(z_i))\} \\ \frac{\partial Net(x_j,y_j,\vec{p})}{\partial y_j} &= \textstyle\sum_i \{a_i \cdot w_{yi}(1-\rho(z_i)) \cdot \rho(z_i)\} \\ \frac{\partial^2 Net(x_j,y_j,\vec{p})}{\partial y_j^2} &= \textstyle\sum_i \{a_i \cdot w_{yi}^2(1-\rho(z_i)) \cdot \rho(z_i) \cdot (1 - 2\cdot\rho(z_i))\}\end{aligned} \tag{8}$$

We have been calculate the derivatives of the Eqs. (6) and (8) with respect to a_i and b_i as:

$$\begin{aligned}
\frac{\partial Net(x_j,y_j,\vec{p})}{\partial a_i} &= \rho(z_i) \\
\frac{\partial^2 Net(x_j,y_j,\vec{p})}{\partial x_j \partial a_i} &= w_{xi} \cdot (1-\rho(z_i)) \cdot \rho(z_i) \\
\frac{\partial^3 Net(x_j,y_j,\vec{p})}{\partial x_j^2 \partial a_i} &= w_{xi}^2 \cdot (1-\rho(z_i)) \cdot \rho(z_i) \cdot (1-2\cdot\rho(z_i)) \\
\frac{\partial^2 Net(x_j,y_j,\vec{p})}{\partial y_j \partial a_i} &= w_{yi} \cdot (1-\rho(z_i)) \cdot \rho(z_i) \\
\frac{\partial^3 Net(x_j,y_j,\vec{p})}{\partial y_j^2 \partial a_i} &= w_{yi}^2 \cdot (1-\rho(z_i)) \cdot \rho(z_i) \cdot (1-2\cdot\rho(z_i))
\end{aligned} \tag{9}$$

and

$$\begin{aligned}
\frac{\partial Net(x_j,y_j,\vec{p})}{\partial b_i} &= a_i \cdot \rho(z_i) \cdot (1-\rho(z_i)) \\
\frac{\partial^2 Net(x_j,y_j,\vec{p})}{\partial x_j \partial b_i} &= a_i \cdot w_{xi} \cdot \rho(z_i \cdot (1-\rho(z_i)) \cdot (1-2\cdot\rho(z_i)) \\
\frac{\partial^3 Net(x_j,y_j,\vec{p})}{\partial x_j^2 \partial b_i} &= a_i \cdot w_{xi}^2 \cdot \rho(z_i) \cdot (1-\rho(z_i)) \cdot (1-2\cdot\rho(z_i)) \\
\frac{\partial^2 Net(x_j,y_j,\vec{p})}{\partial y_j \partial b_i} &= a_i \cdot w_{yi} \cdot \rho(z_i) \cdot (1-\rho(z_i)) \cdot (1-2\cdot\rho(z_i)) \\
\frac{\partial^3 Net(x_j,y_j,\vec{p})}{\partial y_j^2 \partial b_i} &= a_i \cdot w_{yi}^2 \cdot \rho(z_i) \cdot (1-\rho(z_i)) \cdot (1-2\cdot\rho(z_i))
\end{aligned} \tag{10}$$

respectively. We have been calculate the derivatives of the Eqs. (6) and (8) with respect to w_{xi} and w_{yi} as:

$$\begin{aligned}
\frac{\partial Net(x_j,y_j,\vec{p})}{\partial w_{xi}} &= a_i \cdot x_j \cdot \rho(z_i) \cdot (1-\rho(z_i)) \\
\frac{\partial^2 Net(x_j,y_j,\vec{p})}{\partial x_j \partial w_{xi}} &= a_i \cdot \rho(z_i) \cdot (1-\rho(z_i)) \cdot (1 + x \cdot w_x \cdot (1-2\rho(z_i))) \\
\frac{\partial^3 Net(x_j,y_j,\vec{p})}{\partial x_j \partial x_j \partial w_{xi}} &= a_i \cdot w_{ix}\rho(z_i) \cdot (1-\rho(z_i)) \cdot (2 + x \cdot w_x \cdot (1-2\rho(z_i))) \\
\frac{\partial^2 Net(x_j,y_j,\vec{p})}{\partial y_j \partial w_{xi}} &= x_j \cdot a_i \cdot w_{iy} \cdot \rho(z_i) \cdot (1-\rho(z_i)) \cdot (1-2\rho(z_i)) \\
\frac{\partial^3 Net(x_j,y_j,\vec{p})}{\partial y_j \partial y_j \partial w_{xi}} &= x_j \cdot a_i \cdot w_{iy}^2 \cdot \rho(z_i) \cdot (1-\rho(z_i)) \cdot (1-2\rho(z_i))
\end{aligned} \tag{11}$$

and

$$
\begin{aligned}
\frac{\partial Net(x_j,y_j,\vec{p})}{\partial w_{yi}} &= a_i \cdot y_j \cdot \rho(z_i) \cdot (1-\rho(z_i)) \\
\frac{\partial^2 Net(x_j,y_j,\vec{p})}{\partial x_j \partial w_{yi}} &= y_j \cdot a_i \cdot w_{ix} \cdot \rho(z_i) \cdot (1-\rho(z_i)) \cdot (1-2\rho(z_i)) \\
\frac{\partial^3 Net(x_j,y_j,\vec{p})}{\partial x_j \partial x_j \partial w_{yi}} &= y_j \cdot a_i \cdot w_{ix}^2 \cdot \rho(z_i) \cdot (1-\rho(z_i)) \cdot (1-2\rho(z_i)) \\
\frac{\partial^2 Net(x_j,y_j,\vec{p})}{\partial y_j \partial w_{yi}} &= a_i \cdot \rho(z_i) \cdot (1-\rho(z_i)) \cdot (1+y_j \cdot w_y \cdot (1-2\rho(z_i))) \\
\frac{\partial^3 Net(x_j,y_j,\vec{p})}{\partial y_j \partial y_j \partial w_{yi}} &= a_i \cdot w_{iy}\rho(z_i) \cdot (1-\rho(z_i)) \cdot (2+y \cdot w_y \cdot (1-2\rho(z_i)))
\end{aligned}
\tag{12}
$$

respectively.

The definition of the $B(x,y)$ as:

$$\mathbf{B}(x,y) = x \cdot (1-x) \cdot y \cdot (1-y) \tag{13}$$

and we have been obtained the derivatives of the $B(x,y)$ with respect to x and y

$$
\begin{aligned}
\mathbf{B}_x(x,y) &= y \cdot (2x-1) \cdot (1-y) \\
\mathbf{B}_{xx}(x,y) &= 2 \cdot y \cdot (y-1) \\
\mathbf{B}_y(x,y) &= x \cdot (x-1) \cdot (2 \cdot y - 1) \\
\mathbf{B}_{yy}(x,y) &= 2 \cdot x \cdot (x-1).
\end{aligned}
\tag{14}
$$

The definition of the $A(x,y)$ as:

$$
\begin{aligned}
\mathbf{A}(x,y) = &(1-x) \cdot f_0(y) + x \cdot f_1(y) + (1-y) \cdot (g_0(x) - ((1-x) \cdot g_0(0) + x \cdot g_0(1))) \\
&+ y \cdot (g_1(x) - ((1-x) \cdot g_1(0) + x \cdot g_1(1)))
\end{aligned}
\tag{15}
$$

and we have been obtained the derivatives of the $A(x,y)$ with respect to x and y

$$
\begin{aligned}
\mathbf{A}_x(x,y) = &-f_0(y) + f_1(y) - f_0(y) + (1-y) \cdot (g_{0x}(x) - (g_0(0) + g_0(1))) \\
&+ y \cdot (g_{1x}(x) + g_1(0) - g_1(1))
\end{aligned}
\tag{16}
$$

$$
\begin{aligned}
\mathbf{A}_y(x,y) = &x \cdot f_{1y}(y) + (1-x) \cdot f_{0y}(y) - (g_0(x) - (1-x) \cdot g_0(0) - x \cdot g_0(1)) \\
&+ g_1(x) - (1-x) \cdot g_1(0) - x \cdot g_1(1)
\end{aligned}
\tag{17}
$$

$$\mathbf{A}_{xx}(x,y) = (1-y) \cdot g_{0xx}(x) + y \cdot g_{1xx}(x) \tag{18}$$

$$\mathbf{A}_{yy}(x,y) = (1-x) \cdot f_{0yy}(y) + x \cdot f_{1yy}(y) \tag{19}$$

Then the Cost Function and its calculated derivatives given as:

$$E = \sum_i \frac{1}{2} \left\{ -\epsilon \left(\frac{\partial^2 \psi(x_i,y_i)}{\partial x^2} + \frac{\partial^2 \psi(x_i,y_i)}{\partial y^2} \right) + k \left(\frac{\partial \psi(x_i,y_i)}{\partial x} + \frac{\partial \psi(x_i,y_i)}{\partial y} - f(x_i,y_i) \right) \right\}^2 \tag{20}$$

$$\frac{\partial E}{\partial a_i} = \sum_{j=1}^{n} \left(\frac{\partial^2 \psi_t}{\partial x_j \partial x_j} + \frac{\partial^2 \psi_t}{\partial y_j \partial y_j} + \frac{\partial \psi_t}{\partial x_j} + \frac{\partial \psi_t}{\partial y_j} - f(x_j, y_j, \vec{p}) \right) \left(\frac{\partial^3 \psi_t}{\partial a_i \partial x_j \partial x_j} + \frac{\partial^3 \psi_t}{\partial a_i \partial y_j \partial y_j} + \frac{\partial^2 \psi_t}{\partial a_i \partial x_j} + \frac{\partial^2 \psi_t}{\partial a_i \partial y_j} \right) \tag{21}$$

$$\frac{\partial E}{\partial b_i} = \sum_{j=1}^{n} \left(\frac{\partial^2 \psi_t}{\partial x_j \partial x_j} + \frac{\partial^2 \psi_t}{\partial y_j \partial y_j} + \frac{\partial \psi_t}{\partial x_j} + \frac{\partial \psi_t}{\partial y_j} - f(x_j, y_j, \vec{p}) \right) \left(\frac{\partial^3 \psi_t}{\partial b_i \partial x_j \partial x_j} + \frac{\partial^3 \psi_t}{\partial b_i \partial y_j \partial y_j} + \frac{\partial^2 \psi_t}{\partial b_i \partial x_j} + \frac{\partial^2 \psi_t}{\partial b_i \partial y_j} \right) \tag{22}$$

$$\frac{\partial E}{\partial w_{xi}} = \sum_{j=1}^{n} \left(\frac{\partial^2 \psi_t}{\partial x_j \partial x_j} + \frac{\partial^2 \psi_t}{\partial y_j \partial y_j} + \frac{\partial \psi_t}{\partial x_j} + \frac{\partial \psi_t}{\partial y_j} - f(x_j, y_j, \vec{p}) \right) \left(\frac{\partial^3 \psi_t}{\partial w_{xi} \partial x_j \partial x_j} + \frac{\partial^3 \psi_t}{\partial w_{xi} \partial y_j \partial y_j} + \frac{\partial^2 \psi_t}{\partial w_{xi} \partial x_j} + \frac{\partial^2 \psi_t}{\partial w_{xi} \partial y_j} \right) \tag{23}$$

$$\frac{\partial E}{\partial w_{yi}} = \sum_{j=1}^{n} \left(\frac{\partial^2 \psi_t}{\partial x_j \partial x_j} + \frac{\partial^2 \psi_t}{\partial y_j \partial y_j} + \frac{\partial \psi_t}{\partial x_j} + \frac{\partial \psi_t}{\partial y_j} - f(x_j, y_j, \vec{p}) \right) \left(\frac{\partial^3 \psi_t}{\partial w_{yi} \partial x_j \partial x_j} + \frac{\partial^3 \psi_t}{\partial w_{yi} \partial y_j \partial y_j} + \frac{\partial^2 \psi_t}{\partial w_{yi} \partial x_j} + \frac{\partial^2 \psi_t}{\partial w_{yi} \partial y_j} \right) \tag{24}$$

Derivative of cost function according to weights in Gradient Descent method as: $\frac{\partial E}{\partial \vec{p}}$

$$\rho = \frac{1}{1 + e^{(-w_1 * x - w_2 * y - b)}} \tag{25}$$

$$a_i = a_i - \mu \frac{\partial E}{\partial a_i} \tag{26}$$

$$w_{xi} = w_{xi} - \mu \frac{\partial E}{\partial w_{xi}} \tag{27}$$

$$w_{yi} = w_{yi} - \mu \frac{\partial E}{\partial w_{yi}} \tag{28}$$

$$b_i = b_i - \mu \frac{\partial E}{\partial b_i} \tag{29}$$

4. ANN Results

$$-2U_x - 2U_y - U_{xx} - U_{yy} = 4x(y - y^2) + 4y(x - x^2) \tag{30}$$

Solution Domain:

$$x, y \in [0, 1] \tag{31}$$

Exact Solution:

$$u(x, y) = xy(1 - x)(1 - y) \tag{32}$$

One can observe the ANN solution using Tanh activation function, SGD Optimization Technique of the problem (30), in the Figure 1. The ANN solution using Tanh

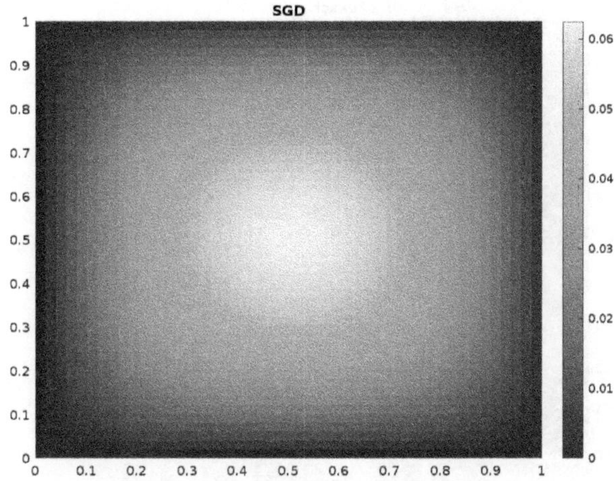

Fig. 1. ANN Solution of the problem (30) using Tanh activation function, SGD optimization technique.

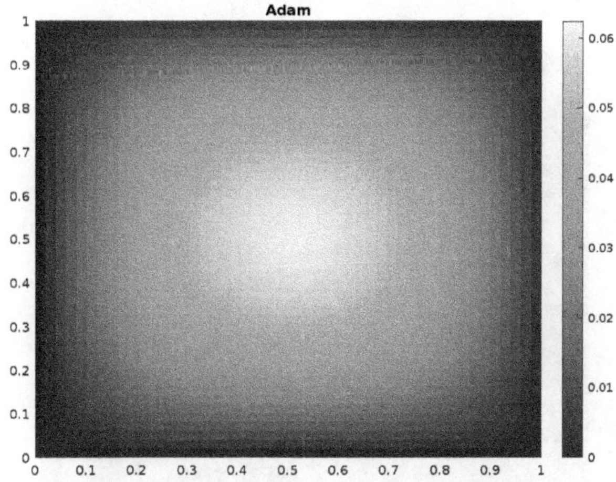

Fig. 2. ANN Solution of the problem (30) using Tanh activation function, Adam optimization technique.

activation function, Adam Optimization Technique of the problem (30) has been given, in the Figure 2.

Figure 3 shows the exact solution of the problem (30).

Then the error between the ANN solution using Tanh activation function with the SGD Optimization Technique and Exact Solution in the Figure 4.

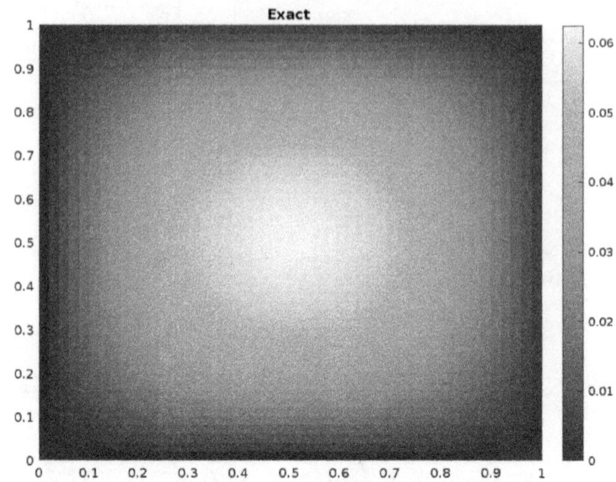

Fig. 3. Exact solution of the problem (30).

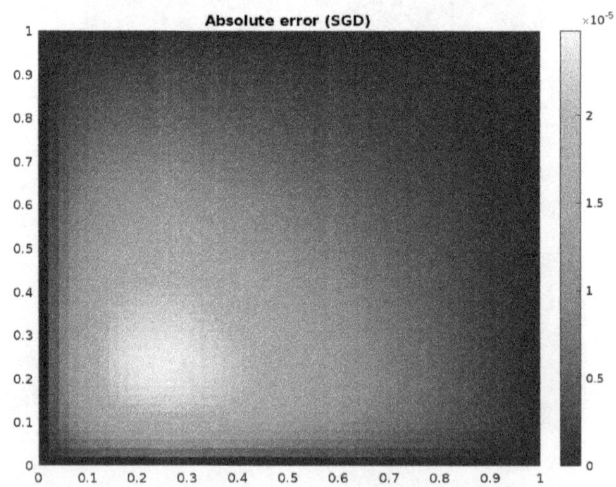

Fig. 4. Error between the ANN solution using Tanh activation function with the SGD Optimization Technique and Exact Solution

One can observe the error between the ANN solution using Tanh activation function with the Adam Optimization Technique and Exact Solution in the Figure 5.

We can say that the error between ANN solution and exact solution smaller than the 10^{-5} using Tanh activation function with the both SGD and Adam Optimization Techniques.

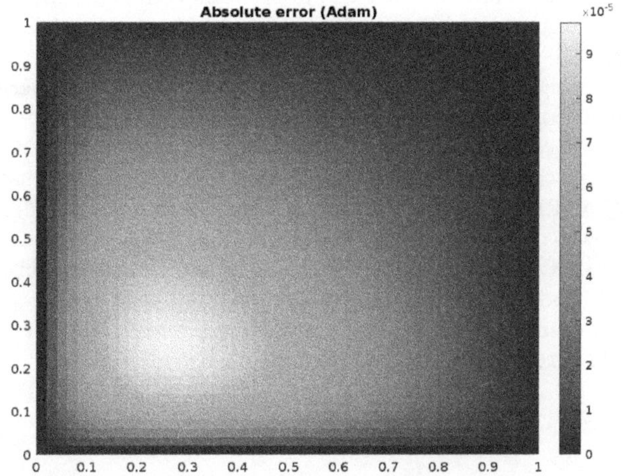

Fig. 5. Error between the ANN solution using Tanh activation function with the Adam Optimization Technique and Exact Solution

References

[1] Lagaris, I. E., Likas, A., & Fotiadis, D. I. (1998). Artificial neural networks for solving ordinary and partial differential equations. *IEEE transactions on neural networks*, **9**(5), 987–1000. `doi:10.1109/72.712178`.

[2] Ramuhalli, P., Udpa, L., & Udpa, S. S. (2005). Finite-element neural networks for solving differential equations. *IEEE transactions on neural networks*, **16**(6), 1381–1392. `doi:10.1109/TNN.2005.857945`.

[3] Mall, S., & Chakraverty, S. (2013). Comparison of artificial neural network architecture in solving ordinary differential equations. *Advances in Artificial Neural Systems*, **2013**, 1–12. `doi:10.1155/2013/181895`.

[4] Berg, J., & Nyström, K. (2018). A unified deep artificial neural network approach to partial differential equations in complex geometries. *Neurocomputing*, **317**, 28–41. `doi:10.1016/j.neucom.2018.06.056`.

[5] Wojciechowski, M. (2012). Solving differential equations by means of feed-forward artificial neural networks. In Artificial Intelligence and Soft Computing: 11th International Conference, ICAISC 2012, Zakopane, Poland, April 29-May 3, 2012, Proceedings, Part I 11, pp. 187–195, *Springer Berlin Heidelberg.*

[6] Rudd, K., & Ferrari, S. (2015). A constrained integration (CINT) approach to solving partial differential equations using artificial neural networks. *Neurocomputing*, **155**, 277–285. `doi:10.1016/j.neucom.2014.11.058`.

[7] Babuška, I., & Szymczak, W. G. (1982). An error analysis for the finite element method applied to convection diffusion problems. *Computer Methods in Applied Mechanics and Engineering*, **31**(1), 19–42. `doi:10.1016/0045-7825(82)90044-5`.

[8] Barrenechea, G. R., John, V., & Knobloch, P. (2024). Finite element methods respecting the discrete maximum principle for convection-diffusion equations. *SIAM Review*, **66**(1), 3–88. `doi:10.1137/22M1488934`.

[9] Li, L., Jiang, Z., & Yin, Z. (2018). Fourth-order compact finite difference method for solving two-dimensional convection–diffusion equation. *Advances in Difference Equations*, 2018, 1–24. doi:10.1186/s13662-018-1652-5.

[10] Romero, L. M., & Benitez, F. G. (2008). A boundary element numerical scheme for the two-dimensional convection–diffusion equation. *International journal for numerical methods in engineering*, **76**(13), 2063–2090. doi:10.1002/nme.2391.

[11] Chai, Z., & Shi, B. (2020). Multiple-relaxation-time lattice Boltzmann method for the Navier-Stokes and nonlinear convection-diffusion equations: Modeling, analysis, and elements. *Physical Review E*, **102**(2), 023306. doi:10.1103/PhysRevE.102.023306.

[12] Chai, Z., Yuan, X., & Shi, B. (2023). Rectangular multiple-relaxation-time lattice Boltzmann method for the Navier-Stokes and nonlinear convection-diffusion equations: General equilibrium and some important issues. *Physical Review E*, **108**(1), 015304. doi:10.1103/PhysRevE.108.015304.

Chapter 4

DRBEM Solutions of Inductionless MHD Flow in a Rectangular Duct with Variably Conducting Walls

Elif Ebren Kaya
The Scientific and Technological Research Council of Turkey, Ankara, Turkey
elifebren90@gmail.com

The two-dimensional laminar, fully-developed magnetohydrodynamic (MHD) flow of a viscous and incompressible fluid is studied in a long pipe of rectangular cross-section under the effect of a vertically applied magnetic field. The induced magnetic field is neglected due to the small magnetic Reynolds number, however, the electric potential is also solved from the divergence of Ohm's law. The governing non-dimensional coupled momentum and electric potential equation (divergence of current density) are solved by using the dual reciprocity boundary element method (DRBEM) with the fundamental solution of Laplace's equation. It has been shown that, the thickness of the boundary layers are controlled by the Hartmann number, however, the wall conductivity determines the structure of the flow.

1. Introduction

The interaction between electrically conducting fluid and a magnetic field has great importance because of its applications such as in the design of MHD generators, pumps, fusion reactors, flowmeters and bearings. The MHD effects originated from the liquid metal flow in a pipe under an applied magnetic field cause some changes on the fluid pressure. For the flow in a duct with conducing walls, the induced currents are closed within the flow domain. However, the currents return through the conducting walls and enhance the MHD pressure drop [1]. Therefore, the electrical conductivity of the duct walls influences the flow behavior greatly [2]. Thus, in order to understand the MHD effects of liquid metal flows better, it is required to model MHD flows through Maxwell's equations and the Navier-Stokes equations under an imposed magnetic field. Since they have many real life applications, there have been several experimental and numerical studies on MHD flows through ducts having different wall conductivity under an effect of the applied magnetic field. The followings give solution of the MHD flow in a rectangular duct under a magnetic field with conducting or non-conducting walls using different numerical methods. Finite element method (FEM) [3–5], boundary element method (BEM) [6–8] have been used mostly for the solutions. Bühler and Horanyi [9] have examined the liquid-metal MHD flows experimentally in a conducting duct. They have considered inductionless transient MHD flow. They identified different flow regimes depending on the Hartmann and Reynolds numbers such as stable flow regime, unstable flow regime and second unstable flow regime. The duct walls conductivities have also an important effect on the flow behavior as investigated in the study [10]. Sterl [11] has also considered MHD flow in rectangular ducts. He has studied MHD flow under an applied magnetic field to examine the effect of the wall conductivities and several Hartmann number values on the flow behavior.

In this paper,we have considered the two-dimensional, laminar, fully-developed MHD flow of a viscous, imcompressible and electrically conducting fluid in a pipe, subjected to a magnetic field. The numerical solutions of MHD flow in a rectangular duct with variably conducting walls under a vertically applied magnetic field are developed based on the DRBEM with the fundamental solution of Laplace equation. The space derivatives in the equations and boundary conditions are computed by using coordinate matrix of DRBEM. The effects of variably conducting walls on the flow behavior are deeply studied and simulated in terms of the velocity and electric potential. It has been seen that as Hartmann number increases the flow is flattened, boundary layers become thinner for all the cases of (non-conducting and variably conducting) conductivity of the duct walls. The flow and electric potential align in the direction of the applied magnetic field and the velocity is higher on the side layers leaving the central region of the duct stagnant. The fluid attempts to form M-shape for variably conducting walls and the magnitude of the flow depends on the conductivity ratio of the side walls.

2. Problem Definition and Mathematical Formulation

We consider the MHD flow in an infinitely long pipe of rectangular cross-section which is placed in a vertically applied magnetic field B_0 (Figure 1). The flow is assumed to be laminar, fully-developed in a rectangular duct. The electrically conducting fluid is subjected to a magnetic field applied vertically on the pipe-axis.

The non-dimensional coupled momentum and electric potential equation (divergence of current density) are given in $(x, y) \in [-1, 1]$ as

$$\begin{aligned} \nabla^2 \omega - Ha^2 \omega &= -1 + Ha^2 \frac{\partial \phi}{\partial x} \\ \nabla^2 \phi &= -\frac{\partial \omega}{\partial x} \end{aligned} \tag{1}$$

where $Ha = B_0 L_0 \sqrt{\sigma}/\sqrt{\mu}$ is the Hartmann number, and L_0, σ and μ are the characteristic length of the duct, electrical conductivity and viscosity of the fluid, respectively.

The no-slip velocity condition is imposed for the flow on the walls of the duct. However, the Neumann or mixed type boundary conditions are used for the electric

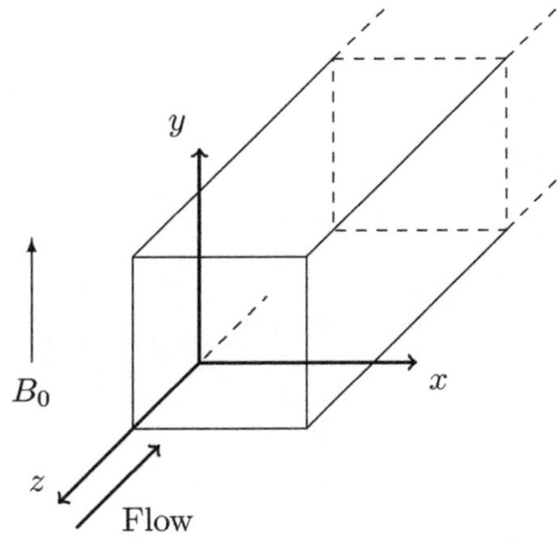

Fig. 1. Rectangular pipe

potential depending on the conductivity of the materials of the walls. Therefore, the boundary conditions can be written as

$$\begin{aligned}
&\omega(x,\pm 1)=\omega(\pm 1,y)=0 \quad \text{no-slip velocity and}\\
&\frac{\partial\phi}{\partial y}(x,\pm 1)=\frac{\partial\phi}{\partial x}(\pm 1,y)=0 \quad \text{non-conducting walls}\\
&\qquad\qquad \text{or}\\
&\omega(x,\pm 1)=\omega(\pm 1,y)=0 \quad \text{no-slip velocity and}\\
&\left.\begin{aligned}
\pm\frac{\partial\phi}{\partial y}(x,\pm 1)&=c\frac{\partial^2\phi}{\partial x^2}(x,\pm 1)\\
\pm\frac{\partial\phi}{\partial x}(\pm 1,y)&=c\frac{\partial^2\phi}{\partial y^2}(\pm 1,y)
\end{aligned}\right\}\ \text{variably conducting walls}
\end{aligned} \tag{2}$$

where c denotes the conductance ratio of the four walls (c_t (top), c_b (bottom), c_l (left), c_r (right)) which can be taken as different values.

3. Application of the DRBEM

The DRBEM procedure is used to solve to the coupled steady MHD Eq. (1) with the fundamental solution of Laplace equation, $u^* = \dfrac{ln(1/r)}{2\pi}$. The terms other than Laplacian are considered as inhomogeneities. Weighting these non-homogenous terms by u^* over Ω and applying Green's second identity twice, the obtained boundary-domain integral equations are

$$\begin{aligned}
c_i\omega_i+\int_\Gamma q^*\omega d\Gamma-\int_\Gamma u^*\frac{\partial\omega}{\partial n}d\Gamma&=-\int_\Omega(-1+Ha^2\omega+Ha^2\frac{\partial\phi}{\partial x})u^*d\Omega\\
c_i\phi_i+\int_\Gamma q^*\phi d\Gamma-\int_\Gamma u^*\frac{\partial\phi}{\partial n}d\Gamma&=-\int_\Omega(-\frac{\partial\omega}{\partial x})u^*d\Omega
\end{aligned} \tag{3}$$

where $q^* = \dfrac{\partial u^*}{\partial n}$, the source point $i \in [1, N]$, N denotes the number of constant boundary elements. The constant c_i equals 0.5 or 1 for boundary solutions or interior computations, respectively.

The domain integrals in Eq. (3) can be converted to boundary integrals, expanding the integrands by radial basis functions $f_j(x,y)$ as

$$\begin{aligned} b_1(x,y) &= -\left(-1 + Ha^2\omega + Ha^2\frac{\partial\phi}{\partial x}\right) = \sum_{j=1}^{N+L} \alpha_j f_j(x,y) \\ b_2(x,y) &= -\left(-\frac{\partial\omega}{\partial x}\right) = \sum_{j=1}^{N+L} \beta_j f_j(x,y) \end{aligned} \tag{4}$$

where $f_j(x,y)$ are connected to the particular solutions $\hat{u}_j$ with $\nabla^2\hat{u}_j = f_j$, the coefficients α_j's and β_j's are undetermined coefficients and L denotes the number of interior nodes.

Substituting $f_j = \nabla^2\hat{u}_j$ in the Eq. (4) and applying Green's second identity twice again to the right hand sides of the Eq. (3), the boundary integral equations are obtained as

$$\begin{aligned} c_i\omega_i + \int_\Gamma q^*\omega d\Gamma - \int_\Gamma u^*\frac{\partial\omega}{\partial n} d\Gamma &= \sum_{j=1}^{N+L} \alpha_j\left(c_i\hat{u}_{ij} + \int_\Gamma q^*\hat{u}_j d\Gamma - \int_\Gamma u^*\hat{q}_j d\Gamma\right) \\ c_i\phi_i + \int_\Gamma q^*\phi d\Gamma - \int_\Gamma u^*\frac{\partial\phi}{\partial n} d\Gamma &= \sum_{j=1}^{N+L} \beta_j\left(c_i\hat{u}_{ij} + \int_\Gamma q^*\hat{u}_j d\Gamma - \int_\Gamma u^*\hat{q}_j d\Gamma\right) \end{aligned} \tag{5}$$

where $\hat{q}_j = \dfrac{\partial\hat{u}_j}{\partial n} = \dfrac{\partial\hat{u}_j}{\partial x}\dfrac{\partial x}{\partial n} + \dfrac{\partial\hat{u}_j}{\partial y}\dfrac{\partial y}{\partial n}$, $i = 1, ..., N$.

Constructing the matrices $\hat{U}$, $\hat{Q}$ and coordinate matrix F by taking the vectors $\hat{u_j}$, $\hat{q_j}$ and $f_{ij} = 1 + r_{ij}$, r_{ij} being the distance from the point i to the point j, as columns respectively, and evaluating all the values at $N + L$ points, the following equations in matrix-vector form are obtained

$$\begin{aligned} H\omega - G\frac{\partial\omega}{\partial n} &= (H\hat{U} - G\hat{Q})F^{-1}\left\{-1 + Ha^2\omega + Ha^2\frac{\partial\phi}{\partial x}\right\} \\ H\phi - G\frac{\partial\phi}{\partial n} &= (H\hat{U} - G\hat{Q})F^{-1}\left\{-\frac{\partial\omega}{\partial x}\right\}. \end{aligned} \tag{6}$$

The components of the H and G matrices for a constant element are given as

$$\begin{aligned} H_{ij} &= c_i\delta_{ij} + \frac{1}{2\pi}\int_{\Gamma_j}\frac{\partial}{\partial n}\left(ln\left(\frac{1}{r}\right)\right)d\Gamma_j, \quad H_{ii} = -\sum_{j=1, j\neq i}^{N} H_{ij} \\ G_{ij} &= \frac{1}{2\pi}\int_{\Gamma_j} ln\left(\frac{1}{r}\right)d\Gamma_j, \quad G_{ii} = \frac{l}{2\pi}\left(ln\left(\frac{2}{l}\right) + 1\right) \end{aligned} \tag{7}$$

where r is the modulus of the distance vector from the point i to element j, δ_{ij} is the Kronecker delta function, l is the length of the constant elements.

When the space derivatives for ω and ϕ are computed by using the coordinate function F as

$$\frac{\partial\phi}{\partial x} = \frac{\partial F}{\partial x}F^{-1}\phi,\ \frac{\partial\omega}{\partial x} = \frac{\partial F}{\partial x}F^{-1}\omega \tag{8}$$

The Eq. in (6) become

$$\begin{aligned} (H - Ha^2K)\omega - G\frac{\partial\omega}{\partial n} - Ha^2K\left(\frac{\partial F}{\partial x}F^{-1}\right)\phi &= K\{-1\} \\ H\phi - G\frac{\partial\phi}{\partial n} + K\left(\frac{\partial F}{\partial x}F^{-1}\right)\omega &= 0 \end{aligned} \tag{9}$$

where $K = (H\hat{U} - G\hat{Q})F^{-1}$.

Enlarged system can be constructed as

$$\begin{bmatrix} H_1 & H_2 \\ H_3 & H_4 \end{bmatrix}\begin{Bmatrix} \omega \\ \phi \end{Bmatrix} - \begin{bmatrix} G & 0 \\ 0 & G \end{bmatrix}\begin{Bmatrix} \frac{\partial\omega}{\partial n} \\ \frac{\partial\phi}{\partial n} \end{Bmatrix} = \begin{Bmatrix} b_1 \\ 0 \end{Bmatrix} \tag{10}$$

where

$$\begin{aligned} H_1 &= H - Ha^2K, \quad H_2 = -Ha^2K\frac{\partial F}{\partial x}F^{-1}, \\ H_3 &= K\frac{\partial F}{\partial x}F^{-1}, \quad H_4 = H, \quad b_1 = K\{-1\}. \end{aligned} \tag{11}$$

The velocity ω and the electric potential ϕ will be obtained at $N + L$ boundary and interior points wherever they are unknowns.

4. Numerical Results and Discussion

The coupled Eq. (10) in matrix-vector form with the entries of the matrices given in Eq. (11) is solved with the different types of boundary conditions given in (2) for Hartmann number values, Ha, ranging as $30, 50, 100$. For $Ha = 30$ and 50, $N = 200$ constant boundary elements are used however $Ha = 100$ requires $N = 300$ boundary elements since the first equation in (1) becomes reaction dominant. Using large N for increasing value of Ha, increases the accuracy of the solution. Since it uses more clustered points close to the walls. Thus it captures boundary layer profiles very well.

From the Table 1 using $N = 200$ constant boundary elements is suitable for numerical computations since there is considerably small difference between the maximum value of the velocity ω and the elapsed time for obtaining numerical results. Also all the integrals in Eq. 6 are first transformed from their intervals on the boundary to the interval $[-1, 1]$. Then, the integrals in Eq. 6 are computed numerically using Gauss-Legendre with 16 points. Moreover, the coefficient matrix is solving the system in Eq. 5 is not sparse, it is a full matrix but there are lots of zeros.

Table 1. The comparison of different constant boundary elements numbers N in terms of the maximum value of velocity ω and the CPU times for $Ha = 30$ for the non-conducting duct walls case

N	ω_{max}	CPU times
120	0.0269	1.002 sec
160	0.0284	3.239 sec
200	0.0295	8.130 sec
240	0.0302	23.082 sec

4.1. *Non-conducting duct walls*

The equations in (10) are solved with no-slip and non-conducting duct walls ($c = 0$, $\frac{\partial \phi}{\partial n} = 0$). The flow behavior is examined for increasing Ha values. The numerical results are shown in Figure 2 in terms of the contour plots of ω, ϕ and ω level curves. It can be observed from Figure 2 that, increasing Ha forms the boundary layers (Hartmann and side layers on perpendicular and parallel walls to the applied magnetic field, respectively) and drops the magnitude of ω. As Ha increases flow is flattened, which is the well-known behavior of MHD flow. The maximum value of the velocities are $\omega_{max} = 0.0295$, $\omega_{max} = 0.0163$ and $\omega_{max} = 0.0082$ corresponding to Ha values as $30, 50, 100$. This means that, applied magnetic field has a slowing down effect on the fluid flow. Thus, the pressure on the fluid drops. This is usually used in controlling the pressure of liquid metals in nuclear power plants. The magnitude of ϕ drops, too. As Ha increases contour plots of ϕ aligns in the direction of applied magnetic field. When Ha is small, the flow is maximum at the centre of the duct and the velocity of the flow drops to zero at the walls, owing to the boundary condition of ω. For large Ha the centre of the duct becomes stagnant for the fluid. This parabolic velocity profile in the level curves is also observed in study [11].

4.2. *Non-conducting side walls and variably conducting Hartmann walls*

The behavior of MHD flow equations under a magnetic field is examined with non-conducting side walls and variably conducting Hartmann walls (i.e $c_l = c_r = 0$ and $c_b = c_t = 0.1$) with no-slip velocity and increasing Ha values. The coupled Eq. (10) with the entries of the matrices given in Eq. (11) are solved iteratively for ω and ϕ using the initial values $\phi^0 = 0$. The numerical results are obtained with a tolerance $\dfrac{||\omega^{n+1} - \omega^n||_\infty}{||\omega^n||_\infty} < 10^{-2}$ and $\dfrac{||\phi^{n+1} - \phi^n||_\infty}{||\phi^n||_\infty} < 10^{-2}$ and ϕ is relaxed using the equation $\phi^{n+1} = (1 - \alpha)\phi^n + \alpha\phi^{n+1}$ with a relaxation parameter taken as $\alpha = 0.01$. The numerical results are shown in Figure 3 in terms of ω, ϕ and ω level curves. The flow aligns in terms of two loops parallel to applied magnetic

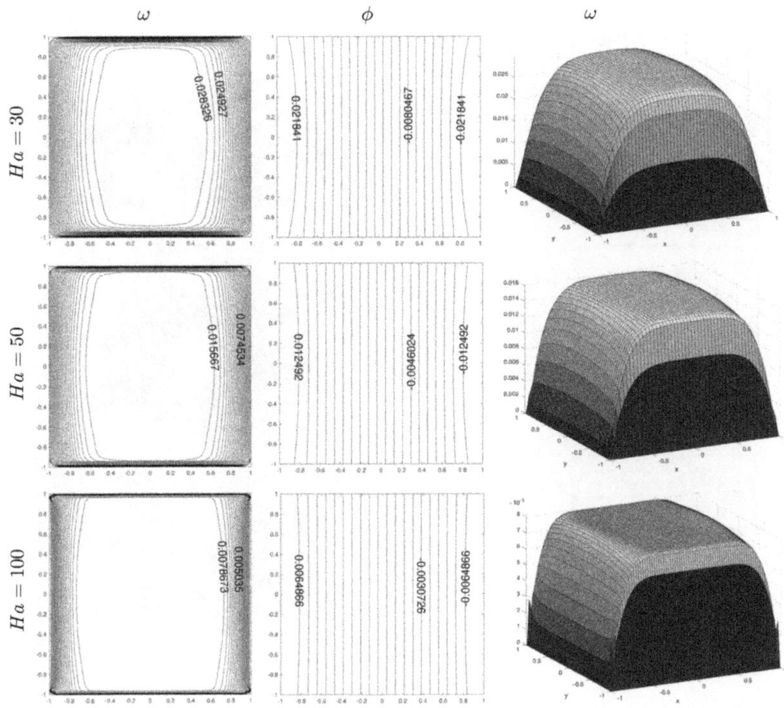

Fig. 2. Velocity and electric potential profiles with no-slip and non-conducting duct walls.

field in front of the side walls. That means, the fluid mainly flows in front of the side walls. As Ha increases the ω and ϕ behaviors stay the same with a decrease in magnitudes. The maximum value of the velocities are $\omega_{max} = 0.0151$, $\omega_{max} = 0.0064$ and $\omega_{max} = 0.0022$ corresponding to Ha values as $30, 50, 100$. An increase in the conductivity of the top and bottom walls causes a drop in ω and ϕ magnitudes compared to the case of non-conducting duct walls. The fluid is nearly stagnant in the centre of the duct. Since higher ω values occurring on the side layers the flow shows M-shape profile as increasing Ha values. The obtained numerical results validate our method by comparing the finite difference results given in [11].

4.3. *Variably conducting walls with the same conductivity ratio*

In addition to the influence of the magnetic field on the flow behavior, the effect of the wall conductance ratio on the flow is analyzed in this section. Therefore, ϕ satisfies mixed type boundary conditions as given in Eq. (2) for variably conducting walls with $c = 0.1$ (i.e $c_t = c_b = c_t = c_r = 0.1$). It requires to solve the MHD equations in coupled form (10) for ω and ϕ iteratively with the initial values $\phi^0 = 0$. The numerical results shown in Figure 4 are obtained with the same tolerance and relaxation parameter used in previous case in Section 4.2. From the Figure 4, it can

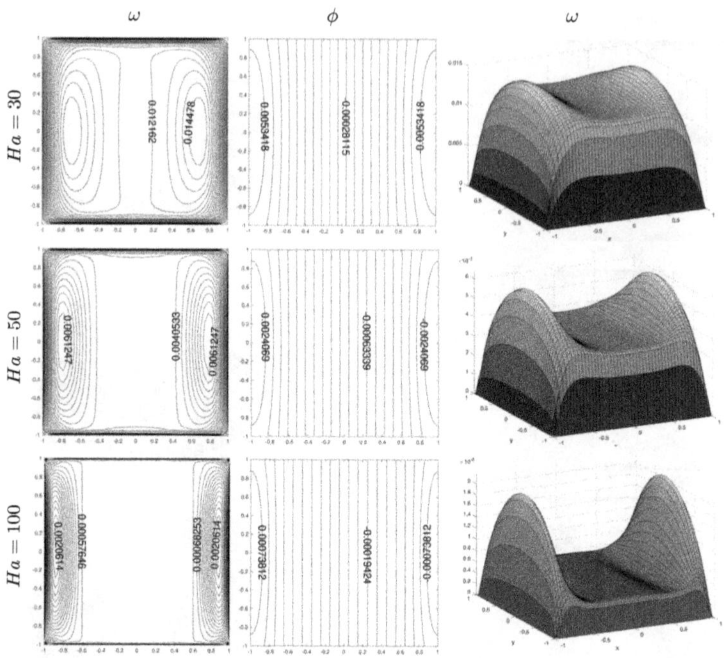

Fig. 3. Velocity and electric potential profiles with no-slip and, non-conducting side walls, variably conducting Hartmann walls.

be observed that, ω and ϕ magnitudes slightly drop compared to the previous case in Section 4.2. The maximum value of the velocities are $\omega_{max} = 0.0136$, $\omega_{max} = 0.0057$ and $\omega_{max} = 0.0019$ corresponding to Ha values as $30, 50, 100$. Furthermore, the formulation of boundary layers and the M-shape of the flow behavior are accelerated by an increase in the conductivity ratio at all walls. Thus, increasing both Ha values and walls conductivity ratios cause the fluid has a slow motion in the pipe.

4.4. *Variably conducting walls with different conductivity ratios*

In this case, the effect of different wall conductivities on the MHD flow behavior are examined for Ha values as $30, 50$ and 100. The three walls of the duct have high electrical conductivity whereas the left wall has a small electrical conductivity such as $c_b = c_t = c_l = 0.1$ and $c_r = 5 \times 10^{-3}$. The MHD flow equations in (10) are solved for ω and ϕ iteratively with the same tolerance, relaxation parameter and initial value of ϕ. The numerical results can be seen from Figure 5. Since the side walls have different conductivities, the flow turbulence is not symmetric which occurs with the effect of increasing Ha values. The maximum value of the velocities are $\omega_{max} = 0.0148$, $\omega_{max} = 0.0060$ and $\omega_{max} = 0.0021$ corresponding to Ha values as $30, 50, 100$. Thus, ω and ϕ magnitudes slightly increase when it compared to the magnitudes in Section 4.3. This difference is due to the drop in

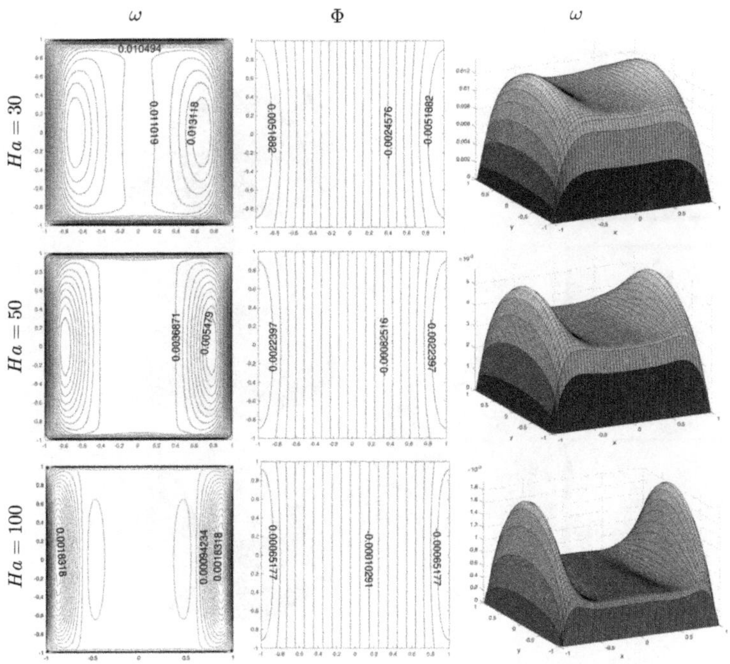

Fig. 4. Velocity and electric potential profiles with no-slip and well-conducting duct walls.

the conductivity on the right side wall. The ω values shows a minimum between core and right side wall distorting slightly the M-shape as in the study [11]. Thus, it can be concluded that, the conductivities of the walls affect the structure of the boundary layers significantly.

5. Conclusion

This paper investigates the steady, fully-developed MHD flow in a pipe of rectangular cross-section (duct) with a vertically applied magnetic field. The effect of the Ha and the electrical conductivity of the walls on the flow are presented. The obtained numerical results by the DRBEM implementation show that Ha controls the thickness of the boundary layers, however the wall conductivity determines the structure of the flow. An increase in the wall conductivities also results in the drop of the magnitudes of ω and ϕ. Instead of the centre of the duct, the maximum value of ω occurs on the non-conducting or conducting side walls causing the M-shape for the flow. As Ha values increases stronger fluctuation on the flow behavior is triggered on the side layers for the case of variably conducting walls. The DRBEM, dealing with all the terms other than Laplacian as inhomogeneity, is used for solving MHD flow equations in terms of ω and ϕ. Furthermore, the variable wall conductivity conditions are easily inserted to the integral equation form on the boundary, which

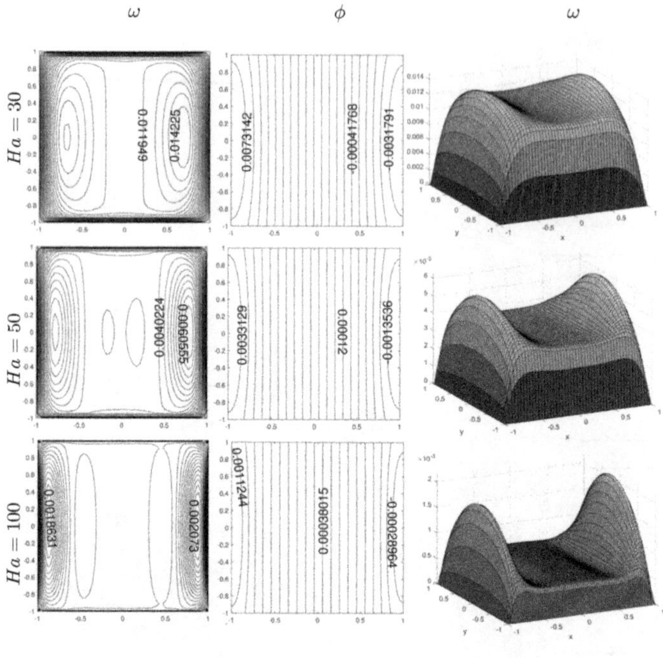

Fig. 5. Velocity and electric potential profiles with different conductivity ratios on the no-slip duct walls. $c_b = c_t = c_l = 0.1$ and $c_r = 5 \times 10^{-3}$.

contains unknown values of ω and ϕ and its normal derivatives. The numerical solutions are obtained by discretization boundary by N elements and arbitrarily chosen L interior nodes resulting small linear system of equations. Thus, it has very low computational cost compared to domain discretization methods.

References

[1] D. Sarma, G. C. Hazarika and P. N. Deka, Numerical study of liquid metal mhd flow through a square duct under the action of strong transverse magnetic field, *International Journal of Computer Applications* **71**, pp. 29–32, (2013).

[2] X. Cai, H. Qiang, S. Dong, J. Lu and D. Wang, Numerical simulations on the fully developed liquid-metal mhd flow at high hartmann numbers in the rectangular duct, *2nd International Conference on Applied Mathematics, Modelling and Statistics Application (AMMSA)* **143**, pp. 68–71, (2018).

[3] S. H. Aydın, I. Nesliturk and M. Tezer-Sezgin, Two-level finite element method with a stabilizing subgrid for the incompressible mhd equations, *International Journal for Numerical Methods in Fluids* **62**, pp. 188–210, (2010).

[4] B. Singh and J. Lal, Finite element method in magnetohydrodynamic channel flow problems, *International Journal for Numerical Methods in Engineering* **18**, pp. 1104–1111, (1982).

[5] R. Scandiuzzi and B. A. Schrefler, Fem in steady mhd duct flow problems, *International Journal for Numerical Methods in Engineering* **30**, pp. 647–659, (1990).

[6] M. Tezer-Sezgin, Boundary element method solution of mhd flow in rectangular duct, *International Journal for Numerical Methods in Fluids* **18**, pp. 937–952, (1994).
[7] M. Tezer-Sezgin and S. Dost, Boundary element method for mhd channel flow with arbitrary wall conductivity, *Applied Mathematical Modelling* **18**, pp. 429–436, (1994).
[8] H. Hosseinzadeh, M. Dehghan and D. Mirzaei, The boundary elements method for magneto-hydrodynamic (mhd) channel flows at high hartmann numbers, *Applied Mathematical Modelling* **37**, pp. 2337–2351, (2013).
[9] L. Bühler and S. Horanyi, Measurements of time-dependent liquid-metal magnetohydrodynamic flows in a flat rectangular duct, *Fusion Engineering and Design*, **84**, pp. 518—521, (2009).
[10] B. S Mazumder, Effect of wall conductances on hydromagnetic flow and heat transfer in a rotating channel, *Acta Mechanica*, **28**, pp. 85–99, (1977).
[11] A. Sterl, Numerical simulation of liquid-metal mhd flows in rectangular ducts, *Journal of Fluid Mechanics*, **216**, pp. 161–191, (1990).

Chapter 5

Robust Constrained Optimization Involving Expensive Function Evaluations with Application to Airfoil Design Optimization

Figen Öztoprak*,‡, Rabia Şener*, Fatma Nur Öter*, İrem Büşra Demiral* and Gorkem Demir†

**Gebze Technical University, Industrial Engineering Department, Kocaeli, Turkiye*
†*Turkish Aerospace (TUSAŞ), Ankara, Turkiye*
‡*figen.topkaya@gtu.edu.tr*

We consider optimization problems that are based on quantities of interest (QoI) computed via computationally expensive routines, and contain uncertainities. In such a setting, each realization of the uncertain QoI requires an execution of the computationally expensive routine (possibly a simulation). Robust formulations of the optimization problem are stated over all possible realizations of those quantities (i.e. over the set of uncertainties) with the aim of generating solutions less sensitive to uncertainties. As QoI are computed via complex routines, the objective/constraints of the robust formulation cannot be stated in closed form. In this work, a new methodology that employs Bayesian optimization to approximately compute robust objective/constraint values at the iterates of a noise-aware sequential optimization algorithm is suggested. The proposed methodology is tested on the application of robust airfoil design optimization under uncertainties in operating conditions, where the realizations are obtained via RANS simulations. The results of our preliminary tests are encouraging in terms of computational efficiency and the quality of the solutions obtained.

1. Introduction

Robust optimization was proposed as a proactive approach to deal with uncertainties. That is, rather than a post-optimality study to see the sensitivity of optimization model output to the perturbations in data, robust optimization suggests incorporating uncertainties to the optimization model so that the obtained solutions are less sensitive to data. A solution to an optimization problem is considered to be *robust* if it stays close to optimal (and almost feasible) for any realization of the uncertain data.

There are different views to formulate a robust counterpart of an optimization problem. The robust formulation can reflect the trade-off in between closeness to optimality and closeness to feasibility via a penalty-type objective function defined over all scenarios regarding uncertainties [1]. If staying feasible is desired for all scenarios, then the straightforward robust formulation contain as many constraints as the possible scenarios. Unlike stochastic optimization, robust optimization does not require assumptions on the distribution of the uncertainties. Instead, it is generally assumed that the scenarios (realizations of the uncertain parameters) lay on a closed and compact set called the *uncertainty set*. When the uncertainty set is continuous, the robust formulation requiring feasibility for all realizations becomes a semi-infinite program.

The infinitely many constraints can be reduced to a finite number of constraints under certain assumptions on the uncertainty set and the constraint functions [2]. For instance, a linear constraint with an ellipsoidal uncertainty set can be reformulated as a second order cone constraint. For optimization problems with more complicated functions, however, such tractable reformulations are not possible. Let us name this case as the *hard case*. An interesting approach for the hard case proposed by Ref. [3] is to consider *local* robustness around an estimation of the uncertain parameters, by employing a local model for the effect of parameter perturbations to function values. Another idea based on localization is to enforce robustness at the local step computation process via restrictions on the search direction [4]. A simple idea to deal with the hard case that is found to be powerful in practice is to solve the robust formulation for certain realizations from the uncertainty set, see if that is feasible for other realizations in that set, and add a new constraint for the violated realization if there are any [5]. On the other hand, an interesting new track employs tensor-based methodologies for obtaining robust solutions via discretization of the uncertain parameters and using low-rank tensor approximations for solving the resulting optimization problems [10].

Beyond the few studies that concentrate on the optimization aspect of the hard case, we observe that the common approach in the literature is to formulate the robust optimization model in terms of statistical quantities. Here, the basic idea is to state the objective/constraints of the robust formulation as a function of the expected value and the variance of the uncertain objective/constraints of the original problem. In different work, the models for distributions of the uncertain objective/constraint functions are obtained via different techniques; e.g. Monte-Carlo [9], polynomial chaos [8], numerical integration [7].

To the best of our knowledge, the closest work to what we suggest in this study is the bilevel optimization framework proposed by Ref. [6]. Similar to what we suggest here (see Section 2), derivative free optimization (DFO) is used in Ref. [6] to approximately solve an inner maximization problem in the space of uncertain parameters which is then employed in an outer level minimization problem in the space of original variables of the problem. However, an ellipsoidal uncertainty set is considered in that study and local quadratic models of the functions (in parameter space) are solved. In particular, local linear models are considered to be too inaccurate although they provide an analytical solution; local convex quadratic models are employed for better accuracy, but since these models do not yield closed-form solutions, a local DFO method is used to solve them.

In this work, we suggest a methodology for robust optimization in the *hard case* mentioned above. Our methodology has a bilevel structure; it solves inner optimization problems in the parameter space to compute the objective/constraints of the robust formulation. Inner optimization must be repeated at each iterate of the outer algorithm because the inner solution depends on the outer iterate (i.e. the propagated uncertainty depends on the values at the variable space). Unlike Ref. [6], we consider global optimization in the parameter space for solving the inner problem, and a method based on local models for solving the outer problem in the variable space. In particular, our methodology employs Bayesian optimization to approximate objective/constraint values at the iterates of a noise-aware sequential optimization algorithm. The key ideas of our methodology are the following: (i) within the optimization process, the (robust) objective/constraints are evaluated only approximately, (ii) uncertainty propagation is modeled by a Gaussian process and sampling is guided by Bayesian optimization to get good approximations of the objective/constraints with less realizations of quantities of interest (QoI) at each iterate of the optimization algorithm, (iii) since approximate evaluations are noisy (i.e. as these are obtained via solving global optimization problems only approximately), the robust optimization problem is solved by a noise-aware optimization algorithm.

The rest of this paper is organized as follows: In Section 2, we explain in detail the methodology we propose, and provide an illustrative example. In Section 3, we present the preliminary results we obtained in solving an airfoil design optimization problem. Finally, we explain our ongoing work as well as the potential future work in Section 4.

2. Proposed Methodology

Let $f : \mathbb{R}^n \to \mathbb{R}$ and $c : \mathbb{R}^n \to \mathbb{R}$ be continuous functions, and consider the following optimization problem.

$$\begin{aligned} \min_{x \in \mathbb{R}} \quad & f(x;p) \\ & c(x;p) \geq 0 \\ & l \leq x \leq u. \end{aligned}$$

Here, $x \in \mathbb{R}^n$ are the variables of the optimization problem whereas $p \in \mathbb{R}^m$ is the vector containing the parameters of the objective function and the constraint function. We state the problem here with a single functional constraint without loss of generality. When p is not known precisely, and it is assumed that $p \in \mathcal{U}_p$ for $\mathcal{U}_p$ closed and compact, the conventional robust formulation of the problem is given with

$$
\begin{aligned}
\min_{x \in \mathbb{R}^n} \quad & \max_{p \in \mathcal{U}} f(x;p) \\
& \min_{p \in \mathcal{U}} c(x;p) \geq 0 \\
& l \leq x \leq u.
\end{aligned} \tag{1}
$$

The solution of the inner optimization problems

$$
\tilde{f}(x) := \max_{p \in \mathcal{U}_p} f(x;p) \qquad \text{and} \qquad \tilde{c}(x) = \min_{p \in \mathcal{U}_p} c(x;p) \tag{2}
$$

depend on x; and thus need to be re-solved each time the variables x are updated. We are interested in the case where (i) closed-form expressions for $\tilde{f}(x)$ and $\tilde{c}(x)$ are not available, (ii) the functions f and c are not necessarily convex in p so that the exact computation of $\tilde{f}(x)$ and $\tilde{c}(x)$ is practically not possible, and (iii) the evaluations of the functions f and c are computationally expensive.

To solve this problem, we propose a bilevel optimization framework which uses Bayesian optimization for the evaluation of $\tilde{f}(x)$ and $\tilde{c}(x)$ in (2) (inner optimization), and uses noise-aware SQP for solving (1) (outer optimization).

2.1. *Inner optimization*

For computing $\tilde{f}$ and $\tilde{c}$ at a given iterate x_k, we need to solve optimization problems (2) in the parameter space. We regard those problems as unconstrained derivative-free optimization problems with expensive function evaluations. Since the set of uncertain parameters generally have small dimension, we consider solving those as global optimization problems with the idea to get better approximations to the true $\tilde{f}(x_k)$ and $\tilde{c}(x_k)$, $k = 0, 1, \ldots$ (as compared to a local optimization approach) over the set $\mathcal{U}_p$ although less precise in terms of local stationarity. With all these motivations, we suggest employing Bayesian optimization as the inner solution procedure.

Bayesian optimization is an iterative procedure that operates based on a Gaussian Process (GP) regression model [11]. Each time a new function evaluation is done, the regression model is updated following the Bayesian approach. Once the model is updated, the point that maximizes the expected improvement is determined, and a new function evaluation is done at this point. The basic procedure is described in Algorithm 1 for solving the maximization problem

$$
\max_{p \in U} \; h(p), \tag{3}
$$

where p is the vector of parameters and h is the QoI so that we are trying to compute the maximum value h can take over all p in the uncertainty set U.

Algorithm 1 (Bayesian optimization framework for solving (3)).

for $t = 0, 1, 2, \ldots$ **do**

Update the probability distribution over h

Choose p_{t+1} maximizing an acquisition function a

$$p_{t+1} = \arg\max_{p \in U} a(p, D_t)$$

Compute the function value $h(p_{t+1})$

$D_{t+1} = D_t \cup (p_{t+1}, h(p_{t+1}))$

end for

2.2. *Outer optimization*

As for the optimization in variable space, we consider the recently proposed noise-aware sequential quadratic optimization algorithm [14], which is a generalization of the conventional sequential quadratic programming (SQP) algorithm for dealing with noisy evaluations. The algorithm applies to the noisy optimization problem with uniformly bounded noise

$$\min_x \ \tilde{f}(x) : \tilde{c}(x) \geq 0,$$

$$|\tilde{f}(x) - \tilde{f}_{noisy}(x)| \leq \epsilon_f, \quad \|\tilde{c}(x) - \tilde{c}_{noisy}(x)\| \leq \epsilon_c,$$

which coincides with (1). In our case, noise arises since we can only approximately compute $\tilde{f}(x_k)$ and $\tilde{c}(x_k)$ at iterates x_k; therefore, $\tilde{f}_{noisy}(x_k)$ and $\tilde{c}_{noisy}(x_k)$ stand for the approximate solutions of the inner optimization problems (2) returned by Bayesian optimization given $x = x_k$. The noise-aware algorithm differs from a regular line search SQP algorithm mainly by a relaxed line search. We give a very rough outline in Algorithm 2. As in conventional SQP, the search direction is computed by solving a local quadratic model of the problem at each iteration, and a merit function is employed for globalization. However, the local model and merit function computations are done based on noisy evaluations, which significantly effects the convergence theory. We refer to Ref. [14] for the details.

Algorithm 2 (Noise-aware SQP framework).

for $k = 0, 1, 2, \ldots$ **do**

Compute direction d_k^{qp} based on noisy evaluations

Update merit function parameter

Compute a steplength α_k via relaxed line search

$x_{k+1} = x_k + \alpha_k d_k^{qp}$

end for

2.3. *Illustrative example*

We adjust the example in Ref. [4] to illustrate the computational efficiency of the proposed methodology. The reference problem is a two-dimensional optimization problem with a nonconvex objective and two nonconvex constraint functions.

Table 1. Optimal solution of problem (4) for varying values of the uncertain parameters p_x and p_y.

p_x	p_y	x^*	y^*	$f(x^*, y^*)$	$c_2(x^*, y^*; 0.9, 1.9)$
0.90	1.90	0.01166	0.81663	4.63785	2.8E-08
0.90	2.50	0.23469	0.23487	0.89855	-10.725
0.90	3.10	0.28503	0.18988	0.83755	-11.389
1.50	1.90	0.02659	0.69878	3.97201	-2.532
1.50	2.50	0.28503	0.18988	0.83755	-11.389
1.50	3.10	0.28503	0.18988	0.83755	-11.389
2.10	1.90	0.09982	0.43603	2.05387	-7.489
2.10	2.50	0.28503	0.18988	0.83755	-11.389
2.10	3.10	0.28503	0.18988	0.83755	-11.389

We consider injecting uncertainty to the parameters of the second constraint. In particular, we consider the following optimization problem.

$$\begin{aligned} \min_{x,y} \; & f(x, y) \\ & c_1(x, y) \leq 0 \\ & c_2(x, y; p_x, p_y) \leq 0, \end{aligned} \tag{4}$$

where

$$c_2(x, y; p_x, p_y) = -(p_x - x)^3 - (y + p_y)^3 + 15.75,$$

and

$$p_x \sim \mathcal{N}(1.5, 0.2), \quad p_y \sim \mathcal{N}(2.5, 0.2).$$

To see sensitivity of the solution to the parameters of the second constraint, we first create a grid of (p_x, p_y) values and solve the problem for varying values of p_x and p_y. In Table 1, we give the optimal solutions obtained at a sample of grid points. We observe that the optimal solution is indeed sensitive to the values of these parameters. We also observe that for the discretized problem over the grid of (p_x, p_y) values, the point $(0.9, 1.9)$ provided the robust solution (a solution that is feasible for all realizations of (p_x, p_y)). In the last column of Table 1, we give the violation of the constraint $c_2(x^*, y^*; 0.9, 1.9)$ when (x^*, y^*) is obtained employing a different choice of (p_x, p_y) but the realization of these parameters occurs as $(0.9, 1.9)$.

Clearly, obtaining an approximate robust solution via such a *multipoint* approach requires too many evaluations of c_2 as the whole optimization process needs to be repeated for each choice of (p_x, p_y). We next consider a common strategy considered in the literature and solve a robust formulation of the problem *based on the mean and the variance of the uncertain constraint.* Employing Monte Carlo estimations to the mean and the variance of c_2, we observed that to get the robust solution in a stable manner we needed a sample size of at least 200 (per iteration). Testing the proposed methodology of Section 2 on this problem, we could get to the same robust solution $(x^*, y^*) = (0.01165988, 0.81663114)$ using only five samples at each outer iteration (three samples in the initial dataset, and two Bayesian Optimization iterations); i.e. with 40 times less evaluations of c_2 per iteration. Indeed, the proposed methodology has required a total of only 120 evaluations of c_2 over the whole optimization process.

3. Implementation to Robust Airfoil Design

We next try our methodology on the application airfoil design optimization. There are different sources of uncertainty regarding the design optimization of airfoils. In our implementation, we consider uncertainties in operating conditions; in particular, we consider uncertainty of the Mach number only. Our setup yields the optimization problem given in (5), which contains a single inequality constraint as well as bounds on the design variables x.

$$\begin{aligned} \min_{x \in \mathbb{R}} \quad & \max_{p \in \mathcal{U}} C_D(x;p) \\ & \min_{p \in \mathcal{U}_p} C_L(x;p) \geq L, \\ & l \leq x \leq u. \end{aligned} \tag{5}$$

Here, the objective is the robust minimization of the *drag coefficient*, subject to the robust satisfaction of the *lift* constraint. Note that the uncertainty of the Mach number parameter p affects both the objective value and the constraint value for a given design. Problem (5) has the same form as (1), and we implemented our suggested solution methodology as described in Section 2.

We tried solving (5) for the RAE2822 test case. We employed the open source CFD solver SU2 [15] for RANS simulations to compute C_D and C_L for a given design x and a given Mach number value p, as well as the computation of adjoints and mesh deformation. We mainly relied on the configuration and the base mesh provided in the tutorial [12] of SU2. Following this reference, we employ a mesh that contains a far-field boundary and a non-slip wall for Navier-Stokes along the airfoil surface. The free-stream temperature is set to 288.15K, the Reynolds number to 6.5×10^6, and the angle of attack to 2.31. The most important modification we did to the reference configuration is the replacement of the equality constraint $C_L(x) = L$ with the inequality constraint in (5). The design variables are 30 Hicks-Henne bump function parameters specified in Ref. [12] (a chord of 1.0 is assumed for the Hicks-Henne bumps). We set $L = 0.3$, and the uncertainty set for the Mach number $\mathcal{U}_p$ to the interval $[0.65, 0.85]$ (recall that a distribution is not required for the conventional robust optimization setting). We set the bounds of the design variables as $l_i = -0.001$ and $u_i = 0.001$, $i = 1, \ldots, 30$.

The proposed methodology was implemented by taking ten Bayesian optimization steps and three samples in the initial set of each inner optimization. For outer optimization, we set $\epsilon_f = \epsilon_c = 10^{-5}$. To get an idea on how Bayesian optimization does global search over the interval $\mathcal{U}_p$, we check the ten iterates p_t of inner optimization in computing $\tilde{C}_D(x_0)$ and $\tilde{C}_L(x_0)$; i.e. in computing the approximate

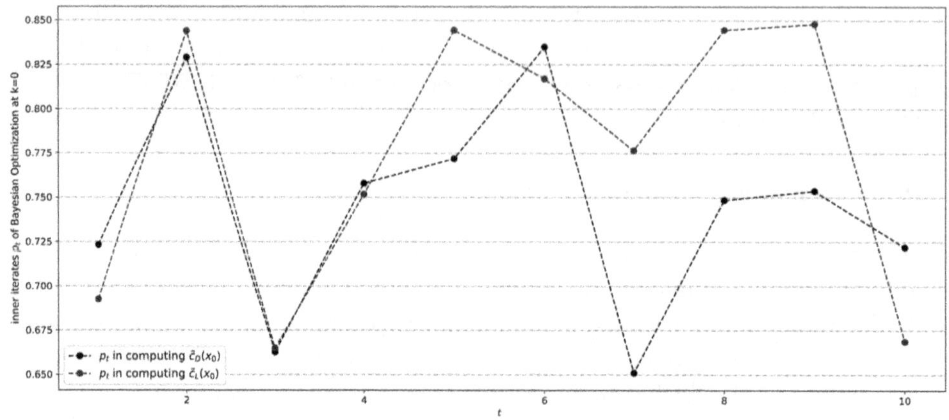

Fig. 1. The iterates p_t of the Bayesian optimization, $t = 1, \ldots, 10$, in evaluating approximately the objective and the constraint values $\tilde{C}_D(x_0)$ and $\tilde{C}_L(x_0)$ at the initial design x_0.

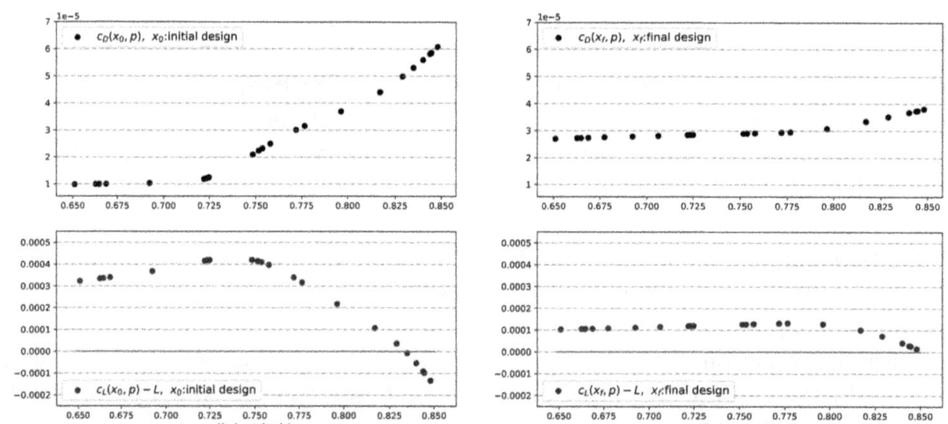

Fig. 2. The objective and the constraint values $C_D(x; p)$ and $(C_L(x; p) - L)$ for different realizations of p at the initial design $x = x_0$ (left) and at the final design $x = x_f$ (right). The upper two plots show values of the objective and the lower two plots show values of the constraint.

values of the objective and the constraint of (5) at the initial design x_0. The corresponding plots are given in Figure 1. We observe that the global search process have covered the whole interval rather than focusing on a part of it as intended.

Next, we look at the output of the outer optimization. A summary is provided in Figure 2. The four plots show samples of the values of drag and lift for different realizations of the Mach number p at the initial and the final points of robust optimization. Overall, we see that the final design is indeed less sensitive to the realization value of the uncertain parameter p. The upper two plots show the objective value $C_D(x; p)$ for different realizations of p. Comparing the initial design (x_0)

with the final one (x_f), the reduction in the variance of the objective can be clearly observed. We also observe that the maximum drag value over the integral $p \in \mathcal{U}_p$ has indeed decreased. The lower two plots show the constraint value $(C_L(x;p) - L)$; that is, the negative values in those plots (the points below the red lines) indicate constraint violation. We observe that the final solution is feasible for all realizations as expected from a robust design.

4. Conclusions and Future Work

In this work, we proposed a new methodology for constrained robust optimization problems involving computationally expensive function evaluations. The results of the preliminary tests are encouraging to do further work. The optimization model (5) is rather incomplete. An intriguing next step is to better assess the practical potential of the proposed methodology by testing it on more realistic airfoil design optimization models. We believe that an interesting extension of the problem is to consider robustness for variations in design variables x, as well as the variations in parameters p; i.e., replacing (2) with

$$\tilde{f}(x) := \max_{p \in \mathcal{U}_p, \delta_x \in \mathcal{U}_\delta} f(x + \delta_x; p) \qquad \text{and} \qquad \tilde{c}(x) = \min_{p \in \mathcal{U}_p, \delta_x \in \mathcal{U}_\delta} c(x + \delta_x; p).$$

Such an extension could model the uncertainties in manufacturing of a design. Finally, we should note that we believe there is a lot to improve in implementation details. The solution process in Section 3 was terminated due to failure of the SU2 RANS solver; so, even the final solution presented in that section can potentially be further improved with a better tuned implementation.

Acknowledgements

This work is supported by TUSAŞ under the LIFT-UP programme with Project No.64ae942b3be22. Computing resources used in this work were provided by the National Center for High Performance Computing of Turkey (UHeM) under grant No.4018112024.

References

[1] J. M. Mulvey, R. J. Vanderbei, and S. A. Zenios, Robust optimization of large-scale systems, *Operations Research*, vol. 43, no. 2, pp. 264–281, (1995).

[2] A. Ben-Tal and A. Nemirovski, Robust optimization–methodology and applications, *Mathematical Programming*, vol. 92, pp. 453–480, (2002).

[3] Y. Zhang, General robust-optimization formulation for nonlinear programming, *Journal of Optimization Theory and Applications*, vol. 132, no. 1, pp. 111–124, (2007).

[4] D. Bertsimas, O. Nohadani, and K. M. Teo, Nonconvex robust optimization for problems with constraints, *INFORMS Journal on Computing*, vol. 22, no. 1, pp. 44–58, (2010).

[5] B. L. Gorissen, I. Yanikoglu, and D. Den Hertog, A practical guide to robust optimization, *Omega*, vol. 53, pp. 124–137, (2015).

[6] P. Kolvenbach, O. Lass, and S. Ulbrich, An approach for robust PDE-constrained optimization with application to shape optimization of electrical engines and of dynamic elastic structures under uncertainty, *Optimization and Engineering*, vol. 19, pp. 697–731, (2018).
[7] D. I. Papadimitriou and C. Papadimitriou, Aerodynamic shape optimization for minimum robust drag and lift reliability constraint, *Aerospace Science and Technology*, vol. 55, pp. 24–33, (2016).
[8] L. W. Cook and J. P. Jarrett, Robust airfoil optimization and the importance of appropriately representing uncertainty, *AIAA Journal*, vol. 55, no. 11, pp. 3925–3939, (2017).
[9] L. W. T. Ng and K. E. Willcox, Multifidelity approaches for optimization under uncertainty, *International Journal for Numerical Methods in Engineering*, vol. 100, no. 10, pp. 746–772, (2014).
[10] S. Garreis and M. Ulbrich, Constrained optimization with low-rank tensors and applications to parametric problems with PDEs, *SIAM Journal on Scientific Computing*, vol. 39, no. 1, pp. A25–A54, (2017).
[11] R. Garnett, *Bayesian optimization.* Cambridge University Press, (2023).
[12] Resources for the tutorial, `https://github.com/su2code/Tutorials/tree/master/design/Turbulent_2D_Constrained_RAE2822`.
[13] National center for HPC. `https://en.uhem.itu.edu.tr/donanim.html`.
[14] F. Oztoprak, R. Byrd and J. Nocedal, Constrained optimization in the presence of noise, *SIAM Journal on Optimization*, vol. 33, no. 3, pp. 2118–2136, (2023).
[15] T. D. Economon, F. Palacios, S. R. Copeland, T. W. Lukaczyk and J. J. Alonso, SU2: An open-source suite for multiphysics simulation and design, *Aiaa Journal*, vol. 54, no. 3, pp. 828–846, (2016).

Chapter 6

The Predictability of Oceanic Circulations via FFT-ANFIS Spectral Adaptive Fuzzy Network

Ali Rıza Alan*,‡ and Cihan Bayındır†,§

*Faculty of Engineering and Natural Sciences, Bahçeşehir University, 34353, Beşiktaş, İstanbul, Turkey

†Faculty of Engineering, İstanbul Technical University, 34469, Sarıyer, İstanbul, Turkey

‡aliriza.alan@bau.edu.tr

§cbayindir@itu.edu.tr

Oceanic currents and circulations are vital components of the marines on Earth for their physical and biological existence [1–6]. In the literature, many different data processing techniques are proposed and employed for their analysis [2–6]. In this paper, we examine the predictability of oceanic circulations using spectral fuzzy logic. More specifically, we propose the possible implementation and usage of the Adaptive Neuro-Fuzzy Inference System (ANFIS) fuzzy network [7, 8] approach for the spectral (FFT) analysis of oceanic circulation data. The data set investigated in this study was acquired by the National Oceanic and Atmospheric Administration (NOAA) between October 2002 and February 2003 in Massachusetts Bay, US. We show that the current speed time series and its spectrum in 2D horizontal directions, namely u and v, can be successfully predicted using the FFT-ANFIS approach. We present prediction error in terms of the coefficient of determination (R^2) and the root-mean-square error (RMSE). The effect of the training data set on the prediction error and the spectral properties of predictions are also analyzed. We show that depending on the temporal and spatial resolution of the data, the prediction times and distances can vary. Depending on their resolution in some cases, they can be very beneficial for the early prediction of the ocean current parameters. Our results can find many important applications including but not limited to predicting the statistics and characteristics of tidal energy variation and controlling the current-induced vibrations of structures in the marine environment.

1. Introduction

Over the course of ocean science's historical evolution, currents and circulations have garnered significant interest as one of the most significant phenomena seen in ocean hydrodynamics [1–11]. Numerous diverse mechanisms can produce these currents and circulations. The aforementioned mechanisms encompass alterations in underwater topography [5], winds [2, 6], tides [3, 4], waves [5], and buoyancy fluxes [4, 6, 12, 13]. The currents can be categorized in a few different ways. They can be categorized as wind-driven, tidal, or thermohaline based on the forcing processes they employ [14]. Another way to categorize currents is based on their depth of occurrence; they can be categorized as surface or subsurface currents. While thermohaline currents are influenced by salinity and temperature, the majority of surface ocean currents are driven by wind. Subsurface thermohaline currents are propelled by denser, saltier water settling to the bottom of the ocean in colder regions. Alternatively, the currents could be categorized based on where they occur; that is, currents that flow a certain distance offshore could be categorized as offshore currents, and currents that flow closer to the shore could be categorized as inshore currents. It would be more reasonable to categorize them as mean or periodic currents from the perspective of time series analysis. Periodic currents are defined as those with cyclically varying speeds and directions at regular intervals. Nonetheless, there are mean portions of oceanic circulations that fluctuate very little or hardly at all throughout time and place.

The marine ecosystem is greatly impacted by ocean circulation and currents. They are among the primary forces behind the earth's mass and heat transport. In addition to carrying chemicals, nutrients, and biological agents, they have the ability to heat or cool the oceans. Numerous technical issues, including greater vibrations, longer transit times, and a higher risk of accidents in important waterways like the Bosphorus, could result from them for both marine traffic and offshore engineering structures [15, 16]. Furthermore, because of their wave-blocking action, they could cause rogue waves to form [17]. They serve as the tidal energy converters' primary driving force [18–20].

Predicting ocean currents and circulation in the marine environment is a crucial issue for the security of maritime operations and structures for a number of reasons, some of which are outlined above. Numerous efforts are undertaken to forecast the various ocean phenomena, such as currents and circulations. In [21, 22], the ocean wave height and energy are predicted using a variety of wave models and time-series techniques. In [23], it is explained how to estimate the near-surface winds using a Kalman filtering technique. For the purpose of forecasting wave height, a transfer function-based approach was put forth in [24]. In [25], the application of artificial neural networks for wave prediction is examined. Real-time wave forecasting in [26] is accomplished with the use of genetic programming techniques. In a recent development, a machine learning method for wave condition prediction is put out in [27]. In [28], a machine learning method is once more suggested for thermocline parameter prediction. In [29], a statistical machine learning methodology is employed to forecast ocean processes. The use of Gaussian processes for the spatiotemporal prediction of ocean currents is covered in [30]. Using deep learning for subgrid parameterization and ocean data inference was another strategy [31].

In this work, we use a deep learning approach to discuss and assess the predictability of the time-varying ocean currents [32]. More precisely, we employ an ocean current data set that NOAA obtained in Massachusetts Bay between October 2002 and February 2003 to function the ANFIS deep learning network.

We investigate how well the ANFIS performed in predicting the ocean current speed data, as well as the spectrum features of the predictions and the root-mean-square (RMS) error. We also look into how predictions are impacted by the training data set. We conclude our results and offer commentary on them.

2. Methodology

2.1. *Review of the ANFIS fuzzy network*

A neural network (NN) is an adaptive learning system that draws inspiration from the organic architecture of the human brain. An input and an output, referred to as neurons, are connected in a section of a neural network (NN). In the process known as training, when NN learns from the training data set, the connections between the neurons of a NN are weighted using appropriate data [9, 10]. The model's outputs are compared to the target values throughout the training phase [9, 10]. By applying this comparison, an iterative procedure reduces the inaccuracy and yields a weighted relationship network. A NN is typically classified into three types based on how the network moves. These are the neural networks that are feedforward (FNN), convolutional (CNN), and recurrent (RNN) [9, 10]. Readers seeking a more thorough discussion of this topic are directed to [9, 10]. In contrast to Aristotelian logic, the fuzzy inference system (FIS) expresses scenarios using a rating scale between 0 and 1 [7]. With one or more inputs connected to one or more conditional rules, this algorithm yields one or more output values [9, 10]. Membership functions are used to represent each variable [7]. Certain mathematical equivalents of verbal utterances are represented by membership functions. A model that combines FIS with NNs is called adaptive neuro-fuzzy inference systems (ANFIS) [8, 33–36]. Similar to FIS, a rule-based relationship exists between input and output values; but, unlike FIS, the neural network algorithm learns the rules adaptively by training on the data set [8]. Unlike the NN structure, this structure associates connections between neurons with membership functions and rules rather than discrete weight numbers. A detailed explanation is provided in [8], which is forwarded to the reader. Throughout this study we use Sugeno type ANFIS system with two trapezoidal membership functions and with 40 epochs.

2.2. *Review of the ocean current data*

In this work, we evaluate the performance of the ANFIS deep learning network for the predictions of ocean current and circulation using an experimental data set. The data set is recorded by NOAA at $42°22'38.3''N, 70°46'54.8''W$ $(42.3773N, 70.7819W)$ in the North Atlantic in the Massachusetts Bay. A VMCM type moored current meter was used to record the experimental data between October 24, 2002, and February 12, 2003. At this particular location, the current meter was placed at a depth of 23.5 m, while the seafloor was 33.7 m deep. Throughout this experiment, the horizontal current velocities, or u and v, are monitored every 3 minutes and 44 seconds. Since the data obtained is very large, only the period between October 24 and November 2, 2002 is used in this study. Readers may visit NOAA's website `https://www.nodc.noaa.gov/gocd/data/a0060062/info/gocd_a0060062_6964vma.html` to view the original data set and further information.

3. Results and Discussion

In this section, we evaluate the ANFIS fuzzy network's performance in predicting the data on ocean current speed. The initial portion of the first component of the original current speed data (u) between October 24 and November 2, 2002, is depicted in Figure 1. For ANFIS, the series' training-to-test ratio is set at 80%–20%. ANFIS's training phase is applied to the first 80% of the first component of the original current speed data (u), and Figure 1 illustrates the contrast between ANFIS predictions and measured data. Figure 2 illustrates the outcome of blind testing the remaining 20% of the same data.

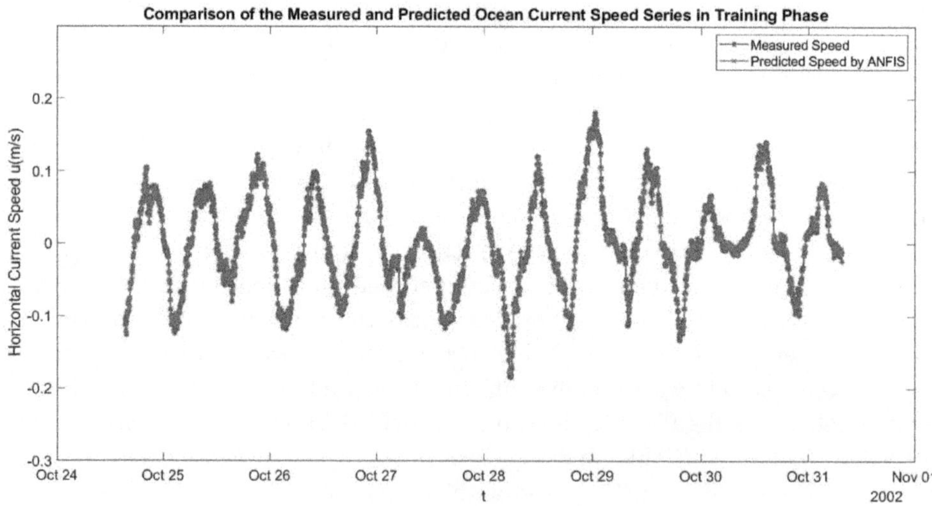

Fig. 1. Observed and measured time series of the first component of the current velocity (u) in the training phase.

In regard to the measured value, the ANFIS results shown in Figures 1 and 2 have a coefficient of determination (R^2) of 0.9859 and a root mean square error (RMSE) of 0.0078 m/s for the training phase, and an R^2 of 0.9890 and an RMSE of 0.0083 m/s for the blind testing phase, respectively. We utilize a spectral approach and obtain the Fourier spectrum via FFT routines to better visualize the matching

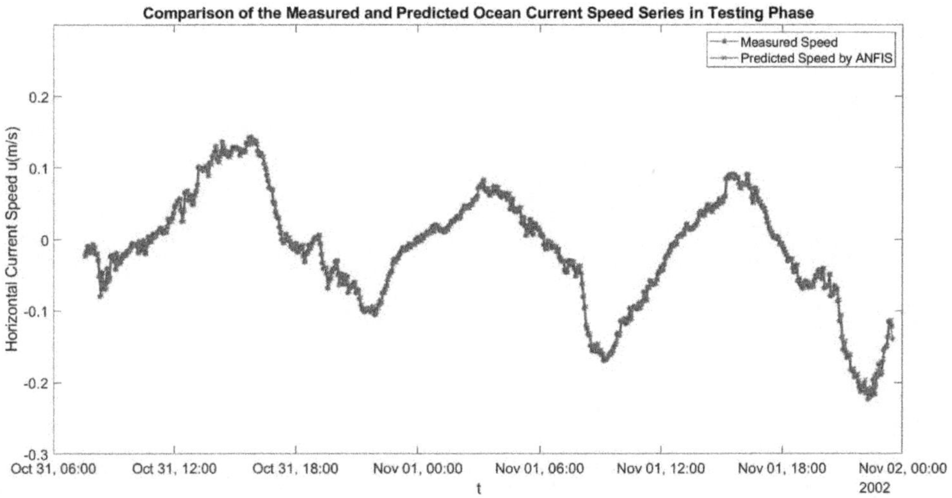

Fig. 2. Observed and measured time series of the first component of the current velocity (u) in the blind testing phase.

of predictions with the measurements. The spectra of the predictions for the first component of the original current speed data (u) are depicted in Figure 3.

Next, we apply the same procedure to the second component of the original current speed data (v) obtained on the same dates. The training-to-test ratio of the series for ANFIS is again 80%–20%. ANFIS's training phase is applied to the first 80% of the second component of the original current speed data (v), and Figure 4 illustrates the contrast between ANFIS predictions and measured data. Figure 5 illustrates the outcome of blind testing the remaining 20% of the same data.

In regard to the obtained value, the ANFIS results shown in Figures 4 and 5 have an R^2 of 0.9706 and an RMSE of 0.0066 m/s for the training phase, and an R^2 of 0.9514 and an RMSE of 0.0076 m/s for the blind testing phase, respectively. We utilize a spectral approach and obtain the Fourier spectrum via FFT routines again to better visualize the matching of predictions with the measurements. The spectra of the predictions for the second component of the original current speed data (v) are depicted in Figure 6.

Given these results, it is reasonable to conclude that ANFIS can forecast the original current speed data's first component (u) and second component (v) with accuracy. If we normalize the peak velocities with RMSE, which are ± 0.3 m/s for u, and ± 0.1 m/s for v, we can state that ANFIS is more successful in predicting u than in predicting v. The present study's prediction time is restricted to a one time step; nonetheless, by employing downsampling and/or additional artificial intelligence approaches, our findings may be extended for improved prediction times.

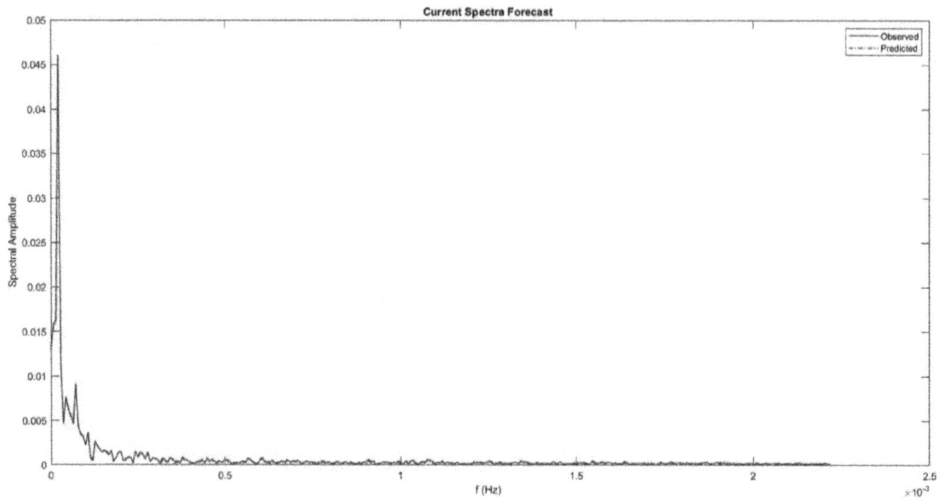

Fig. 3. Comparison of Fourier spectra of time series predictions by ANFIS for the first component of the current velocity (u).

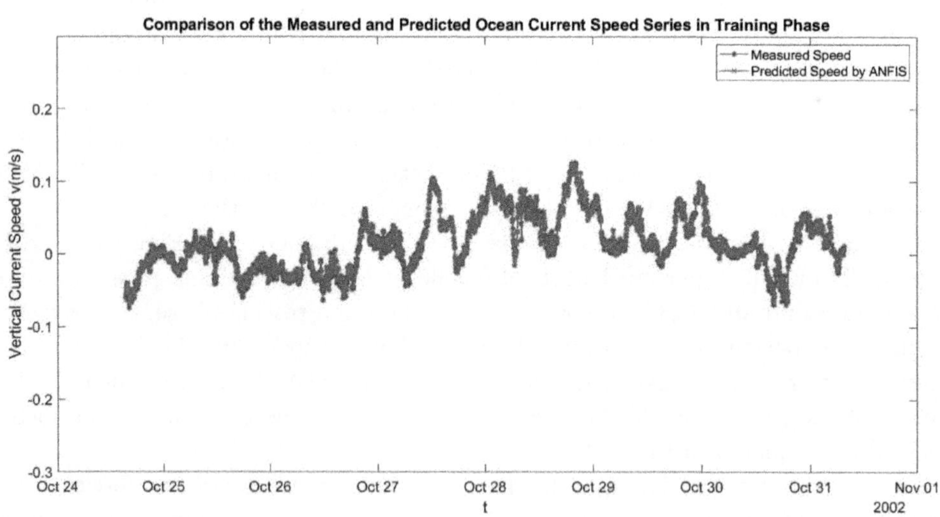

Fig. 4. Observed and measured time series of the second component of the current velocity (v) in the training phase.

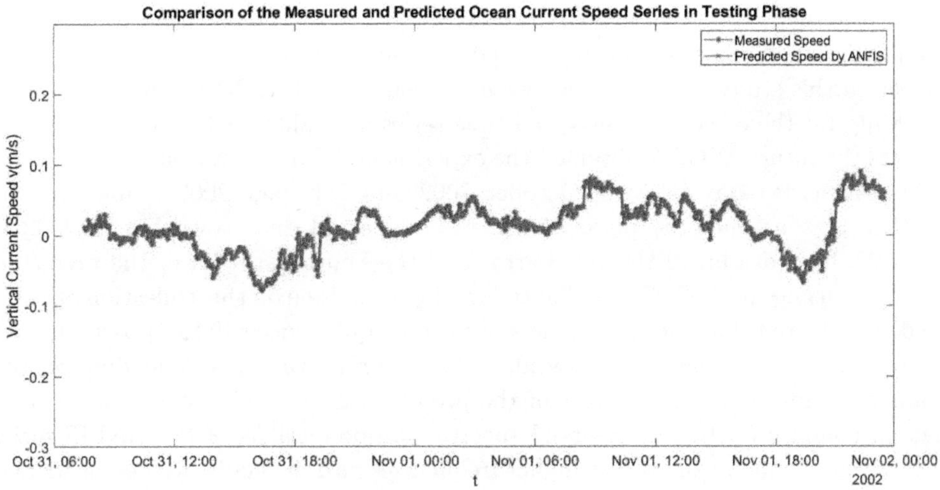

Fig. 5. Observed and measured time series of the second component of the current velocity (v) in the blind testing phase.

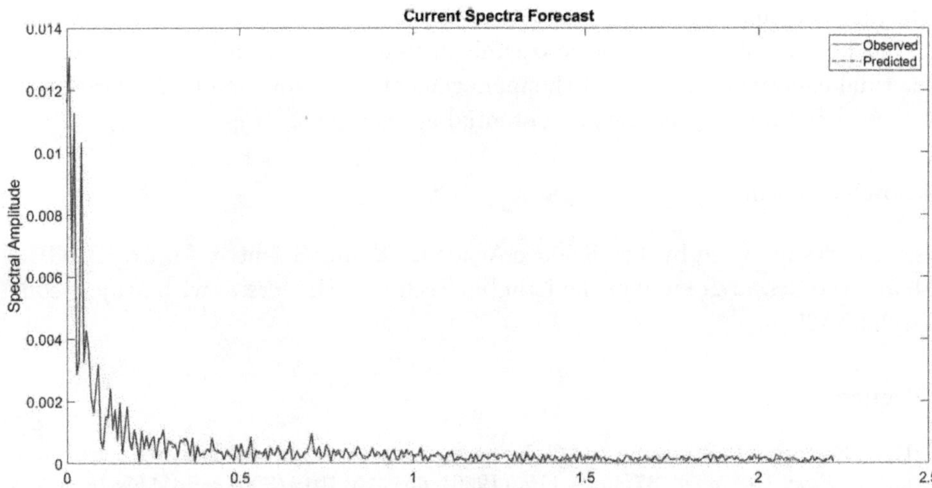

Fig. 6. Comparison of Fourier spectra of time series predictions by ANFIS for the second component of the current velocity (v).

4. Conclusion

The prediction of the ocean current speed time series using deep learning has been covered in this study. In particular, we have examined the ANFIS fuzzy network's suitability for the ocean current speed time series and addressed about how well it predicts the future. NOAA obtained the experimental data set for our investigation in Massachusetts Bay between October 2002 and February 2003. However, for practical reasons only the period between October 24 and November 2, 2002 is used. We have examined the RMS error and the Fourier spectra of the predicted time series using the ANFIS-based deep learning technique on this collection of data. We demonstrated that the ANFIS network can be quite successful in predicting the ocean current speed data and may allow for accurate forecasts depending on the steady-state-in-the-mean behavior of the projected time series. We demonstrated that, in comparison to the observed spectra, the predictions of the ANFIS fuzzy network have some mismatch at higher frequencies. Additionally, we demonstrated that the ocean current time series can be predicted more accurately with longer training sequences. The prediction times can differ substantially based on the data's sample interval. Therefore, the ocean current time series data may be predicted with extremely advantageous prediction times using the ANFIS-based deep learning approach presented in this paper. Other oceanic and atmospheric time-series data, such as the Gulf Stream, North Equatorial, Agulhas currents, and East Wind drift, can be predicted using our methods in a generalized manner. The estimation of the wave blocking point by the chaotic oceanic current and circulation, the prediction of the tidal energy variation, and the management of current-induced vibrations of maritime structures are among the potential applications.

Acknowledgment

This study is supported by The Science Academy-Young Scientist Award (BAGEP) 2024, and the Research Fund of the İstanbul Technical University with project code: MGA-2022-43528.

References

[1] R. G. Peterson, L. Stramma and G. Kortum, Early concepts and charts of ocean circulation, *Prog. Oceanogr.* **37**(1), 1–115, (1996). doi:10.1016/S0079-6611(96)80002-4.
[2] W. Munk, On the wind-driven ocean circulation, *J. Meteor.* **7**(79), 80–93, (1950). doi:10.1175/1520-0469(1950)007<0080:OTWDOC>2.0.CO;2.
[3] B. B. Parker, *Tidal Hydrodynamics* (John Wiley and Sons, New York, 1991).
[4] H. Stommel, A survey of ocean current theory, *Deep Sea Res.* **4**(149), (1957).
[5] M. Tomczak and J. Godfrey, *Regional Oceanography: An Introduction* (Pergamon Press, Oxford, 1994).
[6] H. U. Sverdrup, M. Johnson and R. Fleming, *The Oceans: Their Physics, Chemistry and General Biology* (Prentice Hall, New Jersey, 1942).
[7] L. A. Zadeh, Fuzzy logic, *Computer (Long. Beach. Calif)* **21**(4), 83–93, (1968).

[8] J. Kim and N. Kasabov, HyFIS: Adaptive neuro-fuzzy inference systems and their application to nonlinear dynamical systems, *Neural Networks* **12**(9), 1301–1319, (1999). doi:10.1016/S0893-6080(99)00067-2.

[9] J. Holland, *Adaptation in Natural and Artificial Systems* (MIT Press, Massachusetts, 1975).

[10] D. E. Goldberg, *Genetic Algorithms in Search, Optimization, and Machine Learning* (Addison-Wesley, New York, 1989).

[11] P. L. Richardson, Benjamin Franklin and Timothy Folger's first printed chart of the Gulf Stream, *Science* **207**(4431), 643–645, (1980). doi:10.1126/science.207.4431.643.

[12] G. Wust, Schichtung und Zirkulation des Atlantisches Ozeans, Die Stratosphare, Wissenschaftliche Ergebnisse der Deutschen Atlantischen Expedition auf dem Forschungs- und Vermessungsschiff, *Meteor.* **6**, 1925–1927, (1935).

[13] G. Siedler, J. Church and J. Gould, *Ocean Circulation and Climate: Observing and Modelling the Global Ocean* (Academic Press, 2001).

[14] C. Bayındır, Predicting the ocean currents using deep learning, *TWMS J. App. Eng. Math.* **13**(1), 373–385, (2023).

[15] T. Oguz, E. Özsoy, M. A. Latif, H. I. Sur and Ü. Ünlüata, Modeling of hydraulically controlled exchange flow in the Bosphorus Strait, *J. Phys. Oceanogr.* **20**(7), 945–965, (1990). doi:10.1175/1520-0485(1990)020<0945:MOHCEF>2.0.CO;2.

[16] E. Jarosz, W. J. Teague, J. W. Book and Ş. Beşiktepe, On flow variability in the Bosphorus Strait, *J. Geophys. Res.: Oceans* **116**(C8) (2011). doi:10.1029/2010JC006861.

[17] C. Bayındır, Shapes and statistics of the rogue waves generated by chaotic ocean current, in *The 26th International Ocean and Polar Engineering Conference (ISOPE)*, (2016).

[18] D. Magagna and A. Uihlein, Ocean energy development in Europe: Current status and future perspectives, *Int. J. Mar. Energy* **11**, 84–104, (2015). doi:10.1016/j.ijome.2015.05.001.

[19] A. Uihlein and D. Magagna, Wave and tidal current energy–A review of the current state of research beyond technology, *Renewable Sustainable Energy Rev.* **58**, 1070–1081, (2016). doi:10.1016/j.rser.2015.12.284.

[20] P. A. Lynn, *Electricity from Wave and Tide: An Introduction to Marine Energy* (John Wiley & Sons, 2013).

[21] M. S. Roulston, J. Ellepola, J. von Hardenberg and L. A. Smith, Forecasting wave height probabilities with numerical weather prediction models, *Ocean Eng.* **32**(14–15), 1841–1863, (2005). doi:10.1016/j.oceaneng.2004.11.012.

[22] G. Reikard, P. Pinson and J. R. Bidlot, Forecasting ocean wave energy: The ECMWF wave model and time series methods, *Ocean Eng.* **38**(10), 1089–1099, (2011). doi:10.1016/j.oceaneng.2011.04.009.

[23] A. Malmberg, U. Holst and J. Holst, Forecasting near-surface ocean winds with Kalman filter techniques, *Ocean Eng.* **32**(3–4), 273–291, (2005). doi:10.1016/j.oceaneng.2004.08.005.

[24] P. C. Ho and J. Z. Yim, Wave height forecasting by the transfer function model, *Ocean Eng.* **33**(8–9), 1230–1248, (2006). doi:10.1016/j.oceaneng.2005.09.003.

[25] S. N. Londhe and V. Panchang, One-day wave forecasts based on artificial neural networks, *J. Atmos. Oceanic Technol.* **23**(11), 1593–1603, (2006). doi:10.1175/JTECH1932.1.

[26] S. Gaur and M. C. Deo, Real-time wave forecasting using genetic programming, *Ocean Eng.* **35**(11–12), 1166–1172, (2008). doi:10.1016/j.oceaneng.2008.04.007.

[27] S. C. James, Y. Zhang and F. O'Donncha, A machine learning framework to forecast wave conditions, *Coastal Eng.* **137**, 1–10, (2018). doi:10.1016/j.coastaleng.2018.03.004.

[28] Y. Jiang, Y. Gou, T. Zhang, K. Wang and C. Hu, A machine learning approach to argo data analysis in a thermocline, *Sensors* **17**(10), 2225, (2017). doi:10.3390/s17102225.

[29] G. Hollinger, A. Pereira, V. Ortenzi and G. Sukhatme, Towards improved prediction of ocean processes using statistical machine learning, in *Robotics: Science and Systems Workshop on Robotics for Environmental Monitoring*, (2012).

[30] D. Sarkar, M. A. Osborne and T. A. Adcock, Spatiotemporal prediction of tidal currents using Gaussian processes, *J. Geophys. Res.: Oceans* **124**(4), 2697–2715, (2019).

[31] T. Bolton and L. Zanna, Applications of deep learning to ocean data inference and subgrid parameterization, *J. Adv. Model. Earth Syst.* **11**(1), 376–399, (2019). doi:10.1029/2018MS001472.

[32] D. J. MacKay, *Information Theory, Inference and Learning Algorithms* (Cambridge University Press, 2003).

[33] J.-S.R. Jang, ANFIS: Adaptive-Network-based Fuzzy Inference Systems, *IEEE Trans. Syst. Man Cybern.* **23**(3), 665–685, (1993). doi:10.1109/21.256541.

[34] L. X. Wang, *Adaptive Fuzzy Systems and Control: Design and Stability Analysis* (Englewood Cliffs, NJ: PTR Prentice Hall, 1994).

[35] M. Öztürk and İ. Özkol, Comparison of modified Karnik-Mendel algorithm-based interval type-2 ANFIS and type-1 ANFIS, *Aircr. Eng. Aerosp. Technol.* **93**(10), 1526–1532, (2021). doi:10.1108/AEAT-11-2020-0268.

[36] M. Öztürk, *A Modified ANFIS System for Aerial Vehicles Control*, (PhD Thesis, İstanbul Technical University, 2022).

Chapter 7

Rogue Waves of (3+1)-Dimensional Variable Coefficient Nonlinear Wave Equation

Fatma Nur Kaya Sağlam*,‡ and Zehra Pinar Izgi†

*Department of Information Security Technology, Faculty of Computer and Information Technologies, Cappadocia University, Nevşehir, Turkey

†Department of Mathematics Faculty of Arts and Science, Tekirdağ Namık Kemal University 59030 Tekirdağ, Turkey

‡ftmnrtlp@gmail.com

Rogue waves (RW) are giant waves that occur suddenly and unexpectedly in the ocean. In addition, according to the velocities of the gas bubbles formed by the liquid-gas mixture, RW solutions are obtained by symbolic calculation. In recent years, technological advances and observing tools such as radars, underwater sensors, and satellites have improved the ability to monitor and measure RW. As a result, researchers now have a better understanding of the potential impacts of RW and can take precautions. Nonlinear wave equations (NWEs), which are used in many fields of physics and engineering, such as natural phenomena, gas bubbles, acoustics, optics, hydrodynamics, and plasmas, are modeled with (3+1)-dimensional variable coefficient wave equations, and the complex interactions and nonlinear effects required to explain wave behavior can be explored more effectively. In this study, RW solutions of the (3+1)-dimensional variable/constant coefficient nonlinear wave equation for liquid with gas bubbles are obtained by ansatz-based methods.

1. Introduction

Since many significant natural phenomena and engineering problems are modeled by nonlinear partial differential equations (NLPDEs), exact solutions of NWEs have gained great importance for researchers regarding theory and a wide range of applications [1–12]. Analytical [13–15], semi-analytical [16, 17] and numerical methods [18, 19] have been developed for the solution of these equations, but there is no generalized method for the exact solution. While the concept of the soliton, introduced in the 1950s [20, 21], attracted the attention of researchers in the study of nonlinear phenomena, the concept of the RW has emerged today in optical fibers, Bose-Einstein condensates, gas bubbles, superfluids, and many other fields [22, 23].

RWs, which form extremely RW peaks that appear and disappear suddenly, occur, and persist in the world's major oceans and can occur with or without currents, in deep water, offshore, during storms or general weather conditions, and can cause accidents by affecting the course of ships in the ocean [20]. As a result, while the study of RW has become increasingly important to researchers, multidimensional RW has received little attention in the literature.

NWE are also used to describe the propagation of pressure waves in a mixture of liquid and gas bubbles, taking into account the viscosity and heat transfer of the liquid [24]. Since real nonlinear waves in a liquid containing gas bubbles are multidimensional, they are modeled by (3+1)-dimensional NWE [25]. The use of NWEs with variable coefficients provides a more realistic treatment of the physical phenomena and allows more features to be seen when the medium is heterogeneous and the boundaries are not smooth [26, 27]. Kudryashov *et al.* constructed a model for liquid with gas bubble flow in which all bubbles are assumed to be of the same size, neglecting destruction, interactions, and collisions between bubbles [24].

Gas bubbles can affect, refract, and reflect a wave field in a liquid. This can cause the wavelength to change over time and concentrate in a particular area [28–31]. As a result, RW can be observed, especially in NWEs with varying coefficients. Since gas bubbles can heterogenized the liquid medium, they can affect the fluctuations in the wave field, which can lead to the formation of localized high energy fields. These high-energy fields give rise to RW [28–31]. Furthermore, since these bubbles can interact with other waves in the wave field, they can affect the energy distribution in the wave field, leading to the accumulation of intense energy in some regions and the appearance of RW in these areas [28–31].

While the (3+1)-dimensional NWE used in bubble modeling is mostly a constant coefficient NWE [24], the variable coefficient NWE gives us more realistic results than its constant coefficient equivalent in some physical situations where the environment is not homogeneous and the boundaries are not smooth [20, 26, 27].

$$\alpha(t)(\rho\rho_x)_x + \beta(t)\rho_{xxxx} + \gamma(t)\rho_{xx} + \delta(t)\rho_{yy} + \varsigma(t)\rho_{zz} + \rho_{xt} = 0, \tag{1}$$

where $\rho = \rho(x, y, z, t)$ is the wave amplitude function, the variable coefficients $\alpha(t)$, $\beta(t)$, $\gamma(t)$, $\delta(t)$ and $\varsigma(t)$ are nonlinearity, dispersion, viscosity, perturbation, respectively. Deng and Gao obtained the bilinear form of the equation as well as multisoliton solutions [25]. Zhu and Liu [20] used the Hirota bilinear method to obtain multidimensional RW solutions. Different types of NWE can be obtained by setting different constant values for the parameters $\alpha(t)$, $\beta(t)$, $\gamma(t)$, $\delta(t)$ and $\varsigma(t)$ in Eq. (1).

When,

(i) $\alpha(t) = -6$, $\beta(t) = 1$, $\gamma(t) = 0$, $\delta(t) = \varsigma(t) = 3$,

$$-6(\rho\rho_x)_x + \rho_{xxxx} + 3\rho_{yy} + 3\rho_{zz} + \rho_{xt} = 0, \tag{2}$$

(3+1)-dimensional KP equation handled by Ablowitz and Segur [32].

(ii) $\alpha(t) = \beta(t) = 1$, $\gamma(t) = 0$, $\delta(t) = \varsigma(t) = \frac{1}{2}$,

$$(\rho\rho_x)_x + \rho_{xxxx} + \frac{1}{2}\rho_{yy} + \frac{1}{2}\rho_{zz} + \rho_{xt} = 0, \tag{3}$$

(3+1)-dimensional NWE is obtained [33]. This equation is given to describe the pressure waves in mixing liquid and gas bubbles considering the viscosity of the liquid and the heat transfer. The purpose of this study is to use the Riccati and Bernoulli auxiliary equation method, known as the ansatz-based methods, to obtain RW solutions for the (3+1)-dimensional nonlinear wave equation with variable/constant coefficients (Eq. (1)/Eq. (3)), respectively.

2. Methods

Bernoulli auxiliary equation method and Riccati auxiliary equation method are the most popular ansatz-based methods [11]. Now, let us give you a brief description of the main steps in the process:

(1) Suppose a given NLPDE for $\rho = \rho(x, y, z, t)$, provided by

$$P(\rho, \rho_x, \rho_y, \rho_z, \rho_t, \rho_{xx}, \ldots) = 0. \tag{4}$$

Using the wave transformation $\rho(x, y, z, t) = u(\xi)$, $\xi = x + \eta y + \mu z - ct$, $\mu \neq 0$, $\eta \neq 0$, $c \neq 0$, Eq. (4) is transformed into the given nonlinear ordinary differential equation (NODE),

$$O\left(u, \frac{\partial u}{\partial \xi}, \frac{\partial u}{\partial \xi^2}, \frac{\partial u}{\partial \xi^3}, \ldots\right) = 0. \tag{5}$$

(2) Suppose that Eq. (5) has the exact solution by

$$u(\xi) = \sum_{i=0}^{N} g_i z^i(\xi), \tag{6}$$

where g_i are parameters. Obviously, $z(\xi)$ is the solutions of the Riccati equation

$$\frac{dz}{d\xi} = Pz(\xi) + Qz^2(\xi) + R, \tag{7}$$

or Bernoulli differential equation

$$\frac{dz}{d\xi} = Pz(\xi) + Qz^2(\xi). \tag{8}$$

N is the result of the application of balancing principle to Eq. (4).

(3) Substituting Eq. (6) into Eq. (5), and the system of nonlinear algebraic equations obtained by setting the coefficients of each resulting power equal to zero is solved, as a result we obtain significant and trivial solution sets depending on the parameters P, Q, R, $g_0, g_1, \ldots, g_N$, η, μ and c. After determining these parameters and using the solution of Eq. (7) or Eq. (8) we obtain an explicit-form analytical solution $u(\xi)$.

(4) Then, by substituting the wave transformation, the explicit form of the solutions of Eq. (4) is determined.

3. Solutions

In this section, two cases, the variable coefficient and constant coefficient form of the considered problem will be hold, respectively.

Case (i). To solve of Eq. (1) $\rho = \rho(x, y, z, t)$ is considered where $\xi = x + \eta y + \mu z - ct$, $\mu \neq 0$, $\eta \neq 0$, $c \neq 0$.

As a result, the reduced form is

$$\beta(t)\,u^{(iv)} + (\gamma(t) + \delta(t)\,\eta^2 + \varsigma(t)\,\mu^2 - c)u'' + \alpha(t)\,u'^2 + \alpha(t)\,uu'' = 0. \quad (9)$$

By applying the procedure of the Bernoulli equation method as explained in the previous section, when $Q(\xi) = \frac{\sqrt{-3\beta(t)\alpha(t)g_2}}{6\beta(t)}$ and $P(\xi) = -\frac{\alpha(t)g_1\sqrt{3}}{6\sqrt{-\beta(t)\alpha(t)g_2}}$ are substituted in the equation system and solved, the resulting parameter is given below:

$$g_0 = \frac{-12\delta(t)\eta^2 g_2 - 12\varsigma(t)\mu^2 g_2 + \alpha(t){g_1}^2 - 12\gamma(t)g_2 + 12cg_2}{12\alpha(t)g_2}. \quad (10)$$

Thus, the explicit form of the RW solution is obtained from

$$\rho(x,y,z,t) = \frac{-12\delta(t)\eta^2 g_2 - 12\varsigma(t)\mu^2 g_2 + \alpha(t){g_1}^2 - 12\gamma(t)g_2 + 12cg_2}{12\alpha(t)g_2}$$
$$+ \frac{{g_1}^2}{\left(exp\left(\frac{\alpha(t)g_1(-ct+\eta y+\mu z+x)}{2\sqrt{-3\beta(t)\alpha(t)g_2}}\right)C_1 g_1 - g_2\right)}$$
$$+ \frac{g_2{g_1}^2}{\left(exp\left(\frac{\alpha(t)g_1(-ct+\eta y+\mu z+x)}{2\sqrt{-3\beta(t)\alpha(t)g_2}}\right)C_1 g_1 - g_2\right)^2}. \quad (11)$$

Figure 1 shows the propagation of three-dimensional plots of the solution (Eq. (11)) when $\gamma(t)$, $\delta(t)$, $\varsigma(t)$ and $\alpha(t)$select different functions. When $\gamma(t) = \delta(t) = \varsigma(t) = \cos(t)$, $\alpha(t) = \sin(t)$ with the parameters $c = 0.5$, $\eta = 0.2$, $\mu = 1.2$, $g_1 = -1$, $g_2 = 1$, $y = 2$, $z = 3$, $C_1 = -1.1$, $\beta(t) = -1.7$ a peaked rogue wave solution is shown in Figure 1(a). When $\alpha(t) = \sin(t)$, $\gamma(t) = \delta(t) = \varsigma(t) = t$ with the parameters $c = 0.5$, $\eta = 2$, $\mu = 1.6$, $g_1 = -1$, $g_2 = 0.9$, $y = -1$, $z = 5$, $C_1 = -1.1$, $\beta(t) = -1.7$ of the modulus of the exact wave solution as a peaked rogue wave solution is shown in Figure 1(b).

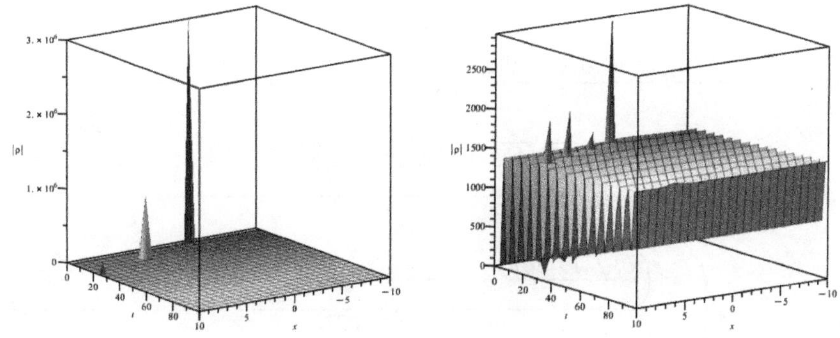

(a) (b)

Fig. 1. RW solution of $|\rho(x,y,z,t)|$ (a) $c = 0.5$, $\eta = 0.2$, $\mu = 1.2$, $g_1 = -1$, $g_2 = 1$, $y = 2$, $z = 3$, $C_1 = -1.1$, $\beta(t) = -1.7$, $\gamma(t) = \delta(t) = \varsigma(t) = \cos(t)$, $\alpha(t) = \sin(t)$ (b) $c = 0.5$, $\eta = 2$, $\mu = 1.6$, $g_1 = -1$, $g_2 = 0.9$, $y = -1$, $z = 5$, $C_1 = -1.1$, $\beta(t) = -1.7$, $\alpha(t) = \sin(t)$, $\gamma(t) = \delta(t) = \varsigma(t) = t$. Figure 1 shows that the speed of the bubbles increases as the $\gamma(t)$, $\delta(t)$, $\varsigma(t)$ parameters changes.

Case (ii). The solution obtained by keeping the coefficients as constants which the constant form of Eq. (1).

$$\alpha(\rho\rho_x)_x + \beta\rho_{xxxx} + \gamma\rho_{xx} + \delta\rho_{yy} + \varsigma\rho_{zz} + \rho_{xt} = 0, \tag{12}$$

where $\rho = \rho(x,y,z,t)$ is the amplitude function of the wave [13]. In order to obtain a solution, for the reduction $\rho = \rho(x,y,z,t)$, $\xi = x + \eta y + \mu z - ct$, $\mu \neq 0$, $\eta \neq 0$, $c \neq 0$ is considered. The wave transformation is substituted into Eq. (12) and reduced to a NODE,

$$\beta u^{(iv)} + (\gamma + \delta\eta^2 + \varsigma\mu^2 - c)u'' + \alpha u'^2 + \alpha u u'' = 0. \tag{13}$$

According to the balancing principle, $u(\xi) = g_0 + \sum_{i=1}^{2} g_i z^i(\xi)$, $g_i(i = 0, 1, \ldots, N)$, is obtained. Where $z(\xi)$ is determined by $\frac{dz}{d\xi} = Pz(\xi) + Qz^2(\xi) + R$. Using the methodical process, numerous solution sets are obtained. Besides the trivial and the constant solutions for the parameters, the following solution set is obtained;

$$g_0 = -\frac{-P^2\alpha g_2 - 8QR\alpha g_2 - 12Q^2c + 12Q^2\delta\eta^2 + 12Q^2\gamma + 12Q^2\varsigma\mu^2}{12Q^2\alpha},$$

$$g_1 = \frac{Pg_2}{Q}, \quad \beta = \frac{-\alpha g_2}{12Q^2}. \tag{14}$$

The solution of Eq. (12) is as follows

$$\rho(x,y,z,t) = -\frac{-P^2\alpha g_2 - 8QR\alpha g_2 - 12Q^2c + 12Q^2\delta\eta^2 + 12Q^2\gamma + 12Q^2\varsigma\mu^2}{12Q^2\alpha}$$
$$+\frac{Pg_2\left(\tan\left(C_1\frac{\sqrt{-P^2+4QR}}{2} + \frac{(-ct+x+\eta y+\mu z)\sqrt{-P^2+4QR)}}{2}\right)\sqrt{-P^2+4QR}-P\right)}{2Q^2}$$
$$+g_2\left(\frac{\tan\left(\frac{C_1\sqrt{-P^2+4QR}}{2} + \frac{(-ct+x+\eta y+\mu z)\sqrt{-P^2+4QR)}}{2}\sqrt{-P^2+4QR}-P)^2\right)}{4Q^2}\right). \tag{15}$$

Figure 2 represents three-dimensional plot of the exact wave solutions (Eq. (15)) with the parame-ters $y = 1$, $z = -1$, $C_1 = 0.9$, $\gamma = 1$, $P = 1$, $Q = -1$, $R = 1.2$, $\alpha = 0.2$, $c = 0.5$, $\delta = -0.7$, $\eta = 2$, $\mu = -1.3$, $\varsigma = 0.6$, $g_2 = 1$.

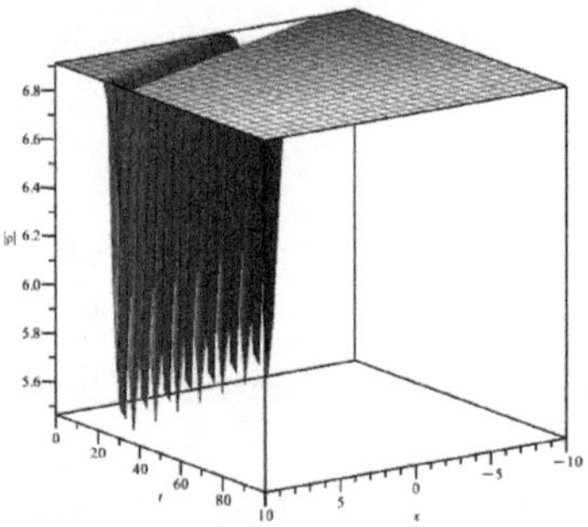

Fig. 2. Three-dimensional plot of the exact wave solution of $|\rho(x,y,z,t)|$ as exact wave form when $y = 1$, $z = -1$, $C_1 = 0.9$, $\gamma = 1$, $P = 1$, $Q = -1$, $R = 1.2$, $\alpha = 0.2$, $c = 0.5$, $\delta = -0.7$, $\eta = 2$, $\mu = -1.3$, $\varsigma = 0.6$, $g_2 = 1$.

Case (iii). In this case, constant coefficient form of the considered problem Eq. (12) is solved via Bernoulli equation method. According to the balancing principle, $u(\xi) = g_0 + \sum_{i=1}^{2} g_i z^i(\xi)$, $g_i(i = 0, 1, \ldots, N)$, is obtained. Where $z(\xi)$ is determined by $\frac{dz}{d\xi} = Pz(\xi) + Qz^2(\xi)$.

Besides the trivial and the constant solutions, the following solution set is obtained;

$$c = \frac{-P\alpha g_1 + 12Q\alpha g_0 + 12Q\delta\eta^2 + 12Q\gamma + 12Q\varsigma\mu^2}{12Q}, \quad \beta = \frac{-\alpha g_1}{12PQ}, \quad g_2 = \frac{Qg_1}{P}. \tag{16}$$

The solution of Eq. (12) is as follows

$$\rho(x,y,z,t) = g_0 + \frac{g_1 P}{\exp\left(-P\left(-ct + x + \eta y + \mu z\right)\right) C_1 P - Q} + \frac{Q g_1 P}{\left(\exp\left(-P\left(-ct\right) + x + \eta y + \mu z\right) C_1 P - Q\right)^2}. \tag{17}$$

Figure 3 represents three-dimensional plot of the exact wave solutions (Eq. (17)) with the parameters $y = 2$, $z = 3$, $C_1 = 0.5$, $\gamma = 1$, $P = 2$, $Q = 1$, $\alpha = 2$, $\delta = 0.5$, $\eta = -1$, $\mu = 0.7$, $\varsigma = 1.2$, $g_0 = -1$, $g_1 = 1$.

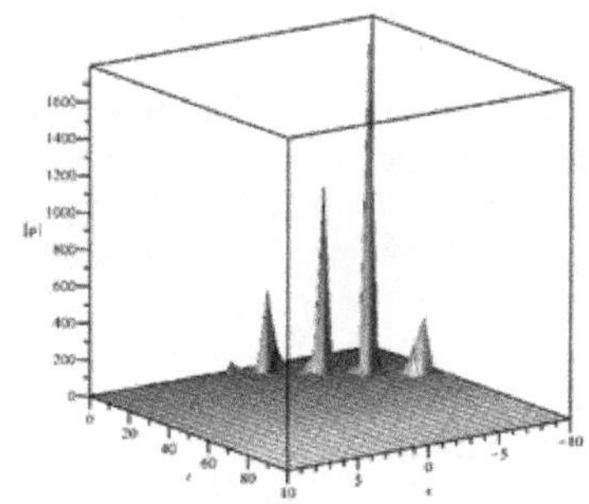

Fig. 3. Three-dimensional plot of the exact wave solution of $|\rho(x,y,z,t)|$ as a peaked rogue wave form when $y = 2$, $z = 3$, $C_1 = 0.5$, $\gamma = 1$, $P = 2$, $Q = 1$, $\alpha = 2$, $\delta = 0.5$, $\eta = -1$, $\mu = 0.7$, $\varsigma = 1.2$, $g_0 = -1$, $g_1 = 1$.

4. Conclusions

We consider the (3+1)-dimensional variable/constant coefficient nonlinear wave equation, which is used to model several nonlinear processes that occur in liquids containing gas bubbles. In the literature, different solution techniques are used to obtain solutions of Eq. (1) with different structures (dark, bright, kink wave solitons, traveling wave solutions, etc.). For Eq. (1) and Eq. (12), which models the gas bubble, the Bernoulli auxiliary equation method and the Riccati auxiliary equation method are used to obtain RW solutions and exact wave solution. From the results shown in Figures 1, 2 and 3, it can be concluded that there are different types of solutions, such as RW and exact wave solutions.

References

[1] J.-G. Liu and W.-H. Zhu, Multiple rogue wave solutions for (2+1)-dimensional Boussinesq equation, *Chinese Journal of Physics*, vol. 67, pp. 492–500, (2020).

[2] X.-Y. Wu, B. Tian, Q.-X. Qu, Y.-Q. Yuan and X.-X. Du, Rogue waves for a (2+1)-dimensional Gross–Pitaevskii equation with time-varying trapping potential in the Bose–Einstein condensate, *Computers & Mathematics with Applications*, vol. 79, no. 4, pp. 1023–1030, (2020).

[3] J.-J. Su, Y.-T. Gao, and C.-C. Ding, Darboux transformations and rogue wave solutions of a generalized AB system for the geophysical flows, *Applied Mathematics Letters*, vol. 88, pp. 201–208, (2019).

[4] S. Kumar and S. K. Dhiman, Lie symmetry analysis, optimal system, exact solutions and dynamics of solitons of a (3+1)-dimensional generalised BKP–Boussinesq equation, *Pramana*, vol. 96, no. 1, p. 31, (2022).

[5] Y. Liu and Y. Yang, Rogue wave solutions for the 3+1-dimensional generalized Camassa–Holm–Kadomtsev–Petviashvili equation, *Chinese Journal of Physics*, vol. 86, pp. 508–514, (2023).

[6] K. Dysthe, H. E. Krogstad, and P. Müller, Oceanic rogue waves, *Annu. Rev. Fluid Mech.*, vol. 40, no. 1, pp. 287–310, (2008).

[7] J. M. Dudley, G. Genty, A. Mussot, A. Chabchoub, and F. Dias, Rogue waves and analogies in optics and oceanography, *Nature Reviews Physics*, vol. 1, no. 11, pp. 675–689, (2019).

[8] S. Singh, L. Kaur, K. Sakkaravarthi, R. Sakthivel, and K. Murugesan, Dynamics of higher-order bright and dark rogue waves in a new (2+1)-dimensional integrable Boussinesq model, *Physica Scripta*, vol. 95, no. 11, p. 115213, (2020).

[9] S. Knobler, D. Liberzon, and F. Fedele, Large waves and navigation hazards of the Eastern Mediterranean Sea, *Scientific Reports*, vol. 12, no. 1, p. 16511, (2022).

[10] F. Fedele, Rogue waves in oceanic turbulence, *Physica D: Nonlinear Phenomena*, vol. 237, no. 14–17, pp. 2127–2131, (2008).

[11] Z. P. Izgi, Rogue waves and solitons of the generalized modified nonlinear Schrödinger equations, *Mathematics and Computers in Simulation*, vol. 208, pp. 535–549, (2023).

[12] M. R. Ali and W.-X. Ma, New exact solutions of Bratu Gelfand model in two dimensions using Lie symmetry analysis, *Chinese Journal of Physics*, vol. 65, pp. 198–206, (2020).

[13] A. Tripathy, S. Sahoo, H. Rezazadeh, and Z. P. Izgi, New optical analytical solutions to the full nonlinearity form of the space–time Fokas–Lenells model of fractional-order, *International Journal of Modern Physics B*, vol. 36, no. 14, p. 2250058, (2022).

[14] M. M. Rashidi and E. Erfani, New analytical method for solving Burgers and nonlinear heat transfer equations and comparison with HAM', *Computer Physics Communications*, vol. 180, no. 9, pp. 1539–1544, (2009).

[15] Z. Pinar Izgi, Analytical studies of Eyring-Powell fluid models, *Multidiscipline Modeling in Materials and Structures*, vol. 19, no. 2, pp. 145–154, (2023).

[16] G. Arora, S. Hussain, R. Kumar, and Y. Mammeri, Semi-Analytical Solution for Condensing Coagulation and Lifshitz-Slyozov Models: Variational Iteration Method, Proceedings of the 2nd International Conference on Nonlinear Dynamics and Applications (ICNDA 2024), Volume 1, pp. 587–598, (2024).

[17] D. J. Tweten, G. M. Lipp, F. A. Khasawneh, and B. P. Mann, On the comparison of semi-analytical methods for the stability analysis of delay differential equations, *Journal of Sound and Vibration*, vol. 331, no. 17, pp. 4057–4071, (2012).

[18] J.-H. He and H. Latifizadeh, A general numerical algorithm for nonlinear differential equations by the variational iteration method, *International Journal of Numerical Methods for Heat & Fluid Flow*, vol. 30, no. 11, pp. 4797–4810, (2020).

[19] B. Wongsaijai, T. Aydemir, T. Ak, and S. Dhawan, Analytical and numerical techniques for initial-boundary value problems of Kolmogorov–Petrovsky–Piskunov equation, *Numerical Methods for Partial Differential Equations*, vol. 40, no. 1, p. e22693, (2024).

[20] J.-G. Liu and W.-H. Zhu, Multiple rogue wave, breather wave and interaction solutions of a generalized (3+1)-dimensional variable-coefficient nonlinear wave equation, *Nonlinear Dynamics*, vol. 103, no. 2, pp. 1841–1850, (2021).

[21] M. S. Osman, New analytical study of water waves described by coupled fractional variant Boussinesq equation in fluid dynamics, *Pramana*, vol. 93, no. 2, p. 26, (2019).
[22] P. A. Clarkson and E. Dowie, Rational solutions of the Boussinesq equation and applications to rogue waves, *Transactions of Mathematics and its Applications*, vol. 1, no. 1, p. tnx003, (2017).
[23] S.-J. Chen, Y.-H. Yin, W.-X. Ma, and X. Lü, Abundant exact solutions and interaction phenomena of the (2+1)-dimensional YTSF equation, *Analysis and Mathematical Physics*, vol. 9, pp. 2329–2344, (2019).
[24] N. A. Kudryashov and D. I. Sinelshchikov, Nonlinear waves in bubbly liquids with consideration for viscosity and heat transfer, *Physics Letters A*, vol. 374, no. 19–20, pp. 2011–2016, (2010).
[25] G.-F. Deng and Y.-T. Gao, Integrability, solitons, periodic and travelling waves of a generalized (3+1)-dimensional variable-coefficient nonlinear-wave equation in liquid with gas bubbles, *The European Physical Journal Plus*, vol. 132, pp. 1–17, (2017).
[26] H.-M. Yin, B. Tian, H.-L. Zhen, J. Chai, and X.-Y. Wu, Bright optical solitons or light bullets for a (3+1)-dimensional generalized nonlinear Schrödinger equation with the distributed coefficients, *Modern Physics Letters B*, vol. 30, no. 25, p. 1650306, (2016).
[27] M. R. Ali and R. Sadat, Lie symmetry analysis, new group invariant for the (3+1)-dimensional and variable coefficients for liquids with gas bubbles models, *Chinese Journal of Physics*, vol. 71, pp. 539–547, (2021).
[28] W. Liu and Y. Zhang, Resonant multiple wave solutions, complexiton solutions and rogue waves of a generalized (3+1)-dimensional nonlinear wave in liquid with gas bubbles, *Waves in Random and Complex Media*, vol. 30, no. 3, pp. 470–480, (2020).
[29] M. Wang, B. Tian, Y. Sun, and Z. Zhang, Lump, mixed lump-stripe and rogue wave-stripe solutions of a (3+1)-dimensional nonlinear wave equation for a liquid with gas bubbles, *Computers & Mathematics with Applications*, vol. 79, no. 3, pp. 576–587, (2020).
[30] X. Tang and Y. Chen, Lumps, breathers, rogue waves and interaction solutions to a (3+1)-dimensional Kudryashov–Sinelshchikov equation, *Modern Physics Letters B*, vol. 34, no. 12, p. 2050117, (2020).
[31] W. Liu and Y. Zhang, Multiple rogue wave solutions for a (3+1)-dimensional Hirota bilinear equation, *Applied Mathematics Letters*, vol. 98, pp. 184–190, (2019).
[32] M. J. Ablowitz and H. Segur, On the evolution of packets of water waves, *Journal of Fluid Mechanics*, vol. 92, no. 4, pp. 691–715, .(1979).
[33] N. A. Kudryashov and D. I. Sinelshchikov, Equation for the three-dimensional nonlinear waves in liquid with gas bubbles, *Physica Scripta*, vol. 85, no. 2, p. 025402, (2012).

Chapter 8

An Optimization Framework Ensuring Steady and Uninterrupted Vehicle Flow

Mahmut Tutam* and Şeyma Emeç†

Department of Industrial Engineering, Erzurum Technical University, Erzurum, 25050, Türkiye

**mahmut.tutam@erzurum.edu.tr*

†*seyma.emec@erzurum.edu.tr*

Despite advancements in technology, ineffective management of vehicle traffic poses a significant hurdle for sustainable and low-emission transportation systems, largely due to issues with road conditions and traffic signals. The former affects rolling resistance, increasing the energy needed to maintain speed, and traction, which impacts the driving efficiency. The latter is a crucial element of roads, improving traffic organization and safety while also restricting vehicle speeds and potentially increasing energy consumption. The goal of this study is to enhance the overall efficiency of electric vehicle operations. To achieve this, a Multi-Stage Decision-Making Framework is developed to optimize the energy consumption by considering road conditions and traffic signals along pre-planned routes. A solution algorithm is designed and implemented within a computational framework, with a small-scale example provided to effectively showcase the model's application and illustrate its practical implications in real-world scenarios. Additionally, movement forces, acceleration/deceleration, and speed profiles for the electric vehicle are determined for this example. Ultimately, the optimal speed profile is established, enabling the electric vehicle to achieve the highest efficiency in energy consumption and performance under the specified constraints.

1. Introduction

As advancements in technology continue to progress, smart systems are increasingly becoming an integral part of daily life by helping people make decisions based on historical or real-time data, leading to more effective and intelligent execution of daily activities [1]. Among these activities, transportation is crucial as it enables people to move around and complete their daily routines. Therefore, coordination between passengers, vehicles, and roads is vital for an efficient, reliable, and sustainable transportation system, which are the essence of smart transportation systems.

Smart transportation systems include the implementation of infrared, wireless networks, sensors and mobile technologies with the objective of ensuring public safety, reducing emissions, promoting environmental friendliness, and improving the efficiency of passenger and freight transportation. Even with all the technological advancements, one of the major challenges faced by smart transportation systems is the ineffective management of vehicle traffic on roadways. Particularly, crucial aspects of roadways are road conditions and traffic signals, which cause frequent acceleration, deceleration, and stopping of vehicles, thereby increasing energy consumption. Therefore, the objective of this study is to develop an optimization model for ensuring a continuous and smooth flow of vehicles.

To give an overview of the paper, we begin by describing the problem in Section 2. Next, we review the literature and position our contribution in Section 3. Section 4 details the methodology used to develop the models and solution algorithm. In Section 5, we present a case study along with the results. The paper ends with a discussion and proposes potential areas for future research.

2. Problem Description

As of 2023, there are about 40 million electric vehicles (EVs) on the roads, with 14 million EVs sold during the year [2]. However, this figure is still relatively small compared to the 1.3 billion internal combustion engine (ICE) vehicles [3]. These numbers suggest that EVs have a significant journey ahead before they can achieve the same level of prevalence as ICE vehicles.

Given the environmental benefits, reduced dependence on oil, and improved energy efficiency, the number of EVs on the road is expected to rise significantly in the near future. However, obstacles such as battery limitations, inadequate charging infrastructure, restricted driving range, and reduced performance in extreme conditions currently hinder EV sales. To address these issues, many automobile manufacturers are focusing on reducing energy consumption to enhance EV range. As a result, most research efforts are directed at optimizing battery usage, either concentrating on improving energy efficiency based on road conditions (while overlooking traffic signals) or traffic signals (while ignoring road conditions).

In this study, a Multi-Stage Decision-Making Framework has been developed to optimize the energy consumption of EVs by considering road conditions and traffic signals on pre-planned routes simultaneously. The model's validity has been tested across various scenarios, and the optimal speed profile has been determined for a small-scale problem.

3. Related Research

A wide range of topics related to EVs is addressed in the research literature. Therefore, we limit our review to the most relevant studies, focusing specifically on road conditions or traffic signals.

A model predictive control framework was introduced to optimize the speed profile of an EV as it approaches a road segment with multiple traffic signals [4].

An optimal speed planning algorithm was created that leverages probabilistic data on traffic signal phases and timing [5]. The probabilistic crossing time of a vehicle at a signalized intersection was expressed in terms of the effective red light duration. To address uncertain crossing times, a data-driven, chance-constrained eco-driving approach designed to enhance both energy efficiency and safety was proposed [6].

A method for optimizing the predictive energy efficiency of an EV was developed by utilizing information on traffic light sequences [7]. An optimization model was proposed to determine the ideal green time and vehicle speed, ensuring that the vehicle passes through the traffic light with the maximum battery state of charge [8]. An optimal crossing strategy for navigating multiple intersections in a crossing task is proposed [9]. The study developed an algorithm to estimate energy consumption for future trips by considering factors such as energy usage, driving behavior, road topography, weather conditions, and traffic conditions. The developed algorithm was validated through 30 driving tests, which included urban, rural, highway, and hilly driving conditions [10].

A comprehensive model is proposed to calculate the energy required for charging EVs for a planned trip [11]. A robust optimal eco-driving control strategy is proposed, considering multi-signal intersections with uncertain traffic signal timings [12]. The study critically analyzes the literature to identify the key factors affecting the remaining driving range [13].

Based on the literature, we conclude that our approach is distinctive. While most studies have predominantly concentrated on energy optimization concerning either road conditions or traffic signals, our methodology simultaneously integrates both factors. Additionally, an algorithm has been developed to enable the proposed model to be applied to small-scale problems. This aims to provide a valuable contribution to the literature.

4. Methodology

We utilize a Multi-Stage Decision-Making Framework to obtain the optimal solution. Initially, we draw on mathematical expressions from the literature to calculate the energy consumption at each state of the stage, specifically determining the velocity, acceleration/deceleration, and movement force for the EV. Note that energy consumption at each state depends on the decisions and values from the states of previous stages. Afterward, we enumerate all possible solutions to identify the optimal one.

4.1. *Notation*

We describe the notation used to develop the mathematical expressions for energy consumption as follows:

f : movement force for the EV (N)
E : energy consumption (kWh)
v : velocity of the vehicle (m/s)
t : travel time (s)
C_d : drag coefficient
A : frontal area of the vehicle (m^2)
ρ_a : air density (kg/m^3)
C_{rr} : rolling resistance coefficient
m : mass of the vehicle (kg)
g : gravitational acceleration (9.81 m/s^2)
θ : inclination angle (radians)
a : acceleration (m/s^2)

4.2. *Mathematical model*

The energy consumption for an EV is calculated based on the summation of the forces acting on the vehicle, which must be overcome over time, and is formulated as

$$E = fvt/3600 \tag{1}$$

The first component of the force is the aerodynamic force (f_{ad}), which is the resistance experienced by an EV due to air flow as it moves and is given by

$$f_{ad} = C_d A v^2 \rho_a / 2 \tag{2}$$

The second component of the force is the rolling resistance force (f_{rr}), which is the resistance encountered by a vehicle's tires as they roll over a surface and is given by

$$f_{rr} = C_{rr} m g \tag{3}$$

Another force is the climbing resistance force (f_{cr}), which is additional resistance encountered by a vehicle when moving uphill/downhill (can be expressed as positive/negative signs) and is calculated by

$$f_{cr} = mg\sin(\theta) \tag{4}$$

Last but not least, the acceleration/deceleration force ($f_{a/d}$) is the required force to change the vehicle's speed (can be expressed as positive/negative signs) and is calculated by

$$f_{a/d} = ma \tag{5}$$

4.3. *Formulation of the optimization model*

The optimization model is designed to minimize energy consumption while adhering to specific time and velocity constraints. It includes parameters that define the vehicle's environment, variables that influence path selection and speed, and constraints that ensure compliance with travel time and speed limits, all aimed at effectively minimizing energy consumption.

Parameters

- distance between traffic lights.
- initial time and wavelength of each traffic light.
- initial speed of the EV.

Variables

- path selection variable.
- velocity for a path between any states of two consecutive stages.
- acceleration/deceleration for a path between any states of two consecutive stages.

Constraints

- passing traffic signals within designated travel times.
- adhering to the specified speed limits

Objective function

- minimize energy consumption.

This model establishes the foundation for the Multi-Stage Decision-Making Framework.

4.4. *Multi-Stage Decision-Making Framework*

We begin by outlining the states and stages of the problem, since the decision-making process involves multiple sequential stages, each dependent on the outcomes of the preceding stage, as shown in Figure 1.

Each traffic signal represents a stage, with the initial stage serving as the starting point. Within each stage, each signal phase represents a state. We calculate the state transfer functions for velocity, acceleration/deceleration, and force for each state based on the states from the preceding stage. Subsequently, energy consumption is calculated.

4.5. *Solution algorithm*

The Multi-Stage Decision-Making Framework model is solved by evaluating all solution alternatives, calculated as $nStates^{nStages-1}$ alternatives, where $nStates$ represents the number of states within each stage, and $nStages$ denotes the number of stages in the problem. For the example given in Figure 1, there are $4^{5-1} = 256$ alternative solutions (see Figure 2). We develop Algorithm 1 to enumerate all potential paths.

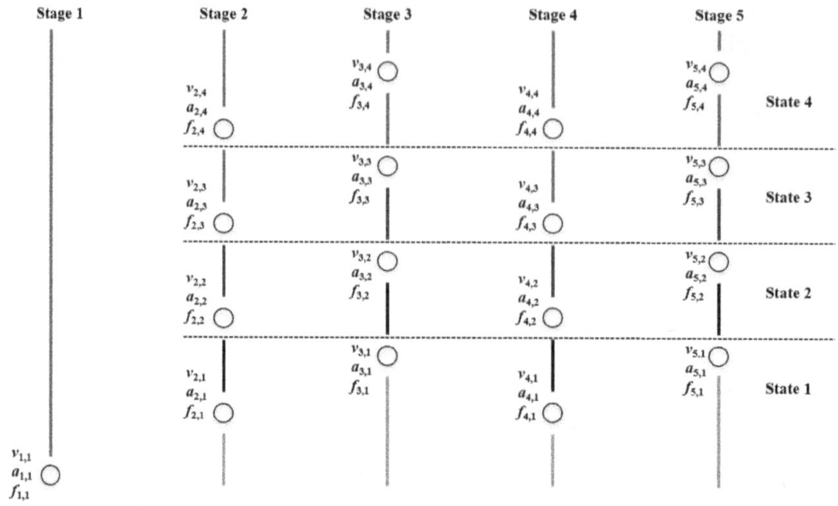

Fig. 1. Multi-stage decision process structure of the problem.

Algorithm 1 (Path Initialization).

initialize allPath with dimensions $(nStates^{nStages-1})$ x $nStages$
set pathNumber to 0
for each combination of Column 2, 3, 4 and 5 from 1 to $nStates$:
    increase pathNumber
    set allPath (pathNumber, :) to [1, Column2, Column3, Column4, Column5]
end

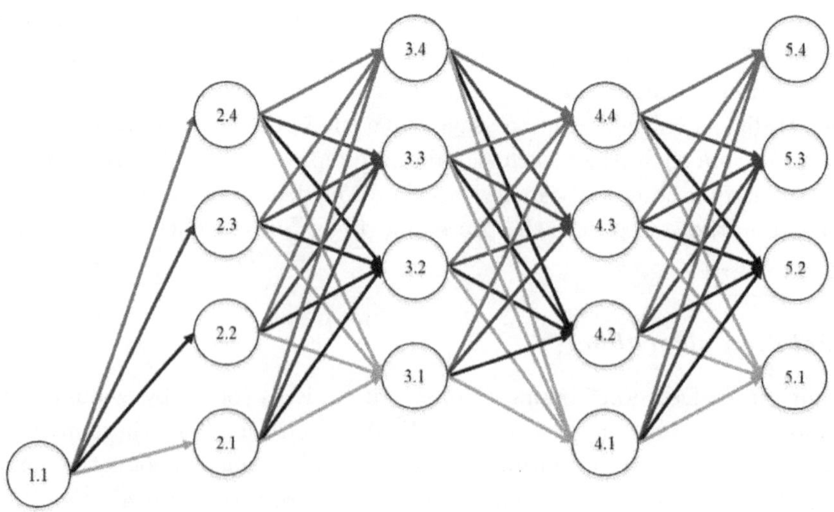

Fig. 2. Potential solution paths for the given example.

For each path, Algorithm 2 calculates velocity, acceleration/deceleration, and energy consumption optimally, then analyzes these results to identify the path minimizing energy consumption.

Algorithm 2 (Energy Calculation).

```
for each path in allPath:
    for each stage:
        calculate velocity [path, stage]
        calculate acceleration [path, stage]
        calculate energy [path, stage]
    end
end
for each path in allPath:
    calculate energyPath [path] as the sum of energy [path, stage].
end
find the minimum value in energyPath
```

In compliance with the minimum and maximum velocity limits based on road conditions, we impose a constraint to ensure adherence to the speed limit.

5. Case Study and Results

We construct a small-scale example to demonstrate the implementation of our models and algorithms. This example involves a problem with five stages and four states, resulting in 256 alternative solutions. We consider the distance between traffic lights to be 1200 meters, with an initial speed of 10 m/s for the EV. The initial time for the first and third traffic lights is 40 seconds, while the second and fourth lights have a base time of 60 seconds, with both having a wavelength of 40 seconds. Additional parameter values used in the example are listed in Table 1.

Table 1. Parameter values used in the example.

Parameter	Value	Unit
C_d	0.356	constant
A	2.841	m^2
ρ_a	1.1985	kg/m^3
C_{rr}	0.0084	constant
m	1745	kg
g	9.81	m/s^2
θ	0	radians

To better understand the application, velocity and acceleration calculation details for routes 1.1-2.1-3.1-4.2-5.2 (feasible) and 1.1-2.2-3.1-4.1-5.1 (infeasible due to time limit) are provided below to illustrate the application.

Starting with $t_{2.1} = 40 > t_{1.1} = 0$ seconds, we compute the new speed and acceleration. The speed at stage 2.1 is

$$v_{2.1} = (2D)/(t_{2.1} - t_{1.1}) - v_{1.1} = (2 \cdot 1200)/(40 - 0) - 10 = 50\,\text{m/s}$$

and the acceleration is

$$a_{2.1} = (v_{2.1} - v_{1.1})/(t_{2.1} - t_{1.1} = (50 - 10)/(40 - 0) = 1\,\text{m/s}^2.$$

At the next stage, $t_{3.1} = 60$ seconds, which is also greater than $t_{2.1} = 40$ seconds, so we calculate the speed and acceleration for this stage as well:

$$v_{3.1} = (2D)/(t_{3.1} - t_{2.1}) - v_{2.1} = (2 \cdot 1200)/(60 - 40) - 50 = 70\,\text{m/s}$$

and

$$a_{3.1} = (v_{3.1} - v_{2.1})/(t_{3.1} - t_{2.1}) = (70 - 50)/(60 - 40) = 1\,\text{m/s}^2.$$

Moving to stage 4.2, $t_{4.2} = 80$ seconds exceeds $t_{3.1} = 60$ seconds, so we compute:

$$v_{4.2} = (2D)/(t_{4.2} - t_{3.1}) - v_{3.1} = (2 \cdot 1200)/(80 - 60) - 70 = 50\,\text{m/s}$$

with

$$a_{4.2} = (v_{4.2} - v_{3.1})/(t_{4.2} - t_{3.1}) = (50 - 70)/(80 - 60) = -1\,\text{m/s}^2.$$

Finally, at stage 5.2, $t_{5.2} = 100$ seconds is greater than $t_{4.2} = 80$ seconds, so the speed and acceleration are calculated as:

$$v_{5.2} = (2D)/(t_{5.2} - t_{4.2}) - v_{4.2} = (2 \cdot 1200)/(100 - 80) - 50 = 70\,\text{m/s}$$

and

$$a_{5.2} = (v_{5.2} - v_{4.2})/(t_{5.2} - t_{4.2}) = (70 - 50)/(100 - 80) = 1\,\text{m/s}^2.$$

For this route, the speed adjustments are 10 m/s, 50 m/s, 70 m/s, 50 m/s, and 70 m/s at each stage, respectively.

In contrast, for route 1.1-2.2-3.1-4.1-5.1, calculations yield different results. With $t_{2.2} = 80$ seconds, which is greater than $t_{1.1} = 0$ seconds, the speed and acceleration at stage 2.2 are calculated as

$$v_{2.2} = (2D)/(t_{2.2} - t_{1.1}) - v_{1.1} = (2 \cdot 1200)/(80 - 0) - 10 = 20\,\text{m/s}$$

and the acceleration is

$$a_{2.2} = (v_{2.2} - v_{1.1})/(t_{2.2} - t_{1.1}) = (20 - 10)/(80 - 0) = 0.125\,\text{m/s}^2.$$

However, $t_{3.1} = 60$ seconds is less than $t_{2.2} = 80$ seconds, which means the EV cannot proceed along this route because it will miss the first wave of the traffic light at stage 3.1. Consequently, this route becomes infeasible as the EV would not pass through the required traffic light waves in the correct sequence.

Implementing the proposed algorithm in the MATLAB environment, we evaluated 256 alternative solutions based on energy consumption.

6. Conclusions and Future Studies

The rise of smart cities is driving the adoption of energy-efficient and eco-friendly transportation options, such as electrical vehicles, which lead to fuel savings and reduced emissions, thereby supporting the development of environmental sustainability. However, the driving range of electrical vehicles remains a major limitation preventing widespread use. In this study, a Multi-Stage Decision-Making Framework is developed to optimize the energy consumption of electrical vehicles based on pre-planned routes and considering both road conditions and traffic signals. Utilizing the developed model, this study aims to enhance both the efficiency of electric vehicle operations. Additionally, a small-scale example is constructed to demonstrate the implementation of our model. A solution algorithm is designed and executed in the MATLAB environment to identify the route with the lowest energy value among the various alternatives generated by the mathematical models. Future studies could assess the validity of the model by applying it to medium- and large-scale problems, as well as real-world scenarios. Furthermore, the developed model could be refined by incorporating additional constraints, such as speed limits. Moreover, designing a more sophisticated algorithm could facilitate the efficient resolution of larger problems within a feasible time frame.

Acknowledgements

The authors would like to acknowledge the Erzurum Technical University Coordination Office for Scientific Research Projects (ETÜ-BAP) for providing funding for this study under project number 2023/006.

References

[1] D. Oladimeji, K. Gupta, N. A. Kose, K. Gundogan, L. Ge, and F. Liang, Smart transportation: An overview of technologies and applications, *Sensors*, vol. 23, no. 8, p. 3880, (2023).

[2] International Energy Agency, Global EV Outlook 2024 Trends in electric cars, reports, (2024). `https://www.iea.org/reports/global-ev-outlook-2024`.

[3] M. Robinet, Fuel for Thought: As the industry goes electric, expect a shakeout among internal-combustion suppliers, (2023). `https://www.spglobal.com/mobility/en/research-analysis/fuel-for-thought-as-the-industry-goes-electric-expect-shakeout.html`.

[4] E. Kural, S. Jones, A. F. Parrilla, and A. Grauers, Traffic light assistant system for optimized energy consumption in an electric vehicle, in *2014 International Conference on Connected Vehicles and Expo (ICCVE)*, 2014, pp. 604–611.

[5] G. Mahler and A. Vahidi, An optimal velocity-planning scheme for vehicle energy efficiency through probabilistic prediction of traffic-signal timing, *IEEE Transactions on Intelligent Transportation Systems*, vol. 15, no. 6, pp. 2516–2523, (2014).

[6] C. Sun, J. Guanetti, F. Borrelli, and S. J. Moura, Optimal eco-driving control of connected and autonomous vehicles through signalized intersections, *IEEE Internet of Things Journal*, vol. 7, no. 5, pp. 3759–3773, (2020).

[7] T. Guan and C. W. Frey, Predictive energy efficiency optimization of an electric vehicle using traffic light sequence information, in *2016 IEEE International Conference on Vehicular Electronics and Safety (ICVES)*, 2016, pp. 1–6.
[8] A. Alturiman and M. Alsabaan, Impact of two-way communication of traffic light signal-to-vehicle on the electric vehicle state of charge, *IEEE Access*, vol. 7, pp. 8570–8581, (2019).
[9] J. Xing, L. Chu, and C. Guo, Optimization of energy consumption based on traffic light constraints and dynamic programming, *Electronics*, vol. 10, no. 18, p. 2295, (2021).
[10] J. Wang, I. Besselink, and H. Nijmeijer, Battery electric vehicle energy consumption prediction for a trip based on route information, *Proceedings of the Institution of Mechanical Engineers, Part D: Journal of Automobile Engineering*, vol. 232, no. 11, pp. 1528–1542, (2018).
[11] Q. T. Tran, L. Roose, C. Vichitpunt, K. Thongmai, and K. Noisopa, A Comprehensive Model to Estimate Electric Vehicle Battery's State of Charge for a Pre-Scheduled Trip Based on Energy Consumption Estimation, *Clean Technologies*, vol. 5, no. 1, pp. 25–37, (2022).
[12] C. Sun, X. Shen, and S. Moura, Robust optimal eco-driving control with uncertain traffic signal timing, in *2018 annual American control conference (ACC)*, 2018, pp. 5548–5553.
[13] M. Smuts, B. Scholtz, and J. Wesson, A critical review of factors influencing the remaining driving range of electric vehicles, in *2017 1st International Conference on Next Generation Computing Applications (NextComp)*, 2017, pp. 196–201.

Chapter 9

Modeling Gradual and Joint Coverage in Location Problems

Mumtaz Karatas[*,§], Levent Eriskin[†,¶] and Ertan Yakici[‡,||]

[*]*Department of Biomedical, Industrial and Human Factors Engineering, Wright State University, Dayton, 45435, USA*

[†]*Department of Industrial Engineering, Piri Reis University, Istanbul, 34940, Turkey*

[‡]*Department of Industrial Engineering, Çankaya University, Ankara, 06790, Turkey*

[§]*mumtaz.karatas@wright.edu*

[¶]*leriskin@pirireis.edu.tr*

[||]*eyakici@cankaya.edu.tr*

Location problems are a core area of research within OR/MS and decision sciences, with diverse applications in logistics, facility location, healthcare, defense, energy and transportation. One important feature of location problems is the coverage of demand or service areas by facilities, which can have significant economic, social, and environmental implications. Conventional models often assume binary or deterministic coverage, where a facility either fully covers a demand point or does not cover it at all. Although this simplification is useful for theoretical derivations, back-of-the-envelope calculations, and performance comparison, it overlooks the nuances and complexities of real-world scenarios.

In this study, we provide an overview of the modeling challenges in location problems that incorporate gradual and joint coverage, where multiple facilities provide partial and cooperative coverage to demand points. Based on previous studies in this domain, we present mathematical formulations, and discuss techniques for linearization and approximation. As an illustrative example, we discuss a capacitated minimal covering location problem (MCLP) adapted from [21], which aims to determine the location and size of undesirable facilities in a given region. We start by introducing the nonlinear formulation that minimizes the sum of demand covered by those undesirable facilities. Subsequently, we introduce three integer linear programming formulations given in [21], two of which involve linear approximations based on a separable programming approach and a tangent line approximation method, while the third involves an exact reformulation of the problem. We also discuss the impact of linearization approximation errors on solution quality and time.

1. Introduction

Over the past several decades, location science has attracted significant attention from researchers and practitioners across various disciplines. As a diverse and complex field, location science encompasses numerous types of location problems, a wide array of solution techniques, and a substantial body of theoretical models. Its development has been largely driven by its connections and interactions with fields like mathematics, geography, logistics, and economics [3].

At the heart of location science lie facility location problems (FLPs), which primarily focus on determining the optimal placement, relocation, or expansion of different types of facilities such as schools [34], fire stations [35], medical centers [36], post offices [37], bank branches [38], warehouses [39], and military stations [33], all while considering various objectives depending on the context. These problems deal with the spatial separation and the interaction of customers and facilities. The high costs associated with constructing, setting up, and relocating facilities make these projects significant long-term investments [4]. Additionally, establishing new facilities involves considering factors like distance between entities, travel times, customer locations and demographics, and long-term strategies. Given the substantial costs of altering facility location decisions, conducting thorough analytical work is crucial. Consequently, these problems play a vital role in strategic decision-making processes for planners in both private and public organizations [5].

There have been several mathematical models, and solution approaches have been proposed to aid planners and decision-makers working in this domain. Some of the prominent survey studies exploring facility location models can refer to foundational works by [6, 6–10] and [11]. More recent studies include contributions by [1, 12–15].

FLPs can take various forms depending on their objectives, constraints, and context. Eiselt and Marianov [16] categorize these problems into minisum, minimax, and covering types based on their objectives.

Minisum problems basically focus on minimizing the total distance between facilities and the demand points in the region. These problems aim to find the best locations for facilities to reduce the overall distance that needs to be traveled or covered. Minimax problems, on the other hand, prioritize balancing service among different demand points. Instead of minimizing the total distance, minimax problems aim to reduce the maximum distance that any demand point has to travel to reach its assigned facility. Covering problems are concerned with ensuring that all demand points are within a specific distance from the nearest facility, which is defined by a fixed threshold. These problems are typically divided into two categories: set covering and maximal covering. The set covering problem focuses on minimizing the total cost of establishing facilities that can cover all demand points. In contrast, the maximal covering problem seeks to maximize the number of demand points covered by a limited number of facilities, especially in situations where covering all demand points is not feasible.

The concept of location covering problems was first introduced by Toregas *et al.* [17] in 1971 and was further formalized by Church and ReVelle [18] in 1974. The basic idea revolves around determining the optimal placement of facilities to ensure that all customers within a predefined "service standard" or "coverage radius" are considered adequately served. In most applications, customers outside this radius are regarded as not being served at all. The specific objectives of these problems vary depending on the context, ranging from minimizing the number of facilities required to ensure coverage to maximizing the population served within a given number of facilities. In particular, set covering problem focuses on minimizing the total cost of establishing facilities that can cover all demand points. In contrast, the maximal covering problem aims to maximize the coverage of demand points covered by limited resources, especially in situations where covering all demand points is not feasible.

All three problem types focus on the optimal placement of facilities in relation to the demand they need to serve. Each of these problems aims to improve how well facilities meet the needs of various locations, though they differ in their specific goals and approaches. All three problems involve a relationship between demand points (places that need service) and facilities (places that provide service). The objective is to optimize this relationship, finding the best possible locations for facilities to serve the demand points effectively.

Minimax, minisum, and covering models are regarded as the foundational formulations from which other location problems have evolved. In addition to these, there exist several other nonstandard location problems that have been of interest to several other researchers. Some of these problem types include, competitive and conditional location problems, location of undesirable or semi-obnoxious facilities, probabilistic or stochastic location problems, and hub location problems, etc.

Modeling gradual and joint coverage poses several challenges to modelers and decision makers, especially when real-world coverage is not binary (either fully covered or not), but rather partial and cooperative. One of the main difficulties lies in representing gradual coverage accurately. Coverage levels decrease as the distance between a facility and a demand point increases, introducing nonlinearities into the model. Capturing this distance-decay accurately (e.g., using exponential or linear decay functions) can make the model harder to solve and often requires approximation or linearization techniques to maintain computational feasibility.

For joint coverage, where multiple facilities contribute to covering a single demand point, the challenge is determining how to combine these partial contributions. Modeling the aggregation of coverage—whether through summation, probabilistic independence, or other methods—requires careful consideration of how facilities interact. The way coverage from different facilities combines (whether it complements or overlaps) is a critical aspect that adds to the complexity. Ensuring that the model reflects these interactions accurately, while also keeping it solvable, is no small task.

Another challenge arises from the nonlinear nature of coverage functions in gradual and joint coverage models. These functions often need to be linearized or approximated to work with optimization solvers. Common techniques, such as piecewise linear approximations or tangent line methods, help but can introduce errors in the solution or complicate the optimization process. The computational complexity of these models increases significantly as the number of demand points and potential facility locations grows. When gradual and joint coverage are included, the problem becomes more computationally expensive. Large-scale problems can quickly overwhelm standard solvers, necessitating the use of heuristics or decomposition methods to manage the size of the problem.

In this paper, we focus on covering problems and discuss formulations, approximation and linearization approaches for problems in the presence of gradual and joint coverage, where multiple facilities provide partial and cooperative coverage to demand points. In Section 2, we first provide the foundations for gradual and joint coverage concepts in location problems in more detail. In Section 3, we introduce some of the recent mathematical formulations, present techniques for linearization and approximations. In particular, we consider the nonlinear mathematical model that was proposed by [21]. Subsequently, we analyze three formulations from the literature which can be solved by commercial solvers. The first two formulations use approximations techniques, while the third involves an exact reformulation of the original problem employing a specialized network mapping technique. Section 4 concludes the paper with future remarks.

2. Gradual and Joint Coverage in Location Problems

In location problems, particularly in the context of facility location and service coverage, the concepts of *Gradual Coverage* and *Joint Coverage* offer more nuanced approaches to defining and optimizing service areas.

Before discussing gradual coverage we find it useful to define the binary coverage. Binary coverage refers to a simplified model in location and detection problems where coverage is treated as an all-or-nothing event. In this model, a point or area is considered fully covered if it falls within a specified range of a facility or detection source, and entirely uncovered if it is outside this range. This creates a clear boundary, typically defined by a coverage radius or detection threshold, within which all customers or targets are assumed to receive adequate service or be successfully detected, while those outside this boundary receive no service or remain undetected. While this model is valuable for analytical derivation due to its simplicity, it often fails to capture the complexities of real-world scenarios. In practice, coverage and detection are rarely absolute; the effectiveness of service or the likelihood of detection usually diminishes gradually with distance rather than dropping off suddenly. Therefore, the binary approximation can be too conservative, underestimating the true potential for coverage or detection in many applications [19].

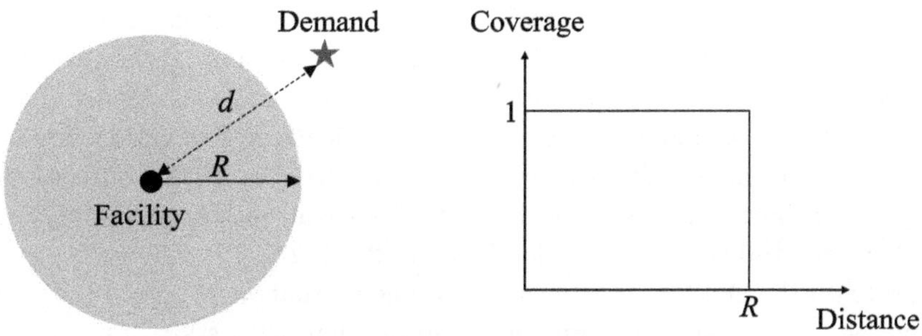

Fig. 1. An illustration of binary coverage concept.

Figure 1 illustrates the concept of binary coverage. In this model, a facility can cover all demand nodes within a range of R. If the distance between the facility and a demand node, d, exceeds R, there is no coverage at all.

Gradual coverage extends the traditional binary notion of coverage by introducing a continuum of coverage levels. Under this concept, the level of service provided to a customer decreases as the distance from the facility increases. Instead of a sharp cutoff at a predefined coverage radius, gradual coverage acknowledges that facilities can still provide some level of service to customers beyond this radius, albeit at a reduced effectiveness. This concept is particularly relevant in situations where the impact or effectiveness of service decays with distance, such as in healthcare facility placement, where proximity to a hospital influences the quality and timeliness of care.

Figure 2 illustrates the concept of gradual coverage. In this model, a facility can fully cover all nodes within a minimum coverage threshold of $\bar{R}$. If the distance between the facility and a demand node exceeds $\bar{\bar{R}}$, there is no coverage at all. Between $\bar{R}$ and $\bar{\bar{R}}$, the coverage proportion takes a value between 0 and 1.

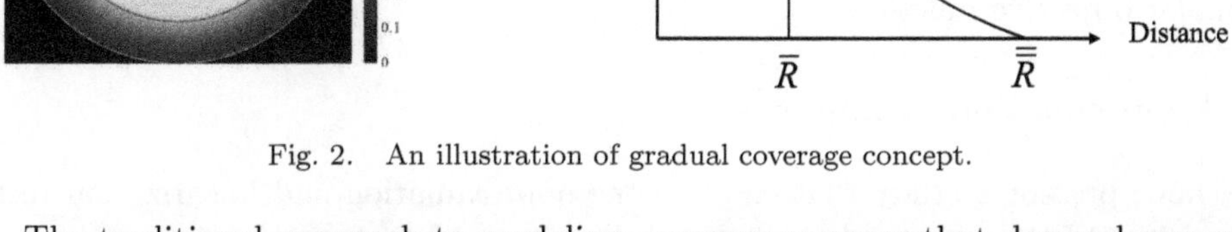

Fig. 2. An illustration of gradual coverage concept.

The traditional approach to modeling coverage assumes that demand can only be met by the nearest facility [20]. Joint coverage refers to the concept where multiple facilities work together to cover a demand point. Rather than a single facility meeting all the demand, several facilities each provide partial coverage, and the total coverage at a demand point is the combined effect of these facilities. This approach is useful when one facility alone cannot fully satisfy the demand or when combining resources from multiple locations enhances service quality or coverage. Joint coverage is commonly used in applications like emergency services or telecommunications, where overlapping coverage from multiple sources enhances reliability and robustness of service delivery. This concept has been implemented in several recent studies some of which include [2, 21–24].

To illustrate the challenges of location problems that incorporate both gradual and joint coverage, consider a simple location problem with a set of demand nodes and candidate facility locations, denoted by $i \in I$ and $j \in J$, respectively. The objective of maximizing overall demand covered in a region by all located facilities can be expressed as follows:

$$\max_{\mathbf{x}} \sum_{i \in I} w_i \left[1 - \prod_{j \in J} (1 - p_{ij} x_j) \right] \tag{1}$$

In Eq. (1), the term $p_{ij}x_j$ represents the partial coverage provided by facility j to demand point i, where p_{ij} is the coverage effectiveness of facility j at demand point i, and x_j is a binary decision variable indicating whether a facility is located at site j. The product $\prod_{j \in J}(1 - p_{ij}x_j)$ captures the probability that demand point i is *not* covered by any facility. Consequently, the expression $\left[1 - \prod_{j \in J}(1 - p_{ij}x_j)\right]$ represents the joint coverage of demand point i by multiple facilities, considering the independent contributions of each. The entire objective function then minimizes the total weighted uncovered demand, with w_i representing the importance (or weight) of demand point i.

Figure 3 illustrates the concept of joint coverage for two facilities. It demonstrates that the collaboration of multiple facilities can significantly improve the total coverage performance. In the figure, the joint coverage is calculated in accordance with Eq. (1), where coverage is provided by multiple facilities independently.

The objective function described above is nonlinear due to the product term in the joint partial coverage expression, making it challenging to solve using commercial optimization solvers, as they often cannot guarantee global optimality for nonlinear problems. To address this issue, in the next section we introduce two linear approximations and an exact reformulation of an illustrative model that employs a similar objective function.

3. Approximation Techniques

We, now, present a study that employs the approximation and linearization techniques in a FLP. In particular, the example we provide is from the work [21] which considers a capacitated minimal covering location problem (MCLP) that aims to determine the location and sizes of multiple undesirable facilities with distance constraints between them. The model has various application areas. For example,

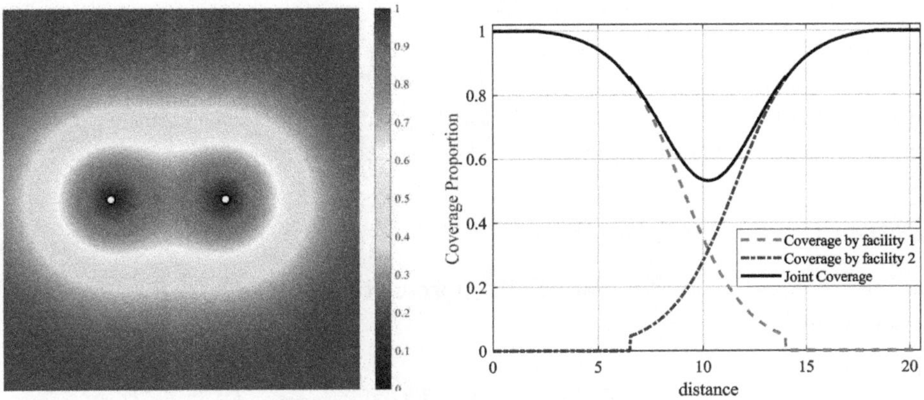

Fig. 3. An illustration of joint coverage concept.

thermal power plants or wastewater purification plants have detrimental effects on both society and the environment. This impact also increases with the size of each facility; for instance, a larger plant contributes more significantly to air or environmental pollution. Therefore, decision-makers face the challenge of determining the optimal design of these facilities to minimize their adverse effects on society. In the following section, for the sake of completeness, we present the full formulations of the MCLP.

Another example of the three approaches used can be found in a more recent study [32]. In this work, the authors consider a hierarchical FLP and aim to determine the optimal layout of a network of facilities to maximize coverage.

3.1. *Problem formulation*

In Section 3.2, we first provide the original nonlinear formulation. In Sections 3.3 and 3.4, we provide the approximation approaches. Finally, in Section 3.5 we present the exact linearization for the original nonlinear model. All models provided in this chapter are borrowed from the literature [21, 32]. Below, we provide a formal description of the common nomenclature used in all formulations throughout this paper:

Let $i \in I$ denote the set of demand nodes. Let $j, j' \in J$ represent the set of candidate facility locations and $s \in S$ be the set of facility size levels. Assuming that each candidate location has a different capacity (area for installing a facility) we introduce the subsets $S_j \subset S$ that represent the set of facility size levels that can be installed at location $j \in J$.

We assume that C represent the service requirement. The capacity of a facility of size $s \in S$ is denoted by c_s. The distance between a demand node $i \in I$ and a candidate facility location $j \in J$ is represented by d_{ij}, while $d_{jj'}$ denotes the distance between candidate locations j and $j' \in J$. A minimum required distance between each pair of located facilities is given by d^{min}. The portion of demand at node $i \in I$ that is covered by a facility of size $s \in S$ located at $j \in J$ is denoted by p_{ijs}, where $p_{ijs} \in [0,1]$. Finally, w_i represents the weight assigned to demand node $i \in I$. The subset $N_j = \{j' \in J \,|\, d_{jj'} \le d^{min}\}, \forall j \in J$ is used to model the minimum distance limitation between located facilities. x_{js} is a binary decision variable indicating whether a facility of size s is located at j.

3.2. *The non-linear formulation*

The Integer Non-Linear Program (INLP) formulation for the MCLP problem with distance constraints is as follows:

$$\textbf{(P):} \min_{\mathbf{x}} \sum_{i \in I} w_i \left[1 - \prod_{j \in J} \prod_{s \in S_j} (1 - p_{ijs} x_{js})\right] \tag{2a}$$

$$\text{subject to} \quad \sum_{j \in J} \sum_{s \in S_j} c_s x_{js} \ge C \tag{2b}$$

$$\sum_{s \in S_j} x_{js} + \sum_{s \in S_{j'}} x_{j's} \le 1, \; \forall j \in J, \forall j' \in N_j \tag{2c}$$

$$x_{js} \in \{0,1\}, \; \forall j \in J, s \in S_j \tag{2d}$$

The objective function (2a) of the INLP aims to minimize the total weighted ill-effect of the installed facilities opened in the region. It is important to note that, similar to the Eq. (1), the individual impact of each facility installed is incorporated into this function to model the joint coverage concept. Constraint (2b) ensures that the combined capacities of opened facilities satisfy the total service requirement. Constraint (2c) stipulates that each facility is assigned a single size and enforces the minimum distance requirement. The variable domain is given by constraint (2d).

3.3. *Linear approximation: Separable programming approach*

This section introduces the Separable Programming Approximation (SPA), which builds upon similar work by [25] and [26]. The SPA employs piecewise linear approximations to address the concave nature of the objective function.

In SPA, the approximation used is typically an under-approximation of the non-linear function, meaning the piecewise linear function lies below or is equal to the actual function. This ensures that the solution obtained is feasible, though not necessarily optimal. SPA leverages the separable nature of the problem, where the objective function can be expressed as a sum of functions depending on subsets of variables, allowing the problem to be solved more effectively by addressing each segment individually.

We note that the implementation of SPA involves considerable effort and algorithms to generate the piecewise linear segments for the nonlinear objective function.

Due to page constraints and for the sake of readability, we will omit the detailed implementation steps here.

Let $J_{is}^{<1} = \{j \mid 0 < p_{ijs} < 1\}$ and $J_{is}^{=1} = \{j \mid p_{ijs} = 1\}$. The implementation of SPA involves defining a new decision variable y_i that can take any value between 0 and 1. Let L_i denote a lower bound for y_i. B_{i1} is the first breakpoint for the demand node i, and R_i is the set of breakpoints used to approximate y_i in $[L_i, 1]$. We also define $\sum_{t \in R_i} \lambda_{it} = 1$ with $\lambda_{it} \geq 0$ for all $t \in R_i$, and R_i may have varying cardinality. Additionally, τ_{it} is a binary decision variable defined for every $i \in I$ and $t \in \{1, 2, \ldots, (|R_i| - 1)\}$. Another decision variable, v_i, corresponds to the probability that node i is not covered by any of the facilities established. Interested readers are referred to [21] for a comprehensive step-by-step explanation of the implementation process. The resulting linear approximation **(SPA)** for the problem **(P)** can be written as:

$$\textbf{(SPA)}: \min_{\mathbf{x},\mathbf{v},\boldsymbol{\tau},\boldsymbol{\lambda}} \sum_{i \in I} w_i(1 - v_i) \tag{3a}$$

subject to (2b) – (2c)

$$\sum_{t \in R_i} \ln(B_{it})\lambda_{it} = \sum_{j \in J_{is}^{<1}} \sum_{s \in S_j} x_{js} \ln(1 - p_{ijs}), \ \forall i \in I \tag{3b}$$

$$v_i \leq \sum_{t \in R_i} B_{it}\lambda_{it}, \ \forall i \in I \tag{3c}$$

$$v_i \leq 1 - x_{js}, \ \forall i \in I, j \in J_{is}^{=1}, s \in S_j \tag{3d}$$

$$\sum_{t \in R_i} \lambda_{it} = 1, \ \forall i \in I \tag{3e}$$

$$\lambda_{i1} \leq \tau_{i1}, \ \forall i \in I \tag{3f}$$

$$\lambda_{it} \leq \tau_{i(t-1)} + \tau_{it}, \ \forall i \in I, t \in \{2, 3, ..., |R_i| - 2\} \tag{3g}$$

$$\lambda_{i|R_i|} \leq \tau_{i(|R_i|-1)}, \ \forall i \in I \tag{3h}$$

$$\sum_{t \in \{1,2,...,(|R_i|-1)\}} \tau_{it} = 1, \ \forall i \in I \tag{3i}$$

$$x_{js} \in \{0, 1\}, \forall j \in J, s \in S_j \tag{3j}$$

$$\tau_{it} \in \{0, 1\}, \forall i \in I, t \in \{1, 2, ..., (|R_i| - 1)\} \tag{3k}$$

$$v_i \geq 0, \forall i \in I \tag{3l}$$

$$\lambda_{it} \geq 0, \forall i \in I, t \in R_i \tag{3m}$$

In the **(SPA)** formulation above, Constraints (3b)–(3e) are used for linearization purposes. Constraint sets (3f)–(3i) are the adjacency constraints and constraint sets (3j)–(3m) declare variable domains.

3.4. *Linear approximation: Tangent line approximation*

The second linear approximation method is the Tangent Line Approximation (**TLA**) [27]. Although similar to the **SPA** in its use of piecewise linear functions to approximate the nonlinear objective function, **TLA** differs by *over*-approximating the objective function with a series of tangent line segments.

By approximating nonlinear functions with linear functions, TLA helps make complex and untractable problems more manageable. This method is especially useful when the objective function or constraints are nonlinear and challenging to address directly. TLA works by approximating a nonlinear function through a series of tangent lines. These tangent lines are calculated at various points on the function, and each line represents the local behavior of the function around its specific point. By combining these tangent lines, TLA provides a piecewise linear approximation of the entire nonlinear function. Different from SPA, TLA typically over-approximates the nonlinear function. This means the piecewise linear function used in TLA lies above the original nonlinear function, ensuring that the approximation provides an upper bound on the objective function. While this makes the problem easier to solve, it can also lead to solutions that are less precise compared to methods that do not over-approximate.

In the TLA formulation, we define a_{il} and b_{il} as the slope and length of the lth line segment, respectively. We also introduce auxiliary binary variables μ_{il}, for all $i \in I$ and $l \in D_i$, to link the decision variable y_{il} and determine the point at which the realization occurs. Similar to SPA, implementing TLA involves tedious steps to generate the linear segments for the objective function. Therefore, we refer the reader to [21] for detailed implementation steps.

$$\textbf{(TLA)} \min_{\mathbf{x},\mathbf{y},\boldsymbol{\mu}} \sum_{i \in I} w_i \sum_{l \in D_i} a_{il} b_{il} y_{il} \tag{4a}$$

$$\text{subject to} \quad (2b) - (2c)$$

$$-\sum_{l \in D_i} b_{il} y_{il} + \sum_{j \in J} \sum_{s \in S_j} \gamma_{ijs} x_{js} = 0, \; \forall i \in I \tag{4b}$$

$$y_{il} \geq y_{i(l+1)}, \; \forall i \in I, l \in \{1, 2, ..., |D_i| - 1\} \tag{4c}$$

$$\mu_{i(l+1)} \leq y_{il} \leq \mu_{il}, \; \forall i \in I, l \in \{1, 2, ..., |D_i| - 1\} \tag{4d}$$

$$x_{js} \in \{0, 1\} \, \forall j \in J, s \in S_j \tag{4e}$$

$$\mu_{il} \in \{0, 1\} \, \forall i \in I, l \in D_i \tag{4f}$$

$$0 \leq y_{il} \leq 1, \; \forall i \in I, l \in D_i \tag{4g}$$

In the **(TLA)** formulation, constraint (4b) is introduced for linearization purposes. Additionally, the auxiliary binary variables $\mu_{il}, \forall i \in I, l \in D_i$, manage the connection of the y_{il} variables through Constraint sets (4c) and (4d). Since the objective function is a minimization, this constraint set ensures that, for a given node i, the y_{il} variables are set to 1 for all line segments preceding the one corresponding to the realized value and 0 for the subsequent line segments. For the particular line segment where the realization occurs, y_{il} takes any value within the range of 0 to 1.

Given that the linearization approximation errors for SAP and TLA are represented by ξ_{SPA} and ξ_{TLA}, respectively, Figure 4 displays the under and overestimations of the two approaches for various approximation values.

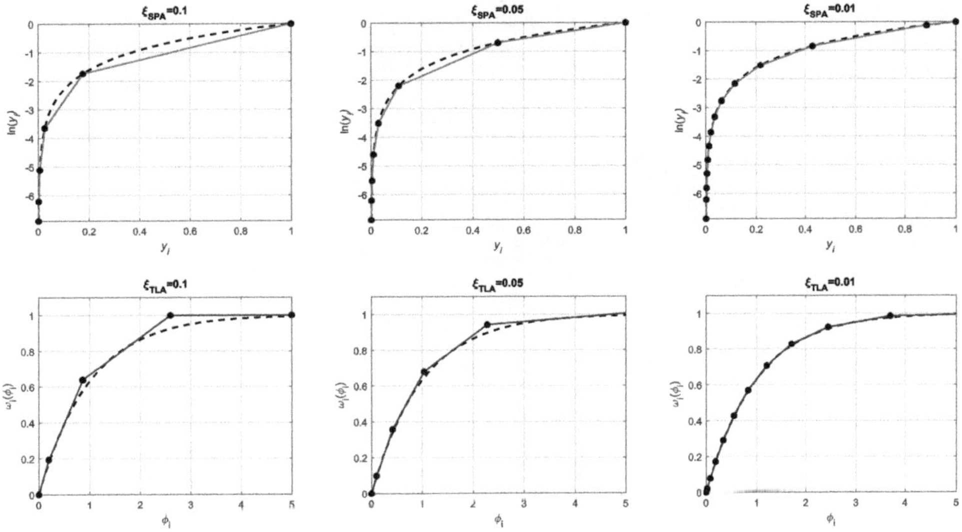

Fig. 4. Piecewise linear line segments used to approximate the objective function for (top) $\xi_{SPA} = \{0.1, 0.05, 0.01\}$ in Model (**SPA**), and (bottom) $\xi_{TLA} = \{0.1, 0.05, 0.01\}$ in Model (**TLA**). Source [21].

3.5. *Exact reformulation: A network mapping technique*

The exact reformulation (EF) which we present here uses a special network mapping approach which was also previously implemented by [31] and later adopted by [30], [28] and [29].

We now present the reformulation of the model (**P1**) using a special network mapping technique as also used by [31] and later adopted by [30], [28] and [29]. The EF approach poses the FLP as a network structure, in which a candidate location is depicted as a node and the proportion of *uncovered* demand is represented by arcs. The general structure of this approach is illustrated in Figure 5.

To implement the EF, we introduce two sets of additional auxiliary parameters: q^+_{ijs} and q^-_{ij}. The parameter q^+_{ijs} defines the scale factors for the uncovered proportion of each (i, j, s) triplet when a facility of size s is located, while q^-_{ij} pertains to the (i, j) pair in the scenario where no facility is located.

Similarly, the decision variables $\pi^+_{i(j,j+1)s}$ and $\pi^-_{i(j,j+1)}$ represent the positive and negative ratios of uncovered demand that is transferred to the adjacent node.

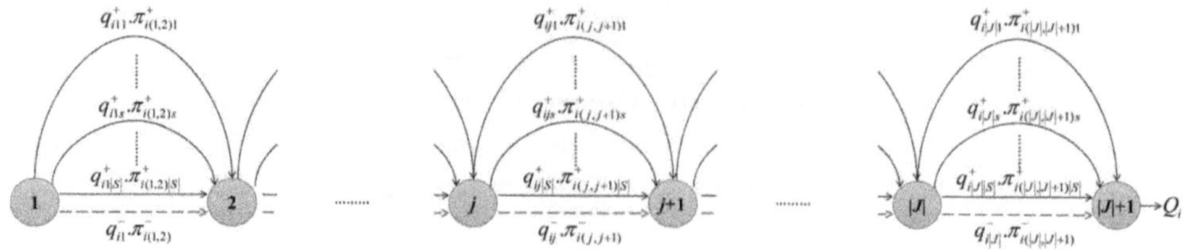

Fig. 5. The network mapping technique used to linearize the original nonlinear model. Each disk represents a candidate facility location, while positive and negative flow arcs are indicated by solid and dashed lines, respectively. Source [21].

Finally, the total ratio of uncovered demand is denoted by $Q_i = 1 - P_i$ for all $i \in I$.
Formulation:

$$\textbf{(EF)} \min_{\mathbf{x},\mathbf{Q},\boldsymbol{\pi}} \sum_{i \in I} w_i \left[1 - Q_i\right] \tag{5a}$$

subject to (2b) – (2c)

$$\pi^-_{i(1,2)} + \sum_{s \in S_j} \pi^+_{i(1,2)s} = 1, \; \forall i \in I \tag{5b}$$

$$q^-_{ij}\pi^-_{i(j,j+1)} + \sum_{s \in S_j} q^+_{ijs}\pi^+_{i(j,j+1)s} = \pi^-_{i(j+1,j+2)} + \sum_{s \in S_j} \pi^+_{i(j+1,j+2)s}, \; \forall i \in I, j \in \{1,2,...,|J|-1\} \tag{5c}$$

$$q^-_{i,|J|}\pi^-_{i,(|J|,|J|+1)} + \sum_{s \in S_j} q^+_{i,|J|,s}\pi^+_{i,(|J|,|J|+1),s} = Q_i, \; \forall i \in I \tag{5d}$$

$$\pi^+_{i(j,j+1)s} \le x_{js}, \; \forall i \in I, j \in \{1,2,...,|J|-1\}, s \in S_j \tag{5e}$$

$$\pi^+_{i(j,j+1)s} \ge \pi^-_{i(j-1,j)} + \sum_{s' \in S_j} \pi^+_{i(j-1,j)s'} - (1 - x_{js}), \; \forall i \in I, j \in \{1,2,...,|J|-1\}, s \in S \tag{5f}$$

$$\pi^-_{i(j,j+1)} \le 1 - \sum_{s \in S} x_{js}, \; \forall i \in I, j \in \{1,2,...,|J|-1\} \tag{5g}$$

$$x_{js} \in \{0,1\}, \; \forall j \in J, s \in S_j \tag{5h}$$

$$Q_i > 0, \; \forall i \in I \tag{5i}$$

$$\pi^+_{i(j,j+1)s} > 0, \; \forall i \in I, j \in J, s \in S_j \tag{5j}$$

$$\pi^-_{i(j,j+1)} > 0, \; \forall i \in I, j \in J \tag{5k}$$

Different from previous formulations, (5a) uses the survival probability Q_i for each demand i. The constraints (5b), (5c), (5d), and (5e) pertain specifically to the EF case. They address flow initialization, flow balancing, uncovered ratio calculation, and flow adjustments when no facility is installed, respectively. Constraints (5f) and (5g) are additional technical constraints designed to ensure the connection between the newly introduced transitive decision variables and the binary location-related decision variables. The remaining constraints specify the variable domains.

Both studies [21] and [32] offer a detailed discussion and experimental results on the performance of the three approaches. Although we do not discuss all the specifics of these comparisons here and refer the reader to those publications, we present some basic findings.

Experiments with various problem sizes show that the linearization approximation errors, ξ_{SPA} and ξ_{TLA}, significantly impact the performance of SPA and TLA. Lower approximation levels require more line segments to represent the concave function, which increases the size of the integer linear problem to be solved. Therefore, before implementing these techniques, planners should seek a balance between speed and quality.

Experiments also indicate that both SPA and TLA perform similarly and can solve small-sized problems in under a second. However, although slightly, TLA generally outperforms SPA in most tests. Since EF involves the introduction of several new decision variables and constraints, it is less efficient compared to the approximations in terms of solution time. Thus, SPA and TLA are effective for solving FLPs with gradual and joint coverage, providing high-quality optimal or near-optimal solutions with an acceptable level of error in the objective function value. However, EF remains a viable alternative if planners require an exact solution without any approximation error.

Moreover, the choice of method should also consider the scalability and complexity of the problem at hand. While SPA and TLA can efficiently handle small to medium-sized FLPs, their performance may degrade as the problem size grows due to the increasing number of line segments required for accurate approximations. In contrast, the EF approach, though slower, may become more attractive for larger problems where the diminishing returns of approximation accuracy no longer justify the increased computational burden. Additionally, in scenarios where the solution quality must be guaranteed or approximation errors could lead to significant operational risks, the exact nature of EF makes it a powerful choice despite its higher computational cost. Ultimately, the decision between using approximation-based methods or exact formulations should be driven by the specific operational requirements and constraints of the problem being addressed.

4. Conclusion

In this study, we examined a FLP that incorporates gradual and joint coverage concepts from location science to more accurately model the location processes observed in many real-world settings. While these extensions enhance the realism of the model, they introduce additional challenges of nonlinearity and intractability. To address these challenges, we discussed two linear approximation techniques, SPA and TLA, as well as an exact reformulation method, the EF.

Both the SPA and TLA are particularly useful in cases where the original problem involves a complex, concave objective function, an exact solution is computationally prohibitive, and a feasible solution that is close to optimal is acceptable for practical purposes. By transforming the problem into a linear form, SPA allows for the use of efficient linear programming solvers, making it easier to obtain solutions to otherwise intractable problems. While these approximation approaches do not guarantee the optimality of the solution, they provide a practical approach for solving complex problems by transforming them into a more manageable linear form. This makes SPA and TLA particularly useful when exact solutions are computationally prohibitive, and a feasible, near-optimal solution is sufficient for practical purposes.

All three approaches constitute a powerful and comprehensive portfolio for addressing FLPs, allowing planners to choose the best balance between solution time and quality. Future research could explore comparing the performance of these approaches against custom heuristics and other metaheuristics. Additionally, future work could investigate the application of these techniques to facility location problems in the continuous domain.

References

[1] E. Bozkaya, L. Eriskin, and M. Karatas, Data analytics during pandemics: A transportation and location planning perspective, *Annals of Operations Research*, vol. 328, no. 1, pp. 193–244, (2023).
[2] L. Erişkin, Point coverage with heterogeneous sensor networks: A robust optimization approach under target location uncertainty, *Computer Networks*, vol. 198, p. 108416, (2021).
[3] G. Laporte, S. Nickel, and F. Saldanha da Gama, *The p-Median Problem*, (Location Science, Springer, 2015), p. 644.
[4] R. Z. Farahani and M. Hekmatfar, *Facility Location: Concepts, Models, Algorithms and Case studies*. Springer Science & Business Media, (2009).
[5] S. H. Owen and M. S. Daskin, Strategic facility location: A review, *European Journal of Operational Research*, vol. 111, no. 3, pp. 423–447, (1998).
[6] R. L. Francis, L. F. McGinnis, and J. A. White, *Facility Layout and Location: An Analytical Approach*, Pearson, (1974).
[7] G. Y. Handler, *Location on Networks Theory and Algorithms*, MIT Press, (1979).
[8] P. B. Mirchandani and R. L. Francis, *Discrete Location Theory*, John Wiley & Sons, New York, (1990).
[9] M. S. Daskin, Network and discrete location models, algorithms and applications, *Location Science*, vol. 1, no. 4, pp. 117–119, (1996).
[10] M. Daskin, Network and discrete location: Models, algorithms and applications, *Journal of the Operational Research Society*, vol. 48, no. 7, pp. 763–764, (1997).
[11] E. Drezner, Facility location: A survey of applications and methods, *Journal of the Operational Research Society*, vol. 47, no. 11, pp. 1421–1421, (1996).
[12] Z. Drezner and H. W. Hamacher, *Facility Location: Applications and Theory*, (Springer Science & Business Media, 2004).
[13] S. Nickel and J. Puerto, *Location Theory: A Unified Approach*, (Springer Science & Business Media, 2006).

[14] R. L. Church and A. T. Murray, *Business Site Selection, Location Analysis, and GIS*, (John Wiley & Sons Hoboken, NJ, 2009).
[15] H. Tozan and M. Karatas, *Operations Research for Military Organizations*, IGI Global, (2018).
[16] H. A. Eiselt and V. Marianov, Pioneering developments in location analysis, in *Foundations of Location Analysis*, (Springer, 2011), pp. 3–22.
[17] C. Toregas, R. Swain, C. ReVelle, and L. Bergman, The location of emergency service facilities, *Operations Research*, vol. 19, no. 6, pp. 1363–1373, (1971).
[18] R. Church and C. R. Velle, The maximal covering location problem, *Papers in Regional Science*, vol. 32, no. 1, pp. 101–118, (1974).
[19] Q. Yang, S. He, J. Li, J. Chen, and Y. Sun, Energy-efficient probabilistic area coverage in wireless sensor networks, *IEEE Transactions on Vehicular Technology*, vol. 64, no. 1, pp. 367–377, (2015).
[20] O. Berman, Z. Drezner, and D. Krass, Discrete cooperative covering problems, *Journal of the Operational Research Society*, vol. 62, no. 11, pp. 2002–2012, (2011).
[21] M. Karatas and L. Eriskin, The minimal covering location and sizing problem in the presence of gradual cooperative coverage, *European Journal of Operational Research*, vol. 295, no. 3, pp. 838–856, (2021).
[22] M. Karatas, A multi-objective facility location problem in the presence of variable gradual coverage performance and cooperative cover, *European Journal of Operational Research*, vol. 262, no. 3, pp. 1040–1051, (2017).
[23] I. Averbakh, O. Berman, D. Krass, J. Kalcsics, and S. Nickel, Cooperative covering problems on networks, *Networks*, vol. 63, no. 4, pp. 334–349, (2014).
[24] T. Drezner and Z. Drezner, The maximin gradual cover location problem, *OR spectrum*, vol. 36, no. 4, pp. 903–921, (2014).
[25] C. Saydam and M. McKnew, Applications and implementation a separable programming approach to expected coverage: An application to ambulance location, *Decision Sciences*, vol. 16, no. 4, pp. 381–398, (1985).
[26] J. D. Camm, S. K. Norman, S. Polasky, and A. R. Solow, Nature reserve site selection to maximize expected species covered, *Operations Research*, vol. 50, no. 6, pp. 946–955, (2002).
[27] R. Aboolian, O. Berman, and D. Krass, Competitive facility location model with concave demand, *European Journal of Operational Research*, vol. 181, no. 2, pp. 598–619, (2007).
[28] M. Karatas and A. Dasci, A two-level facility location and sizing problem for maximal coverage, *Computers & Industrial Engineering*, vol. 139, p. 106204, (2020).
[29] M. Karatas, A multi-objective bi-level location problem for heterogeneous sensor networks with hub-spoke topology, *Computer Networks*, vol. 181, p. 107551, (2020).
[30] J. Salmerón, Deception tactics for network interdiction: A multiobjective approach, *Networks*, vol. 60, no. 1, pp. 45–58, (2012).
[31] D. P. Morton, F. Pan, and K. J. Saeger, Models for nuclear smuggling interdiction, *IIE Transactions*, vol. 39, no. 1, pp. 3–14, (2007).
[32] M. Karatas and L. Eriskin, Linear and piecewise linear formulations for a hierarchical facility location and sizing problem, *Omega*, vol. 118, p. 102850, (2023).
[33] M. Karatas, E. Yakıcı, and A. Dasci, Solving a bi-objective unmanned aircraft system location-allocation problem, *Annals of Operations Research*, vol. 319, no. 2, pp. 1631–1654, (2022).
[34] N. D. Pizzolato and H. B. F. Da Silva, The location of public schools: Evaluation of practical experiences, *International Transactions in Operational Research*, vol. 4, no. 1, pp. 13–22, (1997).

[35] M. A. Badri, A. K. Mortagy, and C. A. Alsayed, A multi-objective model for locating fire stations, *European Journal of Operational Research*, vol. 110, no. 2, pp. 243–260, (1998).

[36] L. Eriskin, M. Karatas, and Y.-J. Zheng, A robust multi-objective model for healthcare resource management and location planning during pandemics, *Annals of Operations Research*, vol. 335, no. 3, pp. 1471–1518, (2024).

[37] M. Tahmasebi, K. Khalili-Damghani, and V. Ghezavati, A customized bi-objective location-routing problem for locating post offices and delivery of post parcels, *Journal of Industrial and Systems Engineering*, vol. 10, no. 1, pp. 1–17, (2017).

[38] P. Miliotis, M. Dimopoulou, and I. Giannikos, A hierarchical location model for locating bank branches in a competitive environment, *International Transactions in Operational Research*, vol. 9, no. 5, pp. 549–565, (2002).

[39] A. A. Kuehn and M. J. Hamburger, A heuristic program for locating warehouses, Management science, vol. 9, no. 4, pp. 643–666, (1963).

Chapter 10

Integrated Gate Assignment Problem, Taxiway Planning Problem and Runway Scheduling Problem: A MIP Model and a Genetic Algorithm

Arif Fıçı, Gözde Zeynep Bolulugil, Furkan Ulutaş and Emine Akyol Özer*

Eskisehir Technical University, Industrial Engineering Department, Eskisehir, Türkiye

** emineakyol@eskisehir.edu.tr*

In recent years, demand for air transportation has been increasing rapidly and airports have difficulties in meeting this demand with their existing capacities. Flight delays are some of the most common problems faced by air traffic managers. One of the main reasons of the flight delays is waiting during ground operations. Efficient taxiway planning can help reduce these delays. However, it is important to integrate gate and runway assignments into taxiway planning. In this study, the taxiway planning problem is addressed together with the gate assignment and runway assignment problems. So, we propose a mixed integer programming model to solve the problem considering safety and operational constraints in airport resources. The objective is to minimize total taxi time (taxi-in and taxi-out times) and waiting time. Because of the NP-hard nature of the problem, the proposed mathematical model may not find feasible solutions for large-size problems. Thus, a genetic algorithm is developed to solve large-size problems. To demonstrate the performance of the proposed methods, we used real data from IGA Istanbul Airport. Experimental results showed that genetic algorithm outperforms the mathematical model in both solution quality and solution time.

1. Introduction

The planning of taxiways is crucial for reducing taxi times, which directly affects fuel consumption, emissions, and flight schedules. Large international airports, with their complex layouts, different aircraft types, operational conditions, and traffic patterns, present significant challenges in this regard. Inefficient taxiway route plan can lead to increased ground delays, higher operating costs for airlines, and a more significant environmental footprint due to unnecessary fuel burn. In this study, we address air-port taxiway planning problem that involves the assignment of gates, runways, and taxi routes.

In recent years, airport taxiway planning problem has gained significant attention in the literature. Several studies focus on integrating aircraft routing with other airport operations. For example, Reference [1] integrates the Ground Routing Problem and the Runway Sequencing Problem to optimize aircraft routing from parking stands to runways and sequence take-offs and landings. They propose a heuristic sequential approach with a new mathematical formulation, which is tested with real data from a major European airport. Consequently, the method significantly reduces total completion and taxi times, proving effective for daily operations. Similarly, Reference [2] tackles airport congestion by integrating arrival and departure management using conflict detection within a predefined route structure. By applying an optimized simulated annealing heuristic and time decomposition, they improve runway assignments and manage aircraft timings. Case studies at Paris Charles De-Gaulle airport reveal a 37% reduction in arrival delays and a 36% reduction in departure delays with the proposed method. Reference [3] address the growing congestion in airside ground traffic by proposing an integrated model that simultaneously handles gate assignment and taxiway planning. The model is formulated as an integer programming problem in a discrete time-space network, considering taxiway conflicts and gate operations. Comparisons with the first-come-first-served method and a heuristic approach show that the integrated model effectively balances resources between gates and taxiing paths. Sensitivity analyses reveal that the number of flight pairs and connection times directly influence gate idle time and aircraft taxi time. Reference [4] addresses airport runway planning and avoid taxiway conflicts by introducing a mathematical model and a hybrid optimization algorithm. This algorithm, which combines particle swarm optimization and ant colony optimization, features a new pheromone initialization and update strategy. Experimental results demonstrate the algorithm's effectiveness in real airport taxiway planning scenarios. Aircraft taxiway planning problem has been the primary focus of many studies in the existing literature. Reference [5] introduces the need for efficient and safe aircraft taxiing guidance in increasingly complex airport environments. They propose an enhanced Q-Learning algorithm, a reinforcement learning method, to autonomously find the shortest taxiing paths. Reference [6] presents a method that combines semantic modeling, reasoning, and real-time planning to enhance decision-making for safe and efficient airport taxiway operations. The main challenge lies in coordinating real-time planning with semantic knowledge to create a system that performs well and can adapt to unforeseen events. The approach is tested on a scenario involving multiple planes moving from the terminal to the same departure runway. Some of studies on the aircraft taxiway planning problem have focused on conflict detection and resolution techniques. For example, Reference [7] considers delays in aircraft taxiing at large airports caused by traffic conflicts. They develop dynamic models for crossing, head-on, and trailing conflicts to deter-mine delay times. By using the Yen algorithm to identify the shortest routes while considering topological networks, conflict-induced delays, and taxiing distributions, the optimized route planning achieves 9.6–12.1% higher precision compared to previous methods. This improvement offers enhanced decision support for airport operations based on environmental parameters and aircraft performance. In another study, Reference [8] focuses on detecting and preventing taxiing conflicts on airport surfaces to ensure operational safety. They outline the types of taxiing conflicts and define conflict areas based on aircraft characteristics. Utilizing Gaussian spatial-temporal regression and real-time taxiing speed data, they quantify the conflict risk and adjust aircraft speed dynamically through a feedback system. Simulations and historical data validate that this method effectively manages conflict risk, keeping it within safe limits and avoiding taxiing conflicts.

As can be seen from the above, the existing literature includes numerous studies on the taxiway planning problem. In some of these studies, gate and runway assignments have been considered independently of taxiways, assuming that these assignments are predetermined. In other studies, taxiway planning problems have been addressed together with either gate planning or runway planning. While these studies provide valuable insights into various aspects of airport taxiway planning, there remains a gap in integrating these approaches comprehensively. This gap is significant, as pointed out in Reference [9], where it is demonstrated that integrated models outperform sequential models in terms of efficiency and effectiveness. To address this gap, this paper aims to optimize gate, runway, and taxiway planning simultaneously.

2. Problem Definition and Mathematical Model

The aim of departure flights is to reach the runway from which they will take off, while arriving flights aim to reach the gate where passengers will disembark . The process of flights moving between gates and runways at an airport is called taxiing, and the paths used for these movements are known as taxiways. The movement of departing flights from the gate to the runway is referred to as taxi-out, while the movement of arriving flights from the runway to the gate is called taxi-in. Airports typically have several taxiways, which may be designated for either taxi-in or taxi-out only. However, especially at larger airports, taxiways are often shared for both movements.

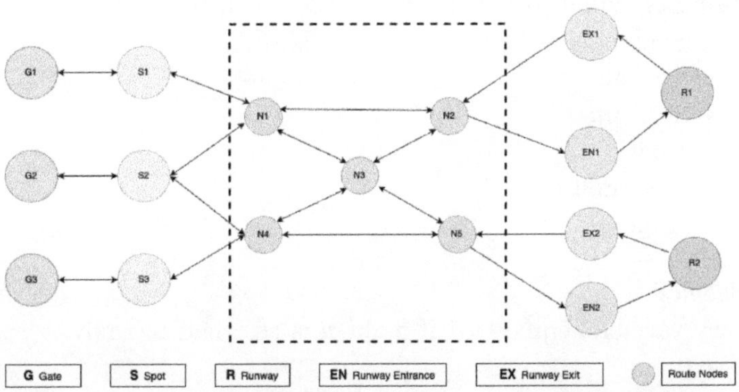

Fig. 1. Airport network.

To better understand the airport taxiway planning problem, an example of airport network structure is illustrated in Figure 1. In this airport, there are three gates, two runways, and six taxiway routes between the runways and gates. For example, a departing flight assigned to gate G1 leaves the airport from runway R2 using the taxi-way route S1-N1-N3-N5-EN2. An arriving flight landing on runway R1 reaches gate G2 by following the route EX1-N2-N3-N4-S3. To ensure safe ground operations at airports, there are waiting points. These include spot waiting points between gates and taxiways, as well as waiting points between taxiways and runway entry/exit points. Flights can wait at these points to avoid conflicts while moving between their assigned runway and gates.

In this study, a Mixed-Integer Programming (MIP) model is developed to minimize total taxi time (taxi-in and taxi-out times) and waiting time considering safety ground operations in airport resources (gates, taxiways. and runways). The notations and model formulation of the airport taxiway planning problem is given below.

Sets and Indices:

N_a: Set of arrival flights, $N_a = \{1,2,...,n_1\}$

N_d: Set of departure flights, $N_d = \{n_1+1, n_1+2,...,n_2\}$

N: Set of all flights, $N = N_a \cup N_d$, indexed by u

G: Set of gates, indexed by g

P_a: Set of runways for arriving flights, $P_a = \{1,2,...,m_1\}$

P_d: Set of runways for departure flights, $P_d = \{m_1+1, m_1+2,...,m_2\}$

P: Set of runways for all flights, $P = P_a \cup P_d$, indexed by p

R: Set of taxiways available for the gate-parking position combination, indexed by r

Parameters:

t_u^s: Scheduled arrival time of arrival flight u

$t_u^{s'}$: Scheduled arrival time to the gate of departure flight u

c_r: The time required to complete taxiway route r

v_{gpr}: 0-1 parameter, equal to 1 if taxiway route r can be used for the gate g and runway p combination 1, otherwise 0.

b^{rw}: Runway usage time

h^{rw}: Runway buffer time

b^{gt}: Gate usage time

h^{tw}: Taxiway buffer time

h^{gt}: Gate buffer time

Decision Variables:

x_{ur}: 0–1 binary variable, equal to 1 if flight u is assigned to taxiway r, otherwise 0.

z_{upg}: 0–1 binary variable, equal to 1 if flight u is assigned to runway p and gate g, otherwise 0.

w_u^{rw}: Waiting time of flight u for runway
t_u^{rwd}: Actual landing time/departure time from runway of flight u
w_u^{tw}: Waiting time of flight u for taxiway
t_u^{tw}: Actual taxiway route starting time for flight u
w_u^{gt}: Waiting time of flight u for gate
t_u^{gta}: Actual arrival time to the gate of flight u
t_u^{gtd}: Actual departure time from gate of flight u
t_u^{rwa}: Actual arrival time to the runway of flight u
$y1_{uj}^{rw}$, $y2_{uj}^{rw}$: 0–1 binary variables used to prevent conflicts on the runway
$y1_{uj}^{tw}$, $y2_{uj}^{tw}$: 0–1 binary variables used to prevent conflicts on the taxiway
$y1_{uj}^{gt}$, $y2_{uj}^{gt}$, $y3_{uj}^{gt}$: 0–1 binary variables used to prevent conflicts at the gate

Objective Function:

$$Min z = \sum_u \sum_g \sum_p \sum_r x_{ur} v_{gpr} c_r + \sum_u w_u^{gt} w_u^{tw} w_u^{rw} \tag{1}$$

The objective function (1) represents the minimization of total taxi times (taxi-in and taxi-out times) and total waiting times. The constraints of mixed integer programming model are explained below:

$$\sum_p \sum_g z_{upg} = 1 \qquad \forall u \in N \tag{2}$$

Constraint (2) ensures that each flight is assigned a gate and a runway.

$$\sum_r x_{ur} = 1 \qquad \forall u \in N \tag{3}$$

Constraint (3) guarantees that each flight is assigned to a taxiway route.

$$\sum_p \sum_g v_{gpr} z_{upg} = x_{ur} \qquad \forall u \in N, \forall r \in R, \tag{4}$$

Constraint (4) ensures that taxiway of a flight should be compatible with their assigned gate-runway combination.

$$t_u^{rwd} = t_u^s + w_u^{rw} \qquad \forall u, u \in N_a \tag{5}$$

$$t_u^{gta} = t_u^{s'} + w_u^{gt} \qquad \forall u, u \in N_d \tag{6}$$

$$t_u^{tw} = t_u^{rwd} + b^{rw} + w_u^{tw} \qquad \forall u, u \in N_a \tag{7}$$

$$t_u^{tw} = t_u^{gta} + b^{gt} + w_u^{tw} \qquad \forall u, u \in N_d \tag{8}$$

$$t_u^{gta} = t_u^{tw} + \sum_r c_r x_{ur} + w_u^{gt} \qquad \forall u, u \in N_a \tag{9}$$

$$t_u^{rwa} = t_u^{tw} + \sum_r c_r x_{ur} + w_u^{rw} \qquad \forall u, u \in N_d \tag{10}$$

$$t_u^{rwd} = t_u^{rwa} + b^{rw} \qquad \forall u, u \in N_d \tag{11}$$

The Constraints (5)–(11) calculate the actual landing time for an arrival flight. Constraint (5), the actual gate arrival time for departure flight; Constraint (6), the actual taxiway starting time for an arrival flight; Constraint (7), the actual taxiway starting time for departure flight; Constraint (8), the actual gate arrival time for arrival flight; Constraint (9), the actual runway arrival time for departure flight; Constraint (10), and the actual takeoff time for departure flight; Constraint (11), taking into account the waiting times.

For safe ground operations, it is essential to prevent conflicts at critical resource points within the airport. Therefore, conflict-free constraints are required for gates, taxiways and runways. These constraints help ensure that flight movements do not overlap or interfere with each other, thereby maintaining efficient and safe operations on the ground.

$$t_u^{rwd} - t_j^{rwd} \geq h^{rw} - M\left(2 - \sum_g z_{upg} - \sum_g z_{jpg} + y1_{uj}^{rw}\right) \quad \forall u, j \in N_a, u \neq j, \forall p \in P \tag{12}$$

$$t_j^{rwd} - t_u^{rwd} \geq h^{rw} - M\left(3 - \sum_g z_{upg} - \sum_g z_{jpg} - y1_{uj}^{rw}\right) \quad \forall u, j \in N_a, u \neq j, \forall p \in P \tag{13}$$

$$t_u^{rwa} - t_j^{rwa} \geq h^{rw} - M\left(2 - \sum_g z_{upg} - \sum_g z_{jpg} + y2_{uj}^{rw}\right) \quad \forall u, j \in N_d, u \neq j, \forall p \in P \tag{14}$$

$$t_j^{rwa} - t_u^{rwa} \geq h^{rw} - M\left(3 - \sum_g z_{upg} - \sum_g z_{jpg} - y2_{uj}^{rw}\right) \quad \forall u, j \in N_d, u \neq j, \forall p \in P \tag{15}$$

Runways: Constraint (12)–(15) state that there should be at least a minimum runway buffer time between two flights which are assigned to the same runway; otherwise, the following flight must wait. Constraints (12) and (13) are formulated for arrival flights, while Constraints (14) and (15) are formulated for departure flights.

$$t_u^{tw} - t_j^{tw} \geq h^{tw} - M(2 - x_{ur} - x_{jr} + y1_{uj}^{tw}) \quad \forall u, j \in N_a, u \neq j, \forall r \in R \tag{16}$$

$$t_j^{tw} - t_u^{tw} \geq h^{tw} - M(3 - x_{ur} - x_{jr} - y1_{uj}^{tw}) \quad \forall u, j \in N_a, u \neq j, \forall r \in R \tag{17}$$

$$t_u^{tw} - t_j^{tw} \geq h^{tw} - M(2 - x_{ur} - x_{jr} + y2_{uj}^{tw}) \quad \forall u, j \in N_d, u \neq j, \forall r \in R \tag{18}$$

$$t_j^{tw} - t_u^{tw} \geq h^{tw} - M(3 - x_{ur} - x_{jr} - y2_{uj}^{tw}) \quad \forall u, j \in N_d, u \neq j, \forall r \in R \tag{19}$$

Taxiways: Constraint (16)–(19) ensure that a time gap, equal to the minimum taxiway buffer time, exists between two arrival/departure flights on the same taxiway route; if not, the following flight should wait before starting its route. Constraints (16) and (17) are formulated for arrival flights, while Constraints (18) and (19) are formulated for departure flights.

$$t_u^{gta} - t_j^{gta} = b^{gt} - M\left(2 - \sum_p z_{upg} - \sum_p z_{jpg} + y1_{uj}^{gt}\right) \quad \forall u, j \in N_a, u \neq j, \forall g \in G \tag{20}$$

$$t_j^{gta} - t_u^{gta} = b^{gt} - M\left(3 - \sum_p z_{upg} - \sum_p z_{jpg} - y1_{uj}^{gt}\right) \quad \forall u, j \in N_a, u \neq j, \forall g \in G \tag{21}$$

$$t_u^{gta} - t_j^{gta} \geq b^{gt} - M\left(2 - \sum_p z_{upg} - \sum_p z_{jpg} + y2_{uj}^{gt}\right) \quad \forall u, j \in N_d, u \neq j, \forall g \in G \tag{22}$$

$$t_j^{gta} - t_u^{gta} \geq b^{gt} - M\left(3 - \sum_p z_{upg} - \sum_p z_{jpg} - y2_{uj}^{gt}\right) \quad \forall u, j \in N_d, u \neq j, \forall g \in G \tag{23}$$

Gates: Constraints (20)–(23) guarantee that there should be gate usage time gap between two flights assigned to the same gate; otherwise, one of the flights should wait before the gate. Similar to runway and taxiway conflict free constraints, Constraints (20) and (21) are formulated for arrival flights, while Constraints (22) and (23) are formulated for departure flights.

$$t_u^{gtd} - t_j^{gta} \geq h^{gt} - M\left(2 - \sum_p z_{upg} - \sum_p z_{jpg} + y3_{uj}^{gt}\right) \quad \forall u, j, g, u \in N_a, j \in N_d \tag{24}$$

$$t_u^{gta} + b^{gt} - t_j^{gtd} \geq h^{gt} - M\left(3 - \sum_p z_{upg} - \sum_p z_{jpg} - y3_{uj}^{gt}\right) \quad \forall u, j, g, u \in N_d, j \in N_a \tag{25}$$

$$t_u^{gta} + b^{gt} - t_j^{gtd} \geq h^{gt} - M\left(2 - \sum_p z_{upg} - \sum_p z_{jpg} + y3_{uj}^{gt}\right) \quad \forall u, j, g, u \in N_d, j \in N_a \tag{26}$$

$$t_u^{gtd} - t_j^{gta} \geq h^{gt} - M\left(3 - \sum_p z_{upg} - \sum_p z_{jpg} - y3_{uj}^{gt}\right) \quad \forall u, j, g, u \in N_a, j \in N_d \tag{27}$$

Constraints (24)–(27) ensure that there should be gate buffer time between an arrival and a departure flight using the same gate.

$$w_u^{rw},\ t_u^{rwd},\ w_u^{tw},\ t_u^{tw},\ w_u^{gt},\ t_u^{gta},\ t_u^{gtd},\ t_u^{rwa} \geq 0 \qquad \forall u \tag{28}$$

$$x_{ur} \in \{0,1\} \qquad \forall u, r \tag{29}$$

$$z_{upg} \in \{0,1\} \qquad \forall u, p, g \tag{30}$$

$$y1_{uj}^{rw},\ y2_{uj}^{rw},\ y1_{uj}^{tw},\ y2_{uj}^{tw},\ y1_{uj}^{gt},\ y2_{uj}^{gt},\ y3_{uj}^{gt} \in \{0,1\} \qquad \forall u, j \tag{31}$$

Constraints (28)–(31) are the variable domains.

3. Genetic Algorithm

Because of the NP-hard nature of the airport taxiway planning problem, a genetic algorithm (GA) was proposed to obtain better solutions in shorter time. The main reason for selecting GA is the good search performance in large and complex solution spaces. The considered problem is a complex problem that involves operational constraints, such as landing and takeoff sequences, taxiway usage, and safety constraints. GAs can achieve successful solutions such complex and constrained optimization problems.

In a classical GA, solutions are encoded as chromosomes, which represent all the values of the decision variables for the problem. An example of chromosome representation for the 4 flights is given in the Figure 2.

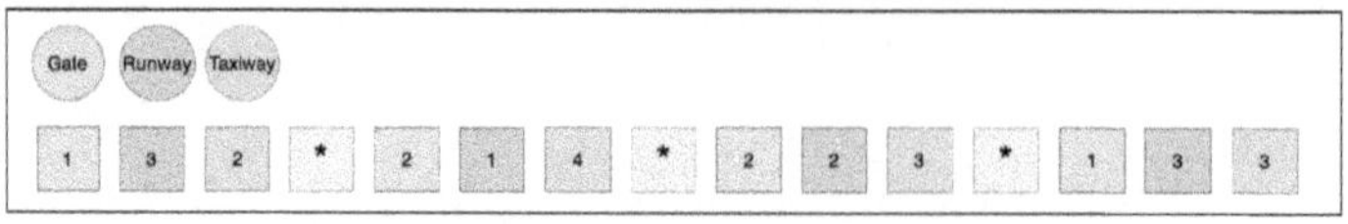

Fig. 2. Chromosome representation.

In Figure 2, the symbol '*' separates flights. The first gene corresponds to the gate number, the second gene to the runway, and the third gene to the taxiway route based on these gate and runway assignments. For instance, the first flight is assigned to gate 1, the third runway, and the second taxiway route.

The fitness values of the chromosomes can be computed using gate, runway and taxiway assignments. In the considered problem, initial population are randomly generated based on Constraint 4. Once an initial population has been formed, selection, crossover and mutation operations are repeatedly performed until number of iterations is reached. The selection process enables individual chromosomes to be chosen for potential inclusion in the next generation. In this study, we used a 2-tournament selection method as a reproduction operator. The crossover process allows the algorithm to combine the best genes from various individuals and the proposed genetic algorithm utilizes one-point crossover operator. The advantage of this operator is that Constraint (4) is still satisfied since only genes in the crossing point of each chromosome are changed. Note that there are no constraints other than Constraint (4) that could lead to an infeasible solution. Since gate, runway, and taxiway assignments are updated, waiting times and route durations are recalculated accordingly. If multiple flights are assigned to the same resource (gate, taxiway, runway), waiting at the resources ensures that the solution remains feasible without violating other constraints. The mutation operator helps to potentially discover better solutions by randomly altering some genes in the chromosome. In the proposed algorithm, a random flight is selected based on mutation rate. A new gate and runway are reassigned to this flight. In addition, taxiway routes are reorganized, and the fitness value of this chromosome are recalculated.

4. Computational Results

To demonstrate the performance of the proposed mathematical model and genetic algorithm, the flight data of IGA Istanbul Airport for one day are used. The mathematical model was implemented in GAMS 24.0.2 using the Cplex solver and the proposed genetic algorithm was coded in Python, running on a 2.8-GHz Intel Core i7 with 4 GB of RAM. The running time of the proposed methods was limited to 3600 s for all the test problems.

Table 1. The test results of proposed solution methods.

	MM				GA						
n_f	$TTRT_{mm}$	TWT_{mm}	z_1	t_1	$TTRT_{ga}$	TWT_{ga}	z_2	t_2	GAP_1	GAP_2	GAP_3
10	150	24	174	1001	160	21	181	0.7	−0.06	0.12	−0.04
20	316	138	454	3600	341	83	424	1.7	−0.08	0.39	0.06
30	494	346	840	3600	529	263	792	2.8	−0.07	0.23	0.06
40	658	619	1277	3600	705	509	1214	3.5	−0.07	0.17	0.05
50	841	1156	1997	3600	894	799	1693	4.5	−0.06	0.30	0.15
60	970	2140	3110	3600	1029	1626	2655	5.9	−0.06	0.24	0.14

The test results of the proposed mathematical model and genetic algorithm based on IGA Istanbul Airport flight data are given in Table 1. The first column in this table shows the number of flights (n^f), with half representing arrival flights and the other half departure flights. The following four columns summarize the results of proposed mathematical model such as total taxi route time (TTRT), total waiting time (TWT), the obtained objective value and solution time in seconds using the GAMS/Cplex solver. Similarly, the next four columns present the results of the genetic algorithm. The last three columns show the results of performance measures.

As can be seen from Table 1, The GAMS/Cplex solver could obtain optimal solution in 3600 s only for $n^f = 10$. For the other test problems, the running time limit is exceeded, and feasible solutions are obtained with the mathematical model. However, the proposed genetic algorithm provides feasible solutions in seconds. It can be clearly observed that GA is remarkably more efficient than the mathematical model regarding running time.

To assess the performance of the mathematical model and the genetic algorithm, we employed three performance measures: Relative gap for total taxi route time (GAP_1), relative gap for total waiting time (GAP_2), and relative gap for total taxi times (taxi-in and taxi-out times) and total waiting times (GAP_3). They are given in Eqs. (32)–(34).

$$GAP_1 = \frac{TTRT_{GAMS} - TTRT_{GA}}{TTRT_{GAMS}} \tag{32}$$

$$GAP_2 = \frac{TWT_{GAMS} - TWT_{GA}}{TWT_{GAMS}} \tag{33}$$

$$GAP_3 = \frac{z_1 - z_2}{z_1} \tag{34}$$

The evaluation of test results based on performance measures are given in the last three columns of Table 1. GAP_1 has a negative value for all test instances since the mathematical models exhibit good performance in reducing total waiting time. However, the proposed genetic algorithm has better results with significantly high GAP_2 and GAP_3 values (except $n^f = 10$). All the GAP_2 values are positive, and it indicates that the GA outperforms the mathematical model in reducing waiting times, particularly in larger test instances. The widest GAP_2 value that is equal to 0.39 is recorded for instance $n^f = 20$. Similarly, all the GAP_3 values (except $n^f = 10$) are also positive, showing that the GA outperforms the mathematical model in minimizing total objective function value. Especially with a large number of flights, GAP_3 reaches peak percentage values (approximately 0.15).

5. Conclusion

In this study, we address airport taxiway planning problem. Unlike studies in the literature, this paper considers the problem integrally by including not only taxiways but also the gates and runways assigned to flights. To solve the problem, a mixed-integer mathematical model is firstly proposed. The objective of the mathematical model is to minimize total taxi time (taxi-in and taxi-out times) and waiting time considering safety regulations. For larger-scale problems, a genetic algorithm is proposed. The performances of the proposed methods are tested using real-life data from IGA Istanbul Airport. It has been demonstrated that the genetic algorithm provides effective solutions in a short time for large-scale problems. As future work, this can be considered as stochastic considering dynamic and real-time data such as unexpected delays. Therefore, stochastic version of the problem can provide efficient airport ground operations.

References

[1] J. Guépet, O. Briant, J.-P. Gayon, and R. Acuna-Agost, Integration of aircraft ground movements and runway operations, *Transportation Research Part E: Logistics and Transportation Review*, vol. 104, pp. 131–149, (2017).

[2] J. Ma, D. Delahaye, M. Sbihi, P. Scala, and M. A. Mujica Mota, Integrated optimization of terminal maneuvering area and airport at the macroscopic level, *Transportation Research Part C: Emerging Technologies*, vol. 98, pp. 338–357, (2019).

[3] Z. G. Jiaming Liu and B. Yu, Optimising Gate assignment and taxiway path in a discrete time–space network: Integrated model and state analysis, *Transportmetrica B: Transport Dynamics*, vol. 11, no. 1, pp. 1–23, (2023).

[4] W. Deng, L. Zhang, X. Zhou, Y. Zhou, Y. Sun, W. Zhu, H. Chen, W. Deng, H. Chen and H. Zhao, Multi-strategy particle swarm and ant colony hybrid optimization for airport taxiway planning problem, *Information Sciences*, vol. 612, pp. 576–593, (2022).

[5] Z. Xiang, H. Sun, and J. Zhang, Application of Improved Q-Learning Algorithm in Dynamic Path Planning for Aircraft at Airports, *IEEE Access*, vol. 11, pp. 107892–107905, (2023).

[6] S. G. Borjigin and M. A. Austin, Multi-Domain Semantic Model and Real-Time Planning Approach to Safe and Efficient Airport Taxiway Operations, *IEEE Transactions on Intelligent Transportation Systems*, vol. 25, no. 7, pp. 7738–7750, (2024).

[7] M. Zhang, S. Liu, and H. Li, Multi-Objective Route Planning for Aircraft Taxiing Under Different Traffic Conflict Types, *Journal of Aerospace Information Systems*, vol. 19, no. 2, pp. 124–142, (2022).

[8] G. Lian, Y. Wang, W. Li, Q. Xu, S. Hao, and T. Wen, Aircraft taxiing conflict avoidance approach based on Gaussian spatial–temporal prediction, *Physica A: Statistical Mechanics and its Applications*, vol. 624, p. 128915, (2023).

[9] C. Yu, D. Zhang, and H. Y. K. Henry Lau, A heuristic approach for solving an integrated gate reassignment and taxi scheduling problem, *Journal of Air Transport Management*, vol. 62, no. C, pp. 189–196, (2017).

Chapter 11

Approximation of Laplace-Steklov EVPs by a Boundary Element Approach

Önder Türk* and Eylem Bahadır
Middle East Technical University, Institute of Applied Mathematics, Ankara, Turkey, 06800
**oturk@metu.edu.tr*

A numerical investigation of a novel eigenvalue problem (EVP), called as the Laplace-Steklov EVP, is considered by the use of a boundary element method. One of the unique features of the problem is that the spectral parameter resides both in the defining partial differential equation and the set of boundary conditions. The discretisation of the EVP is carried out by a dual reciprocity boundary element method (DRBEM). The method is implemented by utilising the fundamental solution of the Laplace equation, and appropriately assuming the rest of the terms as inhomogeneity. Consequently, the first DRBEM approximations of the Laplace-Steklov EVP defined in several two-dimensional configurations are presented in this study to the best of the authors' knowledge.

1. Introduction

Laplace-Steklov eigenvalue problems (EVPs) are defined by special generalisations of Laplace and Steklov EVPs to include the eigenvalue in the differential equation and as well as in at least one of the boundary conditions. Such generalised configurations are at the centre of numerous applications that are of significant physical importance. For instance, the Laplace-Steklov EVP we consider here is associated with the Fourier analysis of a heat problem defined by dynamic boundary conditions representing a heat storing capacity on the boundaries [1, 6].

In general, determining the analytical solutions of eigenproblems is a challenging or even an impossible task using existing analytical techniques, due to several reasons such as geometrically complex problem domains. Therefore, one has to resort to numerical methods to obtain approximate solutions to those problems. There are numerous recent studies on the analysis of Laplace and Steklov eigenproblems by several numerical methods such as finite, boundary, and virtual element methods [9, 10, 12]. However, the only numerical study that tackles the Laplace-Steklov EVP considered in the present study is Ref. [6] in which the problem has Dirichlet, Neumann, and spectral boundary conditions defined on a square domain using a collocation approach, to the best of our knowledge.

In the present study, we mainly aim to propose a novel dual reciprocity boundary element method (DRBEM) formulation to approximate the solution of a Laplace-Steklov EVP. The method has the significant advantage of discretising only the boundary, and thus, providing a potentially suitable framework for the existence of the spectral parameter in the set of boundary conditions. DRBEM has been successfully applied to approximate the solutions of a number of Steklov EVPs in Ref. [12]. In this study, we extend the applicability of DRBEM to handle the novel Laplace-Steklov EVP. The numerical tests are carried out on two different problem domains. Firstly, the problem is considered on a square domain, which allows us to provide an analytical solution to compare with the DRBEM approximations. Secondly, we test our DRBEM formulation on a non-Lipschitz domain, particularly a domain with a cusp. In this latter case, we compare the results of the DRBEM with those obtained by using a finite element method (FEM) with linear Lagrange elements since the analytical solution is not available. In order to achieve this, we include the FEM formulation we follow since we are not aware of its existence in the literature.

2. Mathematical Statement of the Problem

The Laplace-Steklov EVP we consider is defined as: Find eigenvalues $\lambda \in \mathbb{C}$ and eigenfunctions $u \neq 0$ such that

$$\begin{cases} -\Delta u = \mu\lambda u & \text{in } \Omega, \\ u = 0 & \text{on } \Gamma_D, \\ \dfrac{\partial u}{\partial \mathbf{n}} = 0 & \text{on } \Gamma_N, \\ \dfrac{\partial u}{\partial \mathbf{n}} = \rho\lambda u & \text{on } \Gamma_S, \end{cases} \tag{1}$$

where μ and ρ are given parameters, and $\mathbf{n}$ denotes the unit outward normal on the boundary $\Gamma = \Gamma_D \cup \Gamma_N \cup \Gamma_S$ of a bounded domain $\Omega \subset \mathbb{R}^2$.

The case where $\mu = 0$ and $\Gamma_N = \emptyset$ corresponds to a Steklov EVP in which the eigenvalues may represent the squares of the natural vibrational frequencies of an elastic membrane with mass concentrated at the boundary [2]. In the case of $\mu \neq 0$, the problem can be considered as the auxiliary spectral problem associated with a parabolic type initial-boundary value problem defined by Wentzell boundary conditions as given in Ref. [6].

When $\Gamma = \Gamma_S$ with $\mu > 0$, $\rho \geq 0$ or $\mu = 0$, $\rho > 0$, there exists an increasing sequence of infinitely many non-negative eigenvalues of Eq. (1) [3, 5] such that

$$\lambda_0 \leq \lambda_1 \leq \lambda_2 \leq \lambda_3 \leq \cdots . \tag{2}$$

Let us note here that if $\mu = 0$ and $\rho > 0$, $\lambda_0 = 0$, and $\lambda_0 > 0$ otherwise. Moreover, in the case of $\Gamma = \Gamma_S$ with $\mu > 0$ and $\rho < 0$, $\lambda_0 = 0$ is also an eigenvalue, and the problem has infinitely many positive and negative eigenvalues ordered as [3, 4]

$$\cdots \leq \lambda_{-n} \leq \cdots \leq \lambda_{-2} \leq \lambda_{-1} < \lambda_0 = 0 < \lambda_1 \leq \lambda_2 \leq \cdots \leq \lambda_n \leq \cdots . \tag{3}$$

3. Numerical Formulations

In this section, we present the numerical strategies for solving Eq. (1) in detail to some extent. The main objective of the present study is to propose a technique based on boundary elements for approximating the problem at hand, which is given in Section 3.1. As already mentioned, in the absence of analytical solutions, we aim to compare the results obtained from DRBEM by another approach, namely, FEM, whose application to Eq. (1) is introduced in Section 3.2.

3.1. *DRBEM formulation*

The idea is to transform the governing equations into boundary integral equations. To that end, we use the two-dimensional fundamental solution of the Laplace equation [11] given by

$$u^* = \frac{1}{2\pi} \ln\left(\frac{1}{r}\right), \tag{4}$$

as the weight function, where r is the distance from the application point of the concentrated source to any point in the problem domain. Then, the residual of the differential equation together with the ones defined by the boundary conditions in Eq. (1) are weighted with u^*, and its normal derivative $q^* = \partial u^*/\partial n$, respectively, and then forced to vanish in an average sense as follows:

$$\int_\Omega (\Delta u + \mu\lambda u)\, u^* d\Omega = \int_{\Gamma_D} u q^* d\Gamma + \int_{\Gamma_N} q u^* d\Gamma + \int_{\Gamma_S} (\rho\lambda u - q)\, u^* d\Gamma = 0, \tag{5}$$

where by q we denote the normal derivative $\partial u/\partial n$. Next, applying integration by parts twice, collecting the boundary terms together, and using the properties of u^* and q^* [11, 12] results in

$$c_i u_i + \int_\Gamma (u q^* - u^* q) d\Gamma = \lambda \int_\Omega \mu u u^* dx, \tag{6}$$

where the subscript i assigns the source point. The constant c_i is given by $c_i = \theta_i/2\pi$, where θ_i is the internal angle at the source point.

In order to obtain the discrete form of the problem utilising the integral equation given in Eq. (6), we assume that the domain boundary is partitioned with N constant elements. Therefore, we have N boundary nodes and L arbitrarily defined interior nodes within the problem domain for internal investigation. Next, the domain integral in Eq. (6) is substituted by an equivalent boundary integral since we want to discretise only the boundary. To do this, the term μu is approximated by $\sum_{j=1}^{N+L} \alpha_j \varphi_j(x, y)$, where φ_j is a basis function to be described below and α_j is determined by employing a collocation technique. This allows us to represent the solution of the problem as the sum of the Laplacian of a series of particular solutions $\hat{u}_j$ so that $\Delta \hat{u}_j = \varphi_j$. Consequently, the term μu in Eq. (6) is replaced by the approximation $\sum_{j=1}^{N+L} \alpha_j \Delta u_j$.

A reciprocity principle (dual reciprocity) is applied next, and by summing over the boundary elements, the discrete form of Eq. (6) is obtained as

$$c_i u_i + \sum_{k=1}^{N} \int_{\Gamma_k} (uq^* - u^* q)\mathrm{d}s = \lambda \sum_{j=1}^{N+L} \alpha_j \left[c_i \hat{u}_{ij} + \sum_{k=1}^{N} \int_{\Gamma_k} (q^* \hat{u}_j - u^* \hat{q}_j)\mathrm{d}s \right].$$

This discrete equation is expressed in terms of the nodal values by using a radial basis function, considered as $\varphi = 1 + r$ in this study, and integrating over each boundary element. The associated matrix-vector equations in nodal values are given as

$$Hu - Gq = (H\hat{U} - G\hat{Q})F^{-1}(-\lambda\mu u) = -\lambda\mu S u. \tag{7}$$

The matrices $\hat{U}$ and $\hat{Q}$ are formed by using the vectors $\hat{u}_j$ and $\hat{q}_j$ as their respective columns. The entries of the matrices H and G are given by

$$H_{ij} = c_i \delta_{ij} + \frac{1}{2\pi} \int_{\Gamma_j} \frac{\partial}{\partial n} \left(\ln \left(\frac{1}{r} \right) \right) \mathrm{d}\Gamma_j \quad \text{and} \quad G_{ij} = \frac{1}{2\pi} \int_{\Gamma_j} \ln \left(\frac{1}{r} \right) \mathrm{d}\Gamma_j,$$

$i, j = 1, \ldots, N$, where r is the distance from node i to element j, and δ_{ij} is the Kronecker delta function. F is the coordinate matrix whose columns are φ_j, $j = 1, \ldots, N + L$. The second equality in Eq. (7) is written to denote $(H\hat{U} - G\hat{Q})F^{-1}$ by S. Rearranging this equation as

$$Au = Hu + \lambda\mu S u = Gq, \tag{8}$$

reveals the necessity for a partitioned structure. To provide the partitioned structure, we split the set of all boundary and interior nodal points $\mathcal{N}$ as $\mathcal{N} = \mathcal{N}_\mathrm{u} \cup \mathcal{N}_\mathrm{k} \cup \mathcal{N}_\mathrm{s}$. Here $\mathcal{N}_\mathrm{u}$, $\mathcal{N}_\mathrm{k}$, and $\mathcal{N}_\mathrm{s}$ denote the subsets of $\mathcal{N}$ constructed by the nodal values of the subregions, where the solution u is unknown, the Dirichlet condition is imposed on u, and the spectral dependence $q = \lambda\rho u$ is defined, respectively. By introducing the restrictions of the vectors of the nodal values $\eta_\mathrm{p} = \eta(\mathrm{p}_j)_{\mathrm{p}_j \in \mathcal{N}_\mathrm{p}}$ for $\eta = u, q$ where $\mathrm{p} = \mathrm{u}, \mathrm{k}, \mathrm{s}$, the matrix-vector form Eq. (8) can be represented in a partitioned form as

$$\begin{bmatrix} A_\mathrm{uu} & A_\mathrm{uk} & A_\mathrm{us} \\ A_\mathrm{ku} & A_\mathrm{kk} & A_\mathrm{ks} \\ A_\mathrm{su} & A_\mathrm{sk} & A_\mathrm{ss} \end{bmatrix} \begin{bmatrix} u_\mathrm{u} \\ 0 \\ u_\mathrm{s} \end{bmatrix} = \begin{bmatrix} G_\mathrm{uu} & G_\mathrm{uk} & G_\mathrm{us} \\ G_\mathrm{ku} & G_\mathrm{kk} & G_\mathrm{ks} \\ G_\mathrm{su} & G_\mathrm{sk} & G_\mathrm{ss} \end{bmatrix} \begin{bmatrix} 0 \\ q_\mathrm{k} \\ \lambda\rho u_\mathrm{s} \end{bmatrix}. \tag{9}$$

Let us note here that we imposed the condition $q_\mathrm{s} = \lambda\rho u_\mathrm{s}$ in the last entry of the vector in the right-hand side of Eq. (9). This partitioned system is transformed into a standard generalised algebraic eigenvalue problem

$$\widetilde{A}\,\widetilde{f} = \lambda\,\widetilde{G}\,\widetilde{f}, \tag{10}$$

where

$$\widetilde{A} = \begin{bmatrix} A_\mathrm{uu} & -G_\mathrm{uk} & A_\mathrm{us} \\ A_\mathrm{ku} & -G_\mathrm{kk} & A_\mathrm{ks} \\ A_\mathrm{su} & -G_\mathrm{sk} & A_\mathrm{ss} \end{bmatrix}, \quad \widetilde{G} = \rho \begin{bmatrix} 0 & 0 & G_\mathrm{us} \\ 0 & 0 & G_\mathrm{ks} \\ 0 & 0 & G_\mathrm{ss} \end{bmatrix}, \quad \text{and} \quad \widetilde{f} = \begin{bmatrix} u_u \\ q_k \\ u_\mathrm{s} \end{bmatrix}.$$

3.2. *FEM formulation*

In this section, we present a FEM formulation to approximate the problem in Eq. (1) to allow a comparison with the DRBEM approximations when the analytical solution is not known. First, the variational form of the Laplace-Steklov EVP is derived by multiplying both sides of the differential equation given in Eq. (1) by a weight function taken from the space $\mathcal{V} = \{v \in H^1(\Omega) : v|_{\Gamma_D} = 0\}$, where $H^1(\Omega)$ denotes the usual Hilbert space of square-integrable functions whose first order derivatives are also square-integrable on Ω. By applying the divergence theorem and integrating over the domain Ω, we obtain the variational formulation stated as: Find eigenvalues $\lambda \in \mathbb{C}$ and eigenfunctions $0 \neq u \in \mathcal{V}$ such that

$$a(u, v) = \lambda b(u, v) \quad \forall v \in \mathcal{V}, \tag{11}$$

where $a(u, v)$ and $b(u, v)$ are defined as

$$a(u, v) = \int_\Omega \nabla u \cdot \nabla v \mathrm{d}\Omega \quad \text{and} \quad b(u, v) = \int_\Omega \mu u v \mathrm{d}\Omega + \int_{\Gamma_S} \rho u v \mathrm{d}\Gamma.$$

The main idea of the FEM is to approximate the exact eigenfunction u and the associated eigenvalue λ within a finite-dimensional subspace of $\mathcal{V}$. To that end, we assume that Ω_h is constructed by discretising the problem domain Ω from a collection of triangular elements $\mathcal{T}$, where the size of the element partition is denoted by h. On the discrete domain Ω_h, we introduce the standard finite element space $\mathcal{V}_h = \{v \in H^1(\Omega) : v|_\mathcal{T} \in \mathcal{P}_s, \ \forall \mathcal{T} \in \Omega_h\}$, as a finite-dimensional subspace of $\mathcal{V}$, where $\mathcal{P}_s$ denotes the space of piecewise polynomials of degree less than or equal to s defined on $\mathcal{T}$. On this subspace, the discrete problem is stated as: Find $0 \neq u_h \in \mathcal{V}_h$ and $\lambda_h \in \mathbb{C}$ such that

$$a(u_h, v_h) = \lambda_h b(u_h, v_h) \quad \forall v_h \in \mathcal{V}_h. \tag{12}$$

By introducing a basis $\{\phi_i\}|_{i=1}^m$ for $\mathcal{V}_h$, the eigenfunctions can be expressed as $u_h = \sum_{i=1}^m \beta_i \phi_i \in \mathcal{V}_h$. Hence, the discrete problem associated with Eq. (12) reduces to a generalised algebraic eigenvalue system $K\beta = \lambda_h R\beta$. Here $\beta = \{\beta_i\}|_{i=1}^m$ is the vector containing the unknown coefficients β_i. Matrices K and R are the stiffness and mass matrices with the entries given respectively by $k_{ij} = a(\phi_i, \phi_j)$ and $r_{ij} = b(\phi_i, \phi_j)$, $i, j = 1, \ldots, m$. The FEM approximation results provided in the sequel are obtained by solving the associated system when $\mathcal{V}_h$ is constructed with the use of piecewise linear polynomials. The use of linear functions allows us to compute the entries of the stiffness and mass matrices by evaluating the involved integrals analytically. These analytical expressions and more details on FEM applications can be found in standard books on finite elements (e.g., [8]).

4. Numerical Results

In this section, we will present the results we obtain approximating the Laplace-Steklov EVP where $\mu = \rho = 1$. Two different problem domains, namely, a square and a cuspidal domain (to be defined below), are considered. In Section 4.1, the DRBEM results from the square domain are provided together with the corresponding analytical solutions for comparison. In Section 4.2, we present the numerical results when the problem is considered on a domain with a cusp obtained by DRBEM and FEM, the latter providing a means of comparison as no analytical solution is available for this case. In the application of the DRBEM, the boundary integrals are approximated by a Gaussian quadrature using a 16-point formula for the square domain and a 56-point formula for the cuspidal domain following Ref. [12] due to the analogy of the constructions of the matrices involved. All the results we present below are obtained by implementing the described methods using computer programs written in MATLAB. In each case, the resulting generalised matrix eigenvalue problem is solved by its built-in eigensolver.

4.1. *Square domain*

As a first test, we consider the problem given in Eq. (1) on the square domain $\Omega = (0,1)^2$, with a known analytical solution. The problem geometry with the boundary conditions and a sample DRBEM discretisation when $N = L = 8$ are presented in Figure 1. In this configuration, the analytical solutions obtained using a Fourier method [6, 7] are given as

$$u_{n,k} = 2\sin(\sqrt{s_n^k}x)\sin(\sigma_k y),\ k = 1,2,\ldots; n = 0,1,2,\ldots, \tag{13}$$

associated with the eigenvalues $\lambda_{n,k} = \sigma_k^2 + s_n^k$, where $\sigma_k = \pi(2k-1)/2$. The coefficients s_n^k are determined as a root of the non-linear characteristic equation

$$\tan\sqrt{s} = \frac{\sqrt{s}}{\rho(s+\sigma_k^2)}. \tag{14}$$

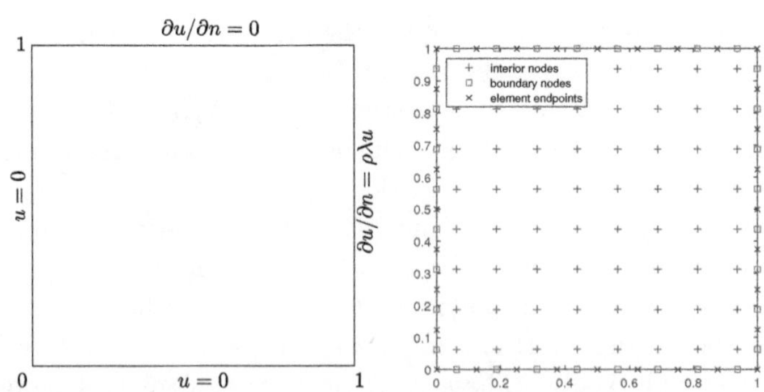

(a) The problem geometry. (b) BEM discretisation.

Fig. 1. The problem geometry with boundary conditions and a sample discretisation with boundary elements where $N = L = 8$.

When the problem is approximated by the DRBEM, specifically when the generalised algebraic eigenvalue system Eq. (10) is solved using $N = 2^\ell$, $\ell = 4,5,6,7$, the first ten eigenvalues obtained together with their rates of convergence towards the reference ones are listed in Table 1. In this test, the reference values are obtained by approximating the roots of the characteristic equation Eq. (14) by Newton's method. From this table, we can deduce that the approximate eigenvalues converge to the reference ones as the number of elements increases with rates greater than or equal to 1.7 near the asymptotic regime. In order to validate that the numerical

Table 1. Computed ten eigenvalues with DRBEM and their rates of convergence.

Ref.	$N = 2^4$	$N = 2^5$	$N = 2^6$	$N = 2^7$
1.3480	1.3526	1.3494 (1.7)	1.3484 (1.7)	1.3481 (1.7)
4.2404	4.2678	4.2495 (1.6)	4.2433 (1.6)	4.2413 (1.7)
7.3699	7.4764	7.4040 (1.6)	7.3804 (1.7)	7.3731 (1.7)
10.5069	10.6977	10.5712 (1.6)	10.5268 (1.7)	10.5129 (1.7)
13.6460	13.9009	13.7797 (0.9)	13.6867 (1.7)	13.6580 (1.8)
13.8935	14.0399	13.8942 (7.8)	13.8936 (3.2)	13.8935 (3.0)
16.7860	17.3119	16.9824 (1.4)	16.8468 (1.7)	16.8039 (1.8)
19.9265	20.7620	20.2524 (1.4)	20.0273 (1.7)	19.9558 (1.8)
23.0672	23.9857	23.4920 (1.1)	23.2025 (1.7)	23.1068 (1.8)
26.2082	27.4157	26.8301 (1.0)	26.4078 (1.6)	26.2663 (1.8)

approach can capture the eigenfunction behaviours qualitatively, the contour plots of the first three DRBEM solutions, as well as those of the corresponding analytical ones denoted by u_1, u_2 and u_3, are provided in Figure 2. From this figure, it can be concluded that the methodology described in Section 3.1 is capable of obtaining accurate eigenfunctions using a moderate number of boundary elements together with a suitable number of inner nodes.

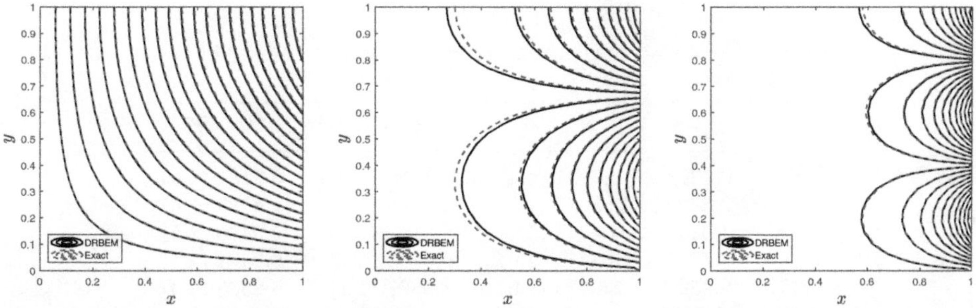

(a) u_1. (b) u_2. (c) u_3.

Fig. 2. DRBEM ($N = L = 24$) approximations and the analytical eigenfunctions u_1, u_2, and u_3.

4.2. *Cuspidal domain*

Now we consider a more challenging configuration, a cuspidal domain defined as $\Omega = \{(x,y) : 0 < x < 1, 0 < y < x^{1.4}\}$, whose boundary is denoted by $\Gamma = \Gamma_D \cup \Gamma_N \cup \Gamma_S$, where $\Gamma_N = \{0 \leq x \leq 1, y = 0\}$, $\Gamma_D = \{x = 1, 0 \leq y \leq 1\}$, and $\Gamma_S = \{0 \leq x \leq 1, y = x^{1.4}\}$. The problem geometry with the boundary conditions is depicted in Figure 3(a). As mentioned before, in this case, the analytical solution is not known; hence the DRBEM results will be compared with the FEM approximations. Sample DRBEM (with $N = 8$) and FEM (with $h = 0.2$) discretisations for this problem domain are provided in Figures 3(b) and 3(c), respectively. The first ten approximate eigenvalues computed using DRBEM and FEM are tabulated together with their rates of convergence to the corresponding reference values in Tables 2 and 3, respectively. The reference values used for both methods are obtained with the use of the FEM strategy on a fine mesh in which $h = 0.001$. These tables reveal the fact that the investigated eigenvalues obtained by both methods converge in similar behaviours to the reference ones. In general, FEM rates are higher than the

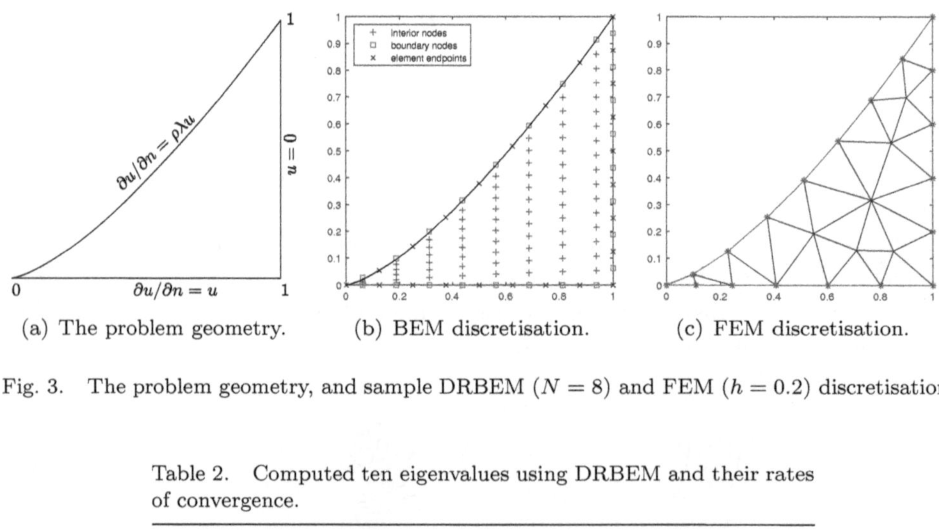

(a) The problem geometry. (b) BEM discretisation. (c) FEM discretisation.

Fig. 3. The problem geometry, and sample DRBEM ($N = 8$) and FEM ($h = 0.2$) discretisations.

Table 2. Computed ten eigenvalues using DRBEM and their rates of convergence.

Ref.	$N = 2^4$	$N = 2^5$	$N = 2^6$	$N = 2^7$
0.7175	0.7298	0.7221 (1.4)	0.7192 (1.5)	0.7181 (1.6)
2.4443	2.5899	2.5081 (1.2)	2.4698 (1.3)	2.4529 (1.6)
4.3188	4.6460	4.4744 (1.1)	4.3858 (1.2)	4.3425 (1.5)
6.1959	6.7797	6.4856 (1.0)	6.3265 (1.1)	6.2434 (1.5)
8.1777	9.0566	8.6175 (1.0)	8.3807 (1.1)	8.2535 (1.4)
10.1963	11.3608	10.7886 (1.0)	10.4736 (1.1)	10.3012 (1.4)
12.2434	13.7625	13.0193 (1.0)	12.6069 (1.1)	12.3826 (1.4)
14.3116	16.1112	15.2627 (0.9)	14.7570 (1.1)	14.4829 (1.4)
16.3963	18.5148	17.5632 (0.9)	16.9390 (1.1)	16.6059 (1.4)
18.4940	20.7515	19.8600 (0.7)	19.1282 (1.1)	18.7388 (1.4)

Table 3. Computed ten eigenvalues using FEM and their rates of convergence.

Ref.	$h = 0.032$	$h = 0.016$	$h = 0.008$	$h = 0.004$
0.7175	0.7188	0.7181 (1.2)	0.7177 (1.5)	0.7176 (1.8)
2.4443	2.4742	2.4581 (1.1)	2.4494 (1.5)	2.4457 (1.8)
4.3188	4.4043	4.3606 (1.0)	4.3346 (1.4)	4.3233 (1.8)
6.1959	6.3789	6.2867 (1.0)	6.2311 (1.4)	6.2060 (1.8)
8.1777	8.4864	8.3276 (1.0)	8.2363 (1.4)	8.1948 (1.8)
10.1963	10.6716	10.4181 (1.1)	10.2826 (1.4)	10.2218 (1.8)
12.2434	12.9312	12.5491 (1.2)	12.3607 (1.4)	12.2785 (1.7)
14.3116	15.2661	14.7154 (1.2)	14.4632 (1.4)	14.3572 (1.7)
16.3963	17.6793	16.9151 (1.3)	16.5860 (1.5)	16.4537 (1.7)
18.4940	20.1681	19.1478 (1.4)	18.7264 (1.5)	18.5642 (1.7)

DRBEM ones, as can be expected since in the former we use linear polynomials whereas the approximations are obtained using constant elements in the latter.

Finally, in Figure 4 we present the contour plots of the first three eigenfunctions obtained using DRBEM and FEM. It is seen that both approaches yield solutions

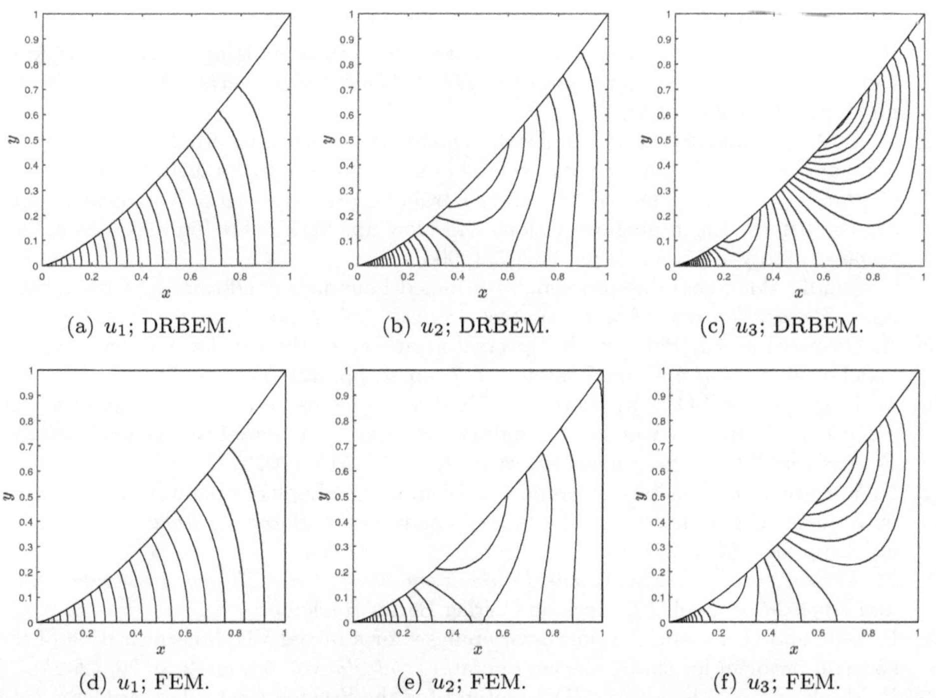

(a) u_1; DRBEM. (b) u_2; DRBEM. (c) u_3; DRBEM.

(d) u_1; FEM. (e) u_2; FEM. (f) u_3; FEM.

Fig. 4. DRBEM ($N = 25$) and FEM ($h = 0.025$) approximations to the eigenfunctions u_1, u_2, and u_3.

that comply very well with each other at the given discretisation levels. It may be important to note that in this example the final systems are of orders 725 for DRBEM and 799 for FEM.

By construction, the DRBEM matrices are denser than the FEM ones. This, however, does not introduce a significant drawback in handling the algebraic systems of moderate sizes which have been shown to be sufficient for obtaining reasonably accurate solutions. Indeed, the results of the example above have been computed in less than ten seconds in average on a standard personal computer. Even though the computing time for constructing the algebraic systems in the DRBEM approach is slightly longer than that of the FEM approach due to the use of quadratures in the former, the computational time of the eigensolver is less than a second in each case.

Acknowledgments

This work has been supported by the Middle East Technical University Scientific Research Projects Coordination Unit under grant number AGEP-705-2023-11411.

References

[1] R. Altmann and C. Zimmer, A second-order bulk–surface splitting for parabolic problems with dynamic boundary conditions, *IMA Journal of Numerical Analysis*, vol. 44, no. 4, pp. 2370–2393, (2023).

[2] M. G. Armentano and A. L. Lombardi, The Steklov eigenvalue problem in a cuspidal domain, *Numerische Mathematik*, vol. 144, no. 2, pp. 237–270, (2020).

[3] C. Bandle, J. Below, and W. Reichel, Parabolic problems with dynamical boundary conditions: Eigenvalue expansions and blow up, *Rendiconti Lincei Matematica e Applicazioni*, vol. 17, no. 1, pp. 35–67, (2006).

[4] C. Bandle, An eigenvalue problem with mixed boundary conditions and trace theorems, *Banach Journal of Mathematical Analysis*, vol. 2, no. 2, pp. 68–75, (2008).

[5] A. Girouard and I. Polterovich, Spectral geometry of the Steklov problem (Survey article), *Journal of Spectral Theory*, vol. 7, no. 2, pp. 321–359, (2017).

[6] M. I. Ismailov and Ö. Türk, Direct and inverse problems for a 2D heat equation with a Dirichlet-Neumann-Wentzell boundary condition, *Communications in Nonlinear Science and Numerical Simulation*, vol. 127, p. 107519, (2023).

[7] N. Y. Kapustin, Basis properties of a problem for the Laplace operator on the square with spectral parameter in a boundary condition, *Differential Equations*, vol. 53, no. 4, pp. 563–565, (2017).

[8] M. Larson and F. Bengzon, *The Finite Element Method: Theory, Implementation, and Applications*, vol. 10. (Springer-Verlag Berlin Heidelberg, 2013).

[9] F. Lepe and G. Rivera, A priori error analysis for a mixed VEM discretization of the spectral problem for the Laplacian operator, *Calcolo*, vol. 58, no. 2, p. 20, (2021).

[10] P. Monk and Y. Zhang, An HDG method for the Steklov eigenvalue problem, *IMA Journal of Numerical Analysis*, vol. 42, no. 3, pp. 1929–1962, (2021).

[11] P. W. Partridge, C. A. Brebbia, and L. C. Wrobel, *The Dual Reciprocity Boundary Element Method*, (Southampton, Boston: Computational Mechanics Publications, 1992).

[12] Ö. Türk, A DRBEM approximation of the Steklov eigenvalue problem, *Engineering Analysis with Boundary Elements*, vol. 122, pp. 232–241, (2021).

Chapter 12

Measuring the Effect of Certain Basic, Social and Academic Activities on Success Using Modern Techniques

Selma Kılınç and Ahmet Şahiner*

Department of Mathematics,
Suleyman Demirel University, Isparta, Turkey
**ahmetsahiner@sdu.edu.tr*

In recent years, the factors affecting academic achievement in education have received considerable attention. Fundamental elements such as environmental and familial support, motivation, passion, technological resources, classroom environment and interpersonal relationships play crucial roles in the measurement and evaluation processes. The integration of digitalization and artificial intelligence in education facilitates the personalization of learning processes, the development of teaching methods and the assessment of various influencing factors. In particular, fuzzy logic offers an effective method for complex and uncertain decision-making problems. This study aims to model the exam success status of high school students using fuzzy logic. It examines the relationship between students' mathematics exam results and variables such as mathematics hopelessness scale, daily study times, play times, homework times, and sleep times. The method employed is the Adaptive Neuro-Fuzzy Inference System (ANFIS) approach, developed by Jang and Sun (1995). This approach combines fuzzy logic and artificial neural networks, using a neural network learning method to adjust system parameters. The study was conducted with 10th and 11th-grade students at Isparta Şehit Ali Hekim Anatolian High School. The data were modeled using MATLAB R2021's Fuzzy Logic Toolbox. The model includes five inputs and one output. When the predicted exam scores were compared with the actual scores, an accuracy rate of 80% was achieved. This approach demonstrates the potential of fuzzy logic to accurately predict academic performance and highlights the importance of various factors.

Keywords: Optimization, Education, Fuzzy Logic Modelling.

1. Introduction

Mathematics has historically occupied a central position in educational systems as an important branch of science. Research has elucidated various factors that influence students mathematical achievement [1] However, traditional statistical methods often fall short in fully addressing the interaction and uncertainty of these factors. This is where fuzzy logic, introduced by Zadeh in 1965, comes into play. Fuzzy logic is a tool developed to cope with uncertainty and work with vague data, allowing for the use of intermediate values rather than precise ones. This enables better modeling of uncertainties in real-life situations [2]. Fuzzy logic is widely used in many fields such as engineering, physics, and education [3]. It is particularly recognized as a significant tool in artificial intelligence, robotics, control systems, economics, and medicine [4]. In artificial intelligence, fuzzy logic is commonly used to handle complex systems and uncertain data [5]. In robotics, it enables robots to respond more flexibly to environmental variations [6]. The process of mathematical modeling involves defining a problem, collecting and analyzing data. Successful modeling requires not only knowledge but also understanding how to use that knowledge. Fuzzy logic has recently been applied in energy conformation problems; [7] studies on optimizing the energy arising from the connection of modules at different angles. In dentistry; [8] studies on age estimation based on pulp measurements. ANFIS (Adaptive Neuro-Fuzzy Inference System) is a combination of artificial neural networks (ANN) and fuzzy inference systems (FIS) [9]. This modeling method is effectively used in solving function approximation problems and adopts a hybrid learning approach to determine the optimal distribution of membership functions. The ANFIS architecture consists of five layers, with each layer defined by node functions composed of several nodes. Inputs in these layers are derived from nodes in preceding layers, facilitating the creation of rules. ANFIS utilizes both artificial neural networks (ANN) and fuzzy logic (FL). This combination makes modeling more systematic and reduces dependency on expert knowledge. ANFIS combines the learning capability of neural networks with the interpretability feature of fuzzy logic [9]. This integration enables automatic generation of fuzzy rules and sets from large datasets up to a certain level of accuracy. ANFIS employs an adaptive backpropagation algorithm to learn the parameters of neural networks [9], allowing it to approximate any real continuous function to any degree of accuracy [10]. ANFIS supports Takagi-Sugeno based systems, where rules are fuzzy if-then rules of Sugeno type, and outputs are obtained using weighted averages. Understanding the relationship between students levels of math hopelessness and their written exam results involves considering several variables. These variables include factors such as daily study time, daily playtime, daily homework time, and daily sleep duration. To comprehend how each of these factors influences math achievement, statistical analyses are typically conducted. Daily Study Time: Study time generally has a positive impact on students' academic achievement. Students who study more often tend to achieve better grades. However, this duration can be either too long or too short. The ideal study time may vary for each student. Daily Playtime: Playtime, especially time spent on computer or mobile games, can affect students' study durations. Playing games for extended periods can lead to distractions and reduced study time. Consequently, excessive gaming generally has a negative impact on academic achievement. Daily Homework Time: Homework time indicates how much time a student dedicates to their studies. Increasing the time spent on homework can help students understand the material better and perform better on exams. Daily Sleep Duration: Getting adequate sleep significantly affects a student's concentration and learning ability. Insufficient sleep can lead to issues such as lack of attention and learning difficulties. Consequently, it can negatively impact exam performance. Math Hopelessness Levels: Math hopelessness levels refer to a student's feelings of failure or anxiety related to mathematics. This can typically affect academic performance. High levels of hopelessness are generally associated with lower performance. In conclusion, when evaluating the factors affecting students' math exam results, it is important to consider all these variables together and understand how they interrelate. Although it is difficult to make generalizations since each student is different, general trends and statistical analyses can help in understanding these relationships. Mathematical achievement is closely associated not only with students knowledge and skills but also with psychological factors. Hope refers to a person's willingness to strive toward desired goals and their perception of achieving those goals [11]. Conversely, hopelessness is based on past negative experiences and leads to anticipation of negative outcomes in the future [12]. Negative attitudes developed toward mathematics by students are one of the reasons for their failure in mathematics [13]. In education; [14] studies on the quantity of academic success based on teachers and students attitudes towards the course, The purpose of this study is to identify factors influencing students mathematical achievement and to create a model using fuzzy logic to predict students written exam results. Variables such as students levels of mathematical hopelessness, daily study durations, daily play durations, daily homework durations, and daily sleep durations will be examined for their effects on written exam results, and a prediction model will be developed. The importance of the study lies in better understanding the factors influencing students' mathematical achievement and in developing strategies to enhance students success by providing information to decision-makers in education. Concepts of hope and hopelessness can influence students attitudes toward mathematics and thus their success. This study aims to investigate the relationship between students' mathematical achievement and their levels of hopelessness using fuzzy logic modeling.

2. Material and Methods

2.1. *Data collection*

When examining the relationships between students mathematics achievement and their attitudes towards mathematics, it was found that these attitudes are associated with academic success. In this study, data collection focused on students study durations, daily playtimes, daily homework durations, and daily sleep durations. Additionally, students were administered the "Mathematics Hopelessness Scale." The "Mathematics Hopelessness Scale" consists of 35 items rated on a scale from 1 to 5. The verbal equivalents of these ratings are "definitely disagree," "disagree," "partially agree," "agree," and "definitely agree." The administered forms were evaluated and scored accordingly. The research was conducted among 10th and 11th grade students at Şehit Ali Hekim Anatolian High School in Isparta province during the 2023-2024 academic year. A total of 58 students participated in the study.

2.2. *Fuzzy Logic Approach*

In a fuzzy set framework, each element in the set is defined by assigning a value that represents its degree of membership in the set. Therefore, the membership of elements in the set can vary. Membership degrees are expressed as real numbers between 0, indicating non-membership, and 1, indicating full membership. A crisp set can be described as a special case of a fuzzy set where membership values are restricted to the values 0 and 1.

Definition 1. The set $A = \{(x, \mu_A(x)) : x \in X\}$ where X is a non-empty set and $\mu_A : X \to [0, 1]$ is a function, is called a fuzzy set on X.

Definition 2. A fuzzy number on $\mathbb{R}^n$ is fuzzy set A on $\mathbb{R}^n$ satisfying the following conditions:[17]

- A is normal fuzzy set, there exists an $x_0 \in \mathbb{R}^n$, such that $\mu_A(x_0) = 1$.
- A is a convex fuzzy set, i.e., for any $x, y \in \mathbb{R}^n$ and $0 \leq \lambda \leq 1, \mu_A(\lambda x + (1-\lambda)y) \geq \min\{\mu_A(x), \mu_A(y)\}$
- $\mu_A : \mathbb{R}^n \to [0, 1]$, is upper semicontinuous, that is, for any $y \in [0, 1], \mu_A^{-1}((-\infty, y))$ is open.
- The closure of $\{x \in \mathbb{R}^n : \mu_A(x) > 0\}$ is compact.

However, since it would be more useful to use certain types of functions in terms of compatibility with the studies, triangular membership function was used in this study. The definitions of two fuzzy number are given below. The general form of triangular membership function can be stated as

$$\mu(x; a, b, m) = \begin{cases} 0, & \text{if } x \leq a \\ \frac{x-a}{m-a}, & \text{if } a < x \leq m \\ \frac{b-x}{b-m}, & \text{if } m < x \leq b \\ 0, & \text{if } x > b \end{cases}$$

where a is the lower boundary, b is the upper boundary, and $a < m < b$. In applications, especially in machinery, planning, and engineering projects, precise data is crucial for comprehensive designs. In such cases, clarifying uncertain data is important. Artificial Intelligence plays a significant role today in fuzzy sets, fuzzy systems, and intelligence research. Results obtained from fuzzy systems are converted into precise data format through fuzzy inference operations, resulting in fuzzy sets. Converting fuzzy data into numerical data is referred to as defuzzification. Fuzzy logic rules are fundamental to fuzzy logic and are employed in modeling systems that deal with uncertainty. These rules establish relationships between inputs and outputs using fuzzy sets and membership functions, aiming to emulate human thought and decision-making processes [15]. Typically structured in an "IF-THEN" format, fuzzy rules specify that a specific outcome will result if certain conditions are met. Each condition and outcome is defined using fuzzy sets [16]. In the fuzzy inference process described by Zadeh (1965), fuzzy rules play a pivotal role and involve three main stages; First, Fuzzification is the process of transforming precise and crisp data into fuzzy sets, where membership degrees represent the degree of belongingness. In this process, exact and uncertain data are converted into linguistic terms or fuzzy sets to facilitate processing within fuzzy logic systems. Second, Rule evaluation involves the activation or assessment of the rule base in fuzzy logic systems. This step calculates the relationships between input values and output values based on fuzzy logic rules and inference mechanisms defined within the system. Each rule's contribution to the input and its impact on the output are evaluated. Third, Defuzzification is the process of converting fuzzy outputs obtained from fuzzy logic systems into crisp and precise numerical values. This process aims to translate fuzzy outputs, often represented as linguistic terms or fuzzy sets, into numerical outputs using methods such as weighted averages or centroid computation. The goal is to obtain a clear and numerical result from fuzzy outputs. These terms describe key steps in the operation and application of fuzzy logic systems, addressing how they handle uncertainty and model complex relationships using linguistic variables and fuzzy sets.

2.3. *ANFIS (Adaptive neuro-fuzzy inference system)*

Successful applications of ANFIS (Figure 1) have been reported in areas such as classification and data analysis.

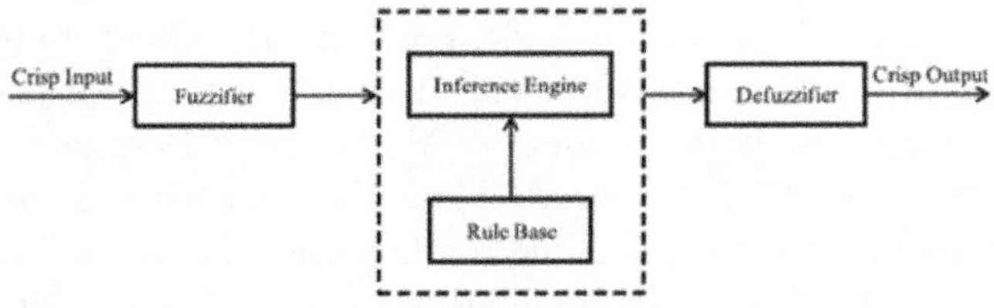

Fig. 1. Block diagram of a neuro-fuzzy (ANFİS) controller.

Firstly, we discuss the first-order Sugeno fuzzy model in ANFIS (Figure 2) with two inputs, p and q, and one output, r.

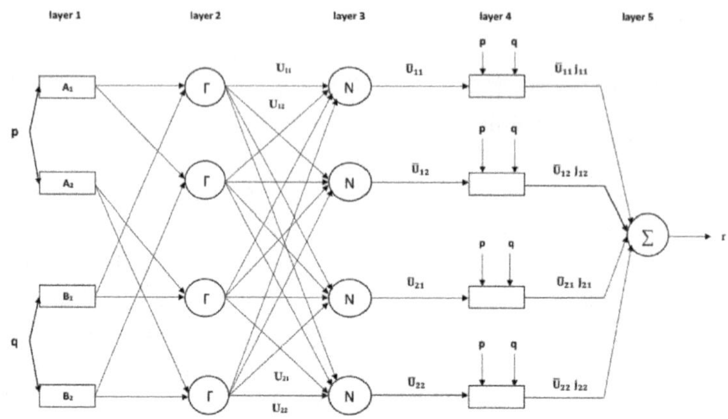

Fig. 2. General ANFIS structure for 2 inputs with 4 rules.

The first-order fuzzy model by Sugeno has four fuzzy if-then rules as follows:

$$R_1: \text{ if } p \text{ is } A_1 \text{ and } q \text{ is } B_1, \text{ then } J_{11} = l_{11}p + m_{11}q + n_{11},$$

$$R_2: \text{ if } p \text{ is } A_1 \text{ and } q \text{ is } B_2, \text{ then } J_{12} = l_{12}p + m_{12}q + n_{12},$$

$$R_3: \text{ if } p \text{ is } A_2 \text{ and } q \text{ is } B_1, \text{ then } J_{21} = l_{21}p + m_{21}q + n_{21},$$

$$R_4: \text{ if } p \text{ is } A_2 \text{ and } q \text{ is } B_2, \text{ then } J_{22} = l_{22}p + m_{22}q + n_{22},$$

Here, A_1, A_2, B_1, and B_2 are fuzzy sets for inputs p and q. $J_{ij}\,(i, j = 1, 2)$ represents outputs within the fuzzy region specified by fuzzy rules, and $l_{ij}, m_{ij}, n_{ij}\,(i, j = 1, 2)$ are design parameters for the inputs p and q, determined during the training process.

3. A New Fuzzy Logic Modelling

In this part of the study, a fuzzy logic-based prediction model was created to evaluate the success status of students in mathematics exams. ANFIS (Adaptive Neuro-Fuzzy Inference System) method was used in the model. ANFIS is a method that combines neural networks and fuzzy logic, providing the ability to learn and adapt from data (Figure 1). The model was constructed using the Sugeno fuzzy model. The ANFIS system consists of five inputs—Mathematics Despair Scale (M), Study Time (ST), Play Time (P), Homework Time (H), and Sleep Time (SL) and 243 fuzzy rules. The output is the predicted mathematics exam grade. The model was created using the MATLAB programming language. These input variables were fuzzified using appropriate membership functions and transformed into linguistic variables. The structure of the fuzzy inference system for two inputs and four rules is shown in Figure 2. After defining membership functions and determining fuzzy rules, the ANFIS training process was carried out using a hybrid learning algorithm. The performance of the model was evaluated by comparing the predicted and actual values. The results of the regression analysis between predicted and actual outputs are presented in Figure 4.

3.1. *Analaysis of Data*

The linguistic expressions on the applied evaluation scale were scored between 1 and 5. Using Matlab's Fuzzy Logic Toolbox, the fuzzy logic modeling step was initiated. During modeling, input variables such as study time, daily play time, daily homework time, daily sleep time, and mathematics hopelessness scale were defined, while student grades were used as outputs. Membership intervals, membership degrees, and membership functions were defined for input and output values to create a rule base. Defuzzification was performed to sharpen the fuzzy values.

3.2. *Construct of the A New Anfis for Data*

If we define the output of the i-th node in the j-th layer as ψ_{ij}, the node functions and operations of each layer in the ANFIS structure are as follows:

Fuzzification: This step involves converting linguistic variables into numerical data. Each linguistic variable is graded and associated with membership functions. The complexity of the model is determined by these membership functions; as the number of membership functions increases, the number of rules increases exponentially.

Inference Rules: The fuzzy inference system (FIS) consists of fuzzy rules that define the relationships between inputs and outputs. Some of the model rules used by ANFIS are depicted in the image below.

Defuzzification: The fuzzy sets obtained from the fuzzy inference operations do not directly correspond to real-life values. Therefore, defuzzification is necessary to convert fuzzy results into numerical data. In this stage, the fuzzy sets are translated into numerical values. The ANFIS architecture of our model is shown in Figure 3.

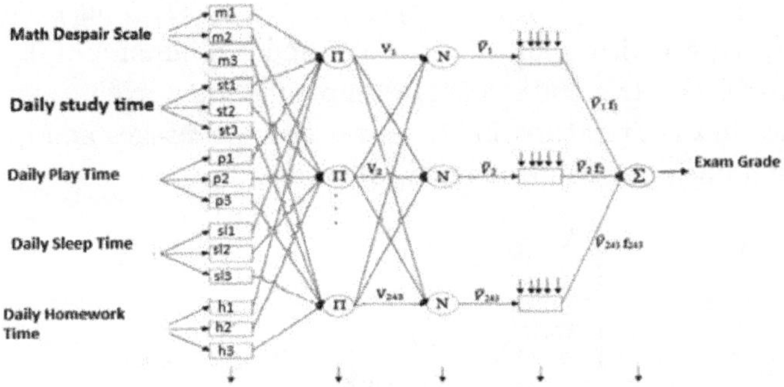

Fig. 3. 243-rule 5-input ANFIS system.

Round and square shapes are used to represent fixed and adaptive nodes, respectively. The model consists of five layers. A detailed explanation of the abstract layers is provided below.

Step 1: *Membership Layer*: The fuzzification layer is the first step of ANFIS, and its primary purpose is to partition the input values into fuzzy sets. This layer defines membership functions using a generalized bell-shaped activation function. Each node's output generates membership degrees based on the input values and the membership function used. This stage employs triangular or bell-shaped membership functions (MF) to organize inputs. Membership functions indicate the degree to which a particular input value belongs to a specific set. This layer is responsible for transmitting the input values, to the next layer. All inputs (i = 1, ..., n), (j = 1, ..., n), (k = 1, ..., n) are assigned to the appropriate fuzzy sets, processed in a meaningful and organized manner. This process ensures accurate and flexible modeling, especially in systems with uncertainty. For example, in a temperature control system, input values are temperature measurements processed by assigning them to fuzzy sets like "low," "medium," and "high." This approach better manages system uncertainty and provides more precise control [15].

$$\psi_{(1,i)} = \mu_{m_i}(M), \quad i = 1,2,3$$

$$\psi_{(1,i)} = \mu_{st_i}(St), \quad i = 1,2,3$$

$$\psi_{(1,i)} = \mu_{p_i}(P), \quad i = 1,2,3$$

$$\psi_{(1,i)} = \mu_{sl_i}(Sl), \quad i = 1,2,3$$

$$\psi_{(1,i)} = \mu_{h_i}(H), \quad i = 1,2,3$$

where M, St, P, Sl, and H are the inputs at the node, and m_i, mst_i, p_i, sl_i, and h_i are the linguistic labels. That is, $\psi_{(1,i)}$ is the membership function, and m_i, mst_i, p_i, sl_i, and h_i represent the degrees of membership of M, St, P, Sl, and H in the membership functions $\mu_{m_i}(M)$, $\mu_{st_i}(St)$, $\mu_{p_i}(P)$, $\mu_{sl_i}(Sl)$, and $\mu_{h_i}(H)$. These are based on an equation with a linear curve of 1 and a minimum of 0.

In this study, the triangular membership function (given below) is the most frequently used in the literature due to its various advantages in approximating the parameters of a linear curve membership function.

$$f(x) = \begin{cases} 0, & x \leq r_1, \\ \frac{x-r_1}{r_2-r_1}, & r_1 < x \leq r_2, \\ \frac{r_3-x}{r_3-r_2}, & r_2 < x \leq r_3, \\ 0, & x \geq r_3. \end{cases} \qquad r_1, r_2, r_3 \in \mathbb{R}$$

Here, r_1, r_2, and r_3 are the parameters that define the membership functions.
Step 2: *Rule Layer*: The rule layer forms the second step of ANFIS, and each node in this layer is fixed, meaning they are non-adaptive nodes. Each node performs the precondition matching of the fuzzy rules and calculates the activation level of each rule. This layer determines the pseudo firing strength of each rule by calculating normalized weights. The output of the layer is obtained by taking the product of

all inputs entering the layer. The number of nodes in the rule layer equals the number of fuzzy rules, and each node represents the rules formed according to the Sugeno fuzzy inference system. The output of each rule node is the product of the membership degrees from the first layer. This process forms the basis of rule-based modeling, determining how the system responds to various inputs and optimizing the overall performance of the system.

$$\psi_{(1,i)} = V_i = m_i(M) \cdot st_i(St) \cdot p_i(P) \cdot sl_i(Sl) \cdot h_i(H), \quad i = 1,2,3$$

Step 3: *Normalization Layer*: The normalization layer constitutes the third step of ANFIS, and each node in this layer is fixed, meaning they are non-adaptive nodes. This layer represents the normalized firing strength, which is obtained by taking the ratio of each output node in the previous layer to the sum of all outputs from the previous layer. Each node accepts all nodes from the rule layer as input values and provides the normalized value for each rule. This layer standardizes the system's output, optimizing the response to the inputs.

$$\psi_{(3,i)} = \bar{V_i} = \frac{V_i}{\sum_{k=1}^{243} V_k}, \quad i = 1,2,\ldots,243$$

Step 4: *Defuzzification*: Every node in this layer is represented by square shape that is every node is an adoptive node. The output of this layer is obtained by taking the product of the normalized firing strength and a first order polynomial. where V_i is the output of Layer (3), and $m_i, st_i, p_i, sl_i, h_i$ are the consequent parameters, and u_i is a constant.

Step 5: *Addition*: The aggregation layer constitutes the fifth step of ANFIS. There is only one node in this layer, labeled Σ. This node performs the summation of all incoming signals (inputs). The overall output of the ANFIS system is obtained by summing the output value of each node in the fourth layer.

$$\psi_{(5,i)} = \sum_i \left(\bar{V_i} f_i\right) = \frac{\sum_i (V_i f_i)}{\sum_i V_i}, \quad i = 1,2,\ldots,243$$

The ANFIS structure is automatically adjusted by least squares estimation and backpropagation algorithm. This algorithm will be used in the next section to develop an ANFIS controller to control various parameters of the induction motor. The flexibility of ANFIS can be used in a wide range of control applications.

4. Result

According to the regression analysis results of our model, which was developed to evaluate the real-world performance of students in mathematics education, an 80% success rate was achieved between the grades received by the students and the results predicted by the model. This finding shows that our model can be used as an effective tool in mathematics education. This approach, developed to support students' mathematics learning processes, has the potential to understand individual learning needs and provide customized instruction accordingly. The importance of using fuzzy logic-based models in mathematics education is once again emphasized with this study. In addition The research we conducted in this study was conducted among a total of 58 students in the 10th and 11th grades at Şehit Ali Hekim Anatolian High School in Isparta province in the 2023–2024 academic year. The data obtained from 10 students randomly selected from these 58 students were used as test data. The data obtained from the remaining 48 students were mathematically modeled using the Adaptive Neuro-Fuzzy Inference System (ANFIS) method. The results obtained from the existing real data for 10 students were compared with the results obtained from the model using regression analysis (Figure 4). As a result, it was shown that they were very close to each other at a rate of 80%. This rate is a measure of the reliability and consistency of our model at a rate of 80%.

In addition, when scientific studies in the field of engineering are examined, it will be seen that new modern methods and modeling techniques are put forward for new world problems and the new data sets they contain. It is also known that the innovation sometimes put forward in such studies is on the development of previously known methods. However, on today, especially in modeling studies on existing data sets in physics, chemistry, biology and educational sciences, studies using the fuzzy inference system are quite few and its popularity has been increasing recently. Especially in our study, the fact that this modeling is done in educational sciences for five inputs using the fuzzy inference system is quite new for the literature.

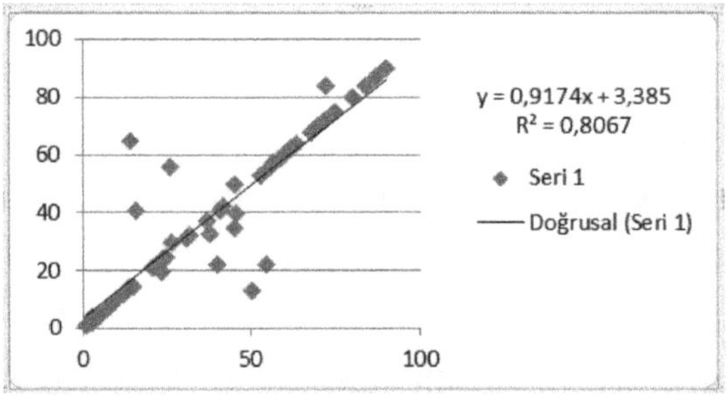

Fig. 4. Regression Analysis.

5. Conclusions

The regression analysis results used in evaluating mathematics achievements reveal the accuracy and reliability of our model. It should be emphasized that fuzzy logic-based models are a flexible and effective tool in terms of understanding the individual differences of students and adjusting the teaching processes according to these differences. These analysis results show the importance of adopting and applying fuzzy logic methods in mathematics education. Future research needs to evaluate the broader applicability and effectiveness of these methods. Such studies can contribute to the development of innovative approaches in mathematics teaching and the improvement of existing practices.

References

[1] S. Yıldırım and A.Işık, (2014). Öğrencilerin Matematik Başarısını Etkileyen Faktörler, *Eğitim ve Bilim*, 39(175), 322–337.

[2] D. Dubois and H. Prade, *Fuzzy Sets and Systems: Theory and Applications.* (New York: *Academic Press*, 1980).

[3] S. Saltan, S. Yıldırım and A. Işık, (2007). Fuzzy Logic Applications in Engineering, *IEEE Transactions on Fuzzy Systems*, 15(1), 45–56.

[4] M. Sakawa, *Fuzzy Sets and Interactive Multiobjective Optimization*, (New York: *Plenum Press*, 1993).

[5] T. Takagi and M. Sugeno, (1985). Fuzzy Identification of Systems and Its Applications to Modeling and Control, *IEEE Transactions on Systems, Man, and Cybernetics*, 15(1), 116–132.

[6] M. Sugeno and T. Takagi, (1983). Fuzzy Identification of Systems and Its Applications to Modeling and Control, *IEEE Transactions on Systems, Man, and Cybernetics*, 15(1), 116–132.

[7] A. Şahiner, R. Akbay and N. Yılmaz (2019). Fuzzy Logic Modeling and Optimization of Academic Achievement of Students, *Yıldız Journal of Educational Research*, 4(1), 85–100.

[8] E. Türk, D. Yıldırım, Ö. Görmez, A. Sahiner and G. Kapusuz (2021). Age estimation on CBCT images of anterior teeth using a fuzzy logic model. In Proceedings of the 1st Genç ORAD Symposium (Abstract paper, p. 102). Gazimağusa, Northern Cyprus: Oral Diagnosis and Maxillofacial Radiology Association (ODMFR).

[9] J. S. R. Jang (1993). ANFIS: Adaptive-network based fuzzy inference systems. *IEEE Trans. on Systems, Man and Cybernetics*, 23, 665–685.

[10] J. S. R. Jang, C. T. Sun and E. Mizutani (1997). *Neuro-fuzzy and soft computing, a computational approach to learning and machine intelligence*, (Prentice Hall, New Jersey, 1997).

[11] C. R. Snyder, S. S. Ilardi, J. Cheavens and S. T. Michael (2000). The Role of Hope in Cognitive-Behavior Therapies, *Cognitive Therapy and Research*, 24(6), 747–762.

[12] N. H. Şahin (2009). Umutsuzluk ve Depresyon: Klinik Bir İnceleme, *Psikiyatri Dergisi*, 21(3), 15–23.

[13] Y. Baykul, *İlköğretimde Matematik Öğretimi*, (Ankara: *Pegem Yayıncılık*, 1990).

[14] A. Şahiner, F. Ucun, G.Kapusuz and N. Yılmaz (2016). Completed Optimised Structure of Threonine Molecule by Fuzzy Logic Modelling, *Zeitschrift für Naturforschung A*, 71(4), 381–386. DOI: 10.1515/zna-2015-0424.

[15] L. A. Zadeh, (1965). Fuzzy Sets, *Information and Control*, 8, 338–353.

[16] G. J Klir and B. Yuan, (1995). *Fuzzy Sets and Fuzzy Logic: Theory and Applications.* (Prentice Hall, New Jersey, 1995).

[17] E. Savas, (2008). On $\bar{\lambda}$ statistically convergent double sequences of fuzzy numbers, *J. Inequal. Appl.*, 147827, 6.

Chapter 13

Variational Iteration Method for Space Fractional Schrödinger Equation

Arzu Guleroglu* and Ece Gunes Zarali†

Department of Mathematics, Trakya University, Edirne, Turkey

* *arzuguleroglu@trakya.edu.tr*

† *ecegunes@trakya.edu.tr*

Fractional Schrödinger equations have attracted much attention from many researchers for their important applications in fractional quantum mechanics. In this study, linear and nonlinear space fractional Schrödinger equations are considered. The variational iteration method is implemented for approximate solutions of linear and nonlinear space fractional Schrödinger equations. Different correctional functionals are constructed with Lagrange multipliers. The correctional functionals are also constructed without handling the space fractional derivative term as a restricted variation. The Lagrange multipliers are identified by the variational theory. Using the identified Lagrange multipliers, different iteration formulas are formed. Example problems for the linear and nonlinear space fractional Schrödinger equations are given to illustrate the efficiency of the variational iteration method. The numerical results show that the variational iteration method is efficient in searching for approximate solutions of the linear and nonlinear space fractional Schrödinger equations. Moreover, depending on the choice of iteration formula, the variational iteration method gives approximate solutions that converge to the exact solution more rapidly.

1. Introduction

Differential equations emerge as mathematical modeling of many problems encountered in science, engineering, and other applied sciences. The equations that best describe most real-world processes are differential equations that involve fractional order derivatives. The basic mathematical model that gives information about a quantum system in mathematical physics is the Schrödinger equation. By expressing the Schrödinger equation using fractional derivatives, space fractional, time fractional, and space-time fractional Schrödinger equations were formulated. The space fractional Schrödinger equation discovered by Laskin [1–3] is the fundamental equation of fractional quantum mechanics. In this paper, we consider the nonlinear space fractional Schrödinger equation (SFSE)

$$\left(i\hbar\frac{\partial}{\partial t}+\gamma_{RZ}D_x^{2\alpha}-V(x,t)-\kappa|\psi(x,t)|^2\right)\psi(x,t)=0,\quad t>0,\, x\in(a,b)\,. \tag{1}$$

Here, ${}_{RZ}D_x^{2\alpha}=-(-\triangle)^\alpha$ is the Riesz fractional derivative $\left(\frac{1}{2}<\alpha\leq 1\right)$, the parameter $\gamma\in\mathbb{R}$, $\gamma>0$ is the group velocity dispersion, $\hbar$ is the Planck constant, V is the arbitrary potential function, κ is the real parameter, and ψ is the solution function. For $\kappa=0$, the linear space fractional Schrödinger equation is recovered.

Various numerical methods have been applied for solving linear and nonlinear SFSEs, such as spectral methods [4, 5], finite element methods [6, 7], and finite difference methods [8, 9]. In the present study, the variational iteration method (VIM) is applied to solve linear and nonlinear SFSE. Different iteration formulas are constructed to provide approximations that converge faster to the exact solution. We present numerical examples to show the performance of the proposed iteration formulas.

2. Preliminaries

Definition 1. [10, 11] The left-sided and right-sided Riemann-Liouville fractional integrals ${}_{RL}I_{a,x}^\alpha$, and ${}_{RL}I_{x,b}^\alpha$ of order $\alpha\in\mathbb{C}$ are defined by

$${}_{RL}I_{a,x}^\alpha u(x)=\frac{1}{\Gamma(\alpha)}\int_a^x\frac{u(\tau)}{(x-\tau)^{1-\alpha}}d\tau,\quad x>a,$$

and

$${}_{RL}I_{x,b}^\alpha u(x)=\frac{1}{\Gamma(\alpha)}\int_x^b\frac{u(\tau)}{(\tau-x)^{1-\alpha}}d\tau,\quad x<b,$$

respectively ($\mathrm{Re}(\alpha)>0$).

Definition 2. [11, 12] The Riesz fractional integration ${}_{RZ}I^\alpha$ is realized via the Riesz potential,

$${}_{RZ}I^\alpha u(x)=\frac{1}{2\Gamma(\alpha)\cos\left(\frac{\alpha\pi}{2}\right)}\int_{-\infty}^{\infty}\frac{1}{|x-\tau|^{1-\alpha}}u(\tau)d\tau,\quad x\in\mathbb{R},0<\alpha<1.$$

Proposition 1. [12] The Riemann-Liouville fractional integral and the Riesz potential are connected as follows

$${}_{RZ}I^\alpha u(x)=\frac{1}{2\cos\left(\frac{\alpha\pi}{2}\right)}\left[{}_{RL}I_{-\infty,x}^\alpha u(x)+{}_{RL}I_{x,\infty}^\alpha u(x)\right],\quad x\in\mathbb{R},0<\alpha<1.$$

Definition 3. [11] The Riemann-Liouville fractional derivatives ${}_{RL}D_{a,x}^\alpha$ and ${}_{RL}D_{x,b}^\alpha$ of order $\alpha\in\mathbb{C}$ are defined by

$${}_{RL}D_{a,x}^\alpha u(x)=\frac{1}{\Gamma(n-\alpha)}\left(\frac{d}{dx}\right)^n\int_a^x\frac{u(\tau)}{(x-\tau)^{\alpha-n+1}}d\tau,\quad x>a,$$

and

$${}_{RL}D_{x,b}^\alpha u(x)=\frac{1}{\Gamma(n-\alpha)}\left(-\frac{d}{dx}\right)^n\int_x^b\frac{u(\tau)}{(\tau-x)^{\alpha-n+1}}d\tau,\quad x<b,$$

respectively, where $\mathrm{Re}(\alpha) \geq 0$, $n = [\mathrm{Re}(\alpha)] + 1$, $[\mathrm{Re}(\alpha)]$ means the integral part of $\mathrm{Re}(\alpha)$.

Definition 4. [13] The Riesz fractional derivative ${}_{RZ}D^{\alpha}$ is defined by

$$ {}_{RZ}D^{\alpha}u(x) = \frac{1}{2\Gamma(-\alpha)\cos\left(\frac{\alpha\pi}{2}\right)} \int_{-\infty}^{\infty} \frac{u(x-\tau)-u(x)}{|\tau|^{1+\alpha}} d\tau, \quad 0<\alpha<2, \alpha \neq 1. $$

Proposition 2. [14] The Riemann-Liouville fractional derivative and Riesz fractional derivative are connected as follows

$$ {}_{RZ}D^{\alpha}u(x) = -\frac{1}{2\cos\left(\frac{\alpha\pi}{2}\right)} \left[{}_{RL}D^{\alpha}_{a,x}u(x) + {}_{RL}D^{\alpha}_{x,b}u(x) \right], \quad a<x<b, $$

where $\alpha \in (n-1, n)$ with $n \in \mathbb{Z}^{+}$.

Definition 5. [15] Let u denote a continuous function. The modified Riemann-Liouville fractional derivative ${}_{MRL}D^{\alpha}_{0,x}$ of order α is defined by

$$ {}_{MRL}D^{\alpha}_{0,x}u(x) = \begin{cases} \frac{1}{\Gamma(-\alpha)} \int_0^x \frac{u(\tau)}{(x-\tau)^{\alpha+1}} d\tau, & \alpha<0 \\ \frac{1}{\Gamma(1-\alpha)} \frac{d}{dx} \int_0^x \frac{u(\tau)-u(0)}{(x-\tau)^{\alpha}} d\tau, & 0<\alpha<1 \\ \frac{1}{\Gamma(n-\alpha)} \frac{d^n}{dx^n} \int_0^x \frac{u(\tau)-u(0)}{(x-\tau)^{\alpha-n}} d\tau, & n \leq \alpha < n+1, n \geq 1. \end{cases} $$

Proposition 3. [15, 16] Let $0<\alpha<1$ and u, w with all their derivatives are continuous on a finite interval $[a, b]$.

$$ \begin{aligned} {}_{MRL}D^{\alpha}\left[u(x)w(x)\right] &= w(x)\,{}_{MRL}D^{\alpha}u(x) + u(x)\,{}_{MRL}D^{\alpha}w(x), \\ {}_{RL}I^{\alpha}_{a,b}\left[w(x)\,{}_{MRL}D^{\alpha}u(x)\right] &= u(x)w(x)\big|_a^b - {}_{RL}I^{\alpha}_{a,b}\left[u(x)\,{}_{MRL}D^{\alpha}w(x)\right]. \end{aligned} $$

3. Variational Iteration Method

J. H. He presented VIM to solve differential equation

$$ (L+N)\omega(t) = u(t) \tag{2} $$

where ω is the unknown solution function, and u is the known analytical function [17]. L and N are linear and nonlinear operators. For Eq. (2), correction functional is expressed as

$$ \omega_{n+1}(t) = \omega_n(t) + \int_0^t \lambda(\tau)\left(L\omega_n(\tau) + N\widetilde{\omega_n}(\tau) - u(\tau)\right) d\tau, \quad n \geq 0 \tag{3} $$

where ω_n is the nth approximate solution, λ is Lagrange multiplier, and $\widetilde{\omega_n}$ is restricted variation, i.e., $\delta\widetilde{\omega_n} = 0$. λ is determined optimally using variational theory. Then, substituting λ into correction functional (3) gives an iteration formula. The approximations ω_n are obtained by using selective function ω_0 and the iteration formula. The solution is

$$ \omega(t) = \lim_{n \to \infty} \omega_n(t). $$

4. Iteration Formulas of VIM for SFSE

For Eq. (1), we construct the correction functionals as follows

$$\psi_{l+1}(x,t) = \psi_l(x,t) + \int_0^t \lambda_I(\tau)\left(i\frac{\partial\psi_l(x,\tau)}{\partial\tau} + \left(\frac{\beta}{2}\left[{}_{MRL}D^{2\alpha}_{a,x} + {}_{MRL}D^{2\alpha}_{x,b}\right]\right.\right.$$
$$\left.\left. - V(x,\tau) - \kappa\left|\widetilde{\psi}_l(x,\tau)\right|^2\right)\widetilde{\psi}_l(x,\tau)\right)d\tau, \tag{4}$$

and

$$\psi_{l+1}(x,t) = \psi_l(x,t) + \frac{1}{\Gamma(\alpha)}\int_0^x \frac{\lambda_{II}(\eta)}{(x-\eta)^{1-\alpha}}\left(\frac{\beta}{2}\left[{}_{MRL}D^{2\alpha}_{a,\eta} + {}_{MRL}D^{2\alpha}_{\eta,b}\right]\psi_l(\eta,t)\right.$$
$$\left. + \left(i\frac{\partial}{\partial t} - V(\eta,t) - \kappa\left|\widetilde{\psi}_l(\eta,t)\right|^2\right)\widetilde{\psi}_l(\eta,t)\right)d\eta, \tag{5}$$

$l \geq 0$, where λ_I and λ_{II} are Lagrange multipliers, $\beta = -\frac{\gamma}{cos(\alpha\pi)}$, $\frac{1}{2} < \alpha < 1$. With known $\delta\widetilde{\psi}_l = 0$, the stationary conditions are

$$1 + i\lambda_I(\tau)|_{\tau=t} = 0,$$
$$\frac{d\lambda_I(\tau)}{d\tau} = 0, \tag{6}$$

and

$$\lambda_{II}(\eta)|_{\eta=x} = 0,$$
$$1 - \frac{\beta}{2}\left[{}_{MRL}D^{\alpha}_{a,\eta} + {}_{MRL}D^{\alpha}_{\eta,b}\right]\lambda_{II}(\eta)\Big|_{\eta=x} = 0, \tag{7}$$
$$\left[{}_{MRL}D^{2\alpha}_{a,\eta} + {}_{MRL}D^{2\alpha}_{\eta,b}\right]\lambda_{II}(\eta) = 0.$$

Hence,

$$\lambda_I(\tau) = i, \lambda_{II}(\eta) = \frac{(\eta-x)^\alpha}{\beta\Gamma(1+\alpha)}. \tag{8}$$

If we substitute the values (8) into functionals (4)–(5), we get the following iteration formulas

Formula I

$$\psi_{l+1}(x,t) = \psi_l(x,t) + i\int_0^t \left(i\frac{\partial}{\partial\tau} + \frac{\beta}{2}\left[{}_{MRL}D^{2\alpha}_{a,x} + {}_{MRL}D^{2\alpha}_{x,b}\right]\right.$$
$$\left. - V(x,\tau) - \kappa|\psi_l(x,\tau)|^2\right)\psi_l(x,\tau)d\tau,\ l \geq 0, \tag{9}$$

and

Formula II

$$\psi_{l+1}(x,t) = \psi_l(x,t) + \frac{1}{\alpha\beta\Gamma^2(\alpha)}\int_0^x \frac{(\eta-x)^\alpha}{(x-\eta)^{1-\alpha}}\left(i\frac{\partial}{\partial t}\right.$$
$$\left. + \frac{\beta}{2}\left[{}_{MRL}D^{2\alpha}_{a,\eta} + {}_{MRL}D^{2\alpha}_{\eta,b}\right] - V(\eta,t) - \kappa|\psi_l(\eta,t)|^2\right)\psi_l(\eta,t)d\eta,\ l \geq 0. \tag{10}$$

The solution is

$$\psi(x,t) = \lim_{l\to\infty}\psi_l(x,t).$$

5. Numerical Examples

We consider numerical examples to test the performance of the presented iteration formulas in solving linear and nonlinear SFSEs.

Example 1. Consider the following problem for linear SFSE

$$\left(i\frac{\partial}{\partial t} - \frac{cos(\alpha\pi)}{2}{}_{RZ}D_x^{2\alpha}\right)\psi(x,t) = 0, (x,t) \in \Omega \tag{11}$$

$$\psi(x,0) = sinx,$$

$$\psi(0,t) = 0, \psi_x(0,t) = e^{-\frac{it}{2}},$$

where $\frac{1}{2} < \alpha \leq 1, \Omega = (0,1) \times (0,1]$.

Formula I

$$\psi_{l+1}(x,t) = \psi_l(x,t) + i\int_0^t \left(i\frac{\partial}{\partial \tau} + \frac{1}{4}\left[{}_{MRL}D_{0,x}^{2\alpha} + {}_{MRL}D_{x,1}^{2\alpha}\right]\right)\psi_l(x,\tau)d\tau,\ l \geq 0. \tag{12}$$

Starting with $\psi_0(x,t) = sin_\alpha(x^\alpha)$ and using Formula I given in Eq. (12), we get the following approximations

$$\psi_1(x,t) = \left(1 - \frac{it}{2}\right) sin_\alpha(x^\alpha),$$

$$\psi_2(x,t) = \left(1 - \frac{it}{2} + \frac{(it)^2}{8}\right) sin_\alpha(x^\alpha),$$

$$\vdots$$

$$\psi_l(x,t) = \left(1 - \frac{it}{2} + \cdots + \frac{(-it)^l}{2^l l!}\right) sin_\alpha(x^\alpha), \tag{13}$$

$l \geq 1$, where

$$sin_\alpha(x^\alpha) = \sum_{l=0}^{\infty} \frac{(-1)^l x^{\alpha(2l+1)}}{\Gamma(1+\alpha(2l+1))}.$$

The closed form of the solution is

$$\psi(x,t) = \lim_{l\to\infty} \psi_l(x,t) = e^{-\frac{it}{2}} sin_\alpha(x^\alpha), \tag{14}$$

which is the exact solution of Eq. (11).

Formula II

$$\psi_{l+1}(x,t) = \psi_l(x,t) + \frac{2}{\alpha\Gamma^2(\alpha)}\int_0^x \frac{(\eta-x)^\alpha}{(x-\eta)^{1-\alpha}}\left(i\frac{\partial}{\partial t} + \frac{1}{4}\left[{}_{MRL}D_{0,\eta}^{2\alpha} + {}_{MRL}D_{\eta,1}^{2\alpha}\right]\right)\psi_l(\eta,t)d\eta,\ l \geq 0. \tag{15}$$

We begin with

$$\psi_0(x,t) = \frac{x^\alpha}{\Gamma(1+\alpha)}e^{-\frac{it}{2}}.$$

By Formula II given in Eq. (15), the following approximations are obtained

$$\psi_1(x,t) = \left(\frac{x^{\alpha}}{\Gamma(1+\alpha)} - \frac{x^{3\alpha}}{\Gamma(1+3\alpha)}\right) e^{-\frac{it}{2}},$$

$$\psi_2(x,t) = \left(\frac{x^{\alpha}}{\Gamma(1+\alpha)} - \frac{x^{3\alpha}}{\Gamma(1+3\alpha)} + \frac{x^{5\alpha}}{\Gamma(1+5\alpha)}\right) e^{-\frac{it}{2}},$$

$$\vdots$$

$$\psi_l(x,t) = \left(\frac{x^{\alpha}}{\Gamma(1+\alpha)} - \frac{x^{3\alpha}}{\Gamma(1+3\alpha)} + \cdots + \frac{(-1)^l x^{\alpha(2l+1)}}{\Gamma(1+\alpha(2l+1))}\right) e^{-\frac{it}{2}}, \qquad (16)$$

$l \geq 1$. Taking the limit of solution (16), we get the exact solution (14).

To verify whether the choice of Formula I given in Eq. (12), and Formula II given in Eq. (15) produces approximations that converge faster to exact solution ψ given in Eq. (14), Figures 1–4 are presented with $\alpha = 0.75$. The approximations ψ_I and ψ_{II} are obtained from approximations ψ_l given in (13) and (16) for $l = 2$, respectively. In Figure 1 (Figure 2), the real (imaginary) parts of the exact solution ψ and approximations ψ_I, ψ_{II} are plotted in (a)–(c). Figure 3 (Figure 4) shows the real (imaginary) parts of the absolute errors $|\psi - \psi_I|$ and $|\psi - \psi_{II}|$ are given in (a)–(b), respectively. When Figures 1–4 are examined, it is seen that approximation ψ_{II} converges to exact solution ψ faster than approximation ψ_I.

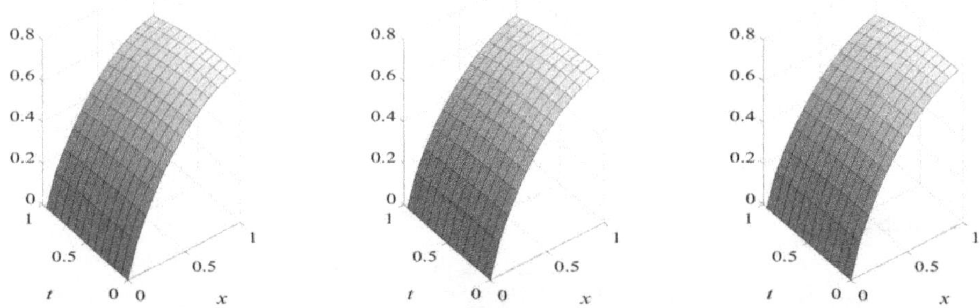

(a) Re[ψ] (b) Re[ψ_I] (c) Re[ψ_{II}]

Fig. 1. Graphs of the real parts of the solutions ψ, ψ_I, and ψ_{II} with $\alpha = 0.75$.

Example 2. Consider the following problem for nonlinear SFSE

$$\left(i\frac{\partial}{\partial t} - \frac{cos(\alpha\pi)}{2}\,{}_{RZ}D_x^{2\alpha} - cos^2 x - |\psi(x,t)|^2\right)\psi(x,t) = 0, (x,t) \in \Omega \qquad (17)$$

$$\psi(x,0) = sinx,$$

$$\psi(0,t) = 0, \psi_x(0,t) = e^{-\frac{3it}{2}},$$

where $\frac{1}{2} < \alpha \leq 1, \Omega = (0,1) \times (0,1]$.

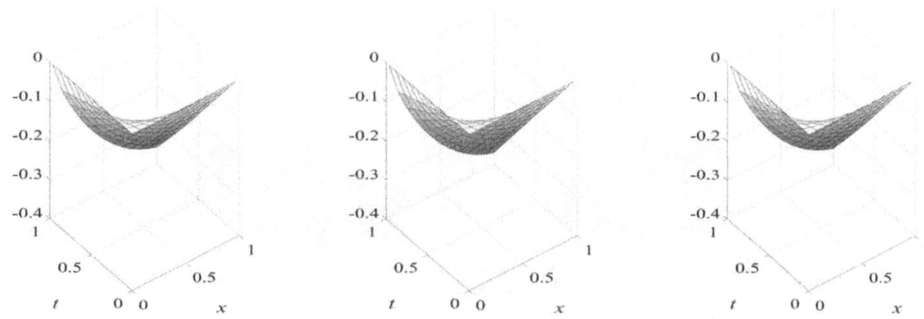

(a) $\mathrm{Im}[\psi]$ (b) $\mathrm{Im}[\psi_I]$ (c) $\mathrm{Im}[\psi_{II}]$

Fig. 2. Graphs of the imaginary parts of the solutions ψ, ψ_I, and ψ_{II} with $\alpha = 0.75$.

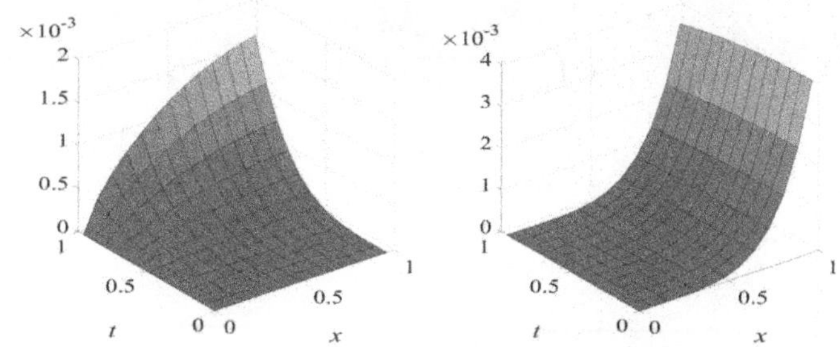

(a) $\mathrm{Re}[|\psi - \psi_I|]$ (b) $\mathrm{Re}[|\psi - \psi_{II}|]$

Fig. 3. Graphs of the real parts of absolute errors $|\psi - \psi_I|$ and $|\psi - \psi_{II}|$ with $\alpha = 0.75$.

Formula I

$$\psi_{l+1}(x,t) = \psi_l(x,t) + i\int_0^t \left(i\frac{\partial}{\partial\tau} + \frac{1}{4}\left[{}_{MRL}D^{2\alpha}_{0,x} + {}_{MRL}D^{2\alpha}_{x,1}\right] - cos^2 x - |\psi_l(x,\tau)|^2 \right) \psi_l(x,\tau)d\tau, \; l \geq 0. \tag{18}$$

Starting with $\psi_0(x,t) = sin_\alpha(x^\alpha)$ and using Formula I given in Eq. (18), we have

$$\psi_1(x,t) = \left(1 - \frac{3it}{2}\right) sin_\alpha(x^\alpha),$$

$$\psi_2(x,t) = \left(1 - \frac{3it}{2} + \frac{(3it)^2}{8}\right) sin_\alpha(x^\alpha),$$

$$\vdots$$

$$\psi_l(x,t) = \left(1 - \frac{3it}{2} + \cdots + \frac{(-3it)^l}{2^l l!}\right) sin_\alpha(x^\alpha), \tag{19}$$

$l \geq 1$. Thus,

$$\psi(x,t) = \lim_{l\to\infty} \psi_l(x,t) = e^{-\frac{3it}{2}} sin_\alpha(x^\alpha), \tag{20}$$

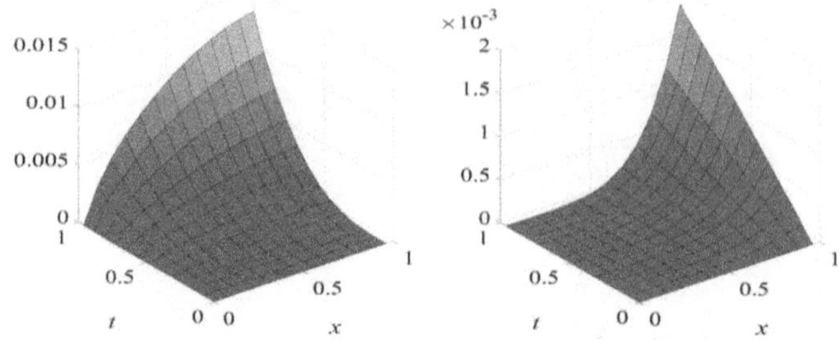

(a) $\mathrm{Im}[|\psi - \psi_I|]$ (b) $\mathrm{Im}[|\psi - \psi_{II}|]$

Fig. 4. Graphs of the imaginary parts of absolute errors $|\psi - \psi_I|$ and $|\psi - \psi_{II}|$ with $\alpha = 0.75$.

which is the exact solution of Eq. (17).

Formula II

$$\psi_{l+1}(x,t) = \psi_l(x,t) + \frac{2}{\alpha\Gamma^2(\alpha)} \int_0^x \frac{(\eta - x)^\alpha}{(x-\eta)^{1-\alpha}} \Bigg(i\frac{\partial}{\partial t} + \frac{1}{4}\left[{}_{MRL}D^{2\alpha}_{0,\eta} + {}_{MRL}D^{2\alpha}_{\eta,1}\right] - cos^2 x - |\psi_l(\eta,t)|^2 \Bigg) \psi_l(\eta,t)d\eta,\ l \geq 0. \tag{21}$$

We start with

$$\psi_0(x,t) = \frac{x^\alpha}{\Gamma(1+\alpha)} e^{-\frac{3it}{2}}.$$

Using Formula II given in Eq. (21), we obtain the following approximate solutions

$$\psi_1(x,t) = \left(\frac{x^\alpha}{\Gamma(1+\alpha)} - \frac{x^{3\alpha}}{\Gamma(1+3\alpha)} \right) e^{-\frac{3it}{2}},$$

$$\psi_2(x,t) = \left(\frac{x^\alpha}{\Gamma(1+\alpha)} - \frac{x^{3\alpha}}{\Gamma(1+3\alpha)} + \frac{x^{5\alpha}}{\Gamma(1+5\alpha)} \right) e^{-\frac{3it}{2}},$$

$$\vdots$$

$$\psi_l(x,t) = \left(\frac{x^\alpha}{\Gamma(1+\alpha)} - \frac{x^{3\alpha}}{\Gamma(1+3\alpha)} + \cdots + \frac{(-1)^l x^{\alpha(2l+1)}}{\Gamma(1+\alpha(2l+1))} \right) e^{-\frac{3it}{2}}, \tag{22}$$

$l \geq 1$. The series solution

$$\psi(x,t) = e^{-\frac{3it}{2}} \sum_{l=0}^{\infty} \frac{(-1)^l x^{\alpha(2l+1)}}{\Gamma(1+\alpha(2l+1))}$$

converge to the exact solution (20) of Eq. (17).

Figures 5–8 are presented to show the performance of Formula I, given in Eq. (18), and Formula II, given in Eq. (21). The approximations ψ_I and ψ_{II} are obtained from approximations ψ_l given in Eqs. (19) and (22) for $l = 2$, respectively. The exact solution ψ given in Eq. (20) and approximations ψ_I and ψ_{II} are considered with $\alpha = 0.75$. In Figure 5 (Figure 6), the graphs of the real (imaginary) parts of the exact solution ψ and approximations ψ_I, ψ_{II} are in (a)–(c). Figure 7 (Figure 8) shows the real (imaginary) parts of the absolute errors $|\psi - \psi_I|$ and $|\psi - \psi_{II}|$ are given in (a)–(b), respectively. Examining Figures 5–8, it is observed that approximation ψ_{II} converges to exact solution ψ faster than approximation ψ_I.

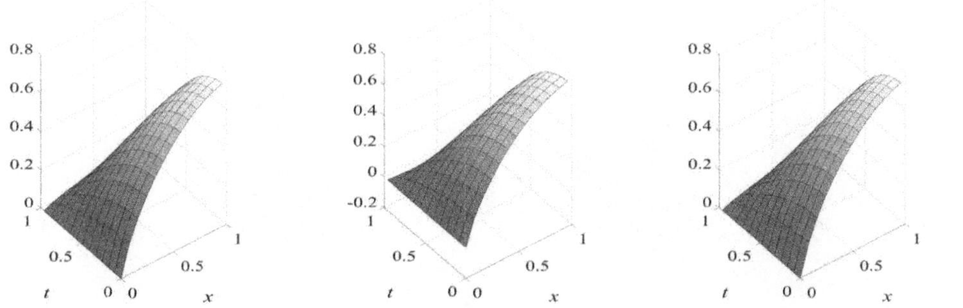

(a) Re[ψ] (b) Re[ψ_I] (c) Re[ψ_{II}]

Fig. 5. Graphs of the real parts of the solutions ψ, ψ_I, and ψ_{II} with $\alpha = 0.75$.

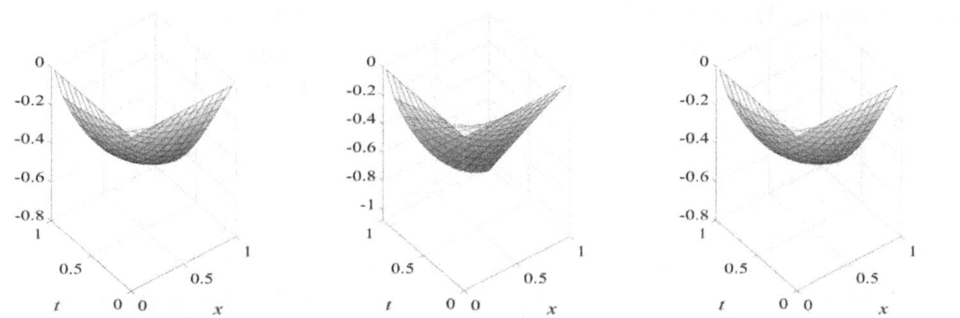

(a) Im[ψ] (b) Im[ψ_I] (c) Im[ψ_{II}]

Fig. 6. Graphs of the imaginary parts of the solutions ψ, ψ_I, and ψ_{II} with $\alpha = 0.75$.

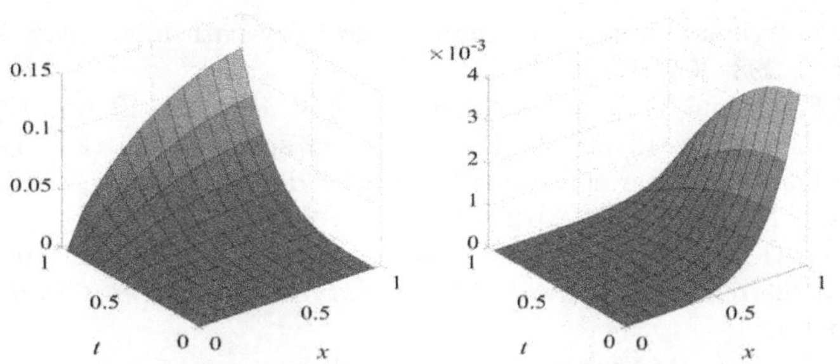

(a) Re[$|\psi - \psi_I|$] (b) Re[$|\psi - \psi_{II}|$]

Fig. 7. Graphs of the real parts of absolute errors $|\psi - \psi_I|$ and $|\psi - \psi_{II}|$ with $\alpha = 0.75$.

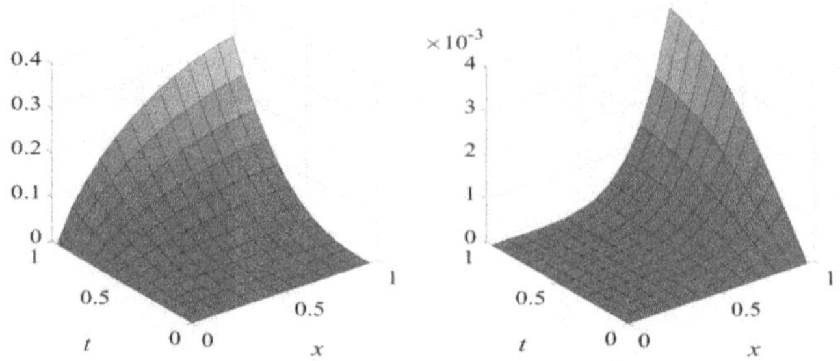

(a) $\mathrm{Im}[|\psi - \psi_I|]$ (b) $\mathrm{Im}[|\psi - \psi_{II}|]$

Fig. 8. Graphs of the imaginary parts of absolute errors $|\psi - \psi_I|$ and $|\psi - \psi_{II}|$ with $\alpha = 0.75$.

6. Conclusion

In this study, VIM is applied to linear and nonlinear SFSEs. Integer and fractional order Lagrange multipliers are introduced. Using these Lagrange multipliers, different iteration formulas are formed for SFSEs. In sample problems of linear and nonlinear SFSEs, approximate solutions that rapidly converge to the exact solution have been obtained by different iteration formulas of the method. It is seen that the approximation obtained by Formula II, having a fractional order Lagrange multiplier, converges to the exact solution ψ more rapidly than the approximation obtained by Formula I, having an integer order Lagrange multiplier. Depending on the choice of iteration formula of VIM for SFSE, approximate solutions that converge faster to the exact solution can be obtained.

References

[1] N. Laskin, Fractional quantum mechanics, *Physical Review E*, **62** (3), 3135–3145, (2000).

[2] N. Laskin, Fractional quantum mechanics and Lévy path integrals, *Physics Letters. A*, **268** (4–6), 298–305, (2000).

[3] N. Laskin, Fractional Schrödinger equation, *Physical Review E*, **66**, 056108, (2002).

[4] Y. Wang, L. Mei, Q. Li and L. Bu, Split-step spectral Galerkin method for the two-dimensional nonlinear space-fractional Schrödinger equation, *Applied Numerical Mathematics*, **136**, 257–278, (2019).

[5] M. Shen and H. Wang, An efficient spectral method for the fractional Schrödinger equation on the real line, *Journal of Computational and Applied Mathematics*, **444**, 115774, (2024).

[6] M. Li, C. Huang and P. Wang, Galerkin finite element method for nonlinear fractional Schrödinger equations, *Numerical Algorithms*, **74**, 499–525, (2017).

[7] M. Li, C. Huang and W. Ming, A relaxation-type Galerkin FEM for nonlinear fractional Schrödinger equations, *Numerical Algorithms*, **83**, 99–124, (2020).

[8] C. Guo and S. Fang, Crank-Nicolson difference scheme for the derivative nonlinear Schrödinger equation with the Riesz space fractional derivative, *Journal of Applied Analysis & Computation*, **11** (3), 1074–1094, (2021).

[9] M. Almushaira, An efficient fourth-order accurate conservative scheme for Riesz space fractional Schrödinger equation with wave operator, *Mathematics and Computers in Simulation*, **210**, 424–447, (2023).

[10] I. Podlubny, *Fractional Differential Equations*, (Academic Press, San Diego, 1999).

[11] A. A. Kilbas, H. M., Srivastava and J. J. Trujillo, *Theory and Applications of Fractional Differential Equations*, vol. **204** (Elsevier, Amsterdam, 2006).

[12] T. M. Atanackovic, S. Pilipovic, B. Stankovic and D. Zorica, *Fractional Calculus with Applications in Mechanics Wave Propagation, Impact and Variational Principles*, (John Wiley and Sons, London, 2014).

[13] V. V. Uchaikin, *Fractional Derivatives for Physicists and Engineers*, (Springer, Berlin, 2013).

[14] C. Changpin Li and M. Cai, *Theory and Numerical Approximations of Fractional Integrals and Derivatives*, (SIAM, Philadelphia, 2019).

[15] G. Jumarie, Modified Riemann Liouville Derivative and Fractional Taylor Series of Non-differentiable Functions Further Results, *Computers & Mathematics with Applications. An International Journal*, **51**, 1367–1376, (2006).

[16] G. Jumarie, Table of Some Basic Fractional Calculus Formulae Derived from a Modified Riemann Liouville Derivative for Non-differentiable Function, *Applied Mathematics Letters*, **22**, 378–385, (2009).

[17] J. H. He, A New Approach to Nonlinear Partial Differential Equations, *Communications in Nonlinear Science and Numerical Simulation*, **2** (4), 230–235, (1997).

Chapter 14

Mathematical Modeling and Analysis of Virus Dynamics: Integrating Immune Response and Multitarget Cells

Taofeek O. Alade*,‡ and Yaqoob Al Rahbi†,§

*National University of Science and Technology, International Maritime College, Science Cluster, Oman

†Faculty of Education and Arts, Sohar University, Oman

‡taofeekalade@gmail.com

§yrahbi@su.edu.om

In this study, we introduce a virus dynamics model incorporating antibody immune response and multitarget cells, offering a comprehensive exploration of the intricate interplay between the immune system and viral infection. The model accounts for two categories of infected cells: latently infected cells and actively infected cells responsible for virus production. Saturated incidence rate is integrated into the model, providing a more nuanced representation of infection dynamics. Ensuring the biological feasibility of our proposed model, we rigorously establish the nonnegativity and boundedness of its solutions. To delve further into the system's behavior, we derive the basic reproduction number $\mathcal{R}_0$, which is crucial for determining the existence and stability of steady states. Our exploration extends to the global stability analysis of these steady states through the construction of suitable Lyapunov functions. To validate the robustness of our theoretical findings, we conduct numerical simulations. These simulations serve as a crucial step in confirming the efficacy and applicability of our proposed model under various conditions. The combination of analytical proofs and numerical simulations enhances the reliability and significance of our research findings.

1. Introduction

In recent years, there has been a significant increase in the development of mathematical models aimed at understanding the complex interactions between various viruses, such as COVID-19, HIV, Dengue, Chikungunya virus and their human target cells (see e.g. [1–13]). By providing a structured framework for analyzing the spread, mutation, and impact of these viruses on human health, mathematical models play a crucial role in advancing the development of effective antiviral drugs. They also offer valuable insights into the behavior of viruses within the host, which is essential for devising targeted therapeutic strategies. Mathematical modeling of viral dynamics is not only fundamental to drug development but also critical for enhancing our ability to predict disease trends. By simulating various scenarios, these models help researchers forecast the trajectory of viral outbreaks, identify potential risks, and formulate timely public health interventions. For instance, during the COVID-19 pandemic, mathematical models were instrumental in projecting infection rates, guiding policy decisions, and evaluating the effectiveness of vaccination campaigns [14].

Moreover, the stability analysis of these models is essential for understanding the long-term dynamics of virus-host interactions. Stability analysis examines whether a viral system will return to equilibrium after a disturbance or whether it will result in persistent infection, viral eradication, or epidemic outbreaks. This type of analysis is critical for identifying the key parameters that influence viral persistence, immune response, and the effectiveness of antiviral strategies. Recent studies have highlighted the importance of such analyses in managing both acute and chronic viral infections, as well as in mitigating the risks of viral resurgence [15, 16].

The basic virus dynamics model involving multitarget cells was initially proposed by Elaiw [17] as follows:

$$\dot{W}_i = \psi_i - d_i W_i - \eta_i W_i Y, \quad i = 1, \ldots, n,$$

$$\dot{I}_i = \eta_i W_i Y - \epsilon_i I_i, \quad i = 1, \ldots, n,$$

$$\dot{Y} = \sum_{i=1}^{n} \alpha_i I_i - bY,$$

In this model, there are n categories of target cells. $W_i(t)$ represents the uninfected target cells, $I_i(t)$ denotes the infected cells, and $Y(t)$ is the concentration of free virus particles. The model assumes that uninfected cells reproduce at a constant rate ψ_i and become infected at a rate $\eta_i W_i Y$, where η_i is the infection rate constant. The death rates of uninfected cells, infected cells, and free virus particles are given by d_i, ϵ_i, and b, respectively. Additionally, infected cells produce free virus particles at a rate $\alpha_i I_i$.

In a subsequent study, Elaiw *et al.* [18] extended this model by incorporating an antibody immune response, represented as:

$$\dot{U} = \lambda + \varpi UY - \delta U,$$

Here, U represents the concentration of antibodies, λ is the production rate of antibodies, ϖUY indicates the growth rate of antibodies following interaction with the virus, and δU is the death rate of antibodies. The introduction of the antibody immune response adds a layer of complexity, providing a mechanism for controlling viral infections. Antibodies are critical proteins that identify and bind to specific antigens, facilitating their removal from the body.

Inspired by the work of [17] and [18], this paper proposes an enhanced virus dynamics model that includes an antibody immune response and multitarget cells. The new model considers both latently and actively infected cells and accounts for a saturated incidence rate, offering a more comprehensive representation of viral dynamics.

In this study, we examine both the basic and global properties of our proposed virus dynamics model. We derive the basic reproduction number, denoted as $\mathcal{R}_0$, and utilize Lyapunov functions to demonstrate the global stability of the model's equilibrium states. Our model not only builds upon but also significantly enhances the frameworks presented in [17] and [18]. A key improvement lies in our modification of the bilinear incidence rate $\eta_i W_i Y$ used in [18]. We replace it with a saturated incidence rate $\frac{\eta_i W_i Y}{1+\theta_i Y}$ for $i = 1, ..., n$.

The bilinear incidence rate, based on the law of mass action, often falls short in accurately capturing the dynamic behavior of viral infections, particularly at high concentrations of the virus [19]. To address this limitation and provide a more realistic representation of viral transmission dynamics, we have incorporated the saturated incidence rate, which better reflects the interactions between viruses and target cells. The model is presented as follows:

$$\dot{W}_i = \psi_i - d_i W_i - \frac{\eta_i W_i Y}{1+\theta_i Y}, \qquad i = 1, ..., n, \tag{1}$$

$$\dot{T}_i = (1-\pi_i)\frac{\eta_i W_i Y}{1+\theta_i Y} - (\beta_i + \rho_i)T_i, \quad i = 1, ..., n, \tag{2}$$

$$\dot{A}_i = \pi_i \frac{\eta_i W_i Y}{1+\theta_i Y} + \rho_i T_i - \epsilon_i A_i, \qquad i = 1, ..., n, \tag{3}$$

$$\dot{Y} = \sum_{i=1}^{n} \mu_i A_i - bY - rUY, \tag{4}$$

$$\dot{U} = \lambda + \varpi UY - \delta U. \tag{5}$$

Here, the within-host virus dynamics model captures the interactions between the virus and n categories of target cells, incorporating both latently and actively infected cells, as well as a saturated incidence rate. Specifically, T_i and A_i, for $i = 1, \ldots, n$, represent the concentrations of latently and actively infected cells, respectively. Actively infected cells produce an average of μ_i virus particles, while free virus particles are neutralized by antibodies at a rate of rUY.

When an antigen is encountered, the antibodies undergo expansion at a constant rate λ, proliferate further at a rate ϖUY, and die at a rate δU.

The model also accounts for the division of infected target cells into two categories: a fraction $(1-\pi_i)$ becomes latently infected cells, while the remaining π_i transforms into actively infected cells, where $0 < \pi_i < 1$. Latently infected cells transition to actively infected cells at a rate of $\rho_i T_i$ and are subject to die at a rate of $\beta_i T_i$.

1.1. *Properties of solutions*

First we show that the solutions of model (1)–(5) are nonnegative and bounded. Let $W = (W_1, W_2, ..., W_n)$, $T = (T_1, T_2, ..., T_n)$ and $A = (A_1, A_2, ..., A_n)$.

Proposition 1. *For system (1)–(5), there exist a positively invariant compact set*

$$\Phi = \{ (\mathbf{W}, \mathbf{T}, \mathbf{A}, Y, U) \in \mathbb{R}^{3n+2}_{\geq 0} : 0 \leq W_i, T_i, A_i \leq N_i, i = 1, ..., n, \\ 0 \leq Y \leq M_1, 0 \leq U \leq M_2 \}.$$

Proof. Since

$$\begin{aligned}
\dot{W}_i\Big|_{W_i=0} &= \psi_i > 0, \\
\dot{T}_i\Big|_{T_i=0} &= (1-\pi_i)\frac{\eta_i W_i Y}{1+\theta_i Y} \geq 0 \quad \text{for all } W_i, Y \geq 0, \\
\dot{A}_i\Big|_{A_i=0} &= \pi_i \frac{\eta_i W_i Y}{1+\theta_i Y} + \rho_i T_i \geq 0 \quad \text{for all } W_i, T_i, Y \geq 0, \\
\dot{Y}\Big|_{Y=0} &= \sum_{i=1}^{n} \mu_i A_i \geq 0 \quad \text{for all } A_i \geq 0, \\
\dot{U}\Big|_{U=0} &= \lambda > 0, \quad i = 1, ..., n,
\end{aligned}$$

Then, $\mathbb{R}^{3n+2}_{\geq 0}$ is positively invariant for system (1)–(5).

We let $S_i(t) = W_i + T_i + A_i$, $i = 1, ..., n$, then

$$\begin{aligned}
\dot{S}_i &= \psi_i - d_i W_i - \frac{\eta_i W_i Y}{1+\theta_i Y} + (1-\pi_i)\frac{\eta_i W_i Y}{1+\theta_i Y} - (\beta_i + \rho_i) T_i + \pi_i \frac{\eta_i W_i Y}{1+\theta_i Y} + \rho_i T_i - \epsilon_i A_i \\
&= \psi_i - d_i W_i - \beta_i T_i - \epsilon_i A_i \\
&\leq \psi_i - \sigma_i S_i,
\end{aligned}$$

where $\sigma_i = \min\{d_i, \beta_i, \epsilon_i\}$. Hence, $0 \leq S_i(t) \leq \frac{\psi_i}{\sigma_i}$ for all $t \geq 0$ if $S_i(0) \leq \frac{\psi_i}{\sigma_i}$. It follows that $0 \leq W_i(t), T_i(t), A_i(t) \leq N_i$ for all $t \geq 0$, if $0 \leq W_i(0) + T_i(0) + A_i(0) \leq N_i$, where $N_i = \frac{\psi_i}{\sigma_i}$. Moreover, we let $G(t) = Y(t) + \frac{r}{\varpi} U(t)$, then

$$\begin{aligned}
G(t) &= \sum_{i=1}^{n} \mu_i A_i - bY + \frac{r}{\varpi}\lambda - \frac{\delta r}{\varpi} U \leq \sum_{i=1}^{n} \mu_i N_i + \frac{r}{\varpi}\lambda - \varphi\left(Y + \frac{r}{\varpi} U\right) \\
&= \sum_{i=1}^{n} \mu_i N_i + \frac{r}{\varpi}\lambda - \varphi G(t),
\end{aligned}$$

where $\varphi = \min\{b, \delta\}$. Then, $G(t) \leq M_1$, $0 \leq Y(t) \leq M_1$ and $0 \leq U(t) \leq M_2$ if $0 \leq Y(0) + \frac{r}{\varpi} U(0) \leq M_1$, where $M_1 = \sum\limits_{i=1}^{n} \frac{\mu_i N_i + \frac{r}{\varpi}\lambda}{\varphi}$ and $M_2 = \frac{\varpi M_1}{r}$.

1.2. *Equilibria*

The basic reproduction number of system (1)–(5) is defined as

$$\mathcal{R}_0 = \sum_{i=1}^{n} \mathcal{R}_{0i} = \sum_{i=1}^{n} \frac{\mu_i \eta_i \psi_i \delta (\beta_i \pi_i + \rho_i)}{d_i \epsilon_i (\beta_i + \rho_i)(b\delta + r\lambda)}.$$

Lemma 1. *(i) If $\mathcal{R}_0 \leq 1$, then the virus-free equilibrium Q_0 is the only equilibrium for the system, (ii) if $\mathcal{R}_0 > 1$, then the system has an endemic equilibrium Q_1 and $Q_1 \in \mathring{\Phi}$, where $\mathring{\Phi}$ is the interior of Φ.*

Proof. To find the equilibria of system (1)–(5) we let its R.H.S be equal zero. Then we get W_i, T_i, A_i and U as functions of Y as:

$$W_i = \frac{\psi_i(1+\theta_i Y)}{d_i(1+a_i Y)}, \quad i = 1, ..., n, \tag{6}$$

$$T_i = \frac{(1-\pi_i)\eta_i W_i Y}{(\beta_i+\rho_i)(1+\theta_i Y)}, \quad i = 1, ..., n, \tag{7}$$

$$A_i = \frac{\eta_i W_i Y(\beta_i \pi_i + \rho_i)}{\epsilon_i(\beta_i+\rho_i)(1+\theta_i Y)}, \quad i = 1, ..., n, \tag{8}$$

$$\sum_{i=1}^{n} \mu_i A_i = (b + rU)\, Y, \tag{9}$$

$$U = \frac{\lambda}{\delta - \varpi Y}, \tag{10}$$

where, $\alpha_i = \frac{\eta_i}{d_i}$ and $a_i = \alpha_i + \theta_i$. From Eqs. (6)–(8) and (10) into Eq. (9) we get

$$\sum_{i=1}^{n} \frac{\mu_i \eta_i \psi_i (\beta_i \pi_i + \rho_i) Y}{d_i \epsilon_i (1+a_i Y)(\beta_i+\rho_i)} - (b+rU)\, Y = 0,$$

$$Y\left(\sum_{i=1}^{n} \frac{\mu_i \eta_i \psi_i \delta (\beta_i \pi_i + \rho_i)}{d_i \epsilon_i\,(1+a_i Y)\,(\beta_i+\rho_i)(b\delta + r\lambda)} - \frac{(b+rU)\,\delta}{(b\delta + r\lambda)}\right) = 0,$$

$$Y\left(\sum_{i=1}^{n} \frac{\mathcal{R}_{0i}}{1+a_i Y} - \frac{\delta\,(\gamma - b\varpi Y)}{\gamma\,(\delta - \varpi Y)}\right) = 0, \tag{11}$$

where, $\gamma = b\delta + r\lambda$. Eq. (11) admits $Y = 0$ as a solution. Substituting $Y = 0$ in Eqs. (6), (7), (8) and (10) we get the virus-free equilibrium $Q_0 = (W^0, T^0, A^0, Y^0, U^0)$, where $W_i^0 = \frac{\psi_i}{d_i}$, $T_i^0 = 0$, $A_i^0 = 0$, $Y^0 = 0$, and $U^0 = \frac{\lambda}{\delta}$. The other possibility of Eq. (11) is

$$\sum_{i=1}^{n} \frac{\mathcal{R}_{0i}\gamma\,(\delta - \varpi Y)}{1+a_i Y} - \delta\,(\gamma - b\varpi Y) = 0, \tag{12}$$

where $Y \neq \frac{\delta}{\varpi}$.

Let us define a function $K(Y)$ as

$$K(Y) = \sum_{i=1}^{n} \frac{\mathcal{R}_{0i}\gamma\,(\delta - \varpi Y)}{1+a_i Y} - \delta\,(\gamma - b\varpi Y).$$

Then we get

$$K(0) = \gamma\delta(\mathcal{R}_0 - 1),$$
$$K\left(\frac{\delta}{\varpi}\right) = -r\delta\lambda.$$

Thus if $\mathcal{R}_0 > 1$, then $K(0) > 0$, $K\left(\frac{\delta}{\varpi}\right) < 0$ and there exists $0 < Y^* < \frac{\delta}{\varpi}$ such that $K(Y^*) = 0$. Substitiuting $Y = Y^*$ in Eqs. (6)–(8) and (10) we get $W_i^* > 0$, $T_i^* > 0$, $A_i^* > 0$ and $U^* > 0$. Then $Q_1 = (W^*, T^*, A^*, Y^*, U^*)$ exists when $\mathcal{R}_0 > 1$.

Clearly $Q_0 \in \Phi$. Now, we show that if $\mathcal{R}_0 > 1$, then $Q_1 \in \mathring{\Phi}$. From Eqs. (6)–(8), we have $d_i W_i^* + \beta_i T_i^* + \epsilon_i A_i^* = \psi_i$, $i = 1, ..., n$. Since $W_i^* > 0, T_i^* > 0$, and $A_i^* > 0$, then

$$d_i W_i^* < \psi_i \Rightarrow W_i^* < \frac{\psi_i}{d_i} \leq N_i,$$
$$\beta_i T_i^* < \psi_i \Rightarrow T_i^* < \frac{\psi_i}{\beta_i} \leq N_i,$$
$$\epsilon_i A_i^* < \psi_i \Rightarrow A_i^* < \frac{\psi_i}{\epsilon_i} \leq N_i.$$

Moreover, from Eqs. (9)–(10) we have

$$\sum_{i=1}^{n} \mu_i A_i^* - bY^* - rY^*U^* + \frac{r}{\varpi}(\lambda + \varpi U^* Y^* - \delta U^*) = 0,$$

$$\Rightarrow bY^* + \frac{\delta r}{\varpi} U^* = \sum_{i=1}^{n} \mu_i A_i^* + \frac{r}{\varpi}\lambda < \sum_{i=1}^{n} \mu_i N_i + \frac{r}{\varpi}\lambda$$
$$\Rightarrow Y^* < \sum_{i=1}^{n} \frac{\mu_i N_i + \frac{r}{\varpi}\lambda}{b} \leq M_1, \quad U_1 < \frac{\varpi}{r} \sum_{i=1}^{n} \frac{\mu_i N_i + \frac{r}{\varpi}\lambda}{\delta} \leq \frac{\varpi M_1}{r} = M_2.$$

It follows that $Q_1 \in \mathring{\Phi}$.

1.3. *Global stability*

In order to investigate the global stability of the two equilibria of system (1)–(5) we will use direct Lyapunov method following the method presented in [20–22]. Let us define

$$H(x) = x - \ln x - 1.$$

Clearly, $H(x) \geq 0$ for $x > 0$ and $H(1) = 0$.

Theorem 1. *Suppose that $\mathcal{R}_0 \leq 1$, then Q_0 is globally asymptotically stable in Φ.*

Proof. We construct the following Lyapunov function:

$$X_0(\mathbf{W}, \mathbf{T}, \mathbf{A}, Y, U) = \sum_{i=1}^{n} y_i \left[W_i^0 H\left(\frac{W_i}{W_i^0}\right) + \frac{\rho_i}{\beta_i \pi_i + \rho_i} T_i + \frac{\beta_i + \rho_i}{\beta_i \pi_i + \rho_i} A_i \right]$$
$$+ Y + \frac{r}{\varpi} U^0 H\left(\frac{U}{U^0}\right),$$

where $y_i = \frac{\mu_i(\beta_i\pi_i+\rho_i)}{\epsilon_i(\beta_i+\rho_i)}$. Note that, $X_0(W,T,A,Y,U) > 0$ for all $W,T,A,Y,U > 0$ and $X_0(W^0,0,0,0,U^0) = 0$. Calculating $\frac{dX_0}{dt}$ along the trajectories of (1)–(5) we get

$$\begin{aligned}
\frac{dX_0}{dt} =& \sum_{i=1}^{n} y_i \left[\left(1-\frac{W_i^0}{W_i}\right)\left(\psi_i - d_iW_i - \frac{\eta_iW_iY}{1+\theta_iY}\right)\right.\\
&+ \frac{\rho_i}{\beta_i\pi_i+\rho_i}\left((1-\pi_i)\frac{\eta_iW_iY}{1+\theta_iY} - (\beta_i+\rho_i)T_i\right) + \left.\frac{\beta_i+\rho_i}{\beta_i\pi_i+\rho_i}\left(\pi_i\frac{\eta_iW_iY}{1+\theta_iY} + \rho_iT_i - \epsilon_iA_i\right)\right]\\
&+ \sum_{i=1}^{n}\mu_iA_i - bY - rUY + \frac{r}{\varpi}\left(1-\frac{U^0}{U}\right)(\lambda + \varpi UY - \delta U)\\
=& \sum_{i=1}^{n} y_id_i\left[\left(1-\frac{W_i^0}{W_i}\right)(W_i^0 - W_i)\right] + \sum_{i=1}^{n}\frac{\mu_i\eta_iW_i^0(\beta_i\pi_i+\rho_i)Y}{\epsilon_i(\beta_i+\rho_i)(1+\theta_iY)} - bY - rU^0Y\\
&+ \frac{r}{\varpi}\left(1-\frac{U^0}{U}\right)(\delta U^0 - \delta U)\\
=& \sum_{i=1}^{n} y_id_i\left(\frac{W_i - W_i^0}{W_i}\right)(W_i^0 - W_i) - \frac{r\delta}{\varpi}\frac{(U-U^0)^2}{U}\\
&+ (b+rU^0)\left(\sum_{i=1}^{n}\frac{\mu_i\eta_i\psi_i(\beta_i\pi_i+\rho_i)}{d_i\epsilon_i(\beta_i+\rho_i)(1+\theta_iY)(b+rU^0)} - 1\right)Y\\
=& -\sum_{i=1}^{n} y_id_i\frac{\left(W_i - W_i^0\right)^2}{W_i} - \frac{r\delta}{\varpi}\frac{(U-U^0)^2}{U} + (b+rU^0)(\mathcal{R}_0 - 1)Y\\
&- (b+rU^0)\sum_{i=1}^{n}\frac{\mathcal{R}_0^{(i)}\theta_iY^2}{1+\theta_iY}.
\end{aligned}$$

Therefore, $\frac{dX_0}{dt} \le 0$ holds when $\mathcal{R}_0 \le 1$. Additionally, $\frac{dX_0}{dt} = 0$ implies that $W_i = W_i^0$, $U = U^0$, and $Y = 0$. Define $D_1 = \{(\mathbf{W},\mathbf{T},\mathbf{A},Y,U) : \frac{dX_0}{dt} = 0\}$, and let D_1 be the largest invariant subset of D_0. The solutions of the system (1)–(5) converge to D_0. For any element in D_0, we have $Y(t) = \dot{Y}(t) = 0$. Consequently, from Eqs. (3) and (4), it follows that $A_i(t) = 0$ and $T_i(t) = 0$. By applying LaSalle's invariance principle [23], we conclude that Q_0 is globally asymptotically stable.

Theorem 2. *Suppose that $\mathcal{R}_0 > 1$, then Q_1 is globally asymptotically stable in $\mathring{\Phi}$.*

Proof. Construct a Lyapunov function X_1 as follows:

$$\begin{aligned}
X_1 =& \sum_{i=1}^{n} y_i\left[W_i^*H\left(\frac{W_i}{W_i^*}\right) + \frac{\rho_i}{\beta_i\pi_i+\rho_i}T_i^*H\left(\frac{T_i}{T_i^*}\right) + \frac{\beta_i+\rho_i}{\beta_i\pi_i+\rho_i}A_i^*H\left(\frac{A_i}{A_i^*}\right)\right]\\
&+ Y^*H\left(\frac{Y}{Y^*}\right) + \frac{r}{\varpi}U^*H\left(\frac{U}{U^*}\right).
\end{aligned}$$

Calculating $\frac{dX_1}{dt}$ along the trajectories of (1)–(5) we get

$$\begin{aligned}\frac{dX_1}{dt} = &\sum_{i=1}^{n} y_i \left[\left(1-\frac{W_i^*}{W_i}\right)\left(\psi_i - d_iW_i - \frac{\eta_i W_i Y}{1+\theta_i Y}\right)\right.\\
&+ \frac{\rho_i}{\beta_i\pi_i+\rho_i}\left(1-\frac{T_i^*}{T_i}\right)\left((1-\pi_i)\frac{\eta_i W_i Y}{1+\theta_i Y} - (\beta_i+\rho_i)T_i\right)\\
&\left.+ \frac{\beta_i+\rho_i}{\beta_i\pi_i+\rho_i}\left(1-\frac{A_i^*}{A_i}\right)\left(\pi_i\frac{\eta_i W_i Y}{1+\theta_i Y} + \rho_i T_i - \epsilon_i A_i\right)\right]\\
&+ \left(1-\frac{Y^*}{Y}\right)\left(\sum_{i=1}^{n}\mu_i A_i - bY - rUY\right) + \frac{r}{\varpi}\left(1-\frac{U^*}{U}\right)(\lambda + \varpi UY - \delta U)\,.\end{aligned}$$

Applying

$$\psi_i = d_iW_i^* + \frac{\eta_i W_i^* Y^*}{1+\theta_i Y^*}, \quad \lambda = \delta U^* - \varpi U^* Y^*,$$

we obtain

$$\begin{aligned}\frac{dX_1}{dt} = y_i &\left[\left(1-\frac{W_i^*}{W_i}\right)(d_iW_i^* - d_iW_i) + \frac{\eta_i W_i^* Y^*}{1+\theta_i Y^*}\left(1-\frac{W_i^*}{W_i}\right) + \frac{\eta_i W_i^* Y}{1+\theta_i Y}\right.\\
&- \frac{\rho_i}{\beta_i\pi_i+\rho_i}(1-\pi_i)\frac{\eta_i W_i Y}{1+\theta_i Y}\frac{T_i^*}{T_i} + \frac{\rho_i(\beta_i+\rho_i)}{\beta_i\pi_i+\rho_i}T_i^* - \frac{\beta_i+\rho_i}{\beta_i\pi_i+\rho_i}\pi_i\frac{\eta_i W_i Y}{1+\theta_i Y}\frac{A_i^*}{A_i}\\
&\left.- \frac{\beta_i+\rho_i}{\beta_i\pi_i+\rho_i}\rho_i T_i\frac{A_i^*}{A_i} + \frac{\beta_i+\rho_i}{\beta_i\pi_i+\rho_i}\epsilon_i A_i^* - \frac{\beta_i+\rho_i}{\beta_i\pi_i+\rho_i}\epsilon_i A_i^*\frac{Y^* A_i}{Y A_i^*}\right]\\
&- bY + bY^* + rUY^* - rU^*Y - rU^*Y^* + rU^*Y^*\left(\frac{U^*}{U}\right)\\
&+ \frac{r}{\varpi}\left(1-\frac{U^*}{U}\right)(\delta U^* - \delta U)\,.\end{aligned}$$

Using the endemic equilibrium conditions

$$(1-\pi_i)\frac{\eta_i W_i^* Y^*}{1+\theta_i Y^*} = (\beta_i+\rho_i)T_i^*, \quad \pi_i\frac{\eta_i W_i^* Y^*}{1+\theta_i Y^*} = \epsilon_i A_i^* - \rho_i T_i^*,$$

$$\sum_{i=1}^{n}\mu_i A_i^* = bY^* + rU^*Y^*,$$

we get

$$\frac{\beta_i+\rho_i}{\beta_i\pi_i+\rho_i}\epsilon_i A_i^* = \frac{\eta_i W_i^* Y^*}{1+\theta_i Y^*} = \frac{\rho_i}{\beta_i\pi_i+\rho_i}(1-\pi_i)\frac{\eta_i W_i^* Y^*}{1+\theta_i Y^*} + \frac{\beta_i+\rho_i}{\beta_i\pi_i+\rho_i}\,\pi_i\frac{\eta_i W_i^* Y^*}{1+\theta_i Y^*},$$

and then

$$\begin{aligned}
\frac{dX_1}{dt} &= \sum_{i=1}^{n} y_i \Bigg[d_i \left(1 - \frac{W_i^*}{W_i}\right)(W_i^* - W_i) + \frac{\rho_i}{\beta_i \pi_i + \rho_i}(1-\pi_i)\frac{\eta_i W_i^* Y^*}{1+\theta_i Y^*} \\
&+ \frac{\beta_i + \rho_i}{\beta_i \pi_i + \rho_i}\pi_i \frac{\eta_i W_i^* Y^*}{1+\theta_i Y^*} - \frac{\rho_i}{\beta_i \pi_i + \rho_i}(1-\pi_i)\frac{\eta_i W_i^* Y^*}{1+\theta_i Y^*}\left(\frac{W_i^*}{W_i}\right) \\
&- \frac{\beta_i + \rho_i}{\beta_i \pi_i + \rho_i}\pi_i \frac{\eta_i W_i^* Y^*}{1+\theta_i Y^*}\left(\frac{W_i^*}{W_i}\right) + \frac{\eta_i W_i^* Y^*}{1+\theta_i Y^*}\frac{(1+\theta_i Y^*)Y}{(1+\theta_i Y)Y^*} - \frac{\eta_i W_i^* Y^*}{1+\theta_i Y^*}\frac{Y}{Y^*} \\
&- \frac{\rho_i}{\beta_i \pi_i + \rho_i}(1-\pi_i)\frac{\eta_i W_i^* Y^*}{1+\theta_i Y^*}\frac{(1+\theta_i Y^*)}{(1+\theta_i Y)}\frac{W_i Y T_i^*}{W_i^* Y^* T_i} + \frac{\rho_i}{\beta_i \pi_i + \rho_i}(1-\pi_i)\frac{\eta_i W_i^* Y^*}{1+\theta_i Y^*} \\
&- \frac{\beta_i + \rho_i}{\beta_i \pi_i + \rho_i}\pi_i \frac{\eta_i W_i^* Y^*}{1+\theta_i Y^*}\frac{(1+\theta_i Y^*)}{(1+\theta_i Y)}\frac{W_i Y A_i^*}{W_i^* Y^* A_i} - \frac{\rho_i}{\beta_i \pi_i + \rho_i}(1-\pi_i)\frac{\eta_i W_i^* Y^*}{1+\theta_i Y^*}\frac{T_i A_i^*}{T_i^* A_i} \\
&+ \frac{\rho_i}{\beta_i \pi_i + \rho_i}(1-\pi_i)\frac{\eta_i W_i^* Y^*}{1+\theta_i Y^*} + \frac{\beta_i + \rho_i}{\beta_i \pi_i + \rho_i}\pi_i \frac{\eta_i W_i^* Y^*}{1+\theta_i Y^*} \\
&- \frac{\rho_i}{\beta_i \pi_i + \rho_i}(1-\pi_i)\frac{\eta_i W_i^* Y^*}{1+\theta_i Y^*}\frac{Y^* A_i}{Y A_i^*} - \frac{\beta_i + \rho_i}{\beta_i \pi_i + \rho_i}\pi_i \frac{\eta_i W_i^* Y^*}{1+\theta_i Y^*}\frac{Y^* A_i}{Y A_i^*} \\
&+ \frac{\rho_i}{\beta_i \pi_i + \rho_i}(1-\pi_i)\frac{\eta_i W_i^* Y^*}{1+\theta_i Y^*} + \frac{\beta_i + \rho_i}{\beta_i \pi_i + \rho_i}\pi_i \frac{\eta_i W_i^* Y^*}{1+\theta_i Y^*}\Bigg] \\
&- 2rU^*Y^* + rUY^* + rU^*Y^*\left(\frac{U^*}{U}\right) - \frac{r\delta}{\varpi}\frac{(U-U^*)^2}{U}. \qquad (13)
\end{aligned}$$

Eq. (13) can be simplified as:

$$\begin{aligned}
\frac{dX_1}{dt} &= \sum_{i=1}^{n} y_i \Bigg[-d_i \frac{(W_i - W_i^*)^2}{W_i} + \frac{\eta_i W_i^* Y^*}{1+\theta_i Y^*}\left(-1 + \frac{(1+\theta_i Y^*)Y}{(1+\theta_i Y)Y^*} - \frac{Y}{Y^*} + \frac{1+\theta_i Y}{1+\theta_i Y^*}\right) \\
&+ \frac{\rho_i}{\beta_i \pi_i + \rho_i}(1-\pi_i)\frac{\eta_i W_i^* Y^*}{1+\theta_i Y^*}\left(5 - \frac{W_i^*}{W_i} - \frac{(1+\theta_i Y^*)}{(1+\theta_i Y)}\frac{W_i Y T_i^*}{W_i^* Y^* T_i} - \frac{T_i A_i^*}{T_i^* A_i} - \frac{Y^* A_i}{Y A_i^*} - \frac{1+\theta_i Y}{1+\theta_i Y^*}\right) \\
&+ \frac{\beta_i + \rho_i}{\beta_i \pi_i + \rho_i}\pi_i \frac{\eta_i W_i^* Y^*}{1+\theta_i Y^*}\left(4 - \frac{W_i^*}{W_i} - \frac{(1+\theta_i Y^*)}{(1+\theta_i Y)}\frac{W_i Y A_i^*}{W_i^* Y^* A_i} - \frac{Y^* A_i}{Y A_i^*} - \frac{1+\theta_i Y}{1+\theta_i Y^*}\right)\Bigg] \\
&- rU^*Y^*\left(2 - \frac{U}{U^*} - \frac{U^*}{U}\right) - \frac{r\delta}{\varpi}\frac{(U-U^*)^2}{U}
\end{aligned}$$

$$\begin{aligned}
\frac{dX_1}{dt} &= \sum_{i=1}^{n} y_i \Bigg[-d_i \frac{(W_i - W_i^*)^2}{W_i} - \frac{\theta_i \eta_i W_i^* (Y - Y^*)^2}{(1+\theta_i Y)(1+\theta_i Y^*)^2} \\
&+ \frac{\rho_i}{\beta_i \pi_i + \rho_i}(1-\pi_i)\frac{\eta_i W_i^* Y^*}{1+\theta_i Y^*}\left(5 - \frac{W_i^*}{W_i} - \frac{(1+\theta_i Y^*)}{(1+\theta_i Y)}\frac{W_i Y T_i^*}{W_i^* Y^* T_i} - \frac{T_i A_i^*}{T_i^* A_i} - \frac{Y^* A_i}{Y A_i^*} - \frac{1+\theta_i Y}{1+\theta_i Y^*}\right) \\
&+ \frac{\beta_i + \rho_i}{\beta_i \pi_i + \rho_i}\pi_i \frac{\eta_i W_i^* Y^*}{1+\theta_i Y^*}\left(4 - \frac{W_i^*}{W_i} - \frac{(1+\theta_i Y^*)}{(1+\theta_i Y)}\frac{W_i Y A_i^*}{W_i^* Y^* A_i} - \frac{Y^* A_i}{Y A_i^*} - \frac{1+\theta_i Y}{1+\theta_i Y^*}\right)\Bigg] \\
&- \frac{r\lambda}{\varpi U^*}\frac{(U-U^*)^2}{U}. \qquad (14)
\end{aligned}$$

We use the following arithmetic mean-geometric mean inequality rule. If $\xi_i \geq 0$, $i = 1, 2, ..., k$, then

$$\frac{1}{k}\sum_{i=1}^{k} \xi_i \geq \sqrt[k]{\prod_{i=1}^{k} \xi_i}, \qquad (15)$$

where equality holding if and only if $\xi_1 = \xi_2 = \cdots = \xi_k$. Thus we get

$$5 \leq \frac{W_i^*}{W_i} + \frac{(1+\theta_i Y^*)}{(1+\theta_i Y)} \frac{W_i Y T_i^*}{W_i^* Y^* T_i} + \frac{T_i A_i^*}{T_i^* A_i} + \frac{Y^* A_i}{Y A_i^*} + \frac{1+\theta_i Y}{1+\theta_i Y^*},$$

$$4 \leq \frac{W_i^*}{W_i} + \frac{(1+\theta_i Y^*)}{(1+\theta_i Y)} \frac{W_i Y A_i^*}{W_i^* Y^* A_i} + \frac{Y^* A_i}{Y A_i^*} + \frac{1+\theta_i Y}{1+\theta_i Y^*}.$$

Then $\frac{dX_1}{dt} \leq 0$ and $\frac{dX_1}{dt} = 0$ if $W_i = W_i^*$, $T_i = T_i^*$, $A_i = A_i^*$, $Y = Y^*$ and $U = U^*$. It follows from LaSalle's invariance principle, Q_1 is globally asymptotically stable in $\mathring{\Phi}$.

1.4. *Numerical simulations*

To carry out numerical simulations for the system (1)–(5), we assume that the virus targets two categories of cells, i.e., $n = 2$. Table 1 lists the values of some of the parameters, while the remaining parameters will be specified below:

Table 1. The data of system (1)–(5).

Parameter	Value	Parameter	Value
ϕ_1	1.826	d_1	0.7979
ϕ_2	3.198	d_2	0.5
$\pi_1 = \pi_2$	0.5	r	0.5964
ϖ	0.5	b	0.4418
$\mu_1 = \mu_2$	2.02	λ	1.402
$\rho_1 = \rho_2$	0.1	δ	1.251
$\epsilon_1 = \epsilon_2$	0.4441	$\beta_1 = \beta_2$	0.5
$\theta_1 = \theta_2$	0.1	η_1	varied
η_2	varied		

Case (I): Effect of π on the dynamical behavior of the system:
We consider the following values $\eta_1 = \eta_2 = 0.035$ and $\theta_1 = \theta_2 = 0.1$. Assume that $\pi = \pi_1 = \pi_2$. The initial values are taken as: $W_1(0) = 2.24, W_2(0) = 6.2, T_1(0) = 0.1, T_2(0) = 0.05, A_1(0) = 0.1, A_2(0) = 0.5, Y(0) = 0.4$ and $U(0) = 1.4$. In Table 2, we have calculated the values of the equilibria and $\mathcal{R}_0$ for different values of π. We can observed that as π is increased then $\mathcal{R}_0$ also increased. Let π^{cr} be the critical value of the parameter π, such that

$$\mathcal{R}_0 = \sum_{i=1}^{2} \frac{\mu_i \eta_i \psi_i \delta(\beta_i \pi^{cr} + \rho_i)}{d_i \epsilon_i (\beta_i + \rho_i)(b\delta + r\lambda)} = 1.$$

Using the data provided in Table 1, we calculate $\pi^{cr} = 0.763593$. As shown in Figure 1, for $0 < \pi \leq 0.763593$, the system's trajectory converges to Q_0, while for $0.763593 < \pi < 1$, the trajectory converges to Q_1. This suggests that the factor $1 - \pi$ functions as a control parameter that can stabilize the system around Q_0. From a biological perspective, $1 - \pi$ can be interpreted as representing the effect of antiviral treatment in achieving viral eradication.

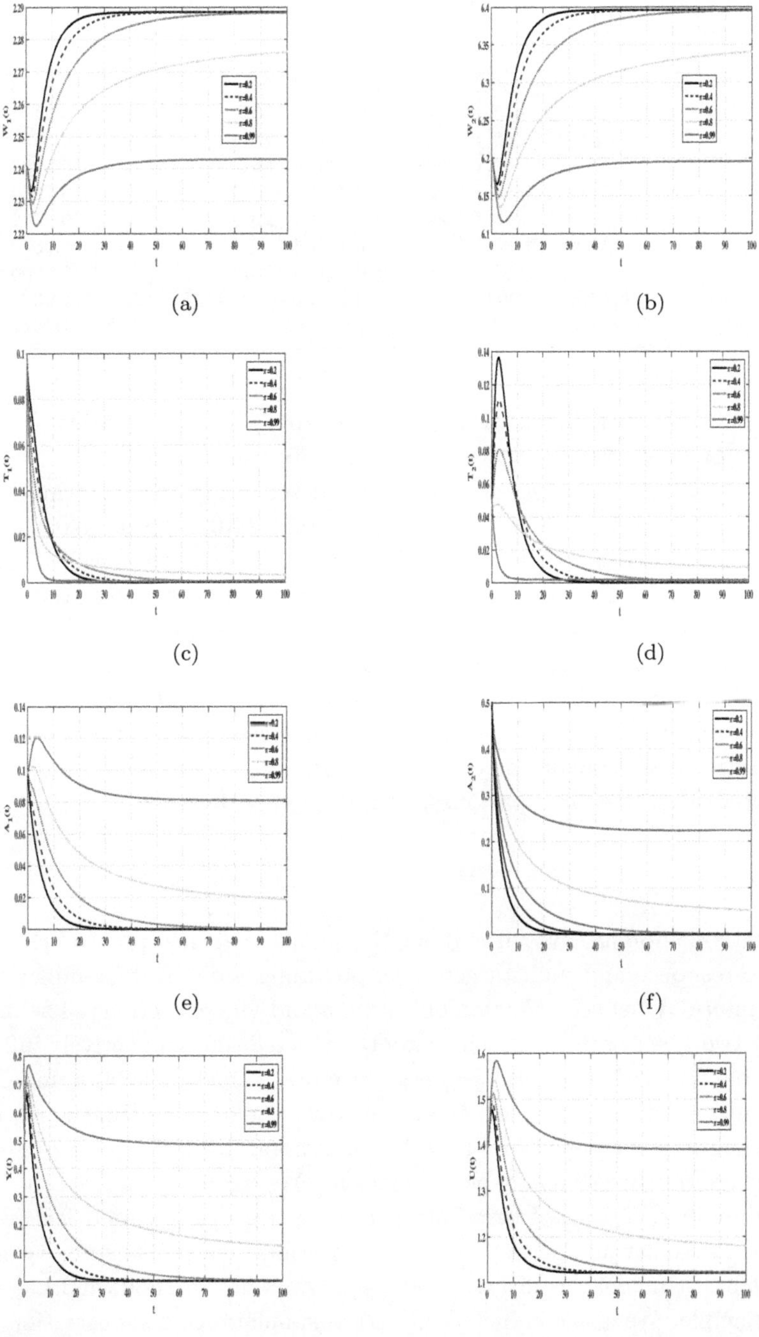

Fig. 1. Effect of π on the dynamical behavior of the system; (a) uninfected target cells 1; (b) uninfected target cells 2; (c) latently infected target cells 1; (d) latently infected target cells 2; (e) actively infected target cells 1; (f) actively infected target cells 2; (g) free virus particles; and (h) antibodies.

Table 2. The values of equilibria, $\mathcal{R}_0$ for model (1)–(5) with different values of π.

π	Equilibria	R_0
0.1	$Q_0 = (2.2885, 6.3960, 0, 0, 0, 0, 0, 1.1207)$	0.3113
0.2	$Q_0 = (2.2885, 6.3960, 0, 0, 0, 0, 0, 1.1207)$	0.4151
0.3	$Q_0 = (2.2885, 6.3960, 0, 0, 0, 0, 0, 1.1207)$	0.5189
0.4	$Q_0 = (2.2885, 6.3960, 0, 0, 0, 0, 0, 1.1207)$	0.6227
0.5	$Q_0 = (2.2885, 6.3960, 0, 0, 0, 0, 0, 1.1207)$	0.7264
0.6	$Q_0 = (2.2885, 6.3960, 0, 0, 0, 0, 0, 1.1207)$	0.8302
0.7	$Q_0 = (2.2885, 6.3960, 0, 0, 0, 0, 0, 1.1207)$	0.9340
0.7636	$Q_0 = (2.2885, 6.3960, 0, 0, 0, 0, 0, 1.1207)$	1.0000
0.8	$Q_1 = (2.2795, 6.3560, 0.0024, 0.0067, 0.0135, 0.0375, 0.0907, 1.1629)$	1.0378
0.9	$Q_1 = (2.2584, 6.2629, 0.004, 0.0111, 0.0495, 0.1374, 0.3131, 1.2810)$	1.1416

Case (II): Impact of Saturation Infection on Virus Dynamics

In this scenario, we assume $\theta = \theta_1 = \theta_2$ and fix the parameters $\pi_1 = \pi_2 = 0.5$ and $\eta_1 = \eta_2 = 0.5$. The simulation is conducted using the following initial conditions: $W_1(0) = 1.4$, $W_2(0) = 4.8$, $T_1(0) = 0.4$, $T_2(0) = 1.0$, $A_1(0) = 1.2$, $A_2(0) = 2.0$, $Y(0) = 1.5$, and $U(0) = 2.5$. The results indicate that the saturation parameter θ does not affect the value of $\mathcal{R}_0$. This finding suggests that while saturation significantly reduces the infection rate, it does not contribute to the complete elimination of the virus from the body.

Figure 2 illustrates the impact of saturation infection. As θ increases, the incidence rate of infection decreases, leading to an increase in the concentrations of uninfected target cells. Concurrently, the concentrations of latently infected cells, actively infected cells, free virus particles, and antibodies decrease.

2. Conclusion

We introduced a comprehensive virus dynamics model that incorporated the antibody immune response and multitarget cells, providing a detailed examination of the complex interplay between the immune system and viral infections. The model accounted for two types of infected cells, latently infected cells and actively infected cells responsible for viral production, offering a more refined understanding of infection dynamics. Additionally, the model integrated a saturated incidence rate, reflecting a more realistic depiction of viral transmission, particularly at high concentrations where transmission may be less efficient. We rigorously derived the basic reproduction number, $\mathcal{R}_0$, a key threshold parameter that played a critical role in determining the stability and persistence of the infection. The global stability of the two equilibria in the model was investigated using Lyapunov methods and LaSalle's invariance principle. We also conducted numerical simulations, demonstrating that both the theoretical and numerical results were consistent. Our analysis highlighted the effects of saturation infection on viral dynamics. The combination of theoretical analysis and numerical simulations validated the model's effectiveness in predicting infection outcomes and informing public health strategies.

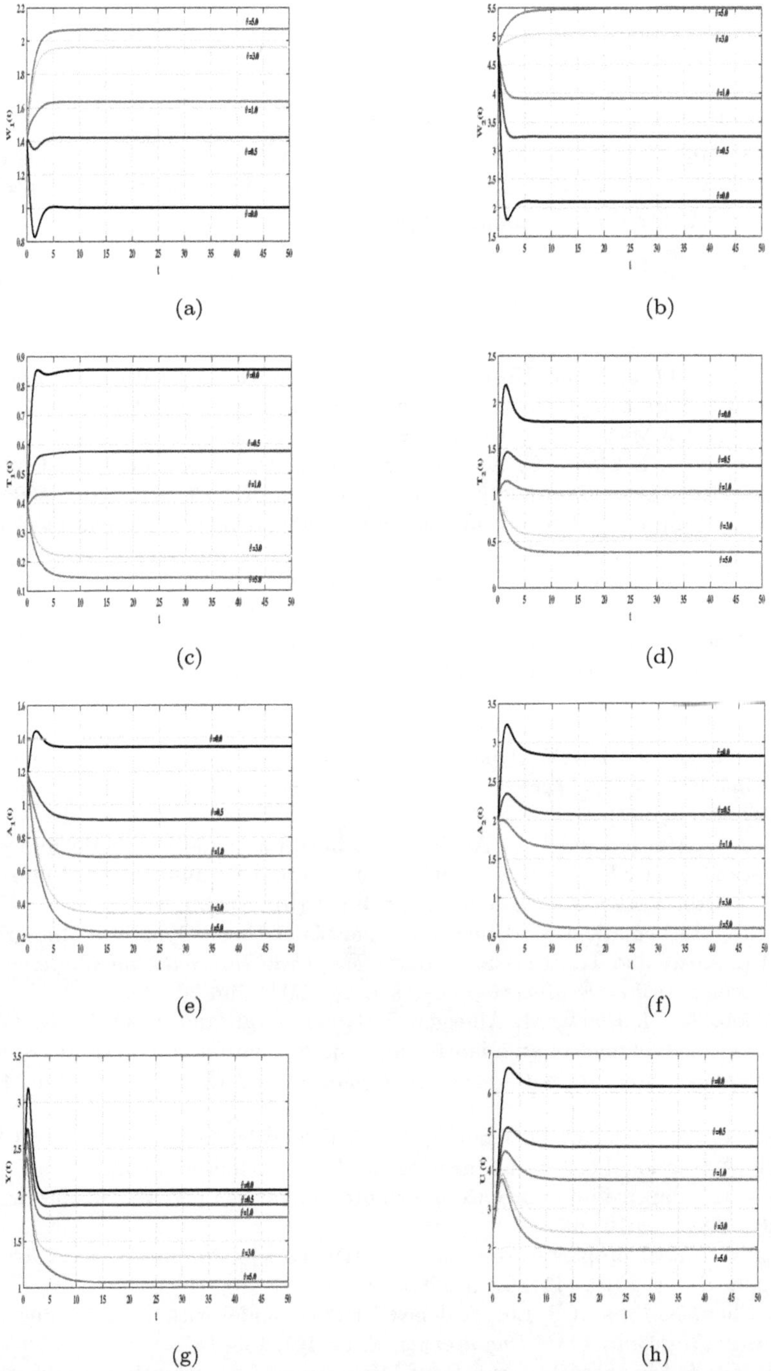

Fig. 2. Effect of the saturation infection on the virus dynamics; (a) uninfected target cells 1; (b) uninfected target cells 2; (c) latently infected target cells 1; (d) latently infected target cells 2; (e) actively infected target cells 1; (f) actively infected target cells 2; (g) free virus particles; and (h) antibodies.

Conflict of Interest

The authors declare no conflict of interest.

Acknowledgements

This study was supported by National University of Science and Technology, (NU) Oman. Research Grant NUFRG/23/IM/0026. The authors, therefore, gratefully acknowledge the NU for their financial support.

References

[1] A. Abidemi and O. J. Peter, Host-vector dynamics of dengue with asymptomatic, isolation and vigilant compartments: Insights from modelling, *The European Physical Journal Plus*, 2023 Mar, 138(3), 1–22.

[2] M. O. Adeniyi, M. I. Ekum, C. Iluno and S. I. Oke, Dynamic model of COVID-19 disease with exploratory data analysis, *Scientific African*, 2020 Sep 1, (9): e00477.

[3] B. S. T. Alkahtani and I. Koca, Fractional stochastic sır model, *Results in Physics*, 24 (2021): 104124.

[4] I. Koca and N. Ozalp, Analysis of a Fractional-Order Couple Model with Acceleration in Feelings, *The Scientific World Journal*, 2013, (1): 730736.

[5] I. Koca, Efficient numerical approach for solving fractional partial differential equations with non-singular kernel derivatives, *Chaos, Solitons & Fractals*, 2018, 116, 278–286.

[6] T. O. Alade, S. Olaniyi, H. A. Idris, Y. Al Rahbi and M. Alnegga, (R2071), Global Stability Analysis of CHIKV Dynamics Model with Adaptive Immunity and Distributed Time Delays, *Applications and Applied Mathematics: An International Journal (AAM)*, 2024, 19(1): 9.

[7] A. Abidemi, M. Alnegga and T. O. Alade, A nonlinear mathematical model for exploring the optimal cost-effective therapeutic strategies and within-host viral infections spread dynamics, *Healthcare Analytics*, 2024 Jun 1 (5): 100321.

[8] T. O. Alade, Mathematical Modeling and Analysis of Virus Dynamics: Integrating Immune Responses and Transmission Modes, *The Third International Conference on Applied Mathematics in Engineering (ICAME'24)*, 2024 Jun 26, 174.

[9] S. A. Ghaleb, A. M. Elaiw, M. Alnegga, E. Ghandourah and T. O Alade, Global stability of virus dynamics of an adaptive immune response with two routes of infection and latency, *International Journal of Dynamics and Control*, 2023 Jun, 11(3): 1002–19.

[10] S. Olaniyi, T. O. Alade, F. M. Chuma, A. W. Ogunsola, O. R. Aderele and S. F. Abimbade, A fractional-order nonlinear model for a within-host chikungunya virus dynamics with adaptive immunity using Caputo derivative operator, *Healthcare Analytics*, 2023 Dec, (4):100205.

[11] A. M. Elaiw, Global properties of a class of HIV models, *Nonlinear Analysis: Real World Applications*, 2010, (11), 2253–2263.

[12] B. Li, Y. Chen, X. Lu and S. Liu, A delayed HIV-1 model with virus waning term, *Mathematical Biosciences and Engineering*, 13 (2016), 135–157.

[13] Y. Wang and X. Liu, Stability and Hopf bifurcation of a within-host chikungunya virus infection model with two delays, *Mathematics and Computers in Simulation*, 138 (2017), 31–48.

[14] A. Abidemi, Z. M. Zainuddin, N. A. Aziz, Impact of control interventions on COVID-19 population dynamics in Malaysia: a mathematical study, *The European Physical Journal Plus*, 136(2), (2021) :237.

[15] T. O. Alade, A. M. Elaiw and S. M. Alsulami, Stability dynamics of a delayed generalized Chikungunya virus infection model, *Journal of Applied Mathematics and Computing*, 65 (2021): 575–595.

[16] S. Marya, S. Harroudi, and K. Allali, Local and global stability of an HCV viral dynamics model with two routes of infection and adaptive immunity, *Computer Methods in Biomechanics and Biomedical Engineering*, 27(11), (2023), 1510–1537.

[17] A. M. Elaiw, Global properties of a class of virus infection models with multitarget cells, *Nonlinear Dynamics*, 69, (2012): 423–435.

[18] A. M. Elaiw, T. O. Taofeek and S. M. Alsulami, Global stability of within-host virus dynamics models with multitarget cells, *Mathematics*, 6(7): (2018): 118.

[19] G. Huang, Y. Takeuchi and W. Ma, Lyapunov functionals for delay differential equations model of viral infections, *SIAM Journal of Applied Mathematics*, 70(7) (2010): 2693–2708.

[20] A. M. Elaiw, T. O. Taofeek and S. M. Alsulami, Analysis of latent CHIKV dynamics model with time delays, *Journal of Computational Analysis and Applications*, 27(1), (2019): 19–36.

[21] A. M. Elaiw, T. O. Taofeek and S. M. Alsulami, Analysis of within-host CHIKV dynamics models with general incidence rate, *International Journal of Biomathematics*, 11(05), (2018): 1850062.

[22] T. O. Alade, A. Abidemi, C. Tunç and S. A Ghaleb, Global stability of generalized within-host chikungunya virus dynamics models, *Applications and Applied Mathematics: An International Journal (AAM)*, 16(1), (2021): 8.

[23] J. K. Hale and S. M. Verduyn Lunel, *Introduction to Functional Differential Equations*, (Springer Verlag, New York, 1993).

Chapter 15

Super Domination Number of Some Transformation Graphs

Yağmur Ceren Güven* and Goksen Bacak-Turan
Department of Mathematics, Manisa Celal Bayar University, Manisa, Turkey
**yagmurcerenguven@gmail.com*

For a given simple graph $G = (V, E)$, a dominating set of G is a subset $D \subseteq V$ such that every vertex in V/D is adjacent to at least one vertex in D. The dominating set problem is the problem of finding a minimum size dominating set, and its corresponding minimum size gives the domination number of G. A dominating set D is called a super dominating set of G if for every vertex $u \in \overline{D} = V/D$, there exists a $v \in D$ such that $N(v) \cap \overline{D} = u$. The super domination number of a graph G, denoted by $\gamma_{sp}(G)$, is the minimum cardinality of a super dominating set. The middle graph $M(G)$ of a graph is obtained from G by inserting a new vertex into every edge of G and joining these new vertices by edges that lie on adjacent edges of G. In this study, the super domination numbers of middle graphs are examined. Some bounds for the super domination number are obtained, and the general results for the super domination number of middle graphs of some graph families are derived.

1. Introduction

A communication network consists of centers and connection lines that enable these centers to communicate with each other. In these communication networks, the centers can be modeled as the vertices of the graph, and the connection lines as the edges of the graph. Graph theory plays a crucial role in mathematics due to its practical applications. The vertices determine the solution to mathematical problems, resulting in faster and more efficient results. The concept of domination has a wide research area in graph theory. Haynes and his colleagues published their studies in this field in a book in 1998 [1]. In 2016, Krishnakumari *et al.* found a lower bound on the number of super domination number of trees and characterised the trees that reached that bound [2]. In 2018, the super domination number of unicyclic graphs was investigated by Alfarisi *et al.* [3]. Zhuang investigated the proportions between super domination and other domination parameters in trees in 2022 [4]. At the same year, Ghanbari and his colleagues presented tight bounds for the super domination number of the neighborhood corona product [7]. They were given the exact number of minimum size super dominating sets for some graph classes.

Also, the super domination number of cactus graphs is studied by Martinez *et al.* in [8]. There are many dominating set applications such as distribution and hierarchy problems that can be modeled with a graph. This paper focuses on the super dominating set problem, which is a variant of the dominating set problem that has recently gained significant attention.

Definition 1 (Lemanska *et al.*, 2015). A subset $D \subseteq V(G)$ is a dominating set if for every vertex $u \in V(G) - D$, there exist $v \in D$ such that u and v are adjacent. The minimum cardinality of a dominating set in G is the domination number of G and is denoted by $\gamma(G)$.

There are types of domination such as total domination, global domination, total global domination, connected domination, super domination, strong domination, k-domination, etc.

Definition 2 (Ghanbari, 2022). A dominating set D of G is called a super dominating set of G, if for every vertex $u \in \overline{D}$, there exists a $v \in D$ such that $N(v) \cap \overline{D} = \{u\}$. In case we say that v super dominates u. The super domination number of G is the cardinality of a minimum size super dominating set of G, denoted by $\gamma_{sp}(G)$.

Corollary 1 (Lemanska *et al.*, 2015). *For any graph G without isolated vetrices,*

$$1 \leq \gamma(G) \leq \frac{n}{2} \leq \gamma_{sp}(G) \leq n-1$$

Example 1. The super domination number of a path graph P_5 is given in detailed below.

Fig. 1. A path graph P_5

If the set $\overline{D} = \{v_1, v_3, v_5\}$ is selected, the set D becomes $D = \{v_2, v_4\}$ since $\overline{D} = V(G)/D$. Set D is the dominating set but not the super dominating set because the vertices in the $\overline{D}$ set have common neighbors. So the $\overline{D}$ set should be reduced. For $\overline{D} = \{v_1, v_4\}$, the set D becomes $D = \{v_2, v_3, v_5\}$.

A set $D \subseteq V(G)$ is a super dominating set of G if D is a dominating set of G and for every $v \in \overline{D}$, there exist $u \in N(v) \cap \overline{D}$. The super domination number of G is

$$\gamma_{sp}(G) = |D| = 3$$

In this paper, the super domination number of middle graphs and the super domination number of generalized transformation graphs are studied.

2. Super Domination Number of Middle Graphs

Definition 3 (Kazemnejad *et al.*, 2020). The middle graph $M(G)$ of a graph G is the graph whose vertex set is $V(G) \cup E(G)$ and two vertices x, y in the vertex set of $M(G)$ are adjacent in $M(G)$ in case one the following holds

(1) $x, y \in E(G)$ and x, y are adjacent in G;

(2) $x \in V(G)$, $y \in E(G)$ and x, y are incident in G.

Example 2. Let the vertices of the path graph P_n be denoted by $V(P_n) = \{v_1, v_3, \ldots, v_{2n-1}\}$ and the edges by $E(P_n) = \{v_2, v_4, \ldots, v_{2n-2}\}$. The middle graph of P_4 is represented in Figure 2.

Fig. 2. A path graph P_4 and its middle graph $M(P_4)$

Example 3. The super domination number of the graph $M(P_4)$ is given below. The more vertices we choose for set D, the fewer vertices are left for the remaining set $\overline{D}$, which gives us the number of domination. For this reason if we select the vertices from $M(P_4)$ as v_1, v_5, this set constitutes the minimum vertex set that defines super domination.

$$\begin{aligned}
\overline{D} &= \{v_1, v_5\} \\
|\overline{D}| &= 2 \\
|D| &= |V(M(P_4))| - |\overline{D}| \\
|D| &= 7 - 2 \\
|D| &= 5
\end{aligned}$$

$\gamma_{sp}(G) \geq \frac{n}{2}$ from the Corollary 1. Clearly D is a minimum set of $M(P_4)$ and when we add one more vertex to the set $\overline{D}$, it contradicts to the definition that the vertices in the set D cannot have common neighbors. Therefore $\gamma_{sp}(M(P_4)) = 5$.

Theorem 1. *Let $M(P_n)$ be the middle graph of P_n for $n \geq 3$. The super domination number of $M(P_n)$ is*

$$\gamma_{sp}(M(P_n)) = \begin{cases} \frac{3n}{2} - 1 & n \equiv 0, 2 \ (mod\, 4) \\ \frac{3n}{2} - \frac{3}{2} & n \equiv 1, 3 \ (mod\, 4) \end{cases}$$

Proof. Let the vertices of $M(P_n)$ be labeled accordingly to the set $V(M(P_n)) = \{v_1, v_2, v_3, \ldots, v_{2n-1}\}$

$$\overline{D} = \{v_1, v_5, v_9, \ldots, v_{4k+1}\}(k \geq 1)$$

$$D = \{v_2, v_3, v_4, v_6, \ldots\}$$

$$D = V(M(P_n))/\overline{D}$$

$$\begin{aligned} |D| &= |V(M(P_n))| - |\overline{D}| \\ &= (2n-1) - \left\lceil \frac{2n-1}{4} \right\rceil \leq 2n-2 \end{aligned}$$

$$\gamma_{sp}(M(P_n)) = (2n-1) - \left\lceil \frac{2n-1}{4} \right\rceil$$

The set D is minimal because when we add one more vertex to the set $\overline{D}$, it contradicts to the definition that the vertices in the set D cannot have common neighbors.

(i) For $n = 4k$

$$\begin{aligned} 2.(4k) - 1 - \left\lceil \frac{2.(4k) - 1}{4} \right\rceil &= 8k - 1 - \left\lceil \frac{8k-1}{4} \right\rceil \\ &= 8k - 1 - 2k = 6k - 1 \\ &= 6.\left(\frac{n}{4}\right) - 1 \\ &= \frac{3n}{2} - 1 \end{aligned}$$

(ii) For $n = 4k + 1$

$$\begin{aligned} 2.(4k+1) - 1 - \left\lceil \frac{2.(4k+1) - 1}{4} \right\rceil &= 8k + 2 - 1 - \left\lceil \frac{8k+2-1}{4} \right\rceil \\ &= 8k + 1 - (2k+1) = 6k \\ &= 6.\left(\frac{n-1}{4}\right) \\ &= \frac{3n}{2} - \frac{3}{2} \end{aligned}$$

(iii) For $n = 4k + 2$

$$\begin{aligned}2.(4k+2)-1-\left\lceil \frac{2.(4k+2)-1}{4}\right\rceil &= 8k+4-1-\left\lceil \frac{8k+3}{4}\right\rceil \\ &= 8k+3-(2k+1) = 6k+2 \\ &= 6.\left(\frac{n-2}{4}\right)+2 \\ &= \frac{3n}{2}-1\end{aligned}$$

(iv) For $n = 4k + 3$

$$\begin{aligned}2.(4k+3)-1-\left\lceil \frac{2.(4k+3)-1}{4}\right\rceil &= 8k+6-1-\left\lceil \frac{8k+5}{4}\right\rceil \\ &= 8k+5-(2k+2) = 6k+3 \\ &= 6.\left(\frac{n-3}{4}\right)+3 \\ &= \frac{3n}{2}-\frac{3}{2}.\end{aligned}$$

Example 4. Let the vertices of the graph $K_{1,n}$ be $V(K_{1,n}) = \{v_0, v_1, v_2, \ldots, v_n\}$ and the edges be $E(K_{1,n}) = \{v_{01}, v_{02}, v_{03}, \ldots, v_{0n}\}$. The graph of $K_{1,n}$ and $M(K_{1,n})$ is represented in Figure 3.

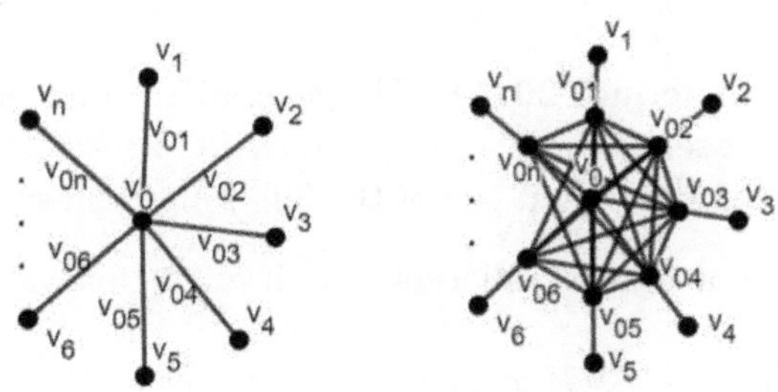

Fig. 3. $K_{1,n}$ and $M(K_{1,n})$

Theorem 2. *Let $M(K_{1,n})$ be the middle graph of $K_{1,n}$ for $n \geq 2$. The super domination number of $M(K_{1,n})$ is*

$$\gamma_{sp}(M(K_{1,n})) = n + 1$$

Proof. To fix the notation, assume $V(K_{1,n}) = \{v_0, v_1, \ldots, v_n\}$ and $E(K_{1,n}) = \{v_{01}, v_{02}, \ldots, v_{0n}\}$. Assume that the set $\overline{D}$ contains the red colored vertices represented in Figure 4 and

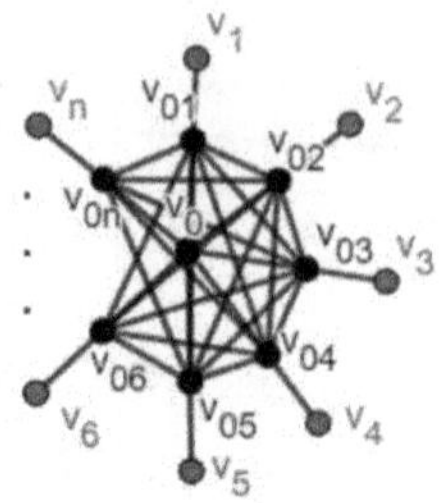

Fig. 4. $M(K_{1,n})$

$$\begin{aligned} D &= \{v_0, v_{01}, v_{02}, \ldots, v_{0n}\} \\ \overline{D} &= \{v_1, v_2, v_3, \ldots, v_n\} \\ D &= V(M(K_{1,n}))/\{v_1, v_2, v_3, \ldots, v_n\} \\ |D| &= 2n + 1 - n = n + 1 \end{aligned}$$

Clearly D is a super dominating set and $|D| = n + 1$. Since $\gamma_{sp}(G) \geq \frac{|V(G)|}{2}$ by Corollary 1 and $|V(M(K_{1,n}))| = 2n + 1$, we have

$$\gamma_{sp}(K_{1,n}) \geq \frac{|V(M(K_{1,n}))|}{2} = \left\lceil \frac{2n+1}{2} \right\rceil = n + 1$$

Thus, D is the minimum set and $\gamma_{sp}(M(K_{1,n})) = |D| = n + 1$.

3. Super Domination Number of Generalized Transformation Graphs

Definition 4 (Basavanagoud, 2015). The generalized transformation graph G^{ab} is a graph whose vertex set is $V(G) \cup E(G)$ and $\alpha, \beta \in V(G^{ab})$. The vertices α and β are adjacent in G^{ab} if and only if one of the followings holds;

(i) For $\alpha, \beta \in V(G)$, α and β are adjacent in G if $a = +$ and α, β are not adjacent in G if $a = -$.

(ii) For $\alpha \in V(G)$ and $\beta \in E(G)$, α and β are incident in G if $b = +$ and α, β are not incident in G if $b = -$.

Example 5. The generalized transformation graph of a path graph P_5 is represented in Figure 5.

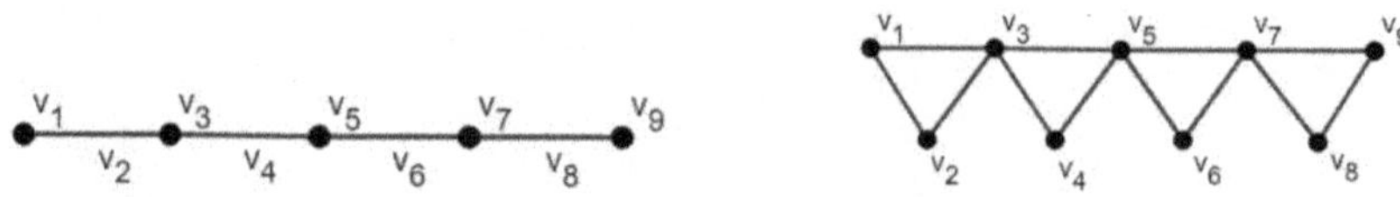

Fig. 5. A path graph P_5 and the generalized transformation graph P_5^{++}

Theorem 3. *Let P_n^{++} be the generalized transformation graph of P_n for $n \geq 3$. The super domination number is*

$$\gamma_{sp}(P_n^{++}) = 2n - 1 - \left\lceil \frac{2n-1}{5} \right\rceil$$

Proof. Let $V(P_n) = \{v_1, v_3, v_5, \ldots, v_{2n-1}\}$ and $E(P_n) = \{v_2, v_4, v_6, \ldots, v_{2n-2}\}$ such that $V(P_n^{++}) = V(P_n) \cup E(P_n)$.

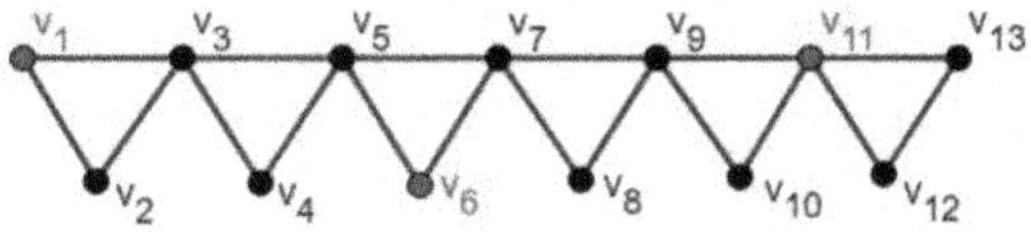

Fig. 6. The generalized transformation graph P_7^{++}

$$\begin{aligned}
\overline{D} &= \{v_1, v_6, v_{11}, \ldots, v_{5k-4}\} \quad (k \geq 1)\\
D &= \{v_2, v_3, v_4, \ldots\}\\
D &= V(P_n^{++})/\overline{D}\\
|D| &= |V(P_n^{++})| - |\overline{D}|\\
|D| &= (2n-1) - \left\lceil \tfrac{2n-1}{5} \right\rceil\\
\gamma_{sp}(P_n^{++}) &= 2n - 1 - \left\lceil \tfrac{2n-1}{5} \right\rceil.
\end{aligned}$$

Clearly D is a super dominating set of P_n^{++}. Because when we add any vertex u to the set D, the distance between u and any vertex v in D is at most two, i.e. u and v have common neighbors in D. This is a contradiction.

Example 6. The generalized transformation graph P_5^{-+} is represented in Figure 7.

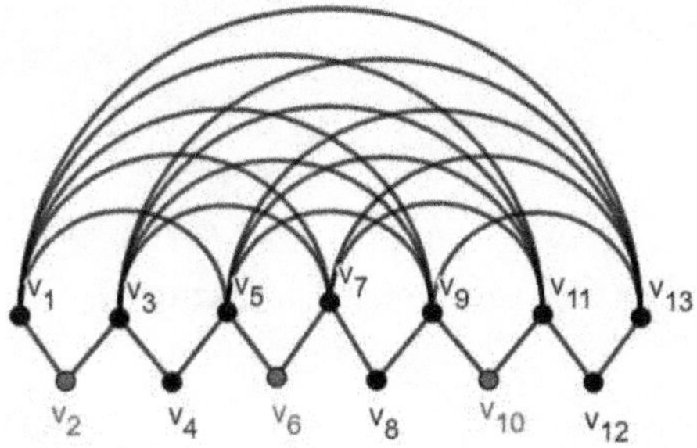

Fig. 7. The generalized transformation graph P_5^{-+}

Theorem 4. *Let P_n^{-+} be the generalized transformation graph of P_n for $n \geq 3$. The super domination number is*

$$\gamma_{sp}(P_n^{-+}) = 2n - 1 - \left\lfloor \frac{n}{2} \right\rfloor.$$

Proof. Assume $V(P_n) = \{v_1, v_3, v_5, \dots, v_{2n-1}\}$ and $E(P_n) = \{v_2, v_4, v_6, \dots, v_{2n}\}$ such that $V(P_n^{-+}) = V(P_n) \cup E(P_n)$ represented in Figure 7.

$$\begin{aligned}
\overline{D} &= \{v_2, v_6, \dots, v_{4k-2}\} \qquad (k \geq 1)\\
D &= \{v_1, v_3, v_5, \dots, v_{2n-1}\}\\
D &= V(P_n^{-+})/\overline{D}\\
|D| &= |V(P_n^{-+})| - |\overline{D}|\\
|D| &= (2n-1) - \left\lfloor \tfrac{n}{2} \right\rfloor\\
\gamma_{sp}(P_n^{-+}) &= 2n - 1 - \left\lfloor \tfrac{n}{2} \right\rfloor.
\end{aligned}$$

Similar to the idea of the proof of the previous theorem, D is a super dominating set of P_n^{-+}.

Example 7. The generalized transformation graph C_5^{++} of a cycle graph C_5 is represented in Figure 8.

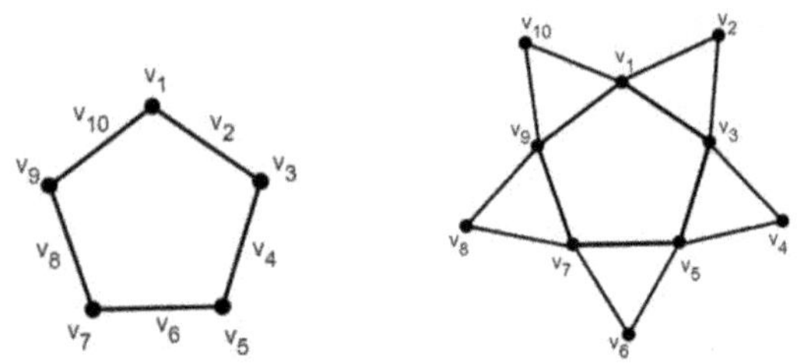

Fig. 8. A cycle graph C_5 and the generalized transformation graph C_5^{++}

Theorem 5. *Let C_n^{++} be the generalized transformation graph of C_n for $n \geq 3$. The super domination number is*

$$\gamma_{sp}(C_n^{++}) = 2n - \left\lfloor \frac{n}{2} \right\rfloor$$

Proof. Let $V(C_n) = \{v_1, v_3, v_5, \dots, v_{2n-1}\}$ and $E(C_n) = \{v_2, v_4, v_6, \dots, v_{2n}\}$ such that $V(C_n^{++}) = V(C_n) \cup E(C_n)$ represented in Figure 9.

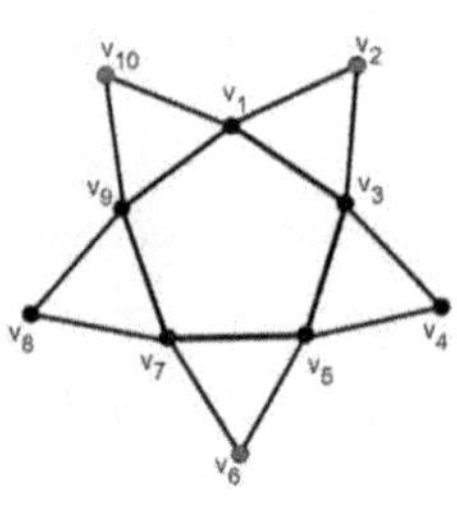

Fig. 9. The generalized transformation graph C_5^{++}

$$\begin{aligned}\overline{D} &= \{v_2, v_6, v_{10}, v_{14}, \ldots, v_{4k-2}\}(k \geq 1)\\ D &= \{v_1, v_3, v_4, v_5, \ldots\}\\ D &= V(P_n^{-+})/\overline{D}\\ |D| &= |V(P_n^{-+})| - |\overline{D}|\\ |D| &= 2n - \lfloor \tfrac{n}{2} \rfloor\end{aligned}$$

$$\gamma_{sp}(C_n^{++}) = 2n - \lfloor \tfrac{n}{2} \rfloor.$$

Example 8. The generalized transformation graph C_5^{-+} is represented in Figure 10.

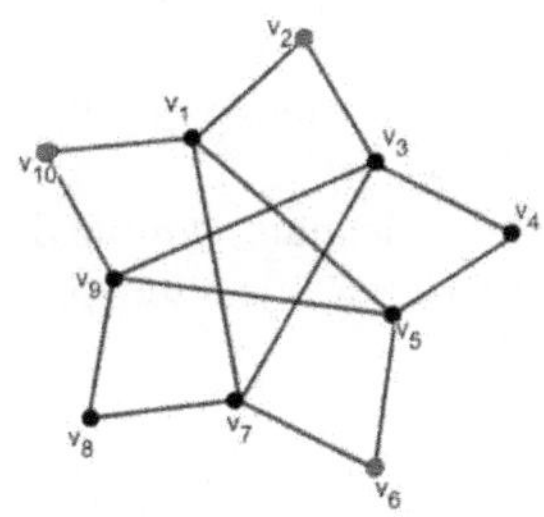

Fig. 10. The generalized transformation graph C_5^{-+} of a cycle graph C_5

Theorem 6. *Let C_n^{-+} be the generalized transformation graph of C_n for $n \geq 3$. The super domination number is*

$$\gamma_{sp}(C_n^{-+}) = 2n - \left\lfloor \frac{n}{2} \right\rfloor$$

Proof. Let $V(C_n) = \{v_1, v_3, v_5, \ldots, v_{2n-1}\}$ and $E(C_n) = \{v_2, v_4, v_6, \ldots, v_{2n}\}$ such that $V(C_n^{-+}) = V(C_n) \cup E(C_n)$ represented in Figure 10.

$$\begin{aligned}\overline{D} &= \{v_2, v_6, v_{10}, v_{14}, \ldots, v_{4k-2}\}(k \geq 1)\\ D &= \{v_1, v_3, v_5, \ldots, v_{2n-1}\}\\ D &= V(P_n^{-+})/\overline{D}\\ |D| &= |V(P_n^{-+})| - |\overline{D}|\\ |D| &= 2n - \lfloor \tfrac{n}{2} \rfloor\end{aligned}$$

$$\gamma_{sp}(C_n^{-+}) = 2n - \lfloor \tfrac{n}{2} \rfloor.$$

Clearly D is a super dominating set of C_n^{-+} similar to the proof of Theorem 5.

4. Conclusions

In this study, a new parameter, super domination number, is considered. The transformation graphs of some classes of graphs are analysed and general formulae giving the super domination number of these graphs are obtained. In future work, it is proposed to apply this parameter to different types of graphs, especially to the transformation graphs of different graphs. The transformation of a graph leads to the loss of its original state, resulting in the formation of a new graph structure. If it is feasible to decode the given graph from the encoded graph in polynomial time, this process can be employed to analyse several structural properties of the original graph in terms of transformation graphs. Since the studies on the super domination number are limited, it is thought that applying this new parameter to transformation graphs and obtaining general results will be an important contribution to the literature.

References

[1] T.W. Haynes, S.T. Hedetniemi, P.J. Slater, 1998, *Fundamentals of Domination in Graphs*, Marcel Dekker, New York.

[2] B. Krishnakumari, Y. B. Venkatakrishnan, Double domination and super domination in trees, *Discrete Mathematics, Algorithms and Applications*, 08(04) (2016): 1650067. doi:10.1142/S1793830916500671.

[3] R. Alfarisi, Dafik, R. Adawiyah, R. M. Prihandini, E. R. Albirri, I. H. Agustin, Super domination number of unicyclic graphs, IOP Conference Series: Earth and Environmental Science, 243 (2019): 012074. doi:10.1088/1755-1315/243/1/012074.

[4] W. Zhuang, Super Domination in Trees, *Graphs and Combinatorics*, 38(1) (2022). doi:10.1007/s00373-021-02409-3.

[5] N. Ghanbari, Some results on the super domination number of a graph, *Discrete Mathematics, Algorithms and Application*, 16(04) (2024): 2350044. doi:10.1142/S1793830923500441.

[6] M. Lemanska, V. Swaminathan , Y. B. Venkatakrishan, R. Zuazua, Super dominating sets in graphs, Proceedings of the National Academy of Sciences, India Section A: Physical Sciences, 85(3) (2015), 353–357. doi:10.1007/s40010-015-0208-2.

[7] N. Ghanbari, G. Jager, T. Lehtila, Super domination: Graphs classes, products and enumeration, *Discrete Applied Mathematics*, 349 (2024): 8–24. doi:10.1016/j.dam.2024.01.039.

[8] A. Cabrera-Martinez, A. C. Peiro, Relating the super domination and 2-domination numbers in cactus graphs, *Open Mathematics*, 2 (2023). doi:10.1515/math-2022-0583.

[9] F. Kazemnejad, B. Pahlavsay, E. Palezzato, M. Torielli, Total Domination Number of Middle Graphs, *Electronic Journal of Graph Theory and Applications*, 10 (1) (2022). doi:10.5614/ejgta.2022.10.1.19.

[10] B. Basavanagoud, I. Gutman, V.R. Desai, Zagreb indices of generalized transformation graphs and their complements, *Kragujevac Journal of Science*, 37(37) (2015): 99–112. doi:10.5937/KgJSci1537099B.

[11] J.A. Rodiguez-Valezquez, D.J. Klein, E. Yi, On the super domination number of graphs, *Communications in Combinatorics and Optimization*, 5(2) (2020): 83–96. doi:10.22049/CCO.2019.26587.1122.

Chapter 16

Approximate Solution to a Non-linear Diffusion Model Implementing the $Sinc_{\pi}(x^n)$ Function

Jordan Hristov

Department of Chemical Engineering, University of Chemical Technology and Metallurgy, Sofia 1756 8 Kliment Ohridsky, Blvd, Bulgaria

jyh@uctm.edu

These are preliminary results on an approximate solution of a non-linear diffusion model with a power-law-type concentration-dependent diffusivity in a semi-infinite domain by the integral-balance method implementing modified function $sinc_{\pi}(x^n)$ function.

1. Introduction

Approximate integral-balance solutions parabolic diffusion models are based on the concept of Goodman [1] about the final penetration depth of the diffusant resulting in a sharp front of the solution propagation [2–6]. It was extensively investigated for years and applied to solutions of practically important problems [5, 7, 8]. The method needs a definition of an appropriate assumed profile and redefinition of the boundary conditions at the solution front (the so-called Goodman's boundary conditions-see Section 2.1). This communication addresses an approximate solution of a temperature (concentration)-dependent thermal (mass) diffusion model (see more details and comments about the model and its properties see [9] and references therein)

$$\frac{\partial u}{\partial t} = \frac{\partial}{\partial x}\left(a\left(u\right)\frac{\partial u}{\partial x}\right) \Rightarrow \frac{\partial u}{\partial t} = \frac{\partial}{\partial x}\left(a_0 u^m \frac{\partial u}{\partial x}\right). \tag{1}$$

The non-linear diffusivity is a power-law function of a scaled temperature (concentration) $D = D_0(u/u_{ref})^m$, $m > 0$ in the case of mass diffusion, where D_0 is the concentration-independent diffusivity) [9]. The model (1), is uniformly parabolic but degenerates for $\theta = 0$ [10] followed by solutions with finite speeds and sharp fronts pertinent to creeping flows [11, 12], non-linear heat conductivity (diffusion) [13], porous media filtration [14].

The idea motivating this work is to build an approximate integral-balance solution to (1) with an assumed profile based on the cardinal $sinc(x)$ (explained in Section 2.2), specifically its modification $Sinc_{\pi}(x^n)$.

2. The background of the new solution

2.1. *Double-integration method*

The double-integration and application of the Leibniz rules yield [9] (for the sake of the clarity of the idea we use the case with $m = 0$) yields

$$\int_0^{\delta}\int_x^{\delta}\frac{\partial u\,(x,t)}{\partial t}dxdx = \int_0^{\delta}\int_x^{\delta}a_0\frac{\partial^2 u\,(x,t)}{\partial x^2}dxdx \Rightarrow \frac{d}{dt}\int_0^{\delta}\int_x^{\delta}u\,(x,t)dxdx = a_0 u\,(0,t) \quad (2)$$

The right-hand side of the integral relation depends on the boundary condition only. However, when $m \neq 0$ a preliminary transform of the diffusion term is needed (the left expression of (3)), and this allows us to present the model in a more convenient form (the right expression of (3))

$$a_0 u^m\frac{\partial u}{\partial x} = \frac{a_0}{m+1}\frac{\partial u^{m+1}}{\partial x} \Rightarrow \frac{\partial u\,(x,t)}{\partial t} = \frac{a_0}{m+1}\frac{\partial u^{m+1}}{\partial x^2} \quad (3)$$

Then, $u(x,t)$ should be replaced in (2) by an assumed profile $u_a = u_a\,(x/\delta\,)$ as a function of the dimensionless ratio (distance) x/δ [9], obeying the so-called Goodman's boundary conditions $u\,(\delta) = \frac{\partial u}{\partial x}\,(\delta) = 0$. This yields a differential equation concerning $\delta\,(t)$ (see in Section 3).

2.2. *sinc(x) functions and a new modification*

2.2.1. *sinc(x) and $sinc_{\pi}(x)$ functions*

The non-normalized $sinc\,(x) = sin\,(x)/x$ function is defined as [15]: $sinc(x) = \frac{sin(x)}{x}$, $x \neq 0$, $sinc(0) = 1$, $sinc(1) = 0$ while the normalized version is: $sinc_{\pi}(\pi x) = \frac{sin(\pi x)}{\pi x}$, $x \neq 0$, $sinc_{\pi}(0) = 1$, $sinc_{\pi}(1) = 0$ (see Figure 1-a). Also, $sinc\,(x = 0) \sim \lim_{x\to 0}\frac{\sin(kx)}{kx} = 1$ for all real $k \neq 0$. These two versions are presented in Figure 1-a). The *sinc* function has a Taylor series expansion, converging for all x [16]

$$sinc\,(x) = \frac{\sin\,(x)}{x} = \sum_{k=0}^{\infty}\frac{(-1)^k\,(x^2)^k}{(2k+1)!} = 1 - \frac{x^2}{3!} + \frac{x^4}{5!} - \frac{x^6}{7!} + \cdots + |0|^{11} \quad (4)$$

For the normalized version we have

$$\frac{\sin\,(\pi x)}{\pi x} = \sum_{k=0}^{\infty}\frac{(-1)^k\,(\pi x)^{2k}}{(2k+1)!} = 1 - \frac{(\pi x)^2}{3!} + \frac{(\pi x)^4}{5!} - \frac{(\pi x)^6}{7!} + \cdots + |0|^{11} \quad (5)$$

2.2.2. *Modified $sinc_{\pi}(x^n)$ function*

For the approximate solution developed next, we introduce a modified version $Sinc_{\pi}\,(x^n)$, defined as

$$Sinc_{\pi}\,(x^n) = \frac{\sin\,(\pi x^n)}{\pi x^n}, \quad n > 0, \quad 0 \leq x \leq 1 \quad (6)$$

Denoting $y = x^n$ we may have a series expansion of (6) such as (5). Moreover, a very attractive feature of $Sinc_\pi\,(x^n)$ is its flexibility in changing the shape as a result of variations of the exponent n: For $n > 1$ all profiles are convex (see Figure 1-b).

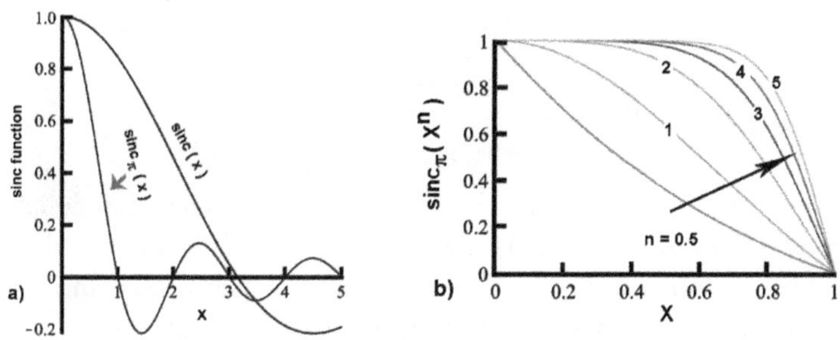

Fig. 1. Comparative plots of $sinc(x)$ functions. (a) $sinc(x)$ and $sinc_\pi(x)$ in the range $0 < x < 1$; (b) convex profiles of $Sinc_\pi(x^n)$ for $n > 1$ and $0 < x < 1$.

3. Approximate solution

We suggest an approximate profile defined through $Sinc_\pi(x^n)$ as

$$u_a = u_s Sinc_\pi\left[\pi\left(\frac{x}{\delta(t)}\right)^n\right] = u_s\frac{\sin(\pi X^n)}{\pi X^n}, \quad 0 \le X = x/\delta \le 1 \tag{7}$$

where $u_s = u_a(x=0)$ because for $x = 0 \Rightarrow x/\delta(t) = X = 0$, and $Sinc_\pi(\pi X^n)_{X\to 0} = 1$. With a Dirichlet boundary condition to Eq. (1) and a semi-infinite medium we get from (7) at $x = 0$ that $u_s = 1$.

The integral balance relation of DIM, expressed through the dimensionless coordinate $X = x/\delta$, yields

$$\frac{d\delta^2}{dt}\underbrace{\left[\int_0^1\left(\int_X^1\frac{\sin(\pi X^n)}{\pi X^n}dX\right)dX\right]}_{S_2(\pi,X)} = \frac{a_0}{m+1} \Rightarrow S_2\frac{d\delta^2}{dt} = \frac{a_0}{m+1} \tag{8}$$

With $S_2(\pi, X)$, expressed trough a truncated series, converging rapidly, we get ($K = 10$ assures very good results in accordance with (5)).

$$S_2(\pi, X) = \sum_{k=0}^{K}\frac{(-1)^k\pi^{2k}}{(2kn+1)(2k+1)!}\left[1 - \frac{1}{(2kn+1)(2kn+2)}\right] \tag{9}$$

The solution of (8) (with $\delta(t=0) = 0$) is

$$\delta = \sqrt{a_0 t}\sqrt{\frac{s_2}{m+1}} \Rightarrow \frac{\delta(t)}{\sqrt{a_0 t}} = \sqrt{\frac{s_2}{m+1}} \Rightarrow \delta^2 = \frac{a_0 t}{S_2(m+1)}, \quad s_2 = 1/S_2 \tag{10}$$

From the definition of $X = x/\delta(t)$ we may express X through the Boltzmann similarity variable $\eta = x/\sqrt{a_0 t}$

$$X = \frac{x}{\sqrt{a_0 t}\sqrt{S_2(m+1)}} = \frac{\eta}{\sqrt{S_2(m+1)}}, \quad \eta = \frac{x}{\sqrt{a_0 t}} \tag{11}$$

Thus, the approximate solution, in terms of η, can be expressed as

$$u_a = \frac{\sin\left(\pi\left(\frac{\eta}{\sqrt{S_2(m+1)}}\right)^n\right)}{\pi\left(\frac{\eta}{\sqrt{S_2(m+1)}}\right)^n} \tag{12}$$

We can see, from Figure 2 that with the increase in n the penetration depth becomes shorter (Figure 2a). The same effect is observed when m is stipulated but n varies (Figure 2b). Hence, both n and m affect the shape of the profile in similar ways and we should see further whether are they interrelated or not.

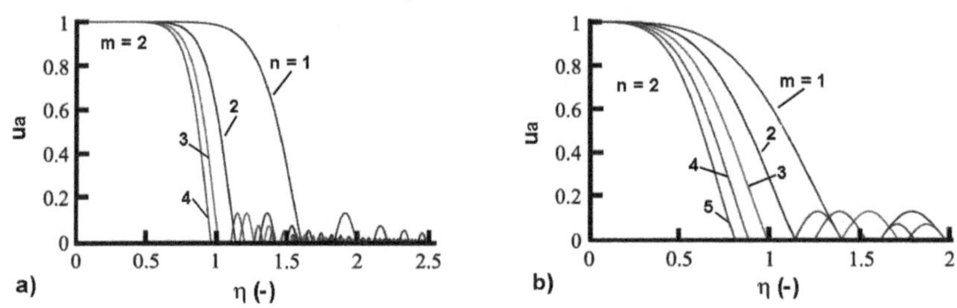

Fig. 2. Approximate solutions (qualitative plots) against the similarity variable η: (a) For a stipulated m and various n; (b) For a stipulated n and various m .
Note: the plots demonstrate oscillatory behaviors beyond the points where the function crosses the abscissa (the fronts of the penetration depths), which are inherent for the $sinc(x)$ function, but this section is outside the zone where the solution is developed

Furthermore, when compared to the solution of Heaslet and Alksne [17], the solution presentations as functions of the dimensionless variable (see Figure 3) indicate that while the approximation errors are unacceptable, the solution needs to be refined (see the next section). The profile shapes are qualitatively suitable. Although Heaslet and Alksne's [17] solution is likewise approximate and was produced using the moment technique, it has been used for years to compare with other approximate solutions that have been developed (see, for instance [9]). Consequently, we require a way to reduce the approximation error.

4. Solution improvement

It was demonstrated in [18] that the concept of constant optimal exponent n of the assumed profile does not work and the approximation errors are too high when compared with reference solutions (the moment solution of Heaslet and Alksne [17] and the parabolic profile solution developed in [9]). Looking for a suitable way to resolve the emerging situation, in the following an idea (conjecture) of a variable exponent is developed.

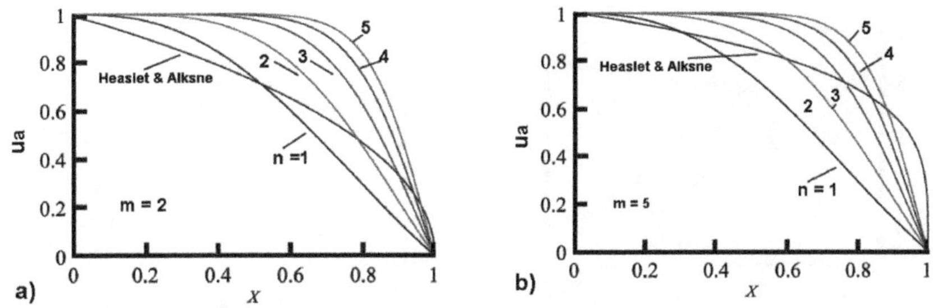

Fig. 3. Qualitative plots of approximate solutions as functions of the dimensionless variable X for various values of the exponent n compared with the solution of Heaslet and Alksne [17]: (a) for $m = 2$ and, (b) for $m = 5$

4.1. *An idea concerning a variable exponent* $n(X)$

We suggest the following $n = f(X)$ functional relationships.

$$\frac{1}{n} = 1 - pX, \quad 0 < p < 1 \tag{13}$$

This approach is mainly driven by intuition and the preceding author's experience in the optimization of integral-balance solutions [19]. It is important to note that, as demonstrated in [9], $1/m$ defines the best exponent of the parabolic profiles. Slightly changing the prefactor p in (13) to $p = \lambda_m/m$ we are able quickly to refine the approximate profiles.

The plots in Figure 4 demonstrate the function $n = f(p, X)$ as a $3D$ smooth surface with a well-defined "valley" along the diagonal $(1, 1) - (00,)$ (Figure 4-a). The correlation of the optimal values of the parameter p (see further in this section) is well-fitted by $1 - p = m^{-1.358}$ (see Figure 5-b) with $R = 0.987$, $P < 0.001$ and standard error of estimate 0.11.

The efficiency of the conjecture (13) is demonstrated by the plots in panels c) and d) of Figure 4. The possibility of changing the plot shape by variations in the parameter p, precisely the variations of exponent n with the independent variable X, leads to successful approximations with errors less than 0.06 (in this specific case we have $m = 4$ and $p = 0.885$, but this is not the optimal value- see the right column in Figure 6 for optimized solutions).

It is worth noting, that for very high values of the exponent n the function $sinc\,(\pi X^n)$ generates an almost rectangular profiles. In Figure 4-b such an ultimate case is presented by the dashed red lines forming the square $[1, 1]$. However, since $n \to \infty$ is only at the point $X = 1$, for $p = 1$, the only vertical asymptote in this case is the line perpendicular to the abscissa at $X = 1$. The generated profile, for $p = 1$, when $X < 1$, however, is not horizontal, since for $n = 1/(1 - X)$, the exponent n is less than infinity and it approaches the extreme limit only at $X = 1$. Therefore, with $0 < p < 1$, the profiles are smooth, and convex and approach the solutions developed by other methods as will demonstrate in what follows.

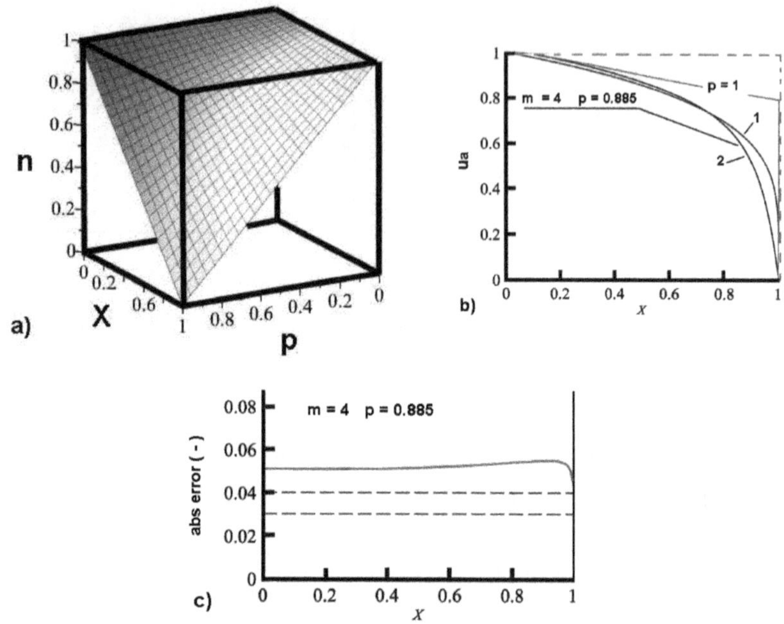

Fig. 4. The function $n = f(p, X)$ and its effect on the approximate solutions: (a) $3D$ presentation of $n = f(p, X)$; (b) Approximate solutions compared with the Heaslet and Alksne solution [17] (left column) (the red lines in the online version); (c) Absolute errors of approximation. The dashed lines mark the common errors attainable by integral-balance solutions [9].

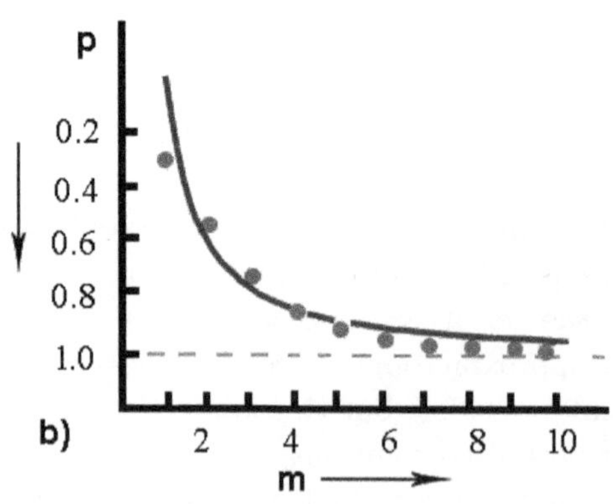

Fig. 5. Correlation of the optimal values of the parameter p

Starting from this initial achievement we are going to define optimal values of p through minimization of the residual function as developed next.

4.2. *Optimization of the solution defining optimal value of p*

The residual function of (1), with $Y = \frac{sin(\pi X^{n(X)}}{\pi X^{n(X)}}$, is defined as (14)

$$R(n,X,m)=\frac{\partial}{\partial t}Y\left(X^{n(X)}\right)-\frac{a_0}{m+1}\frac{\partial^2}{\partial x^2}\left[Y\left(X^{n(X)}\right)\right]^{m+1}\to\min \tag{14}$$

It is not zero, since an approximate solution is used, instead of an exact one, if any. However, it should be minimized over the domain $0\leq X\leq 1$ for a given value of the exponent n. Detailed expressions of $R(n,X,m)$ are available elsewhere [18]. Based on the conjecture (13) the time and space derivatives are

$$\frac{\partial Y}{\partial t}=\frac{p\frac{\partial X(x,t)}{\partial t}\ln(x)}{\left[1-pX(x,t)\right]^2}\left[\cos\left(\pi X^{\frac{1}{1-pX(x,t)}}\right)-\frac{\sin\left(\pi X^{\frac{1}{1-pX(x,t)}}\right)}{\pi X^{\frac{1}{1-pX(x,t)}}}\right] \tag{15}$$

$$\frac{\partial Y}{\partial x}=\left(\frac{p\frac{\partial X(x,t)}{\partial x}\ln(x)}{\left[1-pX(x,t)\right]^2}+\frac{1}{x\left[1-pX(x,t)\right]}\right)\left(\cos\left(\pi X^{\frac{1}{1-pX(x,t)}}\right)-Y(p,X)\right) \tag{16}$$

Derivatives, especially $\frac{\partial^2 Y}{\partial x^2}$ and $\frac{\partial^2 Y^{m+1}}{\partial x^2}$ can be easily developed from (15) and (16) using computer algebra software (Maple, for instance), but due to the limited space, we steer clear of cumbersome terms that everyone can apply. Hence, the profile optimization is a quest for values of the parameter p minimizing the residual function, thus leading to acceptable approximation errors.

The correlation of the optimal values of the parameter p (see further in this section) is well-fitted by $1-p=m^{-1.358}$ (see Figure 5)(with $R=0.987$, $P<0.001$ and standard error of estimate 0.11).

The plots in Figure 6 confirm the feasibility of the conjecture of variable exponent $n(p,X)$. In general, the approximate solutions become more effective for high values of the non-linearity exponent m. For all solutions developed, the errors increase sharply close to the steep fronts of the profiles. It was nothing, that this exists for all approximate solutions due to the very high gradient of the function in this section of the profile (see [9], for instance). The section, at the abscissa, where such a sharp increase of the approximation error starts, moves to the right as the value of m increases. Thus, the solutions with the modified $Sinc_\pi(\pi X^{n(X)})$ function are more effective for high values of m when the model solutions correspond to almost rectangular profiles.

5. Some Comments and Outlines

This short communication just demonstrated the possibility of using the normalized $sinc(\pi x)$ function, precisely its modification $Sinc_\pi(\pi x^n)$, thus allowing the profile to be easily controlled by the value of the exponent n. The conjecture concerning an exponent dependent on the variable $X=x/\delta$ envisages a new trend in the modeling of non-linear diffusion problems by integral methods, thus avoiding emerging problems with the concept of a fixed optimal exponent, commonly applied when parabolic assumed profiles are used. Diffusion models generating almost rectangular concentration profiles are of special interest in modeling moisture penetration

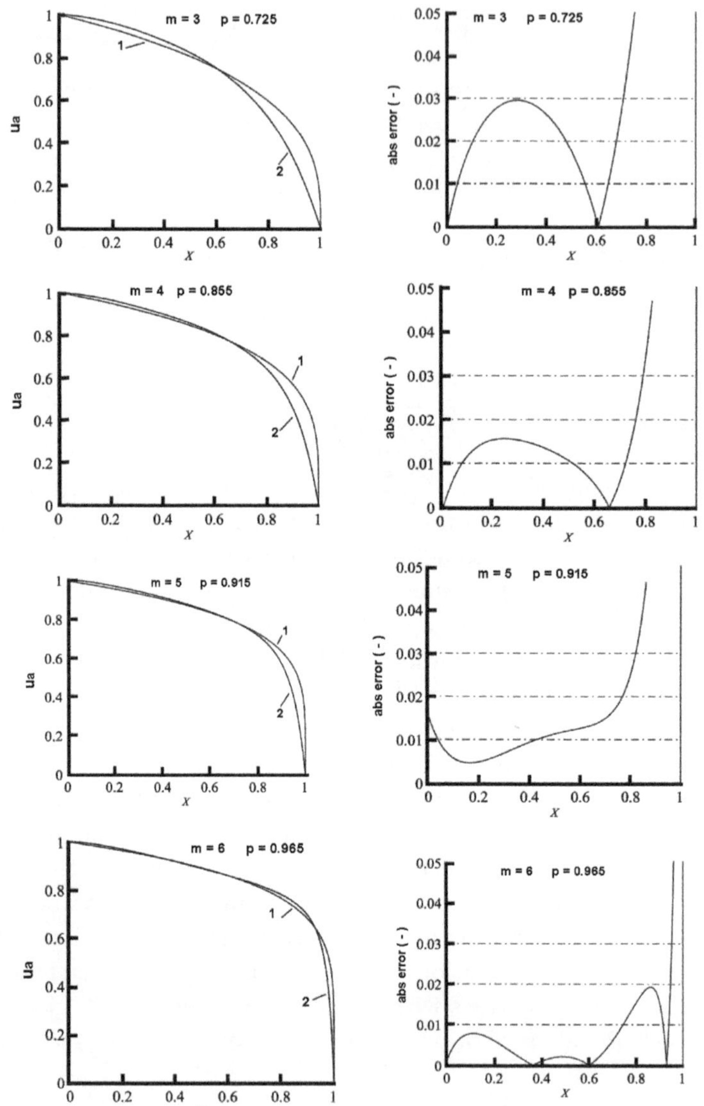

Fig. 6. Approximate solutions (lines 2, blue color in online version) with optimal values of the parameter p. Left column: compared to the solution of Heaslet and Alksne [17] (lines 1, red color in online version); Right column: Errors of approximations.
Note: The dash-dot lines mark the levels of errors commonly attainable by integral-balance solutions (see for instance [9]).

into concretes [20], soils [21], building materials [22, 22–24], adsorption [25], and some cases of magnetic diffusion [26, 27]. Hence, we can foresee the development of novel implementations of the modified $Sinc_{\pi}(\pi X^n)$ function in solutions of strong non-linear diffusion problems emerging in various applied fields.

References

[1] T.R. Goodman, The heat balance integral and its application to problems involving a change of phase, *Trans. ASME.* **80** (2), 335–342, (1958).

[2] A.S. Wood, S. Kutluay, A heat balance integral model of the thermistor, *Int. J. Heat Mass Transfer*, **38** (10), 1831–1840, (1995).

[3] J.K.J. Staggs, Approximate solutions for the pyrolysis of char forming and filled polymers under thermally thick conditions, *Fire and Materials*, **24** (6), 305–308 (2000).

[4] J. Hristov, The heat-balance integral method by a parabolic profile with unspecified exponent: Analysis and benchmark exercises, *Thermal Science*, **13** (2), 27–48 (2009).

[5] J. Hristov, An approximate analytical (integral-balance) solution to a nonlinear heat diffusion equation. *Thermal Science*, **19** (2), 723–733 (2014).

[6] J. Bollati and D. A. Tarzia, Approximate solutions to one-phase Stefan-like problems with space-dependent latent heat, *European J. Appl. Math.*, **32** (2), 337–369 (2021).

[7] E.P. Canzian, F. Santiago, A.V. Brito Lopes, M.R. Barbosa and A.G. Barañano, On the application of the double integral method with quadratic temperature profile for spherical solidification of Lead and Tin metals, *Appl. Thermal Eng.*, **219**, 119528 (2023).

[8] J. Bollati, M.F. Natale, M.F. Semitiel and D.A. Tarzia, A two-phase Stefan problem with power-type temperature dependent thermal conductivity. Existence of a solution by two fixed points and numerical results, *AIMS Math.*, **9** (8), 21189 (2024).

[9] J. Hristov, Integral solutions to transient nonlinear heat (mass) diffusion with a power-law diffusivity: A semi-infinite medium with fixed boundary conditions, *Heat Mass Transfer*, **52** (3), 635–655 (2016).

[10] J. M. Hill, Similarity solutions for nonlinear diffusion-a new integration procedure, *J. Eng. Math.*, **23** (2), 141–155 (1989).

[11] J. Buckmaster, Viscous sheets advancing over dry beds, *J. Fluid Mech.*, **81** (4), 735–756 (1977).

[12] B.M. Marino, L.P. Thomas, R. Gratton, J.A. Diez, and S. Betelu, Waiting-time solutions of a nonlinear diffusion equation: Experimental study of a creeping flow near a waiting front, *Phys. Rev. E*, **54** (3), 2628–2636 (1996).

[13] S.N. Prasad and J.B. Salomon, A new method for analytical solution of a degenerate diffusion equation, *Adv. Water Res.*, **28** (10), 1091–1101 (2005).

[14] D.G. Aronson, The porous medium equation, In eds. A. Fasano and M. Primicerio, Nonlinear Diffusion Problems, *Lecture Notes in Mathematics*, Vol. 1224 (Springer, Berlin, Heidelberg, 1986), pp. 1–46.

[15] D.M. Bradley, Note on Dirichlet sinc integral, *The Am. Sci. Monthly*, **128** (3), 273–274 (2021).

[16] M.E. Selariu, F. Smarandache, M. Nitu, and M. Cardinal functions and integral functions, *Int. J. Geometry*, **1**(1), 54–66 (2012).

[17] M.A. Heaslet and A. Alksne, Diffusion from a fixed surface with concentration-dependent coefficient, *J. Soc. Ind. Appl. Math.*, **9** (4), 584–596 (1961).

[18] J. Hristov, A non-linear diffusion problem with power-law diffusivity: An approximate solution experimenting with modified sinc function, *Math. Mod. Num. Sim. Appl.*, (2024).

[19] J. Hristov, The Heat-Balance Integral: 2. A Parabolic profile with a variable exponent: the concept and numerical experiments, *C.R. Mechanique, Paris*, **340**(7), 493–500 (2012).

[20] D. Lockington, J-Y. Parlange and P. Dux, Sorptivity and the estimation of water penetration into unsaturated concrete, *Mat. Struct.*, **32**, 342–347 (1999).

[21] D. Lockington, Estimating the sorptivity for a wide range of diffusivity dependence on water content, *Transport in Porous Media*, **10** (1), 95–101 (1993).

[22] I. Ioannou, A. Andreou, B. Tsikouras and K. Hatzipanagiotu, Application of the sharp front model to capillary absorption in a vuggy limestone, *Eng. Geol.*, **105** (1–2), 20–23 (2009), doi:10.101/j.enggeo.2008.12.008.

[23] M. Kuntz and P. Lavallee, Experimental evidence and theoretical analysis of anomalous diffusion during water infiltration in porous building materials, *J. Phys. D: Appl. Phys.*, **34** (16), 2547 (2001).

[24] J. Hristov and M. El-Ganaoui, Modelling capillary absorption in building materials with emphasis on the fourth root time law: Time-fractional models, solutions and analysis, In eds P. Agarwal, C. Cattani and S. Momany, *Fractional Differential Equations: Theoretical Aspects and Applications*, (Academic Press, London, UK, 2024), pp. 153–165.

[25] J. Hristov, A new closed-form approximate solution to diffusion with quadratic Fujita's non-linearity: The case of diffusion controlled sorption kinetics relevant to rectangular adsorption isotherms, *Heat Mass Transfer*, **55**, 261–279 (2019).

[26] J. Hristov, Magnetic field diffusion in ferromagnetic materials: Fractional calculus approaches and approximate solutions, *An International Journal of Optimization and Control: Theories & Applications (IJOCTA)*, **11** (3), 1–15 (2021).

[27] B. Xiao, G. Wang, L. Zhao, C. Feng and S. Shu, An exact solution for the magnetic diffusion problem with a step-function resistivity model, *Eur. Phys. J. Plus*, **139**, 305 (2024).

Chapter 17

Harnessing AI for Early Detection and Prediction of Infectious Disease Outbreaks: A Case Study of COVID-19 and Beyond

Oluwaseun Olumide Okundalaye[*,‡] and Necati Özdemir[†,§]

[*] *Adekunle Ajasin University, Akungba-Akoko, Department of Mathematical Sciences, Faculty of Science, Ondo State, Nigeria*

[†] *Balikesir University, Department of Mathematics, Faculty of Science and Art, Balikesir, Türkiye*

[‡] *okundalaye.oluwaseun@aaua.edu.ng*

[§] *nozdemir@balikesir.edu.tr*

The urgent need for accurate and timely infectious disease surveillance was brought to light by the COVID-19 pandemic. By analyzing large datasets and finding patterns that could point to the beginning of epidemics, artificial intelligence (AI). In this chapter, machine learning (ML) and deep learning (DL) models demonstrate the potential for accurate forecasting, supporting public health decision-making processes. Specifically, the COVID-19 pandemic is examined with AI's role in the early detection and prediction of infectious disease outbreaks. This chapter explores AI's methodologies, challenges, and successes in combating infectious diseases. Finally, we discuss the implications for future pandemics, including the need for further research, data integration, and model improvements.

1. Introduction

Throughout history, infectious diseases have posed a constant threat to global health, causing widespread illness and death. Emergencies like the Spanish flu of 1918, the Ebola virus, and the COVID-19 pandemic recently have shown how crucial early detection and prompt response are to controlling outbreaks and reducing the effects on public health [1, 11, 12]. However, in contexts with limited resources, where reporting delays and data gaps are typical, standard surveillance systems frequently fail to deliver timely, correct information [2]. Artificial intelligence (AI) presents novel options for tackling these challenges. AI technologies are being used more and more in the healthcare industry [3]. These technologies, including machine learning (ML), deep learning (DL), and natural language processing (NLP), have demonstrated great promise in improving illness prediction and surveillance [4]. AI-driven tools, such as decision trees and Long Short-Term Memory, were among the first to identify and forecast the spread of the COVID-19 pandemic. These algorithms predicted outbreak areas and directed public health initiatives by utilizing massive datasets, such as travel data, epidemiological records, and even social media trends [5].

The main goal of this chapter is to examine how AI helped in early detection and outbreak prediction during the COVID-19 pandemic. We will discuss predictive modelling techniques, present the results of various AI applications, and look at how these models can be adapted to tackle future challenges in infectious disease prediction.

2. Preliminaries

2.1. *Traditional machine learning models*

Support Vector Machines (SVM): SVM is a supervised learning model used for classification tasks. In the context of infectious diseases, SVMs have been applied to categorize patient outcomes or predict the spread of disease based on epidemiological and demographic data. SVMs are excellent at handling structured data and provide fast, reliable outputs, but they are less effective when it comes to analyzing time-series data or complex patterns involving unstructured data sources like text or social media signals.
Decision Trees and Random Forests: Decision trees split data into subsets based on feature values, making them useful for identifying decision rules that can predict disease outcomes (e.g., predicting the likelihood of an individual contracting a disease based on travel history or exposure). Random forests, which consist of multiple decision trees, enhance accuracy by reducing overfitting, especially in datasets with complex, nonlinear relationships. These models, however, struggle with time-dependent predictions, which are critical in infectious disease forecasting.

2.2. *Deep learning models:*

Recurrent Neural Networks (RNNs): RNNs are a class of artificial neural networks that excel in handling sequential data. They are particularly useful in modelling the progression of infectious diseases over time, as they maintain the memory of previous data points in the sequence, allowing for better predictions of future events based on past trends. A notable limitation of RNNs is their tendency to suffer from the vanishing gradient problem, which makes it challenging for them to capture long-term dependencies.
Long Short-Term Memory Networks (LSTMs): LSTMs are a specialized form of RNNs designed to overcome the limitations of standard RNNs by preserving long-term dependencies in data sequences. LSTMs have been particularly effective in forecasting trends in infectious diseases, such as predicting COVID-19 infection rates over time or identifying potential outbreak hotspots. The ability of LSTMs to maintain historical information over extended periods makes them indispensable for tasks involving time-series prediction and real-time monitoring of epidemics.

2.3. *AI models for disease prediction*

Two main types of AI models for predicting infectious disease outbreaks include deep learning (DL) models and traditional machine learning (ML) models. These models are designed to analyze vast amounts of diverse data, identify patterns, and forecast future outbreak trends. Several of the often-employed AI methods are as follows: Machine Learning Models: Traditional supervised learning models have been used for disease classification and regression tasks. These models include support vector machines (SVM), decision trees, and random forests. Based on labelled datasets, these models are excellent for processing structured data and making rapid predictions [6]. Deep Learning Models: recurrent neural networks (RNNs) and long short-term memory networks (LSTM) networks are two examples of DL models that are particularly good at processing unstructured data and time-series predictions. These models identify long-term connections in the data, which makes them very useful for predicting trends in infectious disease epidemics over time [7].

2.4. *Data sources*

To predict COVID-19 spread, epidemiological data (e.g., daily infection rates, recoveries, deaths) were utilized. These AI systems used a combination of machine learning and deep learning models to parse massive amounts of data for health databases, providing early warnings and valuable predictions regarding the spread of the virus.

3. Model Formulation

3.1. *SEIR model*

The SEIR model is an extension of the traditional Susceptible-Infection-Recovered (SIR) model, with an added compartment for exposed individuals who are infected but not yet infectious [8]. The system of differential equations governing the SEIR model is as follows: S(t): Number of susceptible individuals at time t. E(t): Number of exposed (but not yet infectious) individuals at time t. I(t): Number of infectious individuals at time t. R(t): Number of recovered (or removed) individuals at time t. The equations for the SEIR model are

$$\frac{dS(t)}{dt} = -\beta\frac{S(t)I(t)}{N}, \tag{1}$$

$$\frac{dE(t)}{dt} = \beta\frac{S(t)I(t)}{N} - \sigma E(t), \tag{2}$$

$$\frac{dI(t)}{dt} = \sigma E(t) - \gamma I(t), \tag{3}$$

$$\frac{dR(t)}{dt} = \gamma I(t). \tag{4}$$

Here, β is the transmission rate, σ is the rate of progression from exposure to infectious (inverse of the incubation period), and γ is the recovery rate. N is the total population.

3.2. *Decision tree model*

A Decision Tree is a machine-learning model used for classification or regression tasks. The decision tree recursively splits the data into subsets based on feature values, creating a tree-like structure [9]. The split criterion can be formulated as: Root Node: The starting point where the first decision is made, Internal Nodes: Nodes that represent the features used for splitting the data, and Leaf Nodes: Nodes that represent the outcome (class label or prediction). For a classification decision tree, the general decision-making process can be mathematically formulated as:

$$\text{Split Criterion} = \arg\min_{f,\theta} \left[\frac{n_L}{N} \text{Gini}_L + \frac{n_R}{N} \text{Gini}_R \right], \tag{5}$$

Where f is the feature used for the split, θ is the threshold, and n_L and n_R are the number of samples in the left and right child nodes after the split. N: Total number of samples in the current node. $Gini_L$: $Gini_R$ impurity (or entropy) of the left and right child nodes, respectively

3.3. *LSTM network model*

Long Short-Term Memory (LSTM) is a type of recurrent neural network (RNN) that excels in learning long-term dependencies and time-series data [10]. The general structure of an LSTM cell is defined by the following equations:

$$f_t = \sigma(W_f[h_{t-1}, x_t] + b_f), \tag{6}$$

$$i_t = \sigma(W_i[h_{t-1}, x_t] + b_i), \tag{7}$$

$$C_t = f_t C_{t-1} + i_t \tanh(W_c[h_{t-1}, x_t] + b_C), \tag{8}$$

$$o_t = \sigma(W_o[h_{t-1}, x_t] + b_o), \tag{9}$$

$$h_t = o_t \tanh(C_t), \tag{10}$$

where x_t is the input, h_t is the hidden state, and C_t is the cell state. W_f, W_i, W_c, W_o are weight matrices. b_f, b_i, b_C, b_o are Bias terms, σ is the Sigmoid activation function, and tanh is the Hyperbolic tangent activation function These gates allow the LSTM to maintain information over long sequences, making it particularly effective for time-series prediction tasks like forecasting COVID-19 outbreaks.

4. Numerical Results

4.1. *SEIR model: Graph results*

The blue line represents the number of susceptible individuals over time, the yellow line represents the exposed population (those who are infected but not yet

Table 1. Initial conditions of the SEIR model.

Initial Conditions	**Variable**	**Value**
Total population	N	1000
Initial number of infected individuals	$I(0)$	1
Initial number of exposed individuals $E(0)$	0	
Initial number of recovered individuals	$R(0)$	0
Initial number of susceptible individuals	$S(0)$	999

Table 2. SEIR model parameters.

SEIR Parameters	**Symbol**	**Value**
Transmission rate	β	0.3
Rate of progression (incubation period ~5.2 days)	σ	1/5.2
Recovery rate (infectious period ~14 days)	γ	1/14

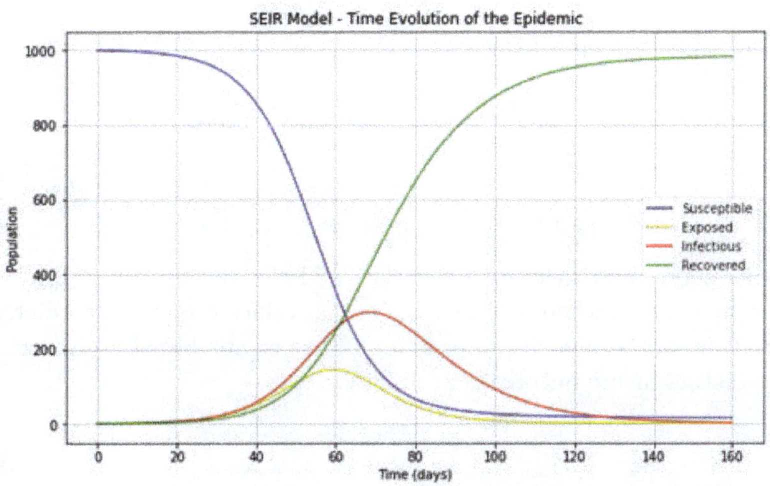

Fig. 1. Time evolution of the epidemic based on the SEIR model.

infectious), the red line represents the infectious population, and the green line represents the recovered population. The graph demonstrates the typical progression of an epidemic, where the number of susceptible individuals decreases as more people are exposed and infected. Over time, the infectious population rises, peaks, and then declines as individuals recover, moving to the recovered compartment.

4.2. *Decision tree: Graph results*

For the Decision Tree, the typical visual result is the tree structure itself, where splits are shown at each node based on features and thresholds.

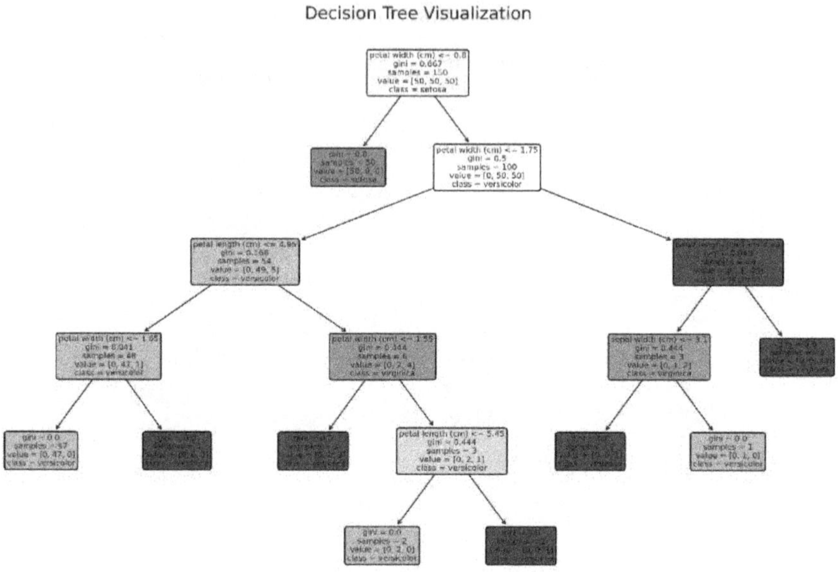

Fig. 2. Decision Tree visualization.

4.3. *LSTM: Graph results*

For the long short-term memory networks (LSTM), the typical graph would be time-series predictions, where you plot the actual values versus the predicted values over time. This can be used for forecasting future trends based on historical data, such as predicting future outbreaks of disease.

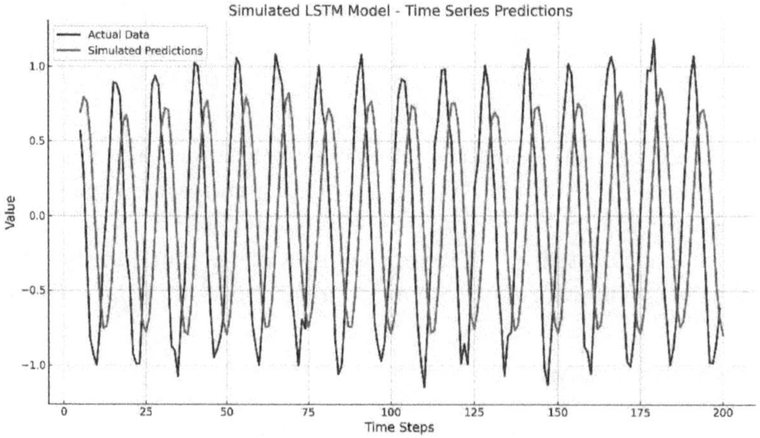

Fig. 3. LSTM model time-series predictions.

4.4. *Evaluation metrics*

Model performance was evaluated using metrics commonly used in machine learning, such as accuracy, precision, recall, and F1-score. These metrics helped assess the models' ability to correctly predict outbreak trends and inform public health interventions. Cross-validation techniques were also applied to ensure that the models were not overfitting to specific datasets and could be generalized to unseen data.

Table 3. Model performance comparison.

Model	Accuracy (%)	Precision (%)	Recall (%)	F1-Score (%)
SEIR Model	85.6	83.2	82.5	82.8
Decision Tree	88.1	85.9	84.2	85.0
LSTM Network	92.4	91.2	89.8	90.5

Table 3 shows a comparison between the model's performance in terms of accuracy, precision, recall, and F1-Score. The results indicated that deep learning models, specifically LSTM networks, outperformed traditional statistical models, the SEIR and Decision Tree model. The LSTM models demonstrated higher accuracy and better handling of time-series data, especially in predicting infection rates and disease progression. The LSTM model's predictive power is particularly evident in the early detection of potential outbreak hotspots. As shown, the LSTM model closely follows the actual infection trend, demonstrating the potential of AI for real-time epidemic forecasting.

5. Discussion

Our numerical results indicate that AI can greatly increase the precision and timeliness of predictions about infectious disease outbreaks. AI models have an advantage over traditional epidemiological models since they can handle large datasets and recognize intricate patterns. This is especially true of deep learning frameworks like LSTM. These models may constantly modify their forecasts by integrating real-time data, which enables public health officials to react to new hazards more skillfully. Still, there are several challenges to overcome before applying AI to disease surveillance. Data gaps and discrepancies in reporting can negatively impact model performance; hence, data quality is a major concern. Furthermore, public health experts frequently find it challenging to comprehend how certain forecasts are created due to the restricted interpretability of AI models. AI models can be modified to forecast the spread of diseases other than COVID-19, including influenza, dengue, and Zika. Because these models provide precise estimates that help allocate resources and influence policy decisions, they can benefit global health initiatives.

6. Conclusion

This chapter demonstrates the critical role AI can play in the early detection and prediction of infectious disease outbreaks. During the COVID-19 pandemic. The application of AI in predicting disease trends, identifying hotspots, and supporting public health decision-making illustrates the transformative potential of these technologies. As we look ahead to future pandemics and disease outbreaks, it is clear that AI will continue to be an essential tool in global health surveillance. To fully realize AI's potential, however, challenges including data quality, model interpretability, and ethical considerations need to be resolved. Future studies should concentrate on strengthening data integration, creating more resilient AI models, and encouraging multidisciplinary cooperation between public health specialists and AI specialists. By doing this, we can develop artificial intelligence (AI) systems that are better suited to safeguard public health in a globalized society. While artificial intelligence (AI) holds great potential for predicting and managing infectious disease outbreaks, several challenges must be addressed to fully realize its capabilities. Key challenges include data quality, model transparency, infrastructure limitations, and ethical considerations. Below is a discussion of these challenges and proposed solutions.

7. Ethical and Legal Considerations

Ethical and privacy concerns present additional challenges in applying AI to public health. AI models often require large datasets, including sensitive personal health information. Without proper regulations, there is a risk of unauthorized access, data misuse, or violations of privacy rights. Moreover, AI models may inadvertently reinforce existing biases, especially if trained on non-representative datasets that do not include certain populations or regions. To address these concerns, governments and international organizations must establish clear ethical frameworks that prioritize privacy, data protection, and informed consent. For instance, AI systems should comply with data protection regulations, such as the General Data Protection Regulation (GDPR) in Europe. Additionally, bias mitigation strategies should be employed to ensure that AI models are trained on diverse datasets that reflect a wide range of populations. Regular audits of AI systems can further ensure that any biases are identified and corrected in real time. Interdisciplinary collaboration is essential for the effective implementation of AI in public health. Successful AI applications require cooperation between AI experts, epidemiologists, public health officials, and policymakers. However, the lack of collaboration between these disciplines can lead to miscommunication and hinder the development of effective AI solutions. Furthermore, while AI offers considerable promise in the early detection and prediction of infectious disease outbreaks, it raises important ethical and legal concerns that must be addressed to ensure its safe and equitable application. Lastly, AI systems rely on vast amounts of data, including sensitive personal health information. Ensuring this data is securely handled and individuals' privacy rights are respected is a fundamental challenge. In many cases, the data used to train AI models is collected from medical records, social media, and travel databases, all raising concerns about unauthorized access and the potential misuse of information. To mitigate these risks, robust data protection policies, such as compliance with the General Data Protection Regulation (GDPR) in the European Union, need to be in place.

References

[1] A. Wilder-Smith and S. Osman, Public health emergencies of international concern: a historic overview, *J. Travel Med.*, 27, 8 (2020).

[2] V. Patil, G. S. Ali, A. Nashte, R. Rautdesai, S. K. Garud, and N. P. Sable, Public Health Policy and Infectious Disease Control: Lessons from Recent Outbreaks, *South East. Eur. J. Public Health*, (2023), 162–170.

[3] Y. K. Dwivedi *et al.*, Artificial Intelligence (AI): Multidisciplinary perspectives on emerging challenges, opportunities, and agenda for research, practice and policy, *Int. J. Inf. Manag.*, 57 (2021): 101994.

[4] M. Y. Jabarulla and H.-N. Lee, A blockchain and artificial intelligence-based, patient-centric healthcare system for combating the COVID-19 pandemic: Opportunities and applications, *Healthcare*, 9(8) (2021): 1019.

[5] C. Comito and C. Pizzuti, Artificial intelligence for forecasting and diagnosing COVID-19 pandemic: A focused review, *Artif. Intell. Med.*, 128, (2022).

[6] S. Uddin, A. Khan, M. E. Hossain, and M. A. Moni, Comparing different supervised machine learning algorithms for disease prediction, *BMC Med. Inform. Decis. Mak.*,. 19, 1 (2019).

[7] D. Zhang, C. Yin, J. Zeng, X. Yuan, and P. Zhang, Combining structured and unstructured data for predictive models: a deep learning approach, *BMC Med. Inform. Decis. Mak.*, 20, (2020).

[8] M. Ghani, Y. Norasia, and W. Ningsih, Dynamics of CoVid-19 Disease in Semarang, Indonesia: Stability, Optimal Control, and Model-Fitting, *Differ. Equ. Dyn. Syst.*, (2023).

[9] M. Zenga and C. Colapinto, Decision tree for classification and forecasting, *Engineering Mathematics and Artificial Intelligence*, CRC Press, (2023), 53–82.

[10] M. Jarrah, Time Series Prediction Analysis of Long Short Term Memory Models, in *6th International Conference on Computing and Informatics (ICCI)*, (2024).

[11] M. Elhia,O. Balatif, L.Boujallal, and M. Rachik. Optimal control problem for a tuberculosis model with multiple infectious compartments and time delays. *An International Journal of Optimization and Control: Theories and Applications (IJOCTA)*, 11(1), 75–91. (2021).

[12] S. Bhatter, S. Kumawat, B. Bhatia, and S. D. Purohit. Analysis of the COVID-19 epidemic with intervention impacts by a fractional operator. *An International Journal of Optimization and Control: Theories and Applications (IJOCTA)*, 14(3), 261–275 (2024).

Chapter 18

A Study of Scattering Theory of Eigenparameter Dependent Impulsive Sturm-Liouville Equation with Matrix Coefficients

Güler Başak Öznur*,‡ and Yelda Aygar†,§

**Department of Mathematics, Gazi University, Ankara, Teknik Okullar, 06500, Turkey*

†Department of Mathematics, Ankara University, Ankara, Tandoğan, 06100, Turkey

‡basakoznur@gazi.edu.tr

§yaygar@ankara.edu.tr

In this study, we set an eigenparameter dependent impulsive boundary value problem for matrix valued Sturm-Liouville equation on the semi axis. The main purpose of this work is to examine the properties of scattering solutions and scattering function of this problem. Firstly, we obtain Jost solution and Jost function of the operator corresponding to the impulsive boundary value problem. Then, we find scattering function by using Jost function and we investigate scattering function's properties. We also get resolvent operator, Green function, the sets of eigenvalues and spectral singularities of this problem. Finally, we present an example to demonstrate the application of our results. The novelty in this study is that it is the first study related to the impulsive Sturm-Liouville problem with matrix coefficients whose boundary conditions depend on spectral parameter.

1. Introduction

In applied mathematics, for the solution of many problems encountered in real life, mathematical models are available and most of these models are expressed with initial or boundary value problems. Operator theory is used to solve these kinds of problems in spectral and scattering analysis. However, processes in which mathematical modeling is used may not always continue uninterruptedly and continuously. Instantaneous and sharp situational changes may be encountered at some stages of the scientific process. Compared with the total process, period of this change is negligible, but the functioning of the system even so changes in nature. These short-term effects are called impulse effects, the differential equations involving these impulsive effects are called impulsive differential equations. Conditions added to the discontinuity in an impulsive equation can be named differently such as interface conditions, transmission conditions, point interactions and jump conditions [1–6]. Recently, many researchers have paid more attention to the boundary value problems with impulsive conditions. Because there are many scientific events in which sudden changes as theoretical physics, atomic physics, radiophysics, electrical technology, mathematical economy, biotechnology processes, ecology, population dynamics, industrial robotics, chemical technology, cruise control of aircraft, control theory, engineering, medicine can be observed [7–9]. These impulsive equations may cause some important results for mathematical theory. Samoilenko and Perestyuk made great contributions to impulsive equations [10, 11]. General theories of impulsive differential equations are given in [11–13], on the other hand, studies on this topic in scattering theory can be cited in [14–20].

In this study, we examine the scattering theory of an impulsive matrix Sturm-Liouville boundary value problem. By using scattering functions and scattering matrices, scattering theory allows analyzing spectrum and properties of an operator. Jost functions are used to define scattering functions. Jost functions have an important role in solving the inverse problem in quantum scattering theory [21–23]. Furthermore, the set of eigenvalues and the set of spectral singularities can be characterized by the zeros of the Jost functions. Although scattering theory of matrix valued operators have been studied with general boundary conditions [24–26], there are only a few studies about spectral and scattering analysis of impulsive matrix Sturm-Liouville boundary value problems [16, 17, 19]. The difference of this study from [16, 17, 19] is that the boundary condition depends on spectral parameter. We will examine the effects of the existing of boundary condition with spectral parameter for an impulsive matrix boundary value problem.

Let us introduce the Hilbert space $\mathcal{L}_2(\mathbb{R}_+, \Upsilon)$ consisting of all matrix valued functions in n-dimensional ($n < \infty$) Euclidian space Υ such that

$$\int_0^\infty ||Y||_\Upsilon^2 \, dx < \infty$$

with the matrix norm

$$||Y||^2_{\mathcal{L}_2(\mathbb{R}_+, \Upsilon)} = \int_0^\infty ||Y||_\Upsilon^2 \, dx,$$

where $\mathbb{R}_+ =: [0, \infty)$. We shall consider an impulsive matrix Sturm-Liouville boundary value problem generated by the following equation

$$-Y'' + Q(x)Y = \lambda^2 Y, \quad x \in [0,1) \cup (1, \infty) \tag{1}$$

with the boundary condition

$$(\zeta_0 + \zeta_1\lambda)\, Y'(0) + (\mu_0 + \mu_1\lambda)\, Y(0) = 0 \tag{2}$$

for $\zeta_0\mu_1 - \zeta_1\mu_0 \neq 0$ and the impulsive conditions

$$Y(1^+) = UY(1^-)$$

$$Y'(1^+) = VY'(1^-), \tag{3}$$

where λ is a spectral parameter and ζ_i, μ_i are real numbers for $i = 0, 1$. Furthermore, $Q = [q_{ij}]_{nxn}$ is a self-adjoint matrix valued function fulfilling the following condition

$$\int_0^\infty x \, ||Q(x)|| \, dx < \infty. \tag{4}$$

Throughout the paper, we will assume that U and V are selfadjoint diagonal matrices in Υ such that all eigenvalues of U and V are positive. We call the impulsive matrix Sturm-Liouville operator $\mathcal{T}$ acting in $\mathcal{L}_2(\mathbb{R}_+, \Upsilon)$ generated by (1)-(3). We shall point out that condition (3) is the impulsive condition of equation (1) and $x = 1$ is the impulsive point for (1)-(3).

In this work, firstly, we obtain Jost solution and Jost function of the operator corresponding to the impulsive boundary value problem. Then, we find scattering function by using Jost function and we investigate scattering function's properties. We also get resolvent operator, Green function and the sets of eigenvalues and spectral singularities of this operator. Finally, we present an example to demonstrate the application of our results. Note that the foundation of the matrix Sturm-Liouville equation (1) remains essential in quantum mechanics. Numerious significant applications of matrix equations have emerged involving particles with internal characteristics. Further exploration of these applications can be found in references [21, 27–29]. In recently work [30], the authors examine the matrix Schrödinger system which is the general form of (1) with a spherically symmetric potential. This approach encompasses all physical phenomena related to matrix Schrödinger equations.

Now, let $\mathcal{G}(x, \lambda)$ and $\mathcal{H}(x, \lambda)$ be the fundamental solutions of (1) in the interval $[0, 1)$ satisfying the following initial conditions

$$\mathcal{G}(0, \lambda) = 0, \quad \mathcal{G}'(0, \lambda) = I$$

and

$$\mathcal{H}(0, \lambda) = I, \quad \mathcal{H}'(0, \lambda) = 0$$

having the representations [27]

$$\mathcal{G}(x, \lambda) = \frac{\sin \lambda x}{\lambda} I + \int_0^x A(x, t) \frac{\sin \lambda t}{\lambda} dt \tag{5}$$

and

$$\mathcal{H}(x, \lambda) = \cos \lambda x I + \int_0^x B(x, t) \cos \lambda t dt, \tag{6}$$

respectively, where the kernels $A(x, t)$ and $B(x, t)$ may be expressed in terms of the potential function Q, and I denotes the identity matrix in Υ.

Lemma 1. *Assume that Y_1 and Y_2 are arbitrary solutions of (1), then the Wronskian*

$$W[Y_1, Y_2] := Y_1 Y_2' - Y_1' Y_2$$

is independent of x [31].

By using Lemma 1, it can be easily found that the Wronskians

$$W\left[\mathcal{G}^T(\lambda), \mathcal{H}(\lambda)\right] = -I, \quad W\left[\mathcal{G}^T(\lambda), \mathcal{G}(\lambda)\right] = 0 \tag{7}$$

and

$$W\left[\mathcal{H}^T(\lambda), \mathcal{G}(\lambda)\right] = I, \quad W\left[\mathcal{H}^T(\lambda), \mathcal{H}(\lambda)\right] = 0 \tag{8}$$

for all $\lambda \in \overline{\mathbb{C}}_+ := \{\lambda \in \mathbb{C} : Im\lambda \geq 0\}$.

In addition, under the condition (4), we show the bounded solution of Eq. (1) fulfilling the condition

$$\lim_{x\to\infty} E(x,\lambda) e^{-i\lambda x} = I, \quad \lambda \in \overline{\mathbb{C}}_+$$

in $(1,\infty)$ by $E(x,\lambda)$. The integral representation of $E(x,\lambda)$ is known from the literature [27]

$$E(x,\lambda) = e^{i\lambda x} I + \int_x^\infty K(x,t) e^{i\lambda t} dt, \quad \lambda \in \overline{\mathbb{C}}_+, \tag{9}$$

where the kernel function $K(x,t)$ may be expressed in terms of Q and is continuously differentiable with respect to x and t. From here, it is easy to see that $E(x,\lambda)$ is analytic with respect to λ on $\mathbb{C}_+ := \{\lambda \in \mathbb{C} : Im\lambda > 0\}$ and continuous up to the real axis. The solution $E(x,\lambda)$ given in (9) is the Jost solution of (1).

On the other hand, under the condition (4), $\widehat{E}(x,\lambda)$ denotes the unbounded solution of Eq. (1) subjecting the condition

$$\lim_{x\to\infty} \widehat{E}(x,\lambda) e^{i\lambda x} = I, \quad \lambda \in \overline{\mathbb{C}}_+.$$

We should remind that if $Y(x,\lambda)$ is the solution of (1), then $Y^T(x,\lambda)$ and $Y^*\left(x,\overline{\lambda}\right)$ will be the solutions for (1), here "T" denotes the transpose and "$*$" denotes the adjoint operator.

2. Jost Function and Scattering Function of $\mathcal{T}$

Firstly, by the help of $\mathcal{G}(x,\lambda)$, $\mathcal{H}(x,\lambda)$ and $E(x,\lambda)$ given in (5), (6) and (9), we construct any solution of (1)–(3) by

$$L(x,\lambda) = \begin{cases} \mathcal{G}(x,\lambda) t_1(\lambda) + \mathcal{H}(x,\lambda) t_2(\lambda) & \text{for } x \in [0,1) \\ E(x,\lambda) & \text{for } x \in (1,\infty) \end{cases} \tag{10}$$

for $\lambda \in \overline{\mathbb{C}}_+$. From the impulsive condition (3), we get

$$U^{-1} E(1,\lambda) = \mathcal{G}(1,\lambda) t_1(\lambda) + \mathcal{H}(1,\lambda) t_2(\lambda) \tag{11}$$

and

$$V^{-1} E'(1,\lambda) = \mathcal{G}'(1,\lambda) t_1(\lambda) + \mathcal{H}'(1,\lambda) t_2(\lambda). \tag{12}$$

By using (7), (8), (11) and (12), we obtain the coefficients $t_1(\lambda)$ and $t_2(\lambda)$ uniquely as

$$t_1(\lambda) = IU^{-1}V^{-1}\left\{U\mathcal{H}^T(1,\lambda)E'(1,\lambda) - V\left[\mathcal{H}^T(1,\lambda)\right]'E(1,\lambda)\right\}$$

and

$$t_2(\lambda) = IU^{-1}V^{-1}\left\{V\left[\mathcal{G}^T(1,\lambda)\right]'E(1,\lambda) - U\mathcal{G}^T(1,\lambda)E'(1,\lambda)\right\}$$

for $\lambda \in \overline{\mathbb{C}}_+$. The function $L(x,\lambda)$ is called the Jost solution of $\mathcal{T}$. In view of (2) and (10), the Jost function of (1)–(3) can be found as

$$L(\lambda) = (\zeta_0 + \zeta_1\lambda)t_1(\lambda) + (\mu_0 + \mu_1\lambda)t_2(\lambda). \tag{13}$$

It is easy to see from (13), $L(\lambda)$ is analytic in $\mathbb{C}_+$ and continuous up to the real axis. For $\lambda \in \mathbb{R}\backslash\{0\}$, Eq. (1) has another solution $M(x,\lambda)$ represented by

$$M(x,\lambda) = \begin{cases} \phi(x,\lambda) & \text{for } x \in [0,1) \\ E(x,\lambda)t_3(\lambda) + E(x,-\lambda)t_4(\lambda) & \text{for } x \in (1,\infty), \end{cases} \tag{14}$$

where $\phi(x,\lambda)$ is the solution of (1) found by

$$\phi(x,\lambda) = (\mu_0 + \mu_1\lambda)\mathcal{G}(x,\lambda) - (\zeta_0 + \zeta_1\lambda)\mathcal{H}(x,\lambda).$$

It is obvious that the solution $\phi(x,\lambda)$ is an entire function of λ. Using (3), we get the following equalities to get the coefficients $t_3(\lambda)$ and $t_4(\lambda)$, we will use same way as finding $t_1(\lambda)$ and $t_2(\lambda)$

$$E(1,\lambda)t_3(\lambda) + E(1,-\lambda)t_4(\lambda) = U\phi(1,\lambda) \tag{15}$$

and

$$E'(1,\lambda)t_3(\lambda) + E'(1,-\lambda)t_4(\lambda) = V\phi'(1,\lambda). \tag{16}$$

By the help of Lemma 1, we calculate the following Wronskians

$$W\left[E^T(\lambda), E(\lambda)\right] = 0, \quad W\left[E^T(\lambda), E(-\lambda)\right] = -2i\lambda I$$

and

$$W\left[E^T(-\lambda), E(\lambda)\right] = 2i\lambda I, \quad W\left[E^T(-\lambda), E(-\lambda)\right] = 0$$

for $\lambda \in \mathbb{R}\backslash\{0\}$. If we use these Wronskians in (15) and (16), the coefficients $t_3(\lambda)$ and $t_4(\lambda)$ are obtained as follows:

$$t_3(\lambda) = -\frac{1}{2i\lambda I}\left[U\left[E^T(1,-\lambda)\right]'\phi(1,\lambda) - VE^T(1,-\lambda)\phi'(1,\lambda)\right]$$

$$t_4(\lambda) = \frac{1}{2i\lambda I}\left[U\left[E^T(1,\lambda)\right]'\phi(1,\lambda) - VE^T(1,\lambda)\phi'(1,\lambda)\right].$$

Since $\mathcal{G}(x,\lambda) = \mathcal{G}(x,-\lambda) = \mathcal{G}^T(x,\lambda)$ and $\mathcal{H}(x,\lambda) = \mathcal{H}(x,-\lambda) = \mathcal{H}^T(x,\lambda)$, the Jost function $L(\lambda)$ and the coefficients $t_3(\lambda)$ and $t_4(\lambda)$ have the following relationship

$$t_4^T(\lambda) = t_3^T(-\lambda) = -\frac{1}{2i\lambda I}UVL(\lambda), \quad \lambda \in \mathbb{R}\backslash\{0\}. \tag{17}$$

As a consequence of Lemma 1 and (17), for all $\lambda \in \mathbb{R}\backslash\{0\}$, the Wronskian of the solutions $L(x,\lambda)$ and $M(x,\lambda)$ is obtained as

$$W\left[L(x,\lambda), M^T(x,\lambda)\right] = \begin{cases} L(\lambda) & \text{for } x \in [0,1) \\ UVL(\lambda) & \text{for } x \in (1,\infty). \end{cases}$$

Theorem 1. *For all* $\lambda \in \mathbb{R}\backslash\{0\}$, $\det L(\lambda) \neq 0$.

Proof. Assume that there exists a $\lambda_0 \in \mathbb{R}\backslash\{0\}$ such that $\det L(\lambda_0) = 0$. We can write from (17)

$$\det t_4^T(\lambda_0) = \det t_3^T(-\lambda_0) = \left(-\frac{1}{2i\lambda_0 I}\right)^n \det U \det V \det L(\lambda_0).$$

By using this equality, we obtain that $\det t_3(\lambda_0) = \det t_4(\lambda_0) = 0$. Since then, there exists a non-zero vector u such that $t_3(\lambda_0) u = 0$ and $t_4(\lambda_0) u = 0$. Then, the solution $M(x,\lambda)$ defined in (14) is equal to zero identically. This gives a trivial solution of (1)-(3) and from here, we get the contradiction, i.e., $\det L(\lambda) \neq 0$ for all $\lambda \in \mathbb{R}\backslash\{0\}$.

Theorem 1 says that the inverse of the function L exists, then we can give the following definition.

Definition 1. The matrix function

$$S(\lambda) = L^{-1}(\lambda) L(-\lambda), \quad \lambda \in \mathbb{R}\backslash\{0\}$$

is called the scattering matrix of the operator $\mathcal{T}$.

Theorem 2. *For all* $\lambda \in \mathbb{R}\backslash\{0\}$, *the scattering matrix satisfies*

$$S(-\lambda) = S^{-1}(\lambda) = S^*(\lambda)$$

and it is an uniter matrix.

Proof. For $\lambda \in \mathbb{R}\backslash\{0\}$, we write directly from Definition 1 that

$$S(-\lambda) = L^{-1}(-\lambda) L(\lambda)$$

and

$$S(\lambda) S(-\lambda) = S(-\lambda) S(\lambda) = I.$$

The last equality gives that

$$S(-\lambda) = S^{-1}(\lambda), \quad \lambda \in \mathbb{R}\backslash\{0\}.$$

Now, let us consider the solutions $L(x,\lambda)$, $L(x,-\lambda)$ and $M(x,\lambda)$ to prove the equation $S^*(\lambda) = S(-\lambda)$, whenever λ belongs to $\mathbb{R}\backslash\{0\}$. Hence, we write

$$M(x,\lambda) = L(x,\lambda)\nu_1 + L(x,-\lambda)\nu_2$$

$$M'(x,\lambda) = L'(x,\lambda)\nu_1 + L'(x,-\lambda)\nu_2,$$

here ν_1, ν_2 are matrices not depending on x and $\lambda \in \mathbb{R}\backslash\{0\}$. From the last equalities, by making the necessary calculations, ν_1 and ν_2 are obtained as follows

$$\nu_1 = -W^{-1}\left[L^*(x,\lambda), L(x,\lambda)\right]\left\{\left[L^*(x,\lambda)\right]' M(x,\lambda) - L^*(x,\lambda) M'(x,\lambda)\right\}$$

and

$$\nu_2 = -W^{-1}\left[L^*(x,-\lambda), L(x,-\lambda)\right]\left\{\left[L^*(x,-\lambda)\right]' M(x,\lambda) - L^*(x,-\lambda) M'(x,\lambda)\right\},$$

respectively. Because of the characteristic features of the impulsive equation, we can write

$$W\left[L^*\left(x,\lambda\right),L\left(x,-\lambda\right)\right]=W\left[L^*\left(x,-\lambda\right),L\left(x,\lambda\right)\right]=0$$

and

$$W\left[L^*\left(x,\lambda\right),L\left(x,\lambda\right)\right]=-W\left[L^*\left(x,-\lambda\right),L\left(x,-\lambda\right)\right]=2i\lambda I.$$

By taking $x=0$ in the expressions for ν_1 and ν_2, we obtain

$$\nu_1=\frac{1}{2i\lambda I}L^*\left(\lambda\right)\quad\text{and}\quad\nu_2=-\frac{1}{2i\lambda I}L^*\left(-\lambda\right).$$

As a consequence, we write

$$M\left(x,\lambda\right)=\frac{1}{2i\lambda I}\left[L\left(x,\lambda\right)L^*\left(\lambda\right)-L\left(x,-\lambda\right)L^*\left(-\lambda\right)\right].\tag{18}$$

Writing $x=0$ in (18) and taking the derivative of (18), we have

$$\left(\zeta_0+\zeta_1\lambda\right)=-\frac{1}{2i\lambda I}\left[t_2\left(\lambda\right)L^*\left(\lambda\right)-t_2\left(-\lambda\right)L^*\left(-\lambda\right)\right]\tag{19}$$

$$\left(\mu_0+\mu_1\lambda\right)=\frac{1}{2i\lambda I}\left[t_1\left(\lambda\right)L^*\left(\lambda\right)-t_1\left(-\lambda\right)L^*\left(-\lambda\right)\right].\tag{20}$$

By the help of (19) and (20), we can obtain

$$L\left(\lambda\right)L^*\left(\lambda\right)=L\left(-\lambda\right)L^*\left(-\lambda\right).\tag{21}$$

From (21), it is seen that

$$L^*\left(\lambda\right)=L^{-1}\left(\lambda\right)L\left(-\lambda\right)L^*\left(-\lambda\right)$$

and

$$L^*\left(\lambda\right)\left[L^*\left(-\lambda\right)\right]^{-1}=L^{-1}\left(\lambda\right)L\left(-\lambda\right).$$

Using Definition 1 and the last equality, we get

$$SS^*=S^*S=I,\qquad ||S||=I,$$

that is, S is uniter. The proof is completed.

3. Resolvent Operator and Eigenvalues of $\mathcal{T}$

In this section, we handle another solution of (1)–(3). This solution is the unbounded solution of $\mathcal{T}$ and let us define it as follows

$$N\left(x,\lambda\right)=\begin{cases}\phi\left(x,\lambda\right) & \text{for } x\in[0,1)\\ E\left(x,\lambda\right)t_5\left(\lambda\right)+\widehat{E}\left(x,\lambda\right)t_6\left(\lambda\right) & \text{for } x\in(1,\infty)\end{cases}\tag{22}$$

for $\lambda\in\overline{\mathbb{C}}_+\backslash\{0\}$. If we apply the impulsive condition (3) to $N\left(x,\lambda\right)$, we find

$$E\left(1,\lambda\right)t_5\left(\lambda\right)+\widehat{E}\left(1,\lambda\right)t_6\left(\lambda\right)=U\phi\left(1,\lambda\right)\tag{23}$$

and

$$E'(1,\lambda)\,t_5(\lambda)+\widehat{E}'(1,\lambda)\,t_6(\lambda)=V\phi'(1,\lambda). \tag{24}$$

By the help of Lemma 1, we write the following Wronskians

$$W\left[\widehat{E}^T(\lambda),E(\lambda)\right]=2i\lambda I,\quad W\left[E^T(\lambda),\widehat{E}(\lambda)\right]=-2i\lambda I$$

and

$$W\left[\widehat{E}^T(\lambda),\widehat{E}(\lambda)\right]=0,\quad W\left[E^T(\lambda),E(\lambda)\right]=0.$$

Using the last Wronskian equalities in (23) and (24), the coefficients $t_5(\lambda)$ and $t_6(\lambda)$ are obtained as

$$t_5(\lambda)=-\frac{1}{2i\lambda I}\left[U\left[\widehat{E}^T(1,\lambda)\right]'\phi(1,\lambda)-V\widehat{E}^T(1,\lambda)\,\phi'(1,\lambda)\right]$$

and

$$t_6(\lambda)=\frac{1}{2i\lambda I}\left[U\left[E^T(1,\lambda)\right]'\phi(1,\lambda)-VE^T(1,\lambda)\,\phi'(1,\lambda)\right],$$

respectively. Note that

$$t_6(\lambda)=-\frac{1}{2i\lambda I}UVL^T(\lambda),\quad \lambda\in\overline{\mathbb{C}}_+\backslash\{0\}.$$

Similar to the Wronskian of the solutions $L(x,\lambda)$ and $M(x,\lambda)$, the Wronskian of the solutions $L(x,\lambda)$ and $N(x,\lambda)$ is found as

$$D(\lambda):=W\left[L(x,\lambda),N^T(x,\lambda)\right]=\begin{cases}L(\lambda) & \text{for } x\in[0,1)\\ UVL(\lambda) & \text{for } x\in(1,\infty)\end{cases} \tag{25}$$

for $\lambda\in\overline{\mathbb{C}}_+\backslash\{0\}$.

Theorem 3. *If the condition* (4) *satisfies, we get*

$$\widetilde{R}_\lambda(\mathcal{T})(f)=\int_0^\infty\widetilde{G}(x,t;\lambda)\,f(t)dt,\quad f\in\mathcal{L}_2(\mathbb{R}_+,\Upsilon),$$

here, $\widetilde{G}(x,t;\lambda)$ *is given by*

$$\widetilde{G}(x,t;\lambda)=\begin{cases}L(x,\lambda)\,D^{-1}(\lambda)\,N^T(t,\lambda) & \text{for } x\in[0,1)\\ N(x,\lambda)\left[D^{-1}(\lambda)\right]^T L^T(t,\lambda) & \text{for } x\in(1,\infty)\end{cases}$$

for $x,t\neq 1$ *and* $\lambda\in\overline{\mathbb{C}}_+\backslash\{0\}$.

Proof. In order to find the resolvent operator and Green function of $\mathcal{T}$, we need to obtain the solutions of the following equation

$$-Y''+Q(x)Y-\lambda^2Y=f(x),\quad x\in[0,1)\cup(1,\infty), \tag{26}$$

where $f\in\mathcal{L}_2(\mathbb{R}_+,\Upsilon)$. From the solutions $L(x,\lambda)$ and $N(x,\lambda)$ defined in (10) and (22), we get the solution of (26)

$$Y(x,\lambda)=L(x,\lambda)\,d_1(x)+N(x,\lambda)\,d_2(x).$$

By the help of the method of variation of parameters, for $\lambda \in \overline{\mathbb{C}}_{+}\backslash\{0\}$, the coefficients $d_1(x)$ and $d_2(x)$ are obtained as

$$d_1(x) = \delta_1 + \int_0^x \frac{N^T(t,\lambda) f(t)}{W[L(x,\lambda), N^T(x,\lambda)]} dt$$

and

$$d_2(x) = \delta_2 + \int_x^\infty \frac{L^T(t,\lambda) f(t)}{W[L^T(x,\lambda), N(x,\lambda)]} dt,$$

respectively, where δ_1 and δ_2 are real numbers. It is known that the solution $Y(x,\lambda)$ is in $\mathcal{L}_2(0,\infty)$ and satisfies the boundary condition (2). Then, we calculate $\delta_1 = \delta_2 = 0$. It completes the proof of Theorem 3.

By using Theorem 3, we define the sets of eigenvalues and spectral singularities of $\mathcal{T}$ by σ_d and σ_{ss} as follows

$$\sigma_d(\mathcal{T}) = \{\mu = \lambda^2 : \lambda \in \mathbb{C}_+ : \det L(\lambda) = 0\}$$

and

$$\sigma_{ss}(\mathcal{T}) = \{\mu = \lambda^2 : \lambda \in \mathbb{R}\backslash\{0\} : \det L(\lambda) = 0\},$$

respectively. It means that the operator $\mathcal{T}$ has no spectral singularity.

Theorem 4. *Assume* (4). *The Jost function* $L(\lambda)$ *satisfies the following asymptotic equation for* $\lambda \in \overline{\mathbb{C}}_+$

$$L(\lambda) = \frac{\zeta_1}{2} i\lambda^2 U^{-1} V^{-1} (U+V)[I + o(1)], \quad |\lambda| \to \infty, \quad \zeta_1 \neq 0.$$

Proof. From (5) and (6), we obtain

$$\mathcal{G}(x,\lambda) = \mathcal{G}^T(x,\lambda), \quad \text{and} \quad \mathcal{H}(x,\lambda) = \mathcal{H}^T(x,\lambda) \tag{27}$$

for $\lambda \in \overline{\mathbb{C}}_+$. By the help of (5),(6) and (27), we write the following equalities

$$\mathcal{G}^T(1,\lambda) = \frac{\sin\lambda}{\lambda} I + \int_0^1 A(1,t) \frac{\sin\lambda t}{\lambda} dt, \tag{28}$$

$$\left[\mathcal{G}^T(1,\lambda)\right]' = \cos\lambda I + A(1,1)\frac{\sin\lambda}{\lambda} + \int_0^1 A_x(1,t)\frac{\sin\lambda t}{\lambda} dt, \tag{29}$$

$$\mathcal{H}^T(1,\lambda) = \cos\lambda I + \int_0^1 B(1,t)\cos\lambda t dt \tag{30}$$

and

$$\left[\mathcal{H}^T(1,\lambda)\right]' = -\lambda\sin\lambda I + B(1,1)\cos\lambda + \int_0^1 B_x(1,t)\cos\lambda t dt. \tag{31}$$

Morever, from (9), we write

$$E(1,\lambda) = e^{i\lambda} I + \int_1^\infty K(1,t)\, e^{i\lambda t} dt \tag{32}$$

and

$$E'(1,\lambda) = i\lambda e^{i\lambda} I - K(1,1)\, e^{i\lambda} + \int_1^\infty K_x(1,t)\, e^{i\lambda t} dt. \tag{33}$$

By using (28)–(33), the following asymptotic equalities are found for $\lambda \in \overline{\mathbb{C}}_+$ whenever $|\lambda| \to \infty$

$$\begin{aligned}
\mathcal{G}^T(1,\lambda) &= \frac{e^{-i\lambda}}{\lambda}\left[\frac{i}{2} I + o(1)\right], \\
\mathcal{H}^T(1,\lambda) &= e^{-i\lambda}\left[\frac{1}{2} I + o(1)\right], \\
\left[\mathcal{G}^T(1,\lambda)\right]' &= e^{-i\lambda}\left[\frac{1}{2} I + O\left(\frac{1}{\lambda}\right)\right], \\
\left[\mathcal{H}^T(1,\lambda)\right]' &= \lambda e^{-i\lambda}\left[-\frac{i}{2} I + O\left(\frac{1}{\lambda}\right)\right]
\end{aligned} \tag{34}$$

and

$$\begin{aligned}
E(1,\lambda) &= e^{i\lambda}\left[I + o(1)\right], \\
E'(1,\lambda) &= e^{i\lambda}\left[i\lambda I + O(1)\right].
\end{aligned} \tag{35}$$

If the asymptotic equations obtained in (34) and (35) are substituted in $t_1(\lambda)$ and $t_2(\lambda)$, we get

$$t_1(\lambda) = \frac{U^{-1}V^{-1}}{2}(U+V)\,\lambda\left[iI + O\left(\frac{1}{\lambda}\right)\right] \tag{36}$$

and

$$t_2(\lambda) = \frac{U^{-1}V^{-1}}{2}(U+V)\left[I + O\left(\frac{1}{\lambda}\right)\right] \tag{37}$$

for $\lambda \in \overline{\mathbb{C}}_+$, $|\lambda| \to \infty$.

By the use of (13), (36) and (37), asymptotic equality of the Jost function of (1)–(3) is obtained as follows

$$L(\lambda) = \frac{\zeta_1}{2} i\lambda^2 U^{-1}V^{-1}(U+V)\left[I + o(1)\right], \quad \lambda \in \overline{\mathbb{C}}_+, \quad |\lambda| \to \infty.$$

This completes the proof.

Example 1. Now, let us consider the following unperturbated impulsive matrix Sturm-Liouville problem of (1)–(3)

$$\begin{gathered} -Y'' = \lambda^2 Y, \quad x \in [0,1) \cup (1,\infty) \\ (\zeta_0 + \zeta_1 \lambda) Y'(0) + (\mu_0 + \mu_1 \lambda) Y(0) = 0 \\ Y(1^+) = UY(1^-) \\ Y'(1^+) = VY'(1^-), \end{gathered} \tag{38}$$

here ζ_i, μ_i are real numbers for $i = 0, 1$ such that $\zeta_0\mu_1 - \zeta_1\mu_0 \neq 0$.
In this example, the solutions $\mathcal{G}(x,\lambda)$, $\mathcal{H}(x,\lambda)$ and $E(x,\lambda)$ are found as follows

$$\mathcal{G}(x,\lambda) = \frac{\sin \lambda x}{\lambda} I, \quad \mathcal{H}(x,\lambda) = \cos \lambda x I, \quad E(x,\lambda) = e^{i\lambda x} I.$$

For $\lambda \in \overline{\mathbb{C}}_+$, we get the Jost solution and Jost function of (38) as

$$L(x,\lambda) = \begin{cases} \dfrac{\sin \lambda x}{\lambda} I t_1(\lambda) + \cos \lambda x I t_2(\lambda) & \text{for } x \in [0,1) \\ e^{i\lambda x} I & \text{for } x \in (1,\infty) \end{cases}$$

and

$$L(\lambda) = (\zeta_0 + \zeta_1 \lambda) t_1(\lambda) + (\mu_0 + \mu_1 \lambda) t_2(\lambda),$$

respectively, where

$$t_1(\lambda) = U^{-1} V^{-1} \lambda e^{i\lambda} (U \cos \lambda i + V \sin \lambda),$$

$$t_2(\lambda) = U^{-1} V^{-1} e^{i\lambda} (V \cos \lambda - U i \sin \lambda).$$

In order to obtain the set of eigenvalues of the Eq. (38), $\det L(\lambda) = 0$ should be examined. Let us take $\zeta_0 = 1$, $\zeta_1 = 0$, $\mu_0 = 0$, $\mu_1 = \lambda$ for ease of calculation. From here, we write

$$\det L(\lambda) = \det U \det V \det [U i \beta_1(\lambda) + V \beta_2(\lambda)],$$

where $\beta_1(\lambda) = \cos \lambda - \sin \lambda$ and $\beta_2(\lambda) = \cos \lambda + \sin \lambda$. It is known that $[u_{jj}]$ and $[v_{jj}]$ are positive real numbers for $1 \leq j \leq n$, then we get

$$[u_{11} i \beta_1(\lambda) + v_{11} \beta_2(\lambda)] [u_{22} i \beta_1(\lambda) + v_{22} \beta_2(\lambda)] ... [u_{nn} i \beta_1(\lambda) + v_{nn} \beta_2(\lambda)] = 0.$$

From here, $u_{jj} i \beta_1(\lambda) + v_{jj} \beta_2(\lambda) = 0$, i.e.,

$$(\mathcal{A}_j - 1) e^{2i\lambda} = \mathcal{A}_j + 1, \quad j \in \{1, 2, ..., n\}, \tag{39}$$

where $\mathcal{A}_j = \dfrac{i v_{jj} - u_{jj}}{i u_{jj} - v_{jj}}$. It follows from (39) that

$$\lambda_k = -\frac{i}{2} \ln \left| \frac{\mathcal{A}_j + 1}{\mathcal{A}_j - 1} \right| + \frac{1}{2} Arg \left(\frac{\mathcal{A}_j + 1}{\mathcal{A}_j - 1} \right) + k\pi, \quad k \in \mathbb{Z}.$$

There appear two special cases:

(i) Assume $0 < \left|\frac{\mathcal{A}_j+1}{\mathcal{A}_j-1}\right| < 1$. Then, it is obtained that

$$\lambda_k = \frac{i}{2}\ln\left|\frac{\mathcal{A}_j+1}{\mathcal{A}_j-1}\right| + \frac{1}{2}Arg\left(\frac{\mathcal{A}_j+1}{\mathcal{A}_j-1}\right) + k\pi, \quad k \in \mathbb{Z}.$$

It implies that $\lambda_k \in \mathbb{C}_+$, $k \in \mathbb{Z}$ are eigenvalues of (38).

(ii) For $\left|\frac{\mathcal{A}_j+1}{\mathcal{A}_j-1}\right| > 1$, we calculate

$$\lambda_k = -\frac{i}{2}\ln\left|\frac{\mathcal{A}_j+1}{\mathcal{A}_j-1}\right| + \frac{1}{2}Arg\left(\frac{\mathcal{A}_j+1}{\mathcal{A}_j-1}\right) + k\pi, \quad k \in \mathbb{Z}.$$

In this case, the problem (38) has no eigenvalues.

References

[1] B. P. Allahverdiev, E. Bairamov and E. Ugurlu, Eigenparameter dependent Sturm–Liouville problems in boundary conditions with transmission conditions, *Journal of Mathematical Analysis and Applications*, **401**(1), 388–396, (2013).

[2] K. Aydemir, H. Olgar, O. S. Mukhtarov and F. Muhtarov, Differential operator equations with interface conditions in modified direct sum spaces, *Filomat*, **32**(3), 921–931, (2018).

[3] G. B. Oznur, Y. Aygar and N. D. Aral, An examination of boundary value transmission problem with quadratic spectral parameter, *Quaestiones Mathematicae*, **46**(5), 871–885, (2023).

[4] G. G. Ozbey, Y. Aygar and G. B. Oznur, Discrete Sturm-Liouville equations with point interaction, *The Journal of Analysis*, 1–13, (2024).

[5] A. S. Ozkan and B. Keskin, Spectral problems for Sturm–Liouville operator with boundary and jump conditions linearly dependent on the eigenparameter, *Inverse Problems in Science and Engineering*, **20**(6), 799–808, (2012).

[6] H. Tas, Y. Kucukevcilioglu and E. Bairamov, Scattering solutions and scattering function of a Klein-Gordon s-wave equation with jump conditions, *Turkish Journal of Mathematics*, **47**(2), 721–731, (2023)

[7] M Cbhoisy, J. F. Guegan and P. Rohani, Dynamics of infectious diseases and pulse vaccination teasing apart the embedded resonance effects, *Physica D: Nonlinear Phenomena*, **223**, 26–35, (2006).

[8] A. D'onofrio, On pulse vaccination strategy in the SIR epidemic model with vertical transmission, *Applied Mathematics Letters*, **18**, 729–732, (2005).

[9] G. Jiang and Q. Lu, Impulsive state feedback control of a predator-prey model, *Journal of Computational and Applied Mathematics*, **200**, 193–207, (2007).

[10] N. A. Perestyuk, V. A. Plotnikov, A. M. Samoilenko and N. V. Skripnik, *Differential Equations with Impulse Effects.* (Walter de Gruyter Co., Berlin, 2011).

[11] A. M. Samoilenko and N. A. Perestyuk, *Impulsive Differential Equations, Volume 14 of Word Scientific Series on Nonlinear Science.* (Series A: Monographs and Treatises, World Scientific Publishing Co., NJ, 1995).

[12] D. D. Bainov and P. S. Simeonov, *Impulsive Differential Equations: Asymptotic Properties of The Solutions.* (World Scientific, 1995).

[13] V. Lakshmikantham, D. D. Bainov and P. S. Simeonov, *Theory of Impulsive Diferential Equations.* (World Scientific Publishing Co., Inc., Teaneck, NJ, 1989).

[14] Y. Aygar, E. Bairamov and G. G. Ozbey, On the spectral and scattering properties of eigenparameter dependent discrete impulsive Sturm-Liouville equations, *Turkish Journal of Mathematics*, **45**(2), 988–1000, (2021).

[15] Y. Aygar and G. G. Ozbey, Scattering analysis of a quantum impulsive boundary value problem with spectral parameter, *Hacettepe Journal of Mathematics and Statistics*, **51**(1), 142–155, (2022).

[16] Y. Aygar and G. B. Oznur, Matrix difference equations with jump conditions and hyperbolic eigenparameter, *Acta Mathematica Universitatis Comenianae*, **91**(2), 149–159, (2022).

[17] E. Bairamov, Y. Aygar and S. Cebesoy, Investigation of spectrum and scattering function of impulsive matrix difference operators, *Filomat*, **33**(5), 1301-1312, (2019).

[18] E. Bairamov, Y. Aygar and G. B. Ozur, Scattering properties of eigenparameter-dependent impulsive Sturm–Liouville equations, *Bulletin of the Malaysian Mathematical Sciences Society*, **13**, 2769–2781, (2020).

[19] S. Cebesoy, E. Bairamov and Y. Aygar, Scattering problems of impulsive Schrödinger equations with matrix coefficients, *Ricerche di Matematica*, **72**(1), 399–415, (2023).

[20] G. B. Oznur and E. Bairamov, Scattering theory of the quadratic eigenparameter depending impulsive Sturm-Liouville equations, *Turkish Journal of Mathematics*, **46**(2), 406–415, (2022).

[21] K. Chadan and P. C. Sabatier, *Inverse Problems in Quantum Scattering Theory.* (Springer, New York, 1977).

[22] B. M. Levitan, *Inverse Sturm-Liouville Problems.* (In: Efimov, O. (ed.) Translated from the Russian, VSP, Zeist, 1987).

[23] V. A. Marchenko, *Sturm-Liouville Operators and Applications, Volume 22 of Operator Theory: Advances And Applications.* (In: Iacob, A. (ed.) Translated from the Russian, Birkhäuser Verlag, Basel, 1986).

[24] Y. Aygar and M. Bohner, Spectral analysis of a matrix valued quantum difference operator, *Dynamic Systems and Applications*, **25**, 29–8, (2016).

[25] E. Bairamov and S. Cebesoy, Spectral singularities of the matrix Schrödinger equations, *Hacettepe Journal of Mathematics and Statistics*, **45**(4), 1007–1014, (2016).

[26] M. Olgun and C. Coskun, Non-selfadjoint matrix Sturm-Liouville operators with spectral singularities, *Applied Mathematics and Computation*, **216**(8), 2271–2275, (2010).

[27] Z. S. Agranovich and V. A. Marchenko, *The Inverse Problem of Scattering Theory.* (Translated from the Russian by B.D. Seckler. Gordon and Breach Science Publishers, New York-London, 1963).

[28] L. D. Faddaev, The inverse problem in the quantum theory of scattering, *Russian Journal of Mathematical Physics*, **4**(1), 72–104, (1963).

[29] R. G. Newton, Connection between the S-matrix and the tensor force, *Physical Review D*, **100**, 412–428, (1955).

[30] T. Aktosun and R. Weder, *Direct and Inverse Scattering For The Matrix Schrödinger Equation.* (Springer, Berlin, 2020).

[31] M. A. Naimark, *Linear Differential Operators.* (Frederick Ungar Publishing Corporation, New York, 1968).

Chapter 19

Optimal Alternative Multiple Testing Procedure for Imputation of Single Cell RNA Sequence Data

Mehmet Ali Kaygusuz[*,‡] and Vilda Purutçuoğlu[†,§]

[*]*Anadolu University, Department of Economics, Eskişehir, 26470, Turkey*
[†]*Middle East Technical University, Department of Statistics, Ankara, 06800, Turkey*
[‡]*makaygusuz1988@gmail.com*
[§]*vpurutcu@metu.edu.tr*

Single-cell RNA-Seq provides transcriptional profiling of thousands of individual cells. This level of throughput analysis enables researchers to understand which genes are expressed and under which quantities they are expressed at the single-cell level as well as how those genes differ across thousands of cells within a heterogeneous sample. On the other hand, scRNA-sequence data often fails to capture expressed genes, leading to the prominent dropout problem. The underlying dropouts cause many challenges in downstream analysis, such as significant increases in noisy observations, power loss in differential expression analysis and uncertainty of gene-to-gene or cell-to-cell relationships. Accordingly, MAGIC (Markov affinity-based graph imputation of cells) is a method for addressing computational difficulties to reduce those technical noises in single-cell data, including under-sampling of mRNA molecules, often termed "dropout" which can severely obscure important gene-gene relationships. In its calculation, this algorithm shares information across similar cells via data diffusion so that the cell count matrix can be denoised and missing transcripts can be filled. But, when the dimension of data raises, the detection of the underlying erroneous signals can be computationally demanding. Hereby, we consider to perform a subsampling algorithm, specifically, principal component analysis, for MAGIC in order to gain from calculation time. Furthermore, we consider to apply the knockoff filtering method to control false discoveries of principle components of MAGIC, resulting in reduction of noise in a single cell RNA data. In our analysis, we use epithelial-to-mesenchymal transition RNA dataset and compare our findings with the ordinary MAGIC algorithm.

1. Introduction

Over the last decade, we have been seeing a dramatic increase in the number of large and highly complex datasets generated from biological experiments, which quantify molecular variables such as gene, protein, and metabolite abundance. Single-cell data which has wide applications in diverse scientific domain such as machine learning, bioinformatics and signal progressing, has those challenges.

The Markov Affinity-based Graph Imputation of Cells (MAGIC) algorithm is particularly developed to analyze the high dimensional data for revealing interactions between thousands of genes in a single cell, and to investigate their changes over time. Basically, MAGIC is a method to reduce technical noise in single-cell data, including under-sampling of mRNA molecules, which is often termed dropout and can rarely obscure important gene-gene relationships. It is commonly used for denoising high-dimensional single-cell RNA sequencing data. In its computation, it shares information across similar cells via data diffusion, to clean the errors in the cell count matrix and to fulfill missing transcripts. In imputing missing expression values, it shares information in similar cells based on the idea of heat diffusion. Here, its key property is to learn the manifold in the data, by using the final graph for smoothing the features and replacing the structure of the data. So, the algorithm conducts this step by creating a Markov transition matrix which is constructed by normalizing the similarity matrix of single cells. Then, in the imputation of a single cell, the weights of other cells are determined by the underlying transition matrix. Indeed, in the study of Van Dijk *et al.* [1] which has been introduced the MAGIC algorithm, this technique is used to impute the principal component analysis (PCA) model with single-cell RNA sequencing data. Then, Amodio *et al.* [2] has proposed autoencoder deep neural network for clustering large amount of single-cell transcripts in a scalable manner while dealing with many patients or conditions simutaneously. Chen *et al.* [3] have adapted robust decomposition algorithm for single-cell RNA sequencing data to suggest computational efficient imputation method. Wang *et al.* [4] has proposed single-cell graph neural networks to provide a hypothesis-free deep learning framework for scRNA-Seq analyses.

On the other hand, multiple testing can be computationally challenging, particularly for high throughout single-cell data. Therefore, the knockoff method is introduced by Barber and Candès [5] as a new statistical multiple-testing paradigm which is used for variable selection in high-dimensional data. It is specifically designed to control the false discovery rate (FDR) while identifying relevant variables. Furthermore, Katsevich and Sabatti [6] has proposed multi-layer knockoff filter for genome-wide studies by using the advantage of group knockoff which is proposed by Dai and Barber [7]. Then, Sesia *et al.* [8] have introduced a novel type of knockoff filter, called knockoff zoom, for causal models with genome-wide studies. More recently, knockoff procedure has been suggested for wavelet classification of bacterial Raman spectra by Chia *et al.* [9] and for PCA with deep neural networks in genomic data by Zhu *et al.* [10].

Furthermore, knockoff procedure for genome-wide studies is proposed with group knockoff, multiple knockoff with application of Alzeheimer`s disease and ghost knockoff for causal models by Chu *et al.* [11], He *et al.* [12, 13] and Ma *et al.* [14], respectively. Moreover, Gablenz *et al.* [15] have combined p-values and knockoff for genome-wide studies with signal models. Görgen *et al.* [16] have proposed weighted and group knockoff procedures for bone recovery rates. Kassani *et al.* [17] have suggested variable selection procedure for causal genetic variants with deep neural networks. Kaygusuz and Purutçuoğlu [18] and Kaygusuz and Purutçuoğlu [19] have investigated deep neural networks for heteregenous effect and conditional independence test for robust feedforward neural networks, respectively.

Hereby, in this study, we bind the idea of MAGIC with knockoff filtering in the analyses of high-dimensional single-cell RNA sequence data. In Section 2, we describe the MAGIC algorithm and and knockoff filtering. In Section 3, we present the analyses which evaluate the gain in accuracy and finally, in Section 4, we conclude our findings and discuss the future direction.

2. Theory and Methods

2.1 MAGIC Algorithm

Single-cell sequencing refers to the sequencing of individual cells to obtain genomic, transcriptomic or multi-omics information at the single-cell level. The review and potential applications of single-cell transcriptomic data over ten years are studied by Goodwin *et al.* [20]. In general, the data which is generated by this process has a much higher resolution than conventional bulk sequencing methods. To deal with this data, PCA is implemented. Simply, PCA is an unsupervised machine learning algorithm which tries to reduce the dimensionality or number of features within a dataset while keeping its information. A common way to compute similarities between vectors is to firstly compute distances and then, apply a kernel function. Furthermore, MAGIC computes a cell-cell distance matrix based on Euclidian distance. In its calculation, distances are converted into an affinity matrix A by using a kernel function which emphasizes close similarities between cells due to the nature of PCA. MAGIC also proposes an adaptive Gaussian kernel, which is why similarity between two cells decreases exponentially with their distance. More specifically, distances below σ which is the standard deviation of the Gaussian kernel, are converted into relatively high affinities, while they are more than standard error. Moreover, we treat A as a weighted adjacency matrix representing the neighbor graph structure where the entry A of i, j, i.e., $A_{(i,j)}$ represents the edge weight between cells i and j. Then, we create a Markov transition matrix M by row normalizing A such that each row sum is 1, presenting the transition probability distribution from a particular cell to every other cell in the data by a single step.

On the other side, MAGIC applies a specific algorithm in order to conduct a kernel density estimate (KDE) which is used to smooth the data and fill in gaps. This special nonparametric approach within the MAGIC algorithm is called DREMI. Shortly, KDE includes a replacement of a smooth kernel, such as Gaussian kernel, at each data point and summation of these Gaussians along a regular grid. Therefore, DREMI aims to normalize the estimate along one of the two dimensions in order to form a conditional density estimate. The crucial steps of DREMI include: (i) Kernel density estimation to compute for two variables x and y, denoted as $p(x, y)$. (ii) Coarse-graining of KDE into larger discrete bins for entropy computation. (iii) Normalization of the coarse-grained KDE in order to calculate $p(x, y)$. (iv) Computation of entropy and mutual information based on the discrete bins.

2.2 Knockoff Procedure

Recently, Barber and Candès [5] has proposed a novel model selection procedure, called the knockoff filtering procedure, by using false discovery rate (FDR) control for low dimensional regression [21, 22]. Then, Candes *et al.* [23] have suggested a conditional randomization test for logistic regression. Moreover, Barber and Candès [24, 25] have proposed robust version of knockoff filter with Kulback-Leibler divergence and for high dimensional regression, separately. More recently, knockoff procedure is suggested to control false discoveries for genome-wide studies by Sesia *et al.* [26, 28] and Brzyski *et al.* [27]. Their selection procedure is based on three stages. The first step is to make knockoff copies of conditional variables. Later, we calculate the importance of variables and finally, the third step is to make the model selection depending on FDR control of type 1 error and to decide which explanatory variables are truly associated with the response variable. While doing this, we bound the FDR as the expected fraction of false discoveries. Therefore, the knockoff filter, as a variable selection method, controls the FDR via the principal component analysis (PCA) when the number of observations is greater than the number of variables. In this approach, shortly, we construct extra copy variables, called knockoff variables, which have certain similar correlation structure as existing variables and allow for FDR control when existing variable selection methods are applied. The main advantages of the knockoff filter over competing methods are that it achieves exact FDR control in finite sample settings, no matter the design matrix, covariates, number of variables, or size of regression coefficients, and that it empirically shows more power than existing methods. The knockoff method works by creating knockoff variables that mimic the original variables in a way that preserves the underlying statistical dependencies. These knockoff variables are constructed to have similar correlation structures with the original variables, but, do not carry any true signal. Then, a variable selection procedure is applied to the combined set of original and knockoff variables, and lastly, a statistical test is performed to distinguish the true signals from the noise.

Furthermore, the most important property of the knock-off filter is the exchangeability property when we work with (X_i, Y_i) pairs obtained by independent and identical distribution, i.e, i.i.d. So, showing $\tilde{X}$ as the knock-off matrix for X variable by $(n \times p)$-dimensional matrix, the ex-changeability property is presented by

$$[X\tilde{X}]_{swap(s)}|y\, Z = [X\tilde{X}]\ |y \tag{1}$$

where both sides of equation have the same distribution and *swap(x)* indicates the interchange between the jth and $(j + p)$th column X_j and $\tilde{X}_j$ for each $j \in S$ where $s \in R^p$ is a positive semi-definite vector and S is the associated sample covariance-variance matrix. For example,

$$(X_1, X_2, X_3, \tilde{X}_1, \tilde{X}_2, \tilde{X}_3)_{swap(2)} = (X_1, \tilde{X}_2, X_3, \tilde{X}_1, X_2, \tilde{X}_3). \tag{2}$$

On the other hand, another property of exchangeability is that the covariance matrix between $(X, \tilde{X})$ must be equal to

$$\mathrm{cov}(X, \tilde{X}) = \begin{bmatrix} \Sigma & \Sigma - diags \\ \Sigma - diags & \Sigma \end{bmatrix} \tag{3}$$

in which Σ stands for the population covariance-variance matrix and *diags* presents the diagonal entries of $\mathrm{cov}(X, \tilde{X})$.

Moreover, we calculate the importance statistics, shown by W_j, before model selection stage. Accordingly, to define the second property of the knock-off filter, we define W_j which depends on the design matrix. So, knockoff copies of $W_j \leq 0$, $W_j \geq 0$ have equal probability for each null j and a threshold t such that

$$\#\{\text{null}\, j : W_j \leq -t\} = \#\{\text{null}\, j : W_j \geq t\}. \tag{4}$$

Finally, the third stage of knockoff filtering procedure involves the model selection by false discovery proportion (FDP) as follows:

$$\mathrm{FDP}(t) = \frac{\#\{null\ j: W_j \geq t\}}{\#\{j: W_j \geq t\}} \text{ and } \widehat{FDP}\,(t) = \frac{\#\{j: W_j \leq -t\}}{\#\{j: W_j \geq t\}}. \tag{5}$$

In Eq. (6), we can use equal probability in order to choose a threshold value with the test statistics W_j. Hereby, we can define a threshold $T > 0$ as below.

$$T = min\{t > 0: \frac{\#\{j: W_j \leq -t\}}{\#\{j: W_j \geq t\}} \leq q\}. \tag{6}$$

where q is the targeted FDR level. Therefore, we can make variable selection with $\hat{S} = \{ j\colon W_j \geq T \}$. Now, we can define FDR at the control level q via $FDR = E(FDP) = \frac{\#\{j: j \in S_0, j:j \in \hat{S}_1\}}{\#\{j:j \in \hat{S}_1\} + 1/q} \leq q$.

3. Numerical Experiments

In the imputation part of the MAGIC algorithm, we use epithelial-to-mesenchymal transition RNA (EMT) dataset which is publically available. This data was collected on Indrop 8–10 days after TGFB-stimulation of HMLE breast cancer cells. Three dimensional scatterplots between canonical EMT genes, namely, E-cadherin, Vimentin, and Fibronectin are shown in Figure 1. In this dataset, cells are subsampled under different degrees by uniformly random iteration (100 iterations each). Then, for each subsampled dataset, we remove any gene that has no expression and imputes the remaining genes by using MAGIC algorithm for the same set of parameters. Afterwards, for each imputed matrix, we calculate the correlation value R^2 per entry against the full dataset. R^2 of re-imputed values on 9,571 genes that are expressed in more than 250 cells are divided into three groups based on expression levels. These are the R^2 of values, cells and genes which are computed for different levels of cell subsampling (with 100 random trials for each level of subsampling). The line plots in Figure 1 show the average and standard deviation of R^2 between full and subsampled data, respectively. Moreover, MAGIC recovers gene-gene interactions in EMT data. Hence, from the plots in Figure 1, it is seen that the scatterplot "before MAGIC" with most cells is near the origin and the scatterplot "after MAGIC" reveals a continuous trajectory. The cells are coloured by the level of Zeb1, which is a key EMT transcription factor increasing through the middle of the mesenchymal branch. Furthermore, to demonstrate MAGIC's ability for revealing the fine structure and temporal ordering of events during cell state transitions, we still choose to examine the epithelial-to-mesenchymal transition.

On the other hand, as seen in Figure 2(a), we profile approximately 7500 HMLE cells (17), 8 and 10 days following TGFβ induction of EMT, collected by using the in-drop microfluidic protocol (2). Here, induction of EMT is asynchronous and each cell progresses along the transition at a different rate. Therefore, on days 8

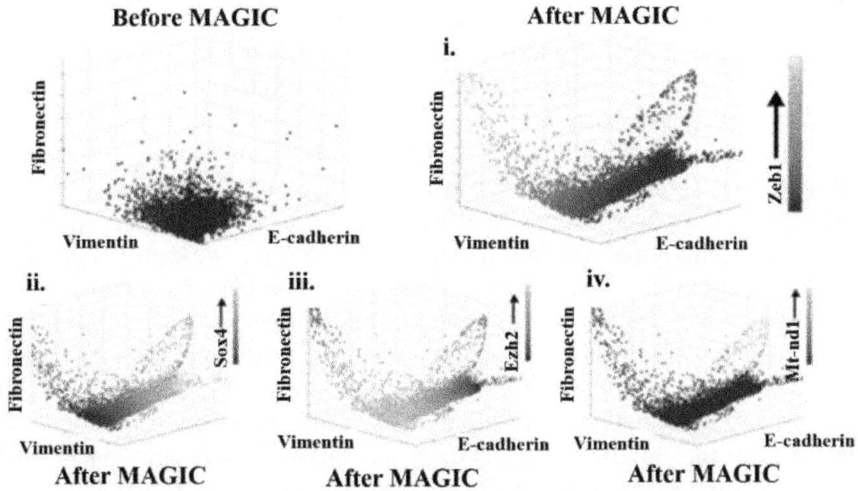

Fig. 1. Before and after MAGIC for EMT data with E-cadherin, Vimentin and Fibronectin genes.

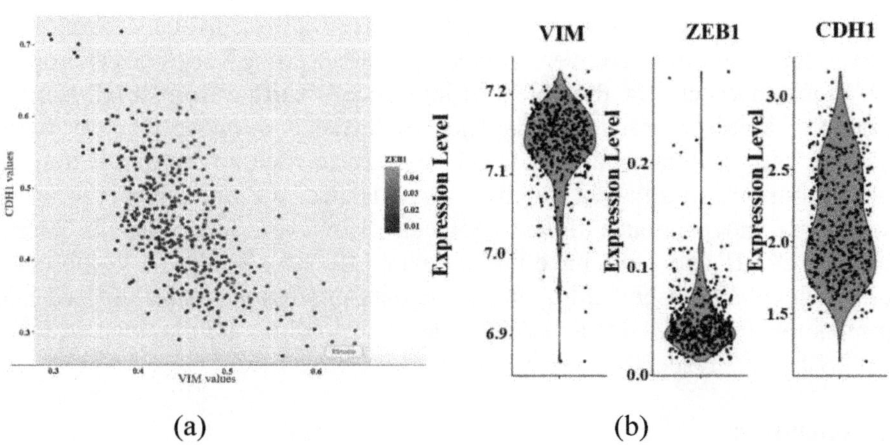

Fig. 2. (a) Scatter plots of Zeb1 and (b) scatter plots of min-max values of VIM, Zeb1 and CDH1.

and 10, cells occupy all phases of EMT. From Figure 2(b), expression of Zeb1, a key transcription factor known to induce EMT, progressively increases as Vimentin and Fibronectin raise and the cells become more mesenchymal. Cdh1 (E-cadherin) are general classes of genes. In CDH1 that is high in the epithelial state, the minimum value is found as 0.2, the mean value is computed as 0.448 and the maximum value is obtained as 0.718. Moreover, the minimum value of VIM is found as 2.46, the mean value is obtained as 3.96 and the maximum value is computed as 5.93. Lastly, the minimum value of ZEB1 is seen as 0.0054, mean value and the maximum values are found as 0.0156 and 0.0468, respectively. According to these statistics, we can say that higher in an intermediate state with MAGIC reveals the known opposite effect between E-cadherin and Vimentin for epithelial-to-mesenchymal transition data. Furthermore, we apply the MAGIC algorithm for epithelial-to-mesenchymal transition data (500 cells and 197 genes) with the sub-sampling procedure in order to find computationally faster selection procedure. As a result, the graph and diffusion operator calculate principal component analyses (PCA) in 0.04 seconds and by using the k-neighborhood algorithm (k-nn) as well as affinity calculation with 0.04 seconds, the total time of calculation after the imputation takes 0.15 seconds with subsampling procedure. On the other hand, the graph and diffusion operator calculate PCA in 0.09 seconds and k-nn in 0.07 seconds and affinities in 0.06 seconds, and thereby, the total time of calculation takes 0.24 seconds without the subsampling procedure.

Additionally, from the multiple testing part of data analysis, we observe that the traditional false discovery rate (FDR) may be too conservative for large-scale biomedical data, especially, when variables are highly correlated. For instance, if a non-null variable X_j is correlated with a null variable X_k, it might be challenging to reject $X_j \perp Y|X_{-j}$. This can be important since highly correlated features can be seen in many settings. To overcome this problem, group knock-off is proposed by [6]. In group knockoffs, the object of inference is shifted from single variables to groups of highly correlated variables. For instance, suppose that we partition p features into g groups and re-order all features in such a way that features of the same group are in adjacent columns of X. Hence, the objective is to test group conditional independence of the hypothesis. However, due to the aforementioned reasons, CDH1 and Zeb1 have the opposite effect for epithelial-to-mesenchymal transition cells. Subsequently, after some data processing procedure, we calculate group knockoff for CDH1 and Zeb1, and their p-values are obtained as 3.87×10^{-8} and 2.70×10^{-7}, respectively, while FDR is found as 0.05.

4. Conclusion

In this study, we have proposed an alternative multiple testing procedure for single-cell RNA-sequencing (scRNA-seq) techniques by using MAGIC (Markov Affinity-based Graph Imputation of Cells) algorithm. This method is a computational method for recovering missing genes by using similar cells in such a way that it can reveal interactions between thousands of genes in a single cell and can detect how they change over time. On the other hand, the knock-off filter method is a statistical technique used for variable selection in high-dimensional data by controlling the false discovery rate (FDR) while identifying relevant variables. But, it can be computationally inefficient for high dimensional single-cell RNA sequence datasets. Hereby, in this study, we have applied group knockoff filter testing procedure and the subsampling procedure together in order to obtain faster dimensional reduction for principal component of the MAGIC algorithm to make imputation, and reduce false discoveries for epithelial-to-mesenchymal transition data. The results have indicated high accuracy with reduced computational demand. Thus, as future study, we consider to apply this testing procedure for auto-encoder neural networks with single-cell datasets or more complex datasets such as Alzheimer`s disease data in order to improve the accuracy without losing from computational efficiency.

Acknowledgement

The authors would like to thank the anonymous referees and editors for their valuable comments which improve the readability of the paper significantly.

References

1. Van Dijk, D., Sharma, R., Nainys, J., Yim, K., Kathail, P., Carr, A. J. and Pe'er, D. (2018). Recovering gene interactions from single-cell data using data diffusion. *Cell*, 174(3), 716–729.
2. Amodio, M., Van Dijk, D., Srinivasan, K., Chen, W. S., Mohsen, H., Moon, K. R. and Krishnaswamy, S. (2019). Exploring single-cell data with deep multitasking neural networks. *Nature Methods*, 16(11), 1139–1145.
3. Chen, C., Wu, C., Wu, L., Wang, X., Deng, M. and Xi, R. (2020). scRMD: imputation for single cell RNA-seq data via robust matrix decomposition. *Bioinformatics*, 36(10), 3156–3161.
4. Wang, J., Ma, A., Chang, Y., Gong, J., Jiang, Y., Qi, R. and Xu, D. (2021). scGNN is a novel graph neural network framework for single-cell RNA-Seq analyses. *Nature Communications*, 12(1), 1882.
5. Barber, R. F. and Candès, E. J. (2015). Controlling the false discovery rate via knockoffs. *Annals of Statistics,* 45(5), 2055–2085.
6. Katsevich, E. and Sabatti, C. (2019). Multilayer knockoff filter: Controlled variable selection at multiple resolutions. *The Annals of Applied Statistics*, 13(1), 1.
7. Dai, R. and Barber, R. (2016). The knockoff filter for FDR control in group-sparse and multitask regression. Proceedings of The 33rd International Conference on Machine Learning, PMLR 48:1851–1859.
8. Sesia, M., Katsevich, E., Bates, S., Candès, E. and Sabatti, C. (2020). Multi-resolution localization of causal variants across the genome. *Nature Communications*, 11(1), 1093.
9. Chia, C., Sesia, M., Ho, C. S., Jeffrey, S. S., Dionne, J., Candes, E. J. and Howe, R. T. (2021). Interpretable classification of bacterial Raman spectra with knockoff wavelets. *IEEE Journal of Biomedical and Health Informatics*, 26(2), 740–748.
10. Zhu, Z., Fan, Y., Kong, Y., Lv, J. and Sun, F. (2021). DeepLINK: Deep learning inference using knockoffs with applications to genomics. *Proceedings of the National Academy of Sciences*, 118(36), e2104683118.
11. Chu, B. B., Gu, J., Chen, Z., Morrison, T., Candes, E., He, Z. and Sabatti, C. (2023). Second-order group knockoffs with applications to GWAS. *arXiv preprint arXiv:2310.15069*.

12. He, Z., Le Guen, Y., Liu, L., Lee, J., Ma, S., Yang, A. C. and Ionita-Laza, I. (2021). Genome-wide analysis of common and rare variants via multiple knockoffs at biobank scale, with an application to Alzheimer disease genetics. *The American Journal of Human Genetics*, 108(12), 2336–2353.
13. He, Z., Liu, L., Belloy, M. E., Le Guen, Y., Sossin, A., Liu, X. and Ionita-Laza, I. (2022). GhostKnockoff inference empowers identification of putative causal variants in genome-wide association studies. *Nature Communications*, 13(1), 7209.
14. Ma, S., Dalgleish, J., Lee, J., Wang, C., Liu, L., Gill, R. and Ionita-Laza, I. (2021). Powerful gene-based testing by integrating long-range chromatin interactions and knockoff genotypes. *Proceedings of the National Academy of Sciences*, 118(47), e2105191118.
15. Gablenz, P. and Sabatti, C. (2024). Catch me if you can: Signal localization with knockoff e-values. *Journal of the Royal Statistical Society Series B: Statistical Methodology*, 87(1), 56–73.
16. Görgen, K., Nazemi, A. and Schienle, M. (2022). Robust knockoffs for controlling false discoveries with an application to bond recovery rates. *arXiv preprint arXiv:2206.06026.*
17. Kassani, P. H., Lu, F., Le Guen, Y., Belloy, M. E. and He, Z. (2022). Deep neural networks with controlled variable selection for the identification of putative causal genetic variants. *Nature Machine Intelligence*, 4(9), 761–771.
18. Kaygusuz, M. A. and Purutçuoğlu, V. (2023a). Conditional independence testing for conformal robust neural networks in outlier detection. In *Proceeding of International Researchers, Statistician, Young Statistician Congress,* 2-5 November 2023, Istanbul, 8 pages.
19. Kaygusuz, M. A. and Purutçuoğlu, V. (2023b). Deep neural networks for heterogeneous effect on biological networks. In *Proceeding of 10th International Congress of Fundamental and Applied Sciences,* ICFAS2023, Istanbul, ID1124, 6 pages.
20. Goodwin, S., McPherson, J. D., and McCombie, W. R. (2016). Coming of age: ten years of next-generation sequencing technologies. *Nature Reviews Genetics*, 17(6), 333–351.
21. Benjamini, Y. and Hochberg, Y. (1995). Controlling the false discovery rate: a practical and powerful approach to multiple testing. *Journal of the Royal Statistical Society: Series B (Methodological)*, 57(1), 289–300.
22. Benjamini, Y. and Yekutieli, D. (2001). The control of the false discovery rate in multiple testing under dependency. *Annals of Statistics*, 1165–1188.
23. Candes, E., Fan, Y., Janson, L. and Lv, J. (2018). Panning for gold: 'Model-X' knockoffs for high dimensional controlled variable selection. *Journal of the Royal Statistical Society Series B: Statistical Methodology*, 80(3), 551–577.
24. Barber, R. F. and Candès, E. J. (2019). A knockoff filter for high-dimensional selective inference. *Annals of Statistics*, 47(5), 2504–2537.
25. Barber, R. F., Candès, E. J., and Samworth, R. J. (2020). Robust inference with knockoffs. The *Annals of Statistics*, 48(3), 1409–1431.
26. Sesia, M., Bates, S., Candès, E., Marchini, J. and Sabatti, C. (2021). False discovery rate control in genome-wide association studies with population structure. *Proceedings of the National Academy of Sciences*, 118(40), e2105841118.
27. Brzyski, D., Peterson, C. B., Sobczyk, P., Candès, E. J., Bogdan, M. and Sabatti, C. (2017). Controlling the rate of GWAS false discoveries. *Genetics*, 205(1), 61–75.
28. Sesia, M., Sabatti, C. and Candès, E. J. (2019). Gene hunting with knockoffs for hidden markov models. *Biomerika*, 106(1), 1–18

Chapter 20

Boole Polynomial Approach for Solving Nonlinear Delay Differential Equations

Hale Gül Dağ* and Kübra Erdem Biçer

Manisa Celal Bayar University, Department of Mathematics
Manisa, 45000 Turkey
**halegul.dag@hotmail.com*

The aim of this study is to develop a numerical method for obtaining approximate solutions of nonlinear delay differential equations with initial-boundary conditions. In this numerical method, Boole polynomials, their derivatives, and collocation points are used. By utilizing the matrix properties of Boole polynomials, the algebraic equation is converted into an augmented matrix equation. The matrix forms of the initial-boundary conditions are applied to this augmented matrix equation, resulting in a new augmented matrix equation. From the solution of this new augmented matrix equation, the unknown Boole coefficients are determined and substituted into the solution. As a result, the solution to the problem is achieved. In this study, examples of exact and approximate solutions are provided to demonstrate the applicability of the Boole matrix method. Approximate solutions of the problems are obtained for various values of N. The Boole solutions are compared with the values obtained by using various numerical methods. Additionally, error analysis is developed to demonstrate the accuracy of the method. The Boole matrix method is implemented using MATLAB computer program codes. The results are compared using tables and figures.

1. Introduction

Newton and Leibniz initiated their research on differential equations in the year 1690. Condorcet conducted studies on delay differential equations in the year 1777. The applications of these equations were employed in the mid-19th century [1]. Delay is observed in many physical, biological, economic, and engineering systems, and is necessary for the facilitation of accurate evaluations of applications [2]. Delay differential equations are encountered in various mathematical modeling contexts within the fields of science and engineering. They are used for examining and researching scientific phenomena [3]. For example, delay differential equations have been utilized in examining simple neural networks with delayed mixed feedback [4], in modeling multiteam prey-predator dynamics [5], in diabetes modeling [6], in mathematically formulating hematopoietic stem cell dynamics [7], and in the mathematical modeling of population dynamics process times [8]. Additionally, delay differential equations have been applied in fields such as economic systems, number theory, control systems, probability, mechanics, electrodynamics, quantum mechanics, biomathematics, and astrophysics [9]. Many methods have been applied to solve delay differential equations. In the literature, numerical methods, such as the variational iteration method [10], Haar wavelet series [11], the Lucas polynomial approach [12], a new operational matrix method based on Taylor polynomials [13], and the Legendre pseudospectral method [14], have been developed for solving delay differential equations [15]. The aim of the study is to develop a method based on Boole polynomials for the numerical solutions of linear and nonlinear delay differential equations.

In this study, Boole matrix method (BMM) is presented for solving the second-order nonlinear delay differential equation given in the form

$$\sum_{k=0}^{m} P_k(x)\, y^{(k)}(x) + \sum_{q=0}^{2}\sum_{p=0}^{q} Q_{pq}(x)\, y^{(p)}(\alpha_p x + \beta_p) y^{(q)}(\lambda_q x + \mu_q) = g(x) \quad (1)$$

with the initial-boundary conditions

$$\sum_{k=0}^{m-1} \left[a_{kj} y^{(k)}(a) + b_{kj} y^{(k)}(b) + c_{kj} y^{(k)}(c)\right] = \lambda_i\,, \quad j = 0, 1, 2 \dots, m-1. \quad (2)$$

The solution of the Eq. (1) is written in the truncated Boole series form

$$y(x) \cong y_N(x) = \sum_{n=0}^{N} a_n R_n(x) \quad (3)$$

where $a_n, n = 0,1,2, \dots, N$ is the unknown Boole coefficients, and $R_n(x), n = 0,1,2, \dots, N$ is the Boole polynomial. The general form and the generated function of Boole polynomials are expressed as follows [16, 17].

$$R_n(x) = \sum_{m=0}^{n} \frac{(-1)^m}{2^m} \binom{x}{n-m} \text{ and } \sum_{n=0}^{\infty} \frac{R_n(x)}{n!} t^n = \frac{2(1+t)^x}{2+t} \quad (4)$$

2. Main Matris Relations and Collocation Method

The matrix form of Boole polynomials is written as

$$\mathbf{R}(x) = \mathbf{X}(x)\mathbf{H}^T \tag{5}$$

where

$$\mathbf{R}(x) = \left[1 \quad x - \frac{1}{2} \quad \cdots \quad \mathbf{R}_N(x)\right], \qquad \mathbf{X}(\mathrm{x}) = [1 \quad x \quad \cdots \quad x^N],$$

$$\mathbf{H} = \begin{bmatrix} 1 & 0 & 0 & \cdots \\ -\frac{1}{2} & 1 & 0 & \cdots \\ \frac{1}{2} & -2 & 1 & \cdots \\ \vdots & \vdots & \vdots & \ddots \end{bmatrix}.$$

According to the relation (5), the matrix form of the solution (3) and its kth derivative are expressed as

$$\begin{aligned} y(x) \cong y_N(x) = \mathbf{R}(x)\mathbf{A} \text{ and } y^{(k)}(x) &\cong y_N^{(k)}(x) = \mathbf{X}^{(\mathrm{k})}(x)\mathbf{H}^{\mathrm{T}}\mathbf{A} \\ &= \mathbf{X}(x)\mathbf{E}^{\mathrm{k}}\mathbf{H}^{\mathrm{T}}\mathbf{A} \end{aligned} \tag{6}$$

where the matrix **E** is the connection between Taylor polynomials and their derivatives, and the matrix **A** is the unknown Boole coefficient [18]. To obtain the terms $y^{(p)}(\alpha_p x + \beta_p)$ and $y^{(q)}(\lambda_q x + \mu_q)$ of the nonlinear part in Eq. (1), x is replaced by $\alpha_p x + \beta_p$ and $\lambda_q x + \mu_q$ in the relation (6), respectively, resulting in the following formulas:

$$y^{(p)}(\alpha_p x + \beta_p) = \mathbf{X}(\alpha_p x + \beta_p)\mathbf{E}^{\mathrm{p}}\mathbf{H}^{\mathrm{T}}\mathbf{A} = \mathbf{X}\,\mathbf{B}(\alpha_p, \beta_p)\mathbf{E}^{\mathrm{p}}\mathbf{H}^{\mathrm{T}}\mathbf{A} \tag{7}$$

$$y^{(p)}(\lambda_q x + \mu_q) = \mathbf{X}(\lambda_q x + \mu_q)\mathbf{E}^{\mathrm{q}}\mathbf{H}^{\mathrm{T}}\mathbf{A} = \mathbf{X}\,\mathbf{B}(\lambda_q, \mu_q)\mathbf{E}^{\mathrm{q}}\mathbf{H}^{\mathrm{T}}\mathbf{A} \tag{8}$$

For $p = 0, q = 0$

$$y(\alpha_0 x + \beta_0)y(\lambda_0 x + \mu_0) = \mathbf{X}(x)\,\mathbf{B}(\alpha_0, \beta_0)\mathbf{H}^{\mathrm{T}}\overline{\mathbf{X}(x)}\,\overline{\mathbf{B}(\lambda_0, \mu_0)}\overline{\mathbf{H}^{\mathrm{T}}}\overline{\mathbf{A}}, \tag{9}$$

where

$$\overline{\mathbf{B}(\lambda_0, \mu_0)} = diag[\mathbf{B}(\lambda_0, \mu_0) \quad \cdots \quad \mathbf{B}(\lambda_0, \mu_0)],$$

$$\mathbf{B}(\alpha_p, \beta_p) = \begin{bmatrix} \binom{0}{0}\alpha_p^0\beta_p^0 & \cdots & \binom{N}{0}\alpha_p^0\beta_p^N \\ \vdots & \ddots & \vdots \\ 0 & \cdots & \binom{N}{N}\alpha_p^N\beta_p^0 \end{bmatrix},$$

$$\mathbf{B}(\lambda_q,\mu_q)=\begin{bmatrix}\binom{0}{0}\lambda_q^0\mu_q^0 & \cdots & \binom{N}{0}\lambda_q^0\mu_q^N\\ \vdots & \ddots & \vdots\\ 0 & \cdots & \binom{N}{N}\lambda_q^N\mu_q^0\end{bmatrix},$$

$$\overline{\mathbf{X}(x)}=diag[\mathbf{X}(x)\ \ \cdots\ \ \mathbf{X}(x)],\qquad \overline{\mathbf{H}^{\mathbf{T}}}=diag[\mathbf{H}^{\mathbf{T}}\ \ \cdots\ \ \mathbf{H}^{\mathbf{T}}].$$

Likewise, the matrix relations (9) are derived for various values of p and q. The relation (6) is written in the Eq. (1) to get

$$\sum_{k=0}^{m} P_k(x)\,\mathbf{X}(x)\mathbf{E}^{\mathrm{k}}\mathbf{H}^{\mathbf{T}}\mathbf{A}+\sum_{q=0}^{2}\sum_{p=0}^{q} Q_{pq}(x)\,y^{(p)}(\alpha_p x+\beta_p)y^{(q)}(\lambda_q x+\mu_q)=g(x)\quad (10)$$

The collocation points are written as

$$x_i=a+\frac{b-a}{N}i,\qquad i=0,1,\dots,N.$$

These points are applied to equation (10) to get

$$\sum_{k=0}^{m} P_k(x_i)\,\mathbf{X}(x_i)\mathbf{E}^{\mathrm{k}}\mathbf{H}^{\mathbf{T}}\mathbf{A}+\sum_{q=0}^{2}\sum_{p=0}^{q} Q_{pq}(x_i)\,y^{(p)}(\alpha_p x_i+\beta_p)y^{(q)}(\lambda_q x_i+\mu_q)=g(x_i)\quad (11)$$

where

$$y^{(p)}(\alpha_p x_i+\beta_p)y^{(q)}(\lambda_q x_i+\mu_q)=\mathbf{Y}^{(\mathrm{p,q})}=\begin{bmatrix}y^{(p)}(\alpha_p x_0+\beta_p)y^{(q)}(\lambda_q x_0+\mu_q)\\ \vdots\\ y^{(p)}(\alpha_p x_N+\beta_p)y^{(q)}(\lambda_q x_N+\mu_q)\end{bmatrix}.$$

For $p=0$ and $q=0$,

$$\mathbf{Y}^{(0,0)}=\begin{bmatrix}\mathbf{X}(x_0)\,\mathbf{B}(\alpha_0,\beta_0)\mathbf{H}^{\mathbf{T}}\overline{\mathbf{X}(x_0)}\ \overline{\mathbf{B}(\lambda_0,\mu_0)\mathbf{H}^{\mathbf{T}}}\\ \vdots\\ \mathbf{X}(x_N)\,\mathbf{B}(\alpha_0,\beta_0)\mathbf{H}^{\mathbf{T}}\overline{\mathbf{X}(x_N)}\ \overline{\mathbf{B}(\lambda_0,\mu_0)\mathbf{H}^{\mathbf{T}}}\end{bmatrix}\begin{bmatrix}a_0\mathbf{A}\\ \vdots\\ a_N\mathbf{A}\end{bmatrix}=\mathbf{X}_{0,0}^{*}\bar{\mathbf{A}}$$

Likewise, this matrix form is obtained for various values of p and q. Thus, relation (11) is written as

$$\sum_{\mathrm{k}=0}^{\mathrm{m}}\mathbf{P}_{\mathrm{k}}\,\mathbf{X}\,\mathbf{E}^{\mathrm{k}}\mathbf{H}^{\mathbf{T}}\mathbf{A}+\sum_{\mathrm{q}=0}^{2}\sum_{\mathrm{p}=0}^{\mathrm{q}}\mathbf{Q}_{\mathrm{pq}}\,\mathbf{X}_{\mathrm{p,q}}^{*}\bar{\mathbf{A}}=\mathbf{G}\quad\Rightarrow\quad \mathbf{WA}+\mathbf{V}\bar{\mathbf{A}}=\mathbf{G}\qquad (12)$$

In this matrix, the $\mathbf{W}=[w_{ij}]$ $i,j=0,1,\dots,N$ matrix is defined for the linear part, and the $\mathbf{V}=[v_{mn}]$, $m=0,1,\dots,N$, $n=0,1,\dots,(N+1)^2-1$ matrix is defined for the nonlinear part. According to the relation (5), the matrix relation of the initial-boundary conditions are written as

$$\sum_{k=0}^{m-1}\left[a_{kj}\mathbf{X}(a)\mathbf{E}^{\mathrm{k}}\mathbf{H}^{\mathrm{T}}+b_{kj}\mathbf{X}(b)\mathbf{E}^{\mathrm{k}}\mathbf{H}^{\mathrm{T}}+c_{kj}\mathbf{X}(c)\mathbf{E}^{\mathrm{k}}\mathbf{H}^{\mathrm{T}}\right]\mathbf{A}=\lambda_i,\Rightarrow \mathbf{UA}+\mathbf{0}^{*}\overline{\mathbf{A}}=\lambda \qquad (13)$$

Finally, in the augmented matrix equation, m rows are removed and replaced with the matrix equation of the conditions, yielding a new augmented matrix equation. The unknown Boole coefficients are obtained from the solution of this matrix and are written in the solution (3). Consequently, the equation (1) is solved.

3. Accuracy of Solution

The accuracy of the solutions is expressed as

$$E(x_r)=\left|\sum_{k=0}^{m}P_k(x_r)\,y^{(k)}(x_r)+\sum_{q=0}^{2}\sum_{p=0}^{q}Q_{pq}(x_r)\,y^{(p)}(\alpha_p x_r+\beta_p)y^{(q)}(\lambda_q x_r+\mu_q)-g(x_r)\right|\cong 0$$

or $E(x_r)\le 10^{-k_r}(-k_r$ is any positive integer) for $x_r\in[a,b], r=0,1,2,\cdots$ (14)

The Boole solution must be satisfied by the Eq. (1) approximately [18]. The absolute error function is written as $e_N(x)=|y(x)-y_N(x)|$ [19]. Also, the residual function is defined as

$$E(x_N)=\sum_{k=0}^{m}P_k(x)\,y_N^{(k)}(x)+\sum_{q=0}^{2}\sum_{p=0}^{q}Q_{pq}(x)\,y_N^{(p)}(\alpha_p x+\beta_p)y_N^{(q)}(\lambda_q x+\mu_q)-g(x) \qquad (15)$$

4. Numerical Example

Example 1. The exact solution of the second-order nonlinear delay differential equation

$$xy^{(2)}(x)-y^{(1)}(x)+y(x)-y^{(2)}(x+1)y^{(1)}(x-1)=x^2-3x+2 \qquad (16)$$

with the initial conditons $y(0)=1,\ y^{(1)}(0)=1$ are investigated for $N=2$. The fundamental matrix relation of the considered equation is expressed as

$$(\mathbf{P}_0\mathbf{X}\mathbf{H}^T+\mathbf{P}_1\mathbf{X}\mathbf{E}^1\mathbf{H}^T+\mathbf{P}_2\mathbf{X}\mathbf{E}^2\mathbf{H}^T)\mathbf{A}+(\mathbf{Q}_{21}\mathbf{X}_{21}^{*})\overline{\mathbf{A}}=\mathbf{G}$$

The augmented matrix form (12) of the Eq. (16) is calculated. The conditions of the problem are written as follows

$$[\mathbf{U}; \mathbf{0}^*: \boldsymbol{\lambda}] = \begin{bmatrix} 1 & -\frac{1}{2} & \frac{1}{2} & ; & 0 & 0 & 0 & 0 & 0 & 0 & 0 & 0 & 0 & : & 1 \\ 0 & 1 & -2 & ; & 0 & 0 & 0 & 0 & 0 & 0 & 0 & 0 & 0 & : & 1 \end{bmatrix}. \quad (17)$$

Finally, the last two rows are deleted from the augmented matrix equation, and the matrix of conditions (17) are written in their place, the new augmented matrix equation is obtained. This matrix are solved and Boole coefficients are found as

$$\mathbf{A} = [2 \quad 3 \quad 1]^T.$$

This matrix written in the solution (3). Therefore, the exact solution of Eq. (16) is obtained as

$$y(x) \cong y_2(x) = x^2 + x + 1.$$

Example 2. In this example, Boole solutions of the second-order nonlinear delay differential equation

$$y^{(3)}(x) = -1 + 2y^2\left(\frac{x}{2}\right) \quad 0 \leq x \leq 1 \quad (18)$$

are investigated with the initial conditions $y(0) = 0, y^{(1)}(0) = 1$, and $y^{(2)}(0) = 0$. The exact solution is $y(x) = sin\,(x)$. The authors in references [20] are solved the problem (18) using the Legendre wavelet method (LWM). For $N = 5, 6$, the absolute error functions of the problem (18) are calculated by BMM. These results are compared with LWM, in Table 1. The values of the residual functions are given in Figure 1, for $N = 5, 6$.

Table 1. Comparison of the absolute errors with other numerical methods.

	Legendre Wavelet Method [20]		Boole Matrix Method	
x_i	$N = 5$	$N = 6$	$N = 5$	$N = 6$
0.1	2.54e-09	5.37e-10	6.3644e-09	2.9547e-09
0.3	2.11e-08	1.59e-09	2.4412e-07	8.2617e-08
0.5	1.21e-07	3.52e-09	6.7786e-07	2.5208e-07
0.7	2.48e-06	6.05e-08	3.6176e-06	4.1971e-07
0.9	3.47e-05	5.08e-06	3.4014e-05	3.7378e-06

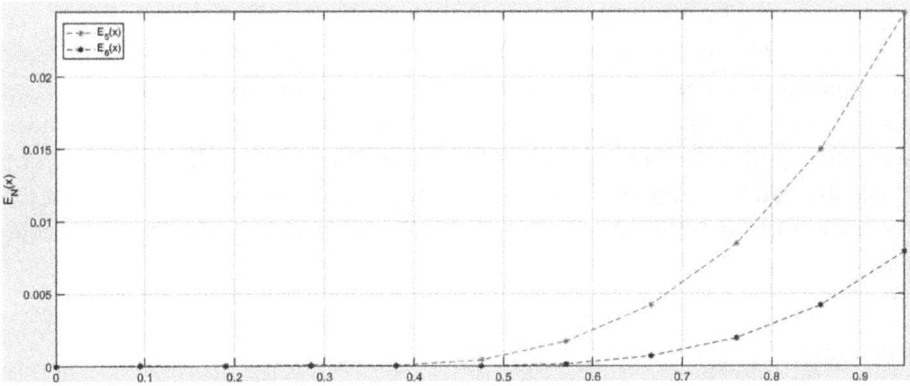

Fig. 1. The comparison of the residual functions for the problem (18) for $N = 5$ and $N = 6$.

Example 3. In this example, the following equation

$$y^{(3)}(x) = -y(x) - y(x - 0.3) + e^{(-x+0.3)} \tag{19}$$

are considered with the initial-boundary conditions $y(0) = 1, y^{(1)}(0) = -1$, and $y^{(2)}(0) = 1$. The exact solution is $y(x) = e^{(-x)}$. In [20], the authors are the problem (19) using the Taylor operation method for $N = 8, 10$. The absolute error functions of the problem (19) are calculated by BMM for $N = 8, 10$. These results are compared with those obtained using the Taylor operation method, in Table 2.

Table 2. Comparison of the absolute errors with other numerical methods.

	Taylor Operation Method [20]		Boole Matrix Method	
x_i	N=8	N=10	N=8	N=10
0.2	6.777e-11	2.313e-13	5.1158e-11	5.5067e-14
0.4	3.312e-10	1.168e-12	2.5057e-10	2.5424e-13
0.6	8.428e-10	2.775e-12	5.9419e-10	5.9419e-13
0.8	1.543e-9	5.125e-12	7.1146e-10	1.0146e-12
1.0	2.491e-9	8.123e-12	2.6861e-08	4.3190e-11

5. Conclusion

In this study, the Boole matrix method is developed to obtain approximate solutions to nonlinear delay differential equations. The method is applied to various numerical examples, demonstrating its effectiveness. Furthermore, the method is adapted to linear delay differential equations. The absolute error functions obtained with the Boole matrix method are compared with those from other methods. In Table 1, the absolute error functions of problem (18) are compared with the Legendre wavelet method. In Table 2, the absolute error functions of problem (19) are compared with the Taylor operation method. The results show that the Boole matrix method is highly applicable and provides accurate solutions for both nonlinear and linear delay differential equations. The versatility and accuracy of the method suggest that it can be a valuable tool for researchers and practitioners dealing with delay differential equations in various fields.

References

1. Z. Sabir, J. L. Guirao, and T. Saeed, Solving a novel designed second order nonlinear Lane–Emden delay differential model using the heuristic techniques, *Applied Soft Computing.* 102 107105, (2021).
2. D. Kumar, S. Upadhyay, S. Singh, and K. N. Rai, Legendre wavelet collocation solution for system of linear and nonlinear delay differential equations, *International Journal of Applied and Computational Mathematics*. 3, 295–310, (2017).
3. N. Sriwastav, A. K. Barnwal, A. M. Wazwaz, and M. Singh, A novel numerical approach and stability analysis for a class of pantograph delay differential equation, *Journal of Computational Science*. 67, 101976, (2023).
4. J. G. Milton, U. van der Heiden, A. Longtin, and M. C. Mackey, Complex dynamics and noise in simple neural networks with delayed mixed feedback, *Biomed. Biochim. Acta*. 49(8–9), 697–707, (1990)
5. P. Kumar and S. Raj, Modelling the multiteam prey-predator Dynamics using the delay differential equation. *Malaysian Journal of Science*, 42–61, (2023).
6. A. Makroglou, I. Karaoustas, J. Li, and Y. Kuang, A review on delay differential equation models in diabetes modeling, II: The insulin therapies and the intracellular activities of β-cells case, *Math Comput Simul*. 1–24, (2011).
7. M. Adimy and F. Crauste, Delay differential equations and autonomous oscillations in hematopoietic stem cell dynamics modeling, *Mathematical Modelling of Natural Phenomena*. 7(6), 1–22, (2012).
8. Y. Kuang (Ed), *Delay differential equations*. (Academic press, New York, 1993).
9. M. M. Bahşi and M. Çevik, Numerical solution of pantograph-type delay differential equations using perturbation-iteration algorithms, *Journal of Applied Mathematics*. 2015(1), (2015).
10. M. M. Khader, Numerical and theoretical treatment for solving linear and nonlinear delay differential equations using variational iteration method, *Arab Journal of Mathematical Sciences*. 19(2), 243–256, (2013).
11. A. Raza and A. Khan, Haar wavelet series solution for solving neutral delay differential equations, *Journal of King Saud University-Science*. 31(4), 1070–1076, (2019).
12. S. Gümgüm, N. Baykuş Savaşaneril, Ö. K. Kürkçü, and M. Sezer, Lucas polynomial solution for neutral differential equations with proportional delays, *TWMS Journal of Applied and Engineering Mathematics*. 10(1), 259–269, (2020).

13. Ş. Yüzbaşı and N. Ismailov, A Taylor operation method for solutions of generalized pantograph type delay differential equations, *Turkish Journal of Mathematics.* 42(2), 395–406, (2018).
14. H. Jafari, M. Mahmoudi, and M. H. Noori Skandari, A new numerical method to solve pantograph delay differential equations with convergence analysis. *Advances in Difference Equations,* 2021(1), 129 (2021).
15. S. K. Vanani and A. Aminataei, On the numerical solution of nonlinear delay differential equations, *Journal of Concrete and Applicable Mathematics.* 8(4), 568–576, (2010).
16. S. Roman, *The Umbral Calculus*, (New York, 1984).
17. K. Jordan, *Calculus of Finite Differences,* (American Mathematical Soc, New York, 1965).
18. K. Erdem Biçer and H. G. Dağ, Numerical solutions of differential equations having cubic nonlinearity using Boole collocation method, *Turkish Journal of Mathematics.* 47(2), 732–745, (2023).
19. W. Adel, K. Erdem Biçer, and M. Sezer, A novel numerical approach for simulating the nonlinear MHD Jeffery–Hamel flow problem, *International Journal of Applied and Computational Mathematics.* 7, 1 (2021).
20. S. Gümgüm, D. Özdek, and G. Özaltun, Legendre wavelet solution of high order nonlinear ordinary delay differential equations. *Turkish Journal of Mathematics*, 43(3), 1339–1352, (2019).

Chapter 21

Optimization of Grinding Conditions during the Clinker Grinding Stage

Yahya Kaya*, Veysel Kobya† and Ali Mardani‡

Department of Civil Engineering, Faculty of Engineering,
Bursa Uludag University, Nilüfer-Bursa, Turkey
**yahyakaya00@gmail.com*
†v.kobya@gmail.com
‡ali.mardani16@gmail.com

Grinding processes are highly energy-intensive, making it crucial to select the right mill and optimize grinding conditions to improve energy efficiency. In cement production, nearly one-third of the total energy is consumed during clinker grinding, so even slight efficiency improvements can yield significant economic and environmental benefits. Key factors like grinding media type and size, mill speed, power, feed size, and quantity greatly influence grinding performance, product size, and energy consumption. While existing studies have extensively explored the effects of ball and feed size, mill speed, and ball shape on grinding efficiency, there is limited research on the combined impact of varying ball sizes, feed sizes, and mill speeds. To address this gap, this study used nine spherical ball sizes (15–65 mm) in different proportions to produce cement under three distinct ball distributions, feed amounts, and mill speeds. The effects of these grinding conditions on efficiency and product properties were assessed using Blaine fineness values and the energy required to achieve the target fineness. Additionally, the Taguchi method and ANOVA were employed to analyze the data, while linear and quadratic regressions were developed to optimize both financial and time outcomes. This comprehensive approach aims to enhance grinding efficiency while minimizing energy consumption and costs.

1. Introduction

Parameters including the type of mill, rotation speed, power, feed size, feed quantity [1], and the type and dimensions of the grinding media [2] influence the properties of clinker. It was understood that there are parameters that directly affect the energy spent in the grinding phase.

It was reported that large-diameter balls cause the feed size to decrease as a result of the impact or impact caused by falling from the shoulder part of the mill during grinding. It was emphasized that the wear effect of small diameter balls during the rotation of the mill causes the grain size to decrease further due to small feeds [3].

It was determined that there is a lack in the literature regarding the synergistic effect of changing parameters such as ball dimensions, feed amount, and mill speed on grinding efficiency.
In the literature review, many experimental studies were found that focused on the effect of parameters such as ball and feed size, mill speed, and ball shape on grinding efficiency and product properties [1–6]. However, the literature addressing the collective influence of varying factors like ball size, feed rate, and mill speed on the efficiency of grinding is scarce. In this context, within the scope of this study, nine spherical balls ranging in size from 15 to 65 mm were used in different ratios to produce cement with six different ball distributions, three different feed rates, and three different mill speeds. The Blaine fineness values of the cements produced at certain speeds and the energy consumed for the target Blaine fineness value were analyzed to investigate the effect of the grinding conditions on grinding efficiency and product properties. In addition, linear and quadratic regressions were developed using Taguchi methods and ANOVA. In this study, the goal is to investigate the interplay between ball distribution, mill speed, and feed amount during the grinding process by analyzing key parameters through both experimental and regression methods. This approach aims to deliver ecological and economic benefits in cement production applications. However, it is important to note that conducting the experiments under laboratory conditions limits the broader applicability of the study's findings.

2. Materials and Methods

In the grinding process of clinker, a laboratory-type mill with a 1.5 kW engine power and a capacity of 5 kg, shown in Figure 1, was used. The mill feed amount (96% clinker + 4% gypsum) was selected as 2, 3 and 4 kg. The grinding process was carried out by applying balls with nine different diameters in three different distributions. The ball size distributions used within the scope of the study are given in Table 1.

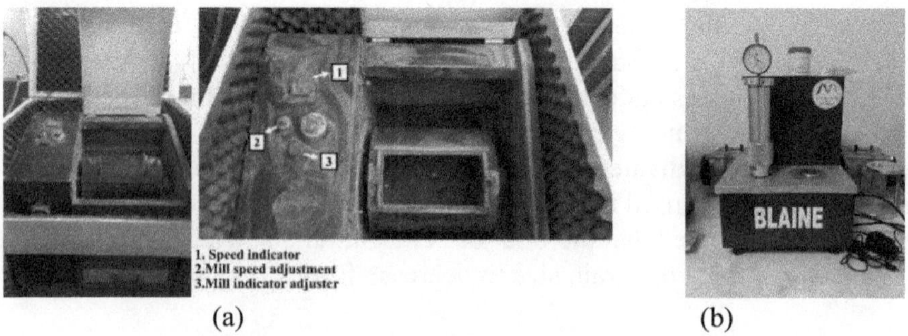

(a) (b)

Fig. 1. a. Laboratory-scale ball mill b. Apparatus for measuring Blaine fineness

Table 1. Ball size distributions

Ball size (mm)	Ball Weight (g/piece)	1. Ball Distribution (piece)	2. Ball Distribution (piece)	3. Ball Distribution (piece)
65	860	4	4	5
55	510	5	5	6
43	260	12	12	12
40	225	7	7	14
37	190	3	8	9
30	110	6	12	22
25	70	5	12	17
20	30	3		
15	16	20	40	50
Total surface area (mm^2)		263093	325673	433097
Total weight (g)		12675	15005	19750

The mill rotation speed was determined as 40, 55, and 70 rpm, which is 50%, 70%, and 90% of the mill critical speed respectively.

The effect of these parameters on grinding efficiency was compared with two different approaches. In the first approach, Blaine fineness values of cements produced after 6000 cycles were measured. In the second approach, the number of grinding cycles that should be applied to obtain the target Blaine fineness value of 3700 ± 100 cm^2/g was determined. The fineness value of the produced cements was determined by the automatic Blaine device.

2.1. *Experimental design with the Taguchi method*

Within the scope of the study, ANOVA and Taguchi methods were applied to determine the optimum combination of grinding parameters. In this study, "smallest is best", shown in Eq. 4 suggested by Mandal *et al.* [8] was used as the objective function.
Here; yi is the data observed in the ith experiment and n refers to the number of observations of the experiment.

The levels of the selected factors are shown in Table 2. To determine the optimum grinding conditions and analyze the effects of grinding parameters, the most suitable orthogonal array was chosen as L9(3^3).

Table 2. Grinding parameters and levels

Parameters	Symbol	Level 1	Level 2	Level 3
Ball Distribution	A	1	2	3
Mass (kg)	B	2	3	4
Speed (rev/min)	C	40	55	70

3. Result and Discussions

When ball distributions are compared, the increase in total ball mass significantly increases the grinding efficiency. Regardless of the ball type, the increase in clinker mass and mill rotation speed negatively affects grinding efficiency.

Table 3. Blaine fineness values after 6000 cycles

Ball Distribution	Mass (kg)	Mill Speed (rev/min)	6000 cycle Blaine fineness (cm^2/g)
Ball Distribution No. 1	2	40	3480
		55	3350
		70	3200
	3	40	2910
		55	2780
		70	2630
	4	40	2500
		55	2370
		70	2220
Ball Distribution No. 2	2	40	3850
		55	3720
		70	3570
	3	40	3280
		55	3150
		70	3000
	4	40	2870
		55	2740
		70	2590
Ball Distribution No. 3	2	40	4130
		55	4000
		70	3850
	3	40	3560
		55	3430
		70	3280
	4	40	3150
		55	3020
		70	2870

Table 4. Energy consumption for target Blaine fineness value.

Ball Distribution	Mass (kg)	Mill Speed (rpm)	Speed for Blaine fineness 3700 cm^2/g	Grinding time for 3700 cm^2/g Blaine fineness	Ampere	Energy consumed for target Blaine fineness value (kWh/ton)
Ball Distribution No. 1	2	40	6387	159.7	1.56	114.17
		55	6613	120.2	1.61	88.72
		70	6850	97.9	1.68	75.35
	3	40	7979	199.5	1.57	95.70
		55	8568	155.8	1.63	77.59
		70	8945	127.8	1.7	66.38
	4	40	9524	238.1	1.58	86.21
		55	10238	186.1	1.65	70.39
		70	11967	171.0	1.72	67.39
Ball Distribution No. 2	2	40	5779	144.5	1.6	105.94
		55	5924	107.7	1.66	81.95
		70	6140	87.7	1.77	71.16
	3	40	6868	171.7	1.61	84.47
		55	7140	129.8	1.67	66.25
		70	7755	110.8	1.74	58.90
	4	40	8758	218.9	1.63	81.78
		55	9631	175.1	1.69	67.82
		70	11085	158.4	1.76	63.87
Ball Distribution No. 3	2	40	4927	123.2	1.66	93.71
		55	5162	93.9	1.72	73.99
		70	5626	80.4	1.79	65.93
	3	40	6368	159.2	1.68	81.73
		55	6710	122.0	1.74	64.86
		70	7328	104.7	1.81	57.90
	4	40	7609	190.2	1.7	74.11
		55	8161	148.4	1.75	59.51
		70	9225	131.4	1.83	55.12

When the energy consumed to obtain 1 ton of product is calculated based on the mass of the milled material, mill rotation speed, and total ball mass increase, it is seen that less energy is consumed. Therefore, it is seen that the most efficient grinding conditions are obtained with ball distribution number 3, 4 kg mass, and 70 rpm speed.

The grinding process takes place when the balls hold on to the mill wall and fall from above. Therefore, the increase in the ball mass used in grinding positively affects the grinding performance.
The mass of the material being ground and the grinding efficiency are inversely related.
Mill speed is important for the grinding process. Because, if the grinding process is carried out above the critical speed, the balls cannot hold on to the mill surface and perform a falling movement [9].

3.1. *Evaluation of analysis results: Analysis of signal-to-noise (S/N) ratio*

With the experimental design, the energy spent in grinding was measured experimentally for each combination of control factors. The obtained data were processed using Taguchi techniques, and the optimization of the measured grinding factors was achieved with signal-to-noise (S/N) ratios.

Table 5. Active parameters according to the Taguchi method

Response table for signal-to-noise ratios				Response table for means			
Level	ball distribution	mass	speed	Level	ball distribution	mass	speed
1	-35.71	-35.33	-35.54	1	61.81	59.28	60.52
2	-34.88	-34.31	-33.91	2	56.3	52.13	49.96
3	-33.88	-33.71	-33.9	3	49.57	48.7	49.63
Delta	1.94	1.63	1.63	Delta	12.68	10.59	10.89
Rank	1	3	2	Rank	1	3	2

3.2. *ANOVA method*

In this study, ANOVA was used to analyze the effects of ball distributions, the mass of the material being ground, and the rotation speed applied in grinding on the energy spent in grinding. The effect ratios of the effective parameters on the grinding performance are shown in Table 6.

Table 6. ANOVA results for consumed energy

Source	Degree of freedom (DoF)	A sum of squares (SS)	Mean square (MS)	F-Value	P-Value	Effect rates (%)
Ball Distribution	5	460.64	230.32	30.61	0.000178	36.89
Mass	2	349.93	174.964	10.04	0.000464	28.02
Speed	2	377.91	75.581	23.25	0.002702	30.26
Error	8	60.2	7.525			4.82
Total	17	1248.67				100.00

3.3. ***Regression analysis***

In this study, the dependent variables were selected as the energy spent in grinding, and the independent variables were ball distribution, mass of the ground material, and mill rotation speed. Regression analysis was used to obtain equations by which the energy spent in grinding could be estimated.

A comparison of the actual test results obtained with the linear regression model and the predicted values is given.

Table 7. Equations by linear regression

Ball Distribution			
1	spent energy	=	97.66 – 5.29 mass – 0.3631 speed
2	spent energy	=	92.14 – 5.29 mass – 0.3631 speed
3	spent energy	=	85.41 – 5.29 mass – 0.3631 speed

The prediction equation for the quadratic regression of energy spent in grinding is given below:

Spent energy = 221.2 − 27.6 mass − 3.85 Speed − 3.42 ball distribution + 1.45 mass^2 + 0.02226 Speed^2 − 0.219 ball distribution × mass + 0.0544 ball distribution × Speed + 0.277 mass × Speed

4. Conclusion

In this study, the effects of grinding parameters such as ball distribution, grinding mass, and rotational speed on grinding performance were investigated. Experimental results are summarized below:

- The increase in the total mass of the ball increases the grinding performance.
- Increasing mill speed led to a general decrease in grinding performance, especially increases up to 90% of the critical speed. This situation negatively affected the shoulder and foot angles of the balls, making it difficult to crush the grains.
- According to the statistical analysis results, it was determined that the most important parameter for energy efficiency was the ball distribution with a contribution of 36.89%.
- The developed quadratic regression models found a very good relationship with high correlation coefficients between the measured and predicted values for the energy spent in grinding.

Acknowledgments
The authors express their appreciation for the assistance provided by the Scientific and Technological Research Council of Turkey (TUBITAK) under Grant No: 222M245. They also recognize the contributions of the Bursa Uludag University Science and Technology Centre (BAP) under grant numbers FGA-2022-1151. The primary author extends gratitude to TUBITAK for the 2211A scholarship received during their doctoral studies.

References

1. Danha, G., Hlabangana, N., and Muzenda, E. (2018). Effect of ball and feed particle size distribution on the milling efficiency of a ball mill: An attainable region approach. *South African Journal of Chemical Engineering*, 25(1), 79–84.
2. Santosh, T., Eswaraiah, C., Soni, R. K., and Kumar, S. (2023). Size reduction performance evaluation of HPGR/ball mill and HPGR/stirred mill for PGE-bearing chromite ore. *Advanced Powder Technology*, 34(1), 103907.
3. Santosh, T., Soni, R. K., Eswaraiah, C., Rao, D. S., and Venugopal, R. (2020). Optimization of stirred mill parameters for fine grinding of PGE-bearing chromite ore. Part. Sci. Technol, 39(6), 663–675.
4. Touil, D., Belaadi, S., and Frances, C. (2008). The specific selection function effect on clinker grinding efficiency in a dry batch ball mill. *International Journal of Mineral Processing*, 87(3–4), 141–145.
5. Kaya, Y., Kobya, V., and Mardani, A. (2024). Evaluation of fresh state, rheological properties, and compressive strength performance of cementitious system with grinding aids. *Journal of Applied Polymer Science*, e55212.
6. Danha, G., Hlabangana, N., and Muzenda, E. (2018). Effect of ball and feed particle size distribution on the milling efficiency of a ball mill: An attainable region approach. *South African Journal of Chemical Engineering*, 25(1), 79–84.

7. Shahbazi, B., Jafari, M., Parian, M., Rosenkranz, J., and Chelgani, S. C. (2020). Study on the impacts of media shapes on the performance of tumbling mills–A review. *Minerals Engineering*, 157, 106490.
8. Mandal, N., Doloi, B., Mondal, B., and Das, R. (2011). Optimization of flank wear using Zirconia Toughened Alumina (ZTA) cutting tool: Taguchi method and Regression analysis. *Measurement*, 44, 2149–2155.
9. Kaya, Y., Kobya, V., Mardani, A., Mardani, N., and Beytekin, H. E. (2024). Effect of Grinding Conditions on Clinker Grinding Efficiency: Ball Size, Mill Rotation Speed, and Feed Rate. *Buildings*, 14(8), 2356.

Chapter 22

Determination of Criteria for Selection of Candidate Sites for Temporary Debris Storage

Esra Aktaş[*,‡] and Ediz Atmaca[†,§]

[*]*Gümüşhane University, Occupational Health and Safety Department, Gümüşhane, 29100, Türkiye*
[†]*Gazi University, Industrial Engineering Department, Ankara, 06050, Türkiye*
[‡]*esra.aktas@gumushane.edu.tr*
[§]*hediz@gazi.edu.tr*

Disasters are difficult-to-control situations that have a profound impact on human life and result in significant material and moral losses. Disasters occur suddenly and frequently without warning, resulting in increased losses both during and after the situation. By making plans and taking the necessary precautions, natural and human hazards can be mitigated or avoided. Disaster management plans are being prepared all over the world, including various actions for disaster preparedness, disaster response and post-disaster. The importance and necessity of disaster management plans cannot be ignored in Türkiye, which has frequently encountered disasters from past to present and suffered great loss of life and property. Disaster management is the process by which society, all organizations, and resources work together to reduce or prevent harm before, during, and after a disaster. In this context, the study aimed to determine the criteria required for the selection of candidate areas for temporary debris storage sites, which are very important after the disaster. The importance weights of the criteria obtained through literature research were determined by soliciting expert opinions via the Analytic Hierarchy Process (AHP) method, which is one of the multi-criteria decision-making methods. The criteria with the highest importance weight are listed.

1. Introduction

Disasters are uncontrollable events that have a profound impact on human life and result in significant material and moral losses. The fact that disasters strike suddenly and frequently without warning has resulted in increased losses during and after the disaster. These losses are physical, psychological, and social. Since the beginning of the twentieth century, natural disasters in Turkey have killed approximately 100 thousand people, injured 175 thousand people, and destroyed or severely damaged approximately 650 thousand houses. Managing these situations with many unknowns becomes extremely difficult and complex. As a result, disaster management has become one of the most popular academic subjects today. Effective and successful disaster management allows for the reduction of loss of life and property, the prompt return of disaster victims to their normal lives, and the rapid repair of the damaged socioeconomic structure. The successful management of activities in the recovery phase, the final stage of disaster management, is critical in ensuring that people return to their normal lives as quickly and effectively as possible. This is because the effectiveness of improvement activities influences intervention efforts.

The ability to respond quickly to road/infrastructure renewal and debris management issues during the recovery phase is closely related to the preparations made prior to the disaster. Debris management emerges as an issue that must be addressed within the first 72 hours and beyond due to the risks it poses and the impact it has on response efforts. Identifying temporary debris storage areas prior to a disaster. It is critical because it is a complex process that considers many criteria, legal limits are established, and disasters cannot be predicted in terms of location and timing. Thus, the performance of specific areas can be evaluated using various scenarios. Furthermore, it will prevent bad decisions from being made in the chaotic environment that will arise when a disaster strikes. Furthermore, identifying the areas is critical for both the environment and human health, as it facilitates intervention efforts by removing debris from roads.

As a result, the study began by thoroughly examining the recovery phase, debris, debris management, and storage areas. Information was provided about debris management activities, as well as the location and importance of storage areas in debris management. Following that, studies in the literature on determining temporary debris storage areas are presented, along with an examination of the methods used and the criteria considered. Thus, the main criteria and sub-criteria to consider when selecting storage areas were identified. The importance weights of the main and sub-criteria were then calculated using the AHP method.

2. Background and Objectives

2.1. *Recovery/reconstruction phase in disaster management*

Disaster management has four basic stages: Preparation, harm reduction, response, and recovery/restructuring. The preparation and harm reduction phases cover the work to be done prior to the disaster, the response phase covers the work to be done during the disaster, and the recovery phase covers the work to be done following the disaster [1]. Figure 1 depicts the disaster management process and its various stages.

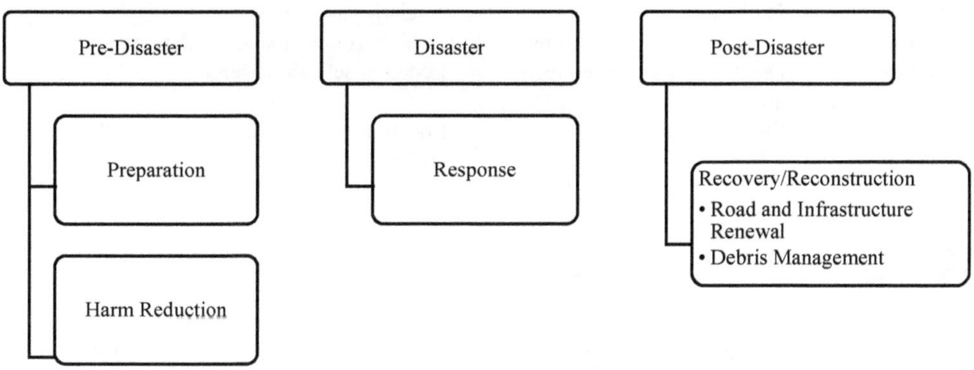

Fig 1. Disaster management process

The recovery phase includes activities that begin immediately following the initial 72-hour response phase and, in some cases, last for years. The goal here is to eliminate the factors that make intervention difficult and to repair the social structure by getting life back to normal as soon as possible. At this stage, debris removal, the proper execution of infrastructure services such as communication, transportation, water, electricity, sewerage, education and health buildings in the disaster area, the construction and strengthening of permanent structures, the development of long-term permanent and temporary settlement alternatives for disaster victims, and the economic and social development of the disaster area, restarting activities, enacting compensatory measures and regulations to revitalize the regional economy, rearranging emergency plans based on lessons learned from previous disasters, strengthening or reconstructing buildings, rearranging city plans, and, as needed, developing the city toward less risky regions in terms of disaster risk, activities are carried out to ensure performance and to develop performance measurement criteria and plans. The main issues at this stage can be summarized as determining the required resources, determining debris collection areas, determining activity performance, and determining road reconstruction resources [2]. Table 1 shows the logistics activity equivalents to the stated problems. During the recovery and reconstruction phase, key logistical activities include addressing problems such as debris removal operations, identifying debris collection areas, and conducting road renewal works. These activities require careful vehicle route and capacity determination, as well as strategic location selection and routing to facilitate efficient collection, transportation, separation, and reuse of debris.

Table 1. Problems and responses in the recovery/reconstruction phase

Phase	Problems	Logistics activities
Recovery / Reconstruction	Debris removal operations	Vehicle-route-capacity determination
	Determining debris collection areas	Location selection, routing
	Road renewal works	Location selection and routing for the collection, transportation, separation, and reuse processes

2.2. *Debris*

"Debris" refers to any collapses, ruins, or debris that occur as a result of a disaster. The size, severity, and location of the disaster all have an impact on the composition and management strategy for the debris that results. As a result, it is critical to determine/predict the severity, location, amount, and composition of debris that may occur during the pre-disaster preparation phase [3]. Thus, it will be ensured that debris removal activities, which are in the final stages of disaster management, are completed successfully, the current situation is correctly analyzed, and the guide and intervention plans to be prepared provide appropriate guidance. Because the type of disaster that may occur is determined by the location and climatic conditions of each region. The debris pile from any disaster is made up of a different mix of debris, as shown in Table 2 [4].

Table 2. Debris heap contents

Debris heap contents	
Vegetative debris or green waste	Putrescible wastes: Such as rotting food
Sediment/soil and rock	Vehicles and vessels
Household hazardous waste	Recyclables: Plastics, metals, etc.
Construction and demolition waste	Electronic and white goods
Industrial and toxic chemicals waste	Human and animal corpses

The types of waste found in the resulting debris pile differ in both type and quantity depending on the type of disaster. This is due to the disaster's variable severity and impact. Furthermore, because the collection-transportation-storage conditions for each type of debris differ, the quantity and diversity of debris generated is critical for improvement efforts. Table 3 depicts the various amounts of debris produced by different disasters. Massive amounts of debris are created during disasters. These amounts also demonstrate the magnitude of the destruction that occurred. The amount of debris produced is 5–15 times the annual waste generated in countries during normal times [5]. Therefore, it also affects the successful implementation of both response phase activities and recovery phase activities.

Table 3. Amounts of debris by disaster type

Disaster Type	**Debris Amount**	**Data Source**
2010 Haiti Earthquake	23–60 million tons	[6]
2008 Sichuan Earthquake	20 million tons	[7]
2005 Hurricane Katrina	76 million cubic meters	[8]
2004 Indian Ocean Tsunami	10 million cubic meters	[9]
1999 Marmara Earthquake	13 million tons	[10]
1995 Hanshin-Awaji Earthquake	15 million tons	[11]

Debris types are collected, sorted, stored, and destroyed in various ways depending on whether they contain harmful/non-harmful/reusable components. This situation makes it difficult to manage (collect and control) and sort large amounts of debris of various types. Table 4 categorizes debris types resulting from various disasters.

Table 4. Different types of disaster and typical debris stream [12]

	Typical debris streams								
Types of disasters	Vegetative	Construction & Demolition (C&D)	Personal Property/ Household Items	Hazardous Waste	Household Hazardous Waste	White Goods	Soil, Mud and Sand	Vehicles and Vessels	Putrescent
Hurricanes / Typhoons	X	X	X	X	X	X	X	X	X
Tsunamis Wildfires	X	X	X	X	X	X	X	X	X
Tornadoes	X	X	X	X	X	X		X	X
Floods	X	X	X	X	X	X	X	X	X
Earthquakes		X	X		X	X	X		
Wildfires	X		X		X	X	X		
Ice Storms	X				X				

Because the hazardous waste in the resulting pile endanger both human health and the environment, safe and clean areas should be designated for disaster victims. At the same time, road closures due to large amounts of debris will complicate intervention efforts such as transporting injured people and aid. As a result, the most important consideration is how to transport/collect the debris, and which types of debris should be prioritized [13].

2.3. *Debris management*

Debris management, one of the two most important activities during the recovery phase, refers to the processes of collecting, transporting, storing, separating, and processing debris from the disaster area. While the work done during the intervention phase continues, it is an activity that begins immediately after the first 72 hours to ensure that disaster victims return to their normal lives as soon as possible and may last for years depending on the severity of the catastrophe. Due to the following advantages of debris management, it is critical to make the necessary preparations prior to the disaster.

- To eliminate many damages caused by debris,
- To accelerate people's return to their normal lives,
- Ensuring rapid intervention, facilitating rapid decision-making,
- Cause less harm to the environment,
- To ensure that post-disaster reconstruction works are carried out at less cost and in a shorter time
- Determining the necessary coordination and communication network for the successful implementation of all these processes.

2.4. *Temporary debris storage sites (TDSS)*

Temporary debris storage areas are defined as areas where debris collected from disaster areas should be temporarily stored, separated, and sometimes defined/determined prior to the disaster, where various processing activities are carried out before being delivered to the centers where it will be final processed [12]. In addition to providing a suitable environment for proper debris separation using appropriate methods, it is critical to identify their location prior to the disaster, as this allows for additional time for these processes. As a result, it will be easier to eliminate substances that are potentially harmful to the environment and human health as soon as possible. The literature generally states that when determining the TDSS, size, location, legal obligations, and environmental factors should all be considered [14]. Figure 2 depicts the debris collection and processing process. The figure depicts the order of temporary storage, processing, and final disposal of the debris. Temporary storage areas appear to be an intermediate step in the debris separation process. Determining temporary debris storage areas prior to a disaster has numerous benefits, including increased flexibility for recovery efforts, simplified recycling and sorting processes, and faster collection efforts. However, it also has disadvantages such as additional costs due to transporting the debris twice, additional costs due to planning, engineering and legal obligations, and cleaning and closing of the areas after all operations regarding the debris are completed. It is possible to eliminate or minimize the issues mentioned as disadvantages by taking them into consideration during pre-disaster planning.

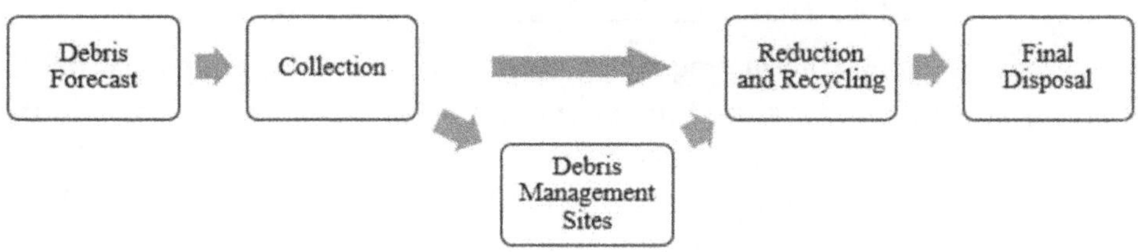

Fig. 2. Debris collection and processing process [12]

3. Material and Methods

The study discusses the problem of identifying candidate areas where debris from a disaster can be stored temporarily. First, the relevant literature and previous studies were reviewed. The criteria that should be considered when determining the storage area were revealed by examining the criteria used in these studies as well as the guides prepared on the subject. The criteria's importance weights were then calculated using the Analytical Hierarchy Process (AHP) method.

3.1. *Determination of criteria*

The literature includes a wide range of criteria for determining temporary debris storage areas. In general, the studies considered size, location, safety, environmental factors, and cost. Table 5 displays the criteria that are generally recommended for use in literature.

Table 5. Studies in literature and criteria

References	Criteria					
	Size	Location	Safety	Cost	Environmental factors	Legal obligations
[12]	X	X	X	X	X	X
[15]	X		X	X	X	
[16]	X			X	X	
[17]	X	X		X	X	X

By reviewing the criteria frequently used in determining temporary debris storage sites and the criteria used in determining solid waste storage areas in the literature, 8 main criteria and 26 sub-criteria for use in the study were developed. The main criteria were determined as "Criteria Related to Debris Structure, Criteria Related to Logistic Infrastructure, Criteria Related to Environmental, Criteria Related to Location, Criteria Related to Safety, Criteria Related to Land Characteristics, Criteria Related to Economic, Criteria Related to Legal Obligations" [4,12–16,18–20].

3.2. *Analytic hierarchy process (AHP)*

The Analytic Hierarchy Process is a mathematical model for measurement and decision-making developed by Thomas L. Saaty of the University of Pennsylvania in the mid- 1970s [21,22]. AHP has been extensively researched in literature and has been used in almost all multi-criteria decision-making applications over the last two decades.

Step 1: A n*n binary comparison matrix is created by evaluating criteria $(C_1, C_2, \ldots, C_N)$ on a 1–9 importance scale (Table 6). In the comparison matrix, the relationship and equality are always present as $a_{ji} = 1/a_{ij}$.

Step 2: To create the normalized comparison matrix, divide each element of the comparison matrix by its column total.

Step 3: Row averages are computed for each row of the normalized comparison matrix. The mean values indicate the relative importance of the criteria.

Step 4: For AHP results to be valid, matrix A must be consistent. In a consistent matrix A, weights w_i (weight vector) are computed. Weights are the vector formed by multiplying and summing the relative importance values calculated in Step 3 with the relevant column of the comparison matrix created in Step 1. The consistency coefficient, represented by the consistency index (CI), was calculated using the Eq. (1). The value is calculated by dividing the weight vector by the appropriate relative importance values.

$$CI = \frac{\lambda_{max} - n}{n - 1} \tag{1}$$

Step 5: The consistency rate (CR) is calculated using the equation CR=CI/RI. The random index (RI) varies with n (the size of the comparison matrix). Table 7 shows the random indicators developed for matrices with sizes ranging from 1 to 15.

Table 6. Saaty's fundamental scale [22].

Intensity of importance	Definition
1	Equal importance
2	Weak
3	Moderate importance
4	Moderate plus
5	Strong importance
6	Strong plus
7	Very strong or demonstrated importance
8	Very, very strong
9	Extreme importance

Table 7. Random index (RI) [23].

n	1	2	3	4	5	6	7	8	9	10	11	12	13	14	15
RI	0.00	0.00	0.58	0.90	1.21	1.24	1.32	1.41	1.45	1.49	1.51	1.48	1.56	1.57	1.59

If the consistency ratio equals zero, it indicates that a completely consistent matrix was obtained. To ensure complete consistency in applications, comparisons should be repeated until CR $\geq$ 0.1.

4. Application and Results

4.1. *Determination of criteria*

Explanations about all main and sub-criteria determined within the scope of the study are shown in Table 8.

Table 8. Main and sub-criteria for determining the temporary debris storage sites.

Main criteria	Sub-criteria	Descriptions
Criteria Related to Debris Structure	Debris Type	The composition of the debris pile, including construction remnants, vegetative waste, hazardous waste, etc.
	Debris Amount	Estimation of the amount of debris generated.
	Debris Transportation Difficulties	Transportation challenges arising from the amount of debris, resource adequacy, and road conditions.
	Hazard Posed by Debris	Dangers stemming from hazardous wastes contained within the debris.
Criteria Related to Logistic Infrastructure	Blocking of Road Networks	Road closures due to the destruction and debris.
	Speed of Debris Clearance on Roads	The rate at which roads closed by debris are cleared.
	Traffic Congestion	Traffic jams caused by closed roads and evacuations.
Criteria Related to Environmental	Distance to Environmentally Sensitive Areas	Areas such as wetlands, habitats of endangered species, and clean water sources should be avoided.
	Distance to Historically Important Areas	Historical and archaeological sites should be avoided.
	Distance to Agricultural Areas	Cultivated agricultural lands should be avoided.
Criteria Related to Location	Distance to Residential Areas	Due to the hazardous composition of the debris, a sufficiently distant area from residential zones should be chosen.
	Distance to Health Centers	Due to the hazardous composition of the debris, a sufficiently distant area from health centers should be chosen.
	Distance to Educational Units	Due to the hazardous composition and structure of the debris, a sufficiently distant area from educational units should be chosen.
	Distance to Disaster Area	The area should be sufficiently distant to avoid aftershocks and new disasters, while also being close enough for ease of transportation.
Criteria Related to Safety	Distance to Power Lines and Power Plants	To prevent new disasters due to the hazardous composition of the debris, the area should be sufficiently distant.
	Distance to Flood Risk Areas	To prevent new incidents at the storage site.
	Distance to Fault Line Areas	Areas with fault lines should be avoided to mitigate the risk of new disasters.
Criteria Related to Land Characteristics	Soil Structure	The land should have soil and topographical features that allow for reuse after the storage process is complete.
	Land Slope	Should be considered with respect to the amount of debris and the transportation process.
	Size	The area should be large enough to store the generated debris.
Criteria Related to Economic	Land Cost	The cost should be affordable by the state budget and funds.
	Funds and Budget	Necessary funds and budgets should be predetermined.
	Equipment-Resource Adequacy	Necessary equipment/resources should be procured in advance based on the estimated amount of debris.
Criteria Related to Legal Obligations	Legal Permits	Necessary laws and legal requirements should be established.
	Regional and Local Regulations Related to the Topic	Regional and local responsibilities should be determined within the framework of regulations/guidelines.

4.2. *Analytic hierarchy process*

A team of five experts from AFAD (Ministry of Interior Disaster and Emergency Management Presidency) evaluated the evaluation matrices created using the determined criteria. After combining and normalizing the pairwise comparison matrices, they were transferred to the EXPERT CHOICE program, where the importance weights were calculated using the AHP method steps for both the main criteria and the sub-criteria based on these criteria. Table 9 shows the results obtained for the main criteria. The results for the sub-criteria are shown in Table 10.

Table 9. Main criteria importance.

Main criteria	**Importance**
Criteria Related to Environmental	**0.426**
Criteria Related to Safety	**0.149**
Criteria Related to Land Characteristics	**0.114**
Criteria Related to Economic	0.074
Criteria Related to Location	0.068
Criteria Related to Legal Obligations	0.067
Criteria Related to Logistic Infrastructure	0.060
Criteria Related to Debris Structure	0.042

Table 10. Main criteria importance.

Main criteria	**Sub-criteria**	**Importance**
Criteria Related to Debris Structure	Debris Type	0.102
	Debris Amount	0.171
	Debris Transportation Difficulties	0.089
	Hazard Posed by Debris	**0.638**
Criteria Related to Logistic Infrastructure	Blocking of Road Networks	**0.491**
	Speed of Debris Clearance on Roads	0.255
	Traffic Congestion	0.254
Criteria Related to Environmental	Distance to Environmentally Sensitive Areas	0.309
	Distance to Historically Important Areas	0.160
	Distance to Agricultural Areas	**0.531**
Criteria Related to Location	Distance to Residential Areas	0.110
	Distance to Health Centers	**0.565**
	Distance to Educational Units	0.136
	Distance to Disaster Area	0.189
Criteria Related to Safety	Distance to Power Lines and Power Plants	**0.472**
	Distance to Flood Risk Areas	0.325
	Distance to Fault Line Areas	0.202
Criteria Related to Land Characteristics	Soil Structure	**0.499**
	Land Slope	0.398
	Size	0.103
Criteria Related to Economic	Land Cost	0.273
	Funds and Budget	0.326
	Equipment-Resource Adequacy	**0.401**
Criteria Related to Legal Obligations	Legal Permits	**0.667**
	Regional and Local Regulations Related to the Topic	0.333

5. Conclusion

The purpose of this study was to identify temporary debris storage sites in the aftermath of a disaster. To this end, criteria have been established to determine the evacuation area. It is critical to identify these areas prior to the disaster in order to avoid making poor decisions in the chaotic aftermath, as well as the hazards posed by debris that endangers both the environment and human health, and the direct impact on intervention efforts. According to the literature, planning should take place prior to the disaster, and there are insufficient studies on the subject. Again, statistical methods are generally preferred, but multiple-criteria decision-making (MCDM) methods, which can consider a wide range of criteria and provide effective and timely solutions, are not. As a result, multi-criteria decision-making techniques were used in the study. Furthermore, because of the methods used in the literature, area selection was based on a small number of criteria; however, the scope of the study was expanded by considering a large number of criteria.

In the study, the importance weights of the criteria identified through literature research were determined using the AHP method, one of the MCDM methods, and expert opinions. According to the results, the Criteria Related to Environmental criterion was identified as the primary criterion with the highest importance weight. The main criteria are then listed in order of importance: Criteria related to safety, criteria related to land characteristics, criteria related to economics, criteria related to location, criteria related to legal obligations, criteria related to logistical infrastructure, and criteria related to debris structure. Sub-criteria for each main criterion are listed in order of importance. Hazard Posed by Debris sub-criteria related to Criteria Related to Debris Structure, Blocking of Road Networks sub-criteria related to Criteria Related to Logistic Infrastructure, Distance to Agricultural Areas sub-criteria related to Criteria Related to Environmental, Distance to Health related to Criteria Related to Location Centers sub criteria, Distance to Power Lines and Power Plants related to Criteria Related to Safety, Soil Structure sub criteria related to Criteria Related to Economic, Equipment-Resource Adequacy sub criteria related to Criteria Related to Legal Obligations Legal Permits sub-criteria were determined as the sub-criteria with the highest importance.

References

1. E. Atmaca, E. Aktaş, and H. N. Öztürk, Evaluated post-disaster and emergency assembly areas using multi-criteria decision-making techniques: A case study of Turkey, *Sustainability*, 15(10), 8350, (2023), doi: 10.3390/su15108350.
2. S. Duran, M. A. Gutierrez, and P. Keskinocak, Pre-positioning of emergency items for CARE international, *Interfaces (Providence)*, 41(3), 223–237, (2011), doi: 10.1287/inte.1100.0526.
3. United States Environmental Protection Agency, Plannig for Natural Disaster Debris, (2019). [Online]. Available: https://www.hsdl.org/?view&did=824828.
4. C. Brown, M. Milke, and E. Seville, Disaster waste management: A review article, *Waste Manag*, 31(6), 1085–1098, (2011), doi: 10.1016/j.wasman.2011.01.027.
5. D. R. Reinhart and P. T. Mccreanor, *Disaster Debris Management – Planning Tools*, (US Environmental Protection Agency Region IV, 1999).
6. The Washington Post, Haiti faces colossal and costly cleanup before it can rebuild, *Chine Civil Engineering Society*, (2010). [Online]. Available: http://cces.net.cn/html/tm/12/18/content/2594.html.
7. Ramzy, The Sichuan Earthquake, Six Months Later, *Time*, (2008). [Online]. Available: https://world.time.com/2008/11/12/the-sichuan-earthquake-six-months-later/.
8. L. Luther, Disaster Debris Removal After Hurricane Katrina : Status and Associated Issues, Congressional Research Service, *Congressoinal Research Service*, (2008).
9. M. Bjerregaard, MSB/UNDP Debris Management Guidelines, January (47), (2010).
10. F. Baycan, Emergency Planning for Disaster Waste: A Proposal Based on The Experience of the Marmara Earthquake in Turkey, *International Conference and Student Competition on post-disaster reconstruction "Planning for reconstruction"*, 90(312), 1–10, (2004).
11. N. Hirayama, T. Shimaoka, T. Fujiwara, T. Okayama, and Y. Kawata, Establishment of disaster debris management based on quantitative estimation using natural hazard maps, *WIT Transactions on Ecology and the Environment,* 140(June 2010), 167–178, (2010), doi: 10.2495/WM100161.
12. Federal Emergency Management Agency, Public assistance. Debris management guide. FEMA-325, (2007). [Online]. Available: www.fema.gov/pdf/government/grant/pa/demagde.pdf%0Ahttp://www.fema.gov/government/grant/pa/demagde.shtm.
13. L. Luther, Managing disaster debris: Overview of regulatory requirements, agency roles, and selected challenges, *Congressional Research Service,* (2010).
14. C. Cheng and R. G. Thompson, Application of boolean logic and GIS for determining suitable locations for Temporary Disaster Waste Management Sites, *International Journal of Disaster Risk Reduction*, 20, 78–92, (2016), doi: 10.1016/j.ijdrr.2016.10.011.
15. UNEP/OCHA Environment Unit, Disaster Waste Management Guidelines, (2011). [Online]. Available: http://www.mendeley.com
16. C. Brown, Solid waste management: A systems approach, University of Canterbury, Department of Civil and Natural Resources Engineering, (2012).
17. S. Grzeda, T. A. Mazzuchi, and S. Sarkani, Temporary disaster debris management site identification using binomial cluster analysis and GIS, *Disasters*, 38(2), 398–419, (2014), doi: 10.1111/disa.12040.

18. Gemitzi, V. A. Tsihrintzis, E. Voudrias, C. Petalas, and G. Stravodimos, Combining geographic information system, multicriteria evaluation techniques and fuzzy logic in siting MSW landfills, *Environmental Geology*, 51(5), 797–811, (2007), doi: 10.1007/s00254-006-0359-1.
19. S. P. Gbanie, P. B. Tengbe, J. S. Momoh, J. Medo, and V. T. S. Kabba, Modelling landfill location using Geographic Information Systems (GIS) and Multi-Criteria Decision Analysis (MCDA): Case study Bo, Southern Sierra Leone, *Applied Geography*, 36, 3–12, (2013), doi: 10.1016/j.apgeog.2012.06.013.
20. Beskese, H. H. Demir, H. K. Ozcan, and H. E. Okten, Landfill site selection using fuzzy AHP and fuzzy TOPSIS: A case study for Istanbul, *Environmental Earth Sciences*, 73(7), 3513–3521, (2015), doi: 10.1007/s12665-014-3635-5.
21. T. L. Saaty, How to Make a Decision: The Analytic Hierarchy Process, *Interfaces*, 24(6) (1994).
22. T. L. Saaty, Decision making with the analytic hierarchy process, *International Journal of Services Science*, 1, 83, (2008), doi: 10.1504/IJSSCI.2008.017590.
23. H. A. Donegan and F.J. Dodd, A note on saaty's random indexes, *Mathematical and Computer Modelling,* 15(10), 135–137, (1991).

Chapter 23

Performance Evaluation of Linear Regression Models Based on Specific Features in the Turkish Second-Hand Car Market

Akın Bedük*,‡ and Özgül Vupa Çilengiroğlu†,§

*The Graduate School of Natural and Applied Sciences,
Dokuz Eylul University, İzmir, Turkey
†Faculty of Sciences - Statistics Department,
Dokuz Eylul University, İzmir, Turkey
‡akinbeduk@gmail.com
§ozgul.vupa@deu.edu.tr

We evaluate the effectiveness of various linear regression models for predicting second-hand car prices in Turkey by analyzing a dataset from the country's leading second-hand car platform, comprising over 17,000 records and 25 attributes, including brand, model, age, mileage, and fuel type. We developed and compared Linear Regression, Ridge, Lasso, and Elastic Net models, finding that while traditional Linear Regression provides a solid baseline, Ridge, Lasso, and Elastic Net significantly enhance prediction accuracy by addressing multicollinearity and improving feature selection. Our results demonstrate the superiority of these advanced models for accurate price estimation in the second-hand car market.

1. Introduction

The automotive industry faces steady growth and fierce competition every year. In the competitive market of second-hand cars, accurate vehicle pricing holds paramount significance for both dealers and customers. Individuals less inclined towards cars and their technical aspects often encounter challenges during the car-buying process. It's common practice to seek help from auto magazines, the Internet, or car dealers to determine a fair price. However, this process is time-consuming and can lead to confusion [1]. Similarly, in the sales process, finding the right balance between ensuring a fair price value and presenting a competitive offer is crucial. Overpricing may prolong the sales process, while underpricing can result in potential losses for sellers [2].

The pricing of second-hand vehicles involves a complex process influenced by various factors. This process can be rendered more comprehensible through supervised machine learning algorithms capable of making predictions using historical data. These algorithms offer solutions to estimate the fair market value by considering the vehicle's characteristics [3].

Various techniques for predicting second-hand vehicle prices have been explored in the literature, with a primary focus on regression analysis. Linear Regression (LR) analysis is a statistical method utilized to estimate the price value of a second-hand vehicle based on multiple predictive variables [4–7]. Determinants include the brand and model of the vehicle, its age, mileage, general condition, etc. By training algorithms on comprehensive datasets of second-hand car sales, predicting the price of a new car based on its features becomes possible through regression analysis [8–11].

In addition to LR, regularization methods such as Ridge, Lasso, and E-NET are also commonly used. Ridge Regression adds a penalty to the loss function based on the squared magnitude of the coefficients, addressing multicollinearity and reducing model complexity [12–17]. Lasso Regression includes a penalty based on the absolute value of the coefficients, encouraging sparsity and feature selection. E-NET combines both Ridge and Lasso penalties, balancing their benefits to handle correlated predictors and enhance model performance. These methods improve prediction accuracy by effectively managing feature interactions and model complexity [18–21].

In this study, models based on price prediction for second-hand vehicle sales were developed using machine learning algorithms and then the performances of these models were compared. One of Turkey's leading second-hand vehicle search platforms was used to train the models. These developed models were found to be the best models that estimate sales prices by considering the vehicle features that affect the price according to the current market. The fact that these models are based on up-to-date data, benefit from different features, offer an innovative approach and provide more precise estimates by considering possible effects and differences is the greatest contribution of the study to the literature. It also enables car buyers to make more conscious purchase or sale decisions.

2. Related Works

There have been many studies in the literature on second-hand car price estimation. In recent years, studies have shown that machine learning is effective in modeling (Table 1). Linear regression methods have been found to generally achieve high accuracy rates. However, the selection of the most appropriate model for second-hand car price estimation may vary depending on the dataset characteristics and analysis objectives.

Table 1. Outline of related works.

Reference	Algorithms used	Results
Satapathy, Vala, Virpariya (2022) [4]	LR, Ridge, Random Forest, Gradient Boosting	92% (Accuracy)
S. Shaprapawad, P. Borugadda and N. Koshika (2023) [7]	LR, Ridge, Lasso, E-Net, Tree Based Methods, Support Vector Machine	94.5% (R-square)
Asghar, M., Mehmood, K., Yasin, S., & Khan, Z. M. (2021) [10]	Lasso, LR	91% (R-square)
Venkatasubbu, P., & Ganesh, M. (2019) [11]	Lasso, Multiple LR, Decision Tree	3.46 (Error Rate)
Praful Rane, Deep Pandya, Dhawal Kotak (2021) [12]	LR, Ridge, Lasso	Not mentioned

3. Methodology and Implementation

Data preprocessing and cleaning is the first and most fundamental step for analysis and interpretation of a model. In this scope, the weekly obtained data was first cleaned, necessary transformations were applied, and the data was recorded. Subsequently, various analyses were conducted for missing and outlier values. The data was then prepared for training machine learning methods. Relationships within the data were analyzed, and its suitability for machine learning methods was assessed through statistical tests. The data was divided into two parts: train and test. All models were trained with the train data, and results were obtained from the test data. Finally, based on the results obtained, the performance values were analyzed, and the most suitable model was selected (Figure 1).

3.1. *Data Collection*

The data for the models to be obtained to predict second-hand car prices were obtained from the last three months of 2023 of Turkey's leading second-hand car search platform, Arabam.com. This data was collected weekly using web scraping techniques implemented with Python scripts. A dataset of over 17,000 records consisted of 25 different features: Ad ID, Ad Date, Price, Make, Series, Model, Km, Transmission, Fuel, Body Type, Warranty, Seller Type, Drive Type, Cylinder Count, Torque, Engine Capacity, Engine Power, Max Power (rpm), Min Power (rpm), Acceleration 0–100 (sec), Max Speed (km/h), Average Fuel Consumption

(L), Year, Changed (indicates whether any part of the car has been modified), and Painted (indicates whether any part of the car has been painted).

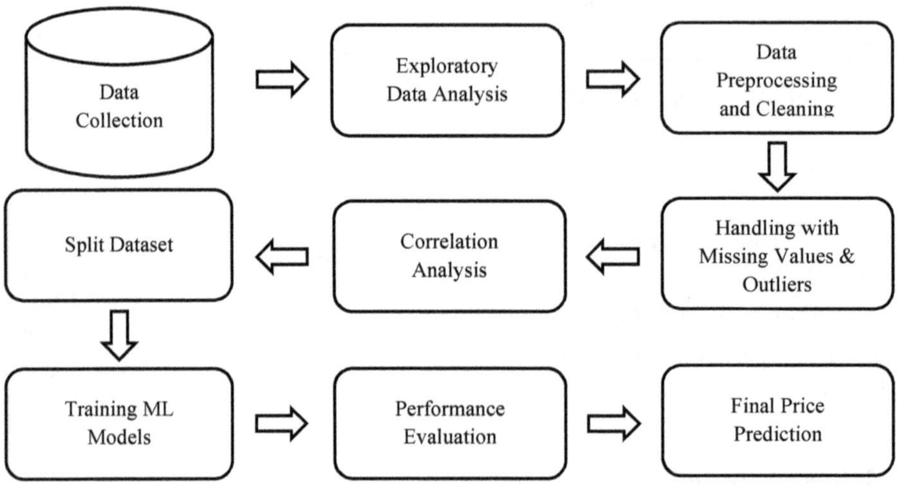

Fig. 1. Block diagram of proposed work.

3.2. *Exploratory data analysis*

In second-hand car price estimation, descriptive statistics were made for continuous data and frequency tables were made for discrete data (Tables 2 and 3). It is noteworthy that from continuous data, vehicle prices vary between 75,000 TL and 3,000,000 TL, mileage varies between 0 and 676,000 kilometers, and their distribution is skewed to the right. These results show that the majority of instruments are concentrated towards the lower end of the spectrum, with a few outliers having extremely high values. Additionally, according to the Shapiro-Wilk normality test, it was determined that continuous data did not come from a normal distribution ($p < 0.05$). It can be seen that there are many unique values in the model and series variables, which are discrete variables. For this reason, models and series were studied with high sample numbers.

3.3. *Data Preprocessing & Cleaning*

Transformation was performed on the raw data set. Erroneous values were identified and corrected, numerical characteristics were distinguished from non-numerical characteristics (age was calculated from the year of production, normality tests were performed, transformations were made for price and mileage variables). This process ensured that the dataset is cleaned and formatted properly before further analysis and modeling.

Table 2. Descriptive statistics of numeric features.

	Price	Km	Torque	Engine capacity	Engine power	Max power	Min power	Acceleration 0 to 100 (sec)	Max speed	Average fuel consumption	Age
mean	702948	169906	196	1475	106	4766	2579	11.8	186	5.3	11.1
std	399136	92314	64	197	27	986	1109	19	17.5	1.28	6.68
min	75000	0	80	875	45	1850	1200	5.8	139	2.8	0
25%	450000	100739	145	1368	90	4000	1750	10.7	172	4.3	6
50%	625000	164000	200	1490	101	4400	2000	11.7	185	5	10
75%	825000	231000	245	1598	120	5750	3800	13	196	6.4	15
max	3994900	676000	500	1999	252	8300	7500	20	250	11.5	34

Table 3. Descriptive statistics of categoric features.

	Make	Series	Model	Transmission	Fuel	Body type
unique	42	245	1833	3	4	8
top	Renault	Clio	1.5 dCi Touch	Manuel	Diesel	Sedan
	Warranty	From	Wheel Drive	Cylinder number	Changed	Painted
unique	2	3	2	3	2	2
top	0	Gallery	Front wheel	4 Cylinder	1	1

3.4. ***Handling with missing values & outliers***

3.4.1. *Handling with missing values*

Missing values in the data set were filled (rows of vehicles whose age was less than 3 years and whose mileage was less than 60,000 km were marked as guaranteed, the others were marked as no warranty, and the parts were separated as without changed/painted parts or having changed/painted parts). This approach not only reduced the impact of missing values but also con-tributed to the overall improvement in data quality, allowing for a more accurate analysis of the relationship between warranty status and other relevant features in the dataset.

3.4.2. *Dealing with outliers*

Box plots, histograms and Q-Q plots were used to identify potential outliers (made for the distribution of price and mileage values, graphs of the transformed price and mileage variables were also obtained) (Figure 2). This process involved meticulous scrutiny of extreme data points, leading to the removal of values significantly deviating from the general trend and thereby im-proving the dataset's integrity. This contributed to the creation of a more robust and reliable dataset for subsequent analyses.

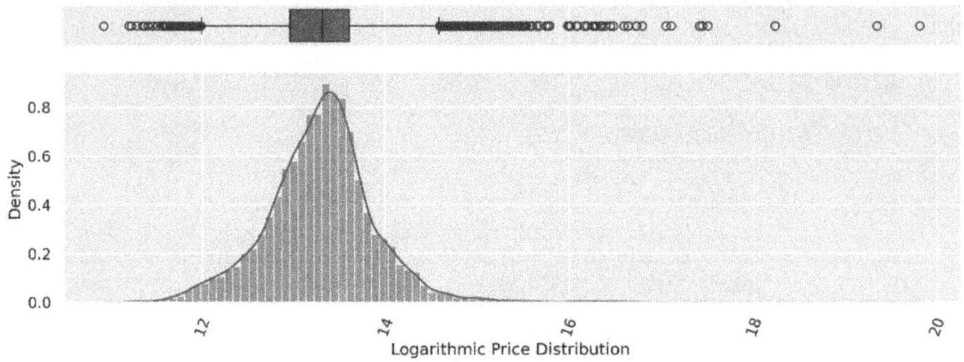

Fig. 2. Price distribution after log transformation.

3.5. *Correlation analysis*

Conducting correlation analysis is crucial for understanding the impact of variables on second-hand car prices. The Pearson correlation analysis of numerical features revealed strong positive correlations between price and maximum speed, torque, and engine power, while negative correlations were found between price and age, mileage, and acceleration. Additionally, multicollinearity was observed among several numerical variables, such as torque, engine capacity, engine power, acceleration, and maximum speed (Figure 3). For categorical variables, significant impacts on price were identified using Spearman correlation coefficients. The top 4 positive and negative correlation values for categorical features are provided in Table 4.

Table 4. Top 4 positive and negative correlation values for categorical features with price.

Top 4 Positive Correlated Features	Correlation	Top 4 Negative Correlated Features	Correlation
Transmission: Semiautomatic	0.4045	Transmission: Manual	−0.6087
Transmission: Automatic	0.3306	Fuel: LPG & Gasoline	−0.4346
Fuel: Gasoline	0.2714	Wheel drive: Front drive	−0.2237
Wheel drive: Others	0.2243	From: Owner	−0.1906

3.6. *Splitting data*

The dataset was divided into two separate subsets, namely the training (70%) and test sets (30%), to train and evaluate machine learning models. The training set was used for model training, while the test set was reserved for evaluating the trained model's performance.

3.7. *Training machine learning models*

In order to predict the prices of second-hand vehicles such as LR, Lasso, Ridge and E-NET models commonly used in the literature were obtained from the training set and their performance on the test set was compared.

3.8. *Performance evaluation*

Metrics that measure the distance between predicted values and actual values are used to determine the best performing model. Of these metrics, Mean Absolute Error (MAE) calculates the average of the absolute differences between actual and predicted values. The Mean Absolute Percentage Error (MAPE) value, which represents the MAE as a percentage of true values, is also frequently used. Mean Square Error (MSE), which includes the average of the squared differences between actual and predicted values, and its root mean Square Error (RMSE) are also performance measures. Finally, the coefficient of determination, denoted R squared, evaluates the proportion of variance in the dependent variable that can be predicted from the independent variables.

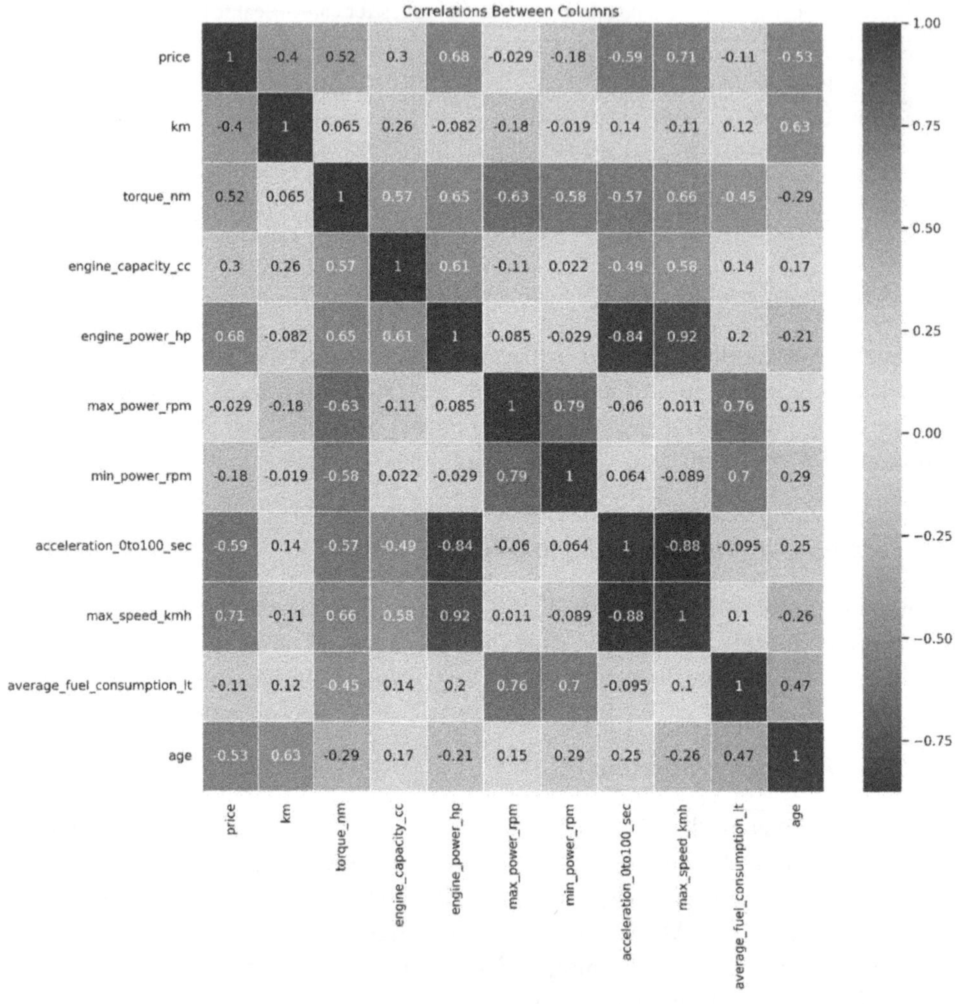

Fig. 3. Correlation analysis with heatmap.

3.9. *Final price prediction*

After comparing the performance metrics of all models, the best-performing models were selected for both the overall dataset and for each brand and series group.

4. Results

This section presents the findings of our study, focusing on the performance of various machine learning models used to predict second-hand car prices. The analysis highlights the strengths and weaknesses of each model based on key evaluation metrics, providing insights into their predictive capabilities and suitability for this task.

4.1. *Performance evaluation of models*

In the second stage of our analysis, we evaluated the error measurement metrics of four different linear regression-based models: LR, Ridge, Lasso, and E-NET. The performance of these models was measured using evaluation metrics, as shown in Table 5.

Among the models tested, the Ridge Regression model demonstrated the highest performance in terms of R-square and RMSE, achieving an R^2 value of 0.91 and an RMSE of 114,747. This indicated that Ridge Regression is the most effective model in capturing the variability in the dataset while maintaining a lower prediction error compared to the other models. Ridge Regression's performance was attributed to its ability to handle multicollinearity by adding a penalty to the coefficients, which reduced overfitting and enhanced generalization on unseen data.

Linear Regression, serving as a baseline model, also showed robust performance with an R^2 value of 0.90 and an RMSE of 117,314. Despite lacking the regularization mechanisms present in Ridge and Lasso models, Linear Regression still managed to provide relatively accurate predictions, suggesting that even a straightforward approach can offer substantial predictive power when the feature set is well-defined and representative of the underlying patterns in the data.

Table 5. Performance evaluation comparisons.

ML Model	R-square	MAE	MAPE	MSE	RMSE
LR	0.90	72,908	0.107	13,762,471,302	117,314
Ridge	0.91	73,669	0.110	13,166,780,418	114,747
Lasso	0.90	71,876	0.104	14,213,989,301	119,222
E-NET	0.71	133,152	0.200	42,179,481,383	205,376

The Lasso Regression model, with an R^2 of 0.90 and a slightly lower MAE (71,876) and MAPE (0.104) compared to Ridge Regression, also performed effectively. Lasso's ability to perform feature selection by shrinking some coefficients to zero was particularly beneficial in handling irrelevant or less significant features in the dataset, which in turn helped in reducing the complexity of the model and improving interpretability.

Elastic Net (E-NET), however, exhibited a significantly lower R^2 value of 0.71, indicating that this model was less effective in capturing the variance in the data compared to the other models. The higher MAE (133,152) and RMSE (205,376) values suggest that the Elastic Net model may have struggled with the complexity of the dataset, potentially due to an overemphasis on the combination of L1 and L2 penalties, which could lead to an inadequate trade-off between feature selection and coefficient shrinkage.

4.2. *Summary of Findings*

In summary, the results indicated that while all models, except Elastic Net, exhibited strong predictive performance, Ridge Regression was found to be the best-performing model in this study. Its superior handling of multicollinearity and its ability to maintain stable predictions made it a valuable tool for predicting second-hand car prices. Furthermore, the low MAPE values across most models suggested that the predictions are reliable, allowing for accurate price estimates, which is essential for enhancing decision-making in the second-hand car market in Turkey.

These findings highlighted the importance of model selection and regularization techniques in improving the accuracy and robustness of predictive models. By effectively managing the complexity of the dataset, the models developed in this study were shown to offer significant potential for application in real-world scenarios, contributing to a more efficient and transparent second-hand car market.

5. Conclusion

The primary objective of this study was to develop models capable of accurately predicting prices for second-hand car sales and to compare their performances. Among the models, the Ridge Regression model was found to have achieved an R-squared value of 0.91, indicating a high level of accuracy. These results demonstrate a strong potential for effective pricing in the second-hand car market.

Comparing our results with previous studies, the Ridge Regression model demonstrated competitive performance. For instance, Satapathy *et al.* [4] reported an accuracy of 92% using various models, including Ridge Regression, while Shaprapawad *et al.* [7] achieved an R-squared value of 94.5% using a combination of linear regression and tree-based methods. Similarly, Asghar *et al.* [10] obtained an R-squared value of 91% with Lasso and Linear Regression, which aligns closely with our findings. However, the error rate of 3.46 reported by Venkatasubbu and Ganesh [11] using Multiple Linear Regression and Decision Trees demonstrates how model performance can vary significantly depending on algorithm choice and dataset characteristics.

Our study contributes uniquely to the field by focusing on the Turkish second-hand car market using a 2023 dataset of over 17,000 records, which includes a diverse range of modern vehicles and a wide array of features such as brand, model, age, mileage, and fuel type. This comprehensive, up-to-date dataset enabled us to more accurately capture current market dynamics, enhancing the model's predictive power. These findings suggest that our model is well-suited for the Turkish market and potentially adaptable to other regions, offering a valuable tool for stakeholders in the second-hand car industry.

To further improve model accuracy and reliability, we recommend testing the developed model on a larger and more diverse dataset. This would help ensure transparent pricing processes in the second-hand car market. Additionally, generalizing the model to consider the unique conditions of different regional markets could enable more precise and localized predictions. Keeping the model updated with current data is crucial for maintaining its relevance and adaptability to changing market conditions.

Despite the promising results, this study faced some limitations, such as the availability and quality of data, which may affect the generalizability of the findings. Future research should focus on incorporating additional features, exploring different modeling techniques, and applying the model to other markets or regions to validate and extend the findings. Additionally, future studies could investigate brand-series factors by analyzing price prediction models based on specific vehicle brand and series. This approach might provide more tailored pricing predictions, reflecting the unique characteristics and market demand for different vehicle types.

The practical implications of this study are significant for stakeholders in the second-hand car market, including car dealers, buyers, and pricing analysts. Integrating the developed model into industry platforms, such as mobile apps or online marketplaces, could provide real-time pricing recommendations, enhancing the transparency and reliability of the trading environment. Continuous data collection and the inclusion of diverse data sources are crucial for improving model accuracy over time.

In conclusion, this study represents a significant step towards establishing fair and accurate pricing processes in the second-hand car market. By leveraging the developed model, the industry could foster a more transparent and reliable trading environment, ultimately leading to a more efficient and equitable market-place.

References

1. B. Hemendiran and P. N. Renjith, Predicting the Prices of the Used Cars using Machine Learning for Resale, in Proc. *IEEE Int. Students' Conf. Electrical, Electronics and Computer Science*, Bhopal, 2023, pp. 1–5.
2. C. Jin, Price Prediction of Used Cars Using Machine Learning, in Proc. *IEEE Int. Conf. Emergency Science and Information Technology*, Chongqing, 2021, pp. 223–230.
3. C. V. Narayana, C. L. Likhitha, S. Bademiya, and K. Kusumanjali, Machine Learning Techniques to Predict the Price of Used Cars: Predictive Analytics in Retail Business, in Proc. *IEEE Int. Conf. Electronics and Sustainable Communication Systems*, Coimbatore, 2021, pp. 1680–1687.
4. S. K. Satapathy, R. Vala, and S. Virpariya, An Automated Car Price Prediction System Using Effective Machine Learning Techniques, in Proc. IEEE Int. Conf. *Computational Intelligence and Sustainable Engineering Solutions*, Greater Noida, 2022, pp. 402–408.
5. F. Ecer, Forecasting of Second-Hand Automobile Prices and Identification of Price Determinants in Turkey, *Anadolu University Journal of Social Sciences 13(4)*, 2013, 101–112.
6. C. V. Narayana, N. O. G. Madhuri, A. NagaSindhu, M. Aksha, and C. Naveen, Second Sale Car Price Prediction using Machine Learning Algorithm, in Proc. *IEEE Int. Conf. Communication and Electronics Systems*, Coimbatore, 2022, pp. 1171–1177.
7. S. Shaprapawad, P. Borugadda, and N. Koshika, Car Price Prediction: An Application of Machine Learning, in Proc. *IEEE Int. Conf. Inventive Computation Technologies*, Lalitpur, 2023, pp. 242–248.
8. Y. Yamasari, N. Rochmawati, R. E. Putra, A. Qoiriah, Y. Asmunin, and W. Yustanti, Predicting the Students Performance using Regularization-based Linear Regression, in Proc. *IEEE Int. Conf. Vocational Education and Electrical Engineering*, Surabaya, 2021, pp. 1–5.
9. R. Sutaria, and R. Jain, Auto-Price Forecast: An Analysis of Car Value Trends, in Proc. *IEEE Int. Conf. Emerging Technology*, Belgaum, 2023, pp. 1–6.
10. A. Muhammad, K. Mehmood, S. Yasin, and Z. M. Khan, Used Cars Price Prediction Using Machine Learning with Optimal Features, *Pakistan Journal of Engineering and Technology 4(2)*, 2021,113–119.
11. M. Ganesh and P. Venkatasubbu, Used Cars Price Prediction using Supervised Learning Techniques, *International Journal of Engineering and Advanced Technology 9*, 2019, 216–223.
12. P. Rane, D. Pandya, and D. Kotak, Used Car Price Prediction, in Proc. *International Research Journal of Engineering and Technology* 8(4), 2021.
13. S. Han, J. Qu, J. Song, and Z. Liu, Second-hand Car Price Prediction Based on a Mixed-Weighted Regression Model, in Proc. *IEEE Int. Conf. Big Data Analytics*, Guangzhou, 2022, pp. 90–95.

14. Y. Li, Y. Li, and Y. Liu, Research on Used Car Price Prediction Based on Random Forest and LightGBM, in Proc. *IEEE Int. Conf. Data Science and Computer Application*, Dalian, 2022, pp. 539–543.
15. F. Abdullah, M. Rahman, M. Shidujaman, M. Hasan, and M. Habib, Machine Learning Modeling for Reconditioned Car Selling Price Prediction, in Proc. *SPIE 12543, Applications of Machine Learning 2023*, Bellingham, 2023, pp. 100.
16. M. Babaoglu, A. Coskuncay, and T. Aydin, Modeling Automobile Sales in Turkiye with Regression-Based Machine Learning Algorithms, Journal of Data Applications 1, 2023, 19–33.
17. M. Bilen, Predicting Used Car Prices with Heuristic Algorithms and Creating a New Dataset, *Journal of Multidisciplinary Developments 6(1)*, 2021, pp 29–43.
18. Ö. Çelik and U. Osmanoğlu, Prediction of the Prices of Second-Hand Cars, *European Journal of Science and Technology*, 2019, pp. 77–83.
19. F. Uysal, Comparative Analysis of Various Machine Learning and Deep Learning Approaches for Car Resale Price Prediction in the Turkish Market, *NOHU Journal of Engineering Sciences 13(1)*, 2024, pp. 342–349.
20. S. Kumar, D. Kaur, and A. Parvez, Prediction of Car Prices with Machine Learning, in Proc. *IEEE Int. Conf. Communication and Computing Research*, 2022, pp.1-4.
21. R. Swarnkar and R. Sawant, S. P. H. R., Multiple Linear Regression Algorithm-Based Car Price Prediction, in Proc. *IEEE Int. Conf. Artificial Intelligence and Smart Energy*, Coimbatore, India, 2023, pp. 675–681.

Chapter 24

Mixed Element Network Structure Determination via Artificial Neural Networks

Metin Şengül*,‡ and Atilla Özmen†,§

**Electrical and Electronics Engineering, İstanbul Kültür University, İstanbul, Turkey*

†Electrical and Electronics Engineering, Kadir Has University, İstanbul, Turkey

‡m.sengul@iku.edu.tr

§aozmen@khas.edu.tr

At high frequencies, the connection points between lumped elements are considered as distributed circuit elements. In this case, mixed element circuits are obtained and these circuits are described in terms of two-variable functions. After designing a mixed element network, from the obtained two-variable network functions, network structure (connection order of lumped and distributed elements) must be determined. In literature, only some of the possible structures can be determined. But, it is possible to determine all of the mixed element network structures with 100% accuracy by means of the proposed artificial neural network (ANN) based approach here.

1. Introduction

Especially, at high frequencies, the connection points between lumped elements destroy the performance of the circuit. To be able to solve this problem, the effect of these connection points must be considered during design process, and they must be considered as circuit elements. In this case, the circuit will contain two different type of circuit elements, lumped and distributed elements.

Transcendental or multivariable functions are utilized in the description of mixed element networks. In the first approach, single variable transcendental functions are used. In the second approach, characterization of mixed lumped and distributed element networks is based on Richards transformation, ($\lambda = \tanh(p\tau)$, on the imaginary axis the transformation can be expressed as $\lambda = j\Omega = jtan(\omega\tau)$, where τ is the commensurate delay of the distributed elements, namely transmission lines) [1]. Generalization of this approach to mixed lumped and distributed element networks yields multivariable synthesis procedures, where the Richards variable $\lambda = \Sigma + j\Omega$ is utilized for distributed element section and the original frequency variable $p = \sigma + j\omega$ is utilized for lumped element section. As a result, mixed element networks are described in terms of two frequency variables [1, 2].

Unfortunately, the problems associated with the approximation and synthesis of mixed element networks has not been solved yet, namely a complete design theory for mixed lumped and distributed element networks still does not exist.

A special and very useful network configuration has particular interest in most existing studies, in which lumped reactances and cascaded ideal commensurate transmission lines (Unit Elements, UEs) are used. So, this mixed element structure consists of cascaded lossless lumped elements and unit elements. At this step, let us briefly explain the two-variable description of the mixed element circuits mentioned above.

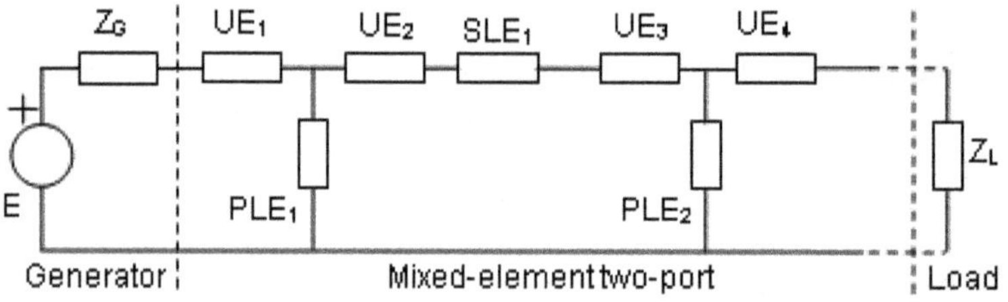

Fig. 1. Mixed-element two-port.

In this work, since low-pass mixed element networks are considered, PLE_i and SLE_i seen in Figure 1 are parallel connected capacitors and series connected inductors, respectively. *UEi* are the unit elements representing the connection points between lumped elements.

In this work, two-variable scattering parameters are utilized to define the mixed element networks, which are expressed via two-variable polynomials g, h, f as follows [3–5]:

$$S(p,\lambda) = \begin{bmatrix} S_{11}(p,\lambda) & S_{12}(p,\lambda) \\ S_{21}(p,\lambda) & S_{22}(p,\lambda) \end{bmatrix} = \begin{bmatrix} \frac{h(p,\lambda)}{g(p,\lambda)} & \frac{\mu f(-p,-\lambda)}{g(p,\lambda)} \\ \frac{f(p,\lambda)}{g(p,\lambda)} & \frac{-\mu h(-p,-\lambda)}{g(p,\lambda)} \end{bmatrix} \tag{1}$$

where $|\mu| = 1$.

As mentioned above, the variables of the polynomials in (1) are p and λ, where λ is the Richards variable related with unit elements, and p is the complex frequency variable related with lumped elements.

In the representation given in (1), the polynomial $g(p,\lambda)$ is a real coefficient scattering Hurwitz polynomial. The degree of the polynomial is $n_p + n_\lambda$ and $g(p,\lambda)$ can be expressed as $g(p,\lambda) = \mathbf{P}^T \Lambda_g \boldsymbol{\lambda} = \boldsymbol{\lambda}^T \Lambda_g^T \mathbf{P}$ where

$$\Lambda_g = \begin{bmatrix} g_{00} & g_{01} & \cdots & g_{0n_\lambda} \\ g_{10} & g_{11} & \cdots & \vdots \\ \vdots & \vdots & \ddots & \vdots \\ g_{n_p 0} & \cdots & \cdots & g_{n_p n_\lambda} \end{bmatrix}, \tag{2a}$$

$$\boldsymbol{P}^T = [1 \quad p \quad p^2 \quad \cdots \quad p^{n_p}], \tag{2b}$$

$$\boldsymbol{\lambda}^T = [1 \quad \lambda \quad \lambda^2 \quad \cdots \quad \lambda^{n_\lambda}]. \tag{2c}$$

where n_p and n_λ are the numbers of lumped elements and unit elements, respectively.

Since the polynomial $g(p,\lambda)$ has two variables, its partial degrees can be expressed as the highest power of a nonzero-coefficient variable, i.e., $n_p = \deg_p g(p,\lambda)$, $n_\lambda = \deg_\lambda g(p,\lambda)$.

In a similar manner, the coefficients of polynomial $h(p,\lambda)$ are real and its degree is $n_p + n_\lambda$ and the polynomial can be expressed as $h(p,\lambda) = \mathbf{P}^T \Lambda_h \boldsymbol{\lambda} = \boldsymbol{\lambda}^T \Lambda_h^T \mathbf{P}$, where

$$\Lambda_h = \begin{bmatrix} h_{00} & h_{01} & \cdots & h_{0n_\lambda} \\ h_{10} & h_{11} & \cdots & \vdots \\ \vdots & \vdots & \ddots & \vdots \\ h_{n_p 0} & \cdots & \cdots & h_{n_p n_\lambda} \end{bmatrix}. \tag{3}$$

The real coefficient polynomial $f(p,\lambda)$ is constructed via the mixed element networks' transmission zeros and it can be expressed as

$$f(p,\lambda) = f(p) f(\lambda) \tag{4}$$

where the polynomials $f(\lambda)$ and $f(p)$ contain the transmission zeros of the UE and lumped element sections, respectively.

The polynomial $f(p)$ will be either an even or an odd polynomial in p, if only the real frequency transmission zeros related with lossless lumped element section are considered. If unit elements are cascade connected, the polynomial $f(\lambda)$ is expressed as

$$f(\lambda) = (1 - \lambda^2)^{n_\lambda/2}. \quad (5)$$

A practical form of $f(p,\lambda)$ can be obtained as seen in (6), if the finite imaginary axis zeros except the zeros at DC are disregarded,

$$f(p,\lambda) = p^k(1 - \lambda^2)^{n_\lambda/2} \quad (6)$$

where k is the number of transmission zeros at DC.

During the design of a mixed element network, to be able to obtain a realizable network, two-variable polynomials must satisfy the following expression [4, 5]

$$g(p,\lambda)g(-p,-\lambda) = h(p,\lambda)h(-p,-\lambda) + f(p,\lambda)f(-p,-\lambda). \quad (7)$$

In the design procedures utilizing (7), the polynomial $f(p,\lambda)$ is formed by means of the transmission zeros of the two-port network, and the coefficients of the polynomial $h(p,\lambda)$ is initialized by the designer as +/-1. Therefore, the right hand side of (7) is formed, and it is equal to the following expression

$$g(p,\lambda)g(-p,-\lambda) = G(p^2,\lambda^2) = h(p,\lambda)h(-p,-\lambda) + f(p,\lambda)f(-p,-\lambda) \quad (8)$$

where $G(p^2,\lambda^2)$ is an even polynomial. Then, the polynomial $g(p,\lambda)$ is easily constructed by using the left half plane roots of the obtained even polynomial.

After satisfying (7) or (8) and completing the design, the designer will get the coefficient matrices given in (2a) and (3). Then, by means of these matrices, component values must be calculated, and mixed element network structure must be obtained. But in literature only some of the possible network structures are determined [4–7]. In this work, an ANN based approach is proposed to determine all of the possible mixed element network structures from the coefficient matrices (2a) and (3).

In the following section, ANN used in this work is briefly explained, and then an example is given to show the performance of the proposed approach.

2. Example: Low Pass Mixed Element Network

Table 1. All possible low-pass mixed element network structures with degree four.

	Network structures
1	Series L, parallel C, UE, UE
2	Series L, UE, parallel C, UE
3	Series L, UE, UE, parallel C
4	UE, series L, parallel C, UE
5	UE, series L, UE, parallel C
6	UE, UE, series L, parallel C
7	Parallel C, series L, UE, UE
8	Parallel C, UE, series L, UE
9	Parallel C, UE, UE, series L
10	UE, parallel C, series L, UE
11	UE, parallel C, UE, series L
12	UE, UE, parallel C, series L

As an example, a low-pass mixed element network with two lumped and two unit elements is selected. So, the degree of the mixed element structure is four, and the number of possible network structures is 12. In the literature, only four of them can be identified. The list of all possible low-pass mixed element network structures with degree four can be found in Table 1.

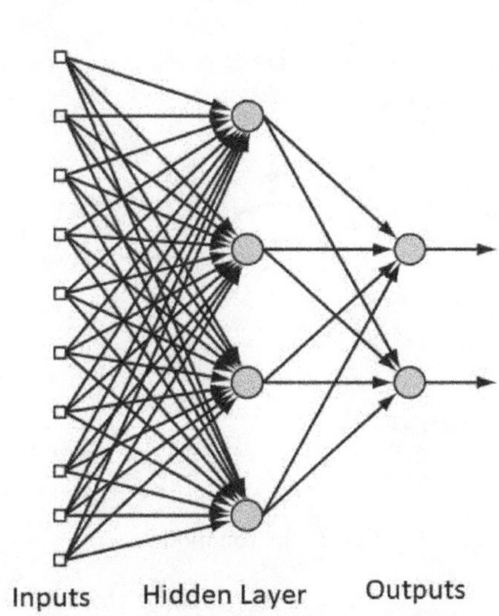

Fig. 2. ANN structure.

In this example, a simple multilayer perceptron (MLP) structure with one hidden layer containing 36 neurons is used. The general structure of the proposed ANN is given in Figure 2. As shown in this figure, each neuron has multiple inputs and only one output. Neuron model finds the weighted sum of its inputs and then applies this sum to the corresponding activation functions to find the output of the neuron [8–9].

A typical neuron model, consisting of a summing node and an activation function, is shown in Figure 3. The activation function is represented by $f(x)$, which introduces non-linearity into the neuron's output. The overall output of the neuron is determined by the following expression

$$y = f(\sum_{i=1}^{n} x_i w_i + b). \tag{9}$$

In (9),

- $x_1, x_2, x_3, \ldots, x_n$ are the input signals to the neuron,
- $w_1, w_2, w_3, \ldots, w_n$ are the corresponding weight coefficients,
- b is the bias term, which allows the neuron to shift activation threshold,
- n is the total number of inputs to the neuron,
- y is the output of the neuron.

The summation computes the weighted sum of all inputs and adds the bias term b. Then, it is passed through the activation function to produce the neuron's output y.

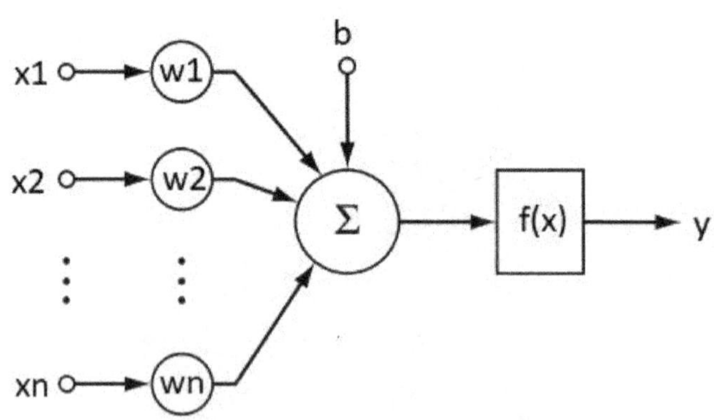

Fig. 3. Neuron model.

Rectified Linear Unit (ReLU) and linear functions are used as activation functions for the hidden and output layers, respectively. ReLU function mostly is used in deep neural network structures. It is shown that ReLU activation function improves the training of neural network and provides significant performance when compared to other activation functions [10]. The ReLU function is a linear function if its input is greater than zero, otherwise it gives zero as shown in Figure 4.

The ReLU function is mathematically defined as follows

$$f(x) = \max(0, x). \tag{10}$$

In (10), x represents the input value to the function. The $\max(0, x)$ expression returns the greater of 0 and x. In other words, if x is positive, the function returns x; otherwise, it returns 0.

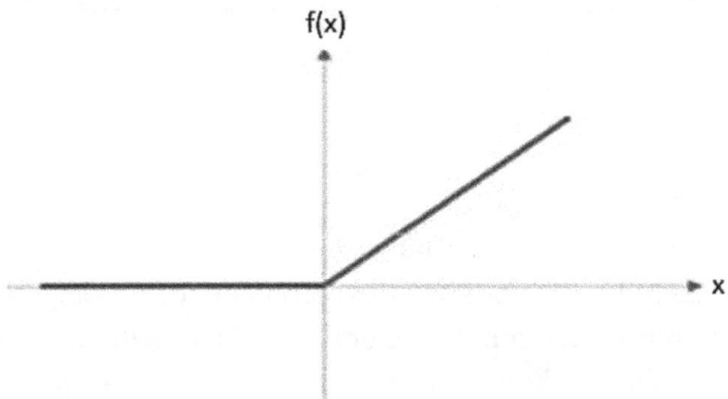

Fig. 4. ReLU activation function.

Λ_h and Λ_g coefficient matrices are arranged as a single column vector with a size of 18. For each of 12 circuit topology, 10000 input vectors are created to feed the proposed ANN. Each input vector is obtained from the uniformly distributed circuit component values between 0 and 1. In total, 120000 column vectors are obtained for all possible circuit topologies. The input data is divided into three parts as 70% for train, 15% for validation and 15% for testing. Since, the signs of Λ_h and Λ_g coefficients characterize the circuit topology better than the coefficients themselves, these coefficients are not used directly to feed ANN. The following signum function is used to obtain only sign of the coefficients

$$\operatorname{sgn}(x) = \begin{cases} +1, & x > 0 \\ 0, & x = 0. \\ -1, & x < 0 \end{cases} \tag{11}$$

ANN schematic diagram used to determine the mixed element network structure is given in Figure 5.

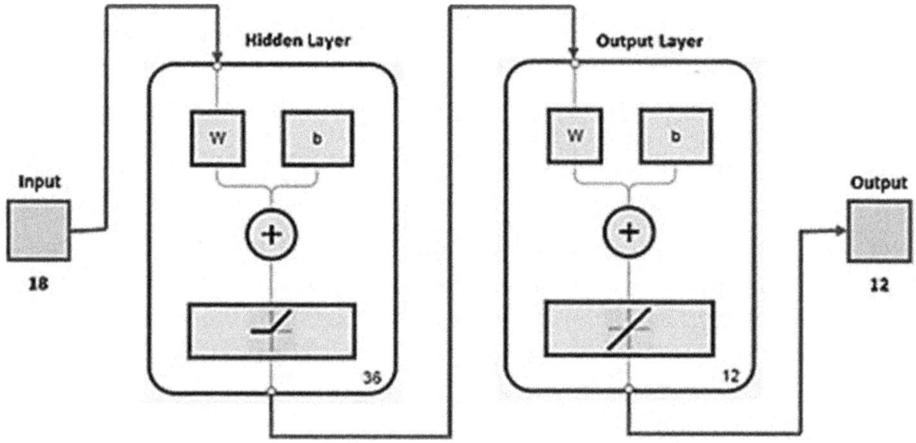

Fig. 5. Used ANN schematic diagram.

Since the determination of circuit topologies from Λ_h and Λ_g coefficient matrices is considered as a classification problem, the used ANN's output is designed as a column vector containing only 0s and 1s with a size of 12 for each circuit topology. Table 2 shows the output column vectors and corresponding circuit structures.

Table 2. Circuit structures and corresponding ANN output column vectors.

UE-SL-UE-PC	**UE-PC-UE-SL**	**SL-PC-UE-UE**	**PC-SL-UE-UE**	**UE-UE-PC-SL**	**PC-UE-UE-SL**	**SL-UE-UE-PC**	**UE-UE-SL-PC**	**SL-UE-PC-UE**	**PC-UE-SL-UE**	**UE-PC-SL-UE**	**UE-SL-PC-UE**
1	0	0	0	0	0	0	0	0	0	0	0
0	1	0	0	0	0	0	0	0	0	0	0
0	0	1	0	0	0	0	0	0	0	0	0
0	0	0	1	0	0	0	0	0	0	0	0
0	0	0	0	1	0	0	0	0	0	0	0
0	0	0	0	0	1	0	0	0	0	0	0
0	0	0	0	0	0	1	0	0	0	0	0
0	0	0	0	0	0	0	1	0	0	0	0
0	0	0	0	0	0	0	0	1	0	0	0
0	0	0	0	0	0	0	0	0	1	0	0
0	0	0	0	0	0	0	0	0	0	1	0
0	0	0	0	0	0	0	0	0	0	0	1

The Levenberg-Marquardt (LM) backpropagation algorithm, which is well-suited for small and medium-sized optimization problems, is used to train the proposed ANN to find the weight coefficients. When compared to other learning algorithms, LM has faster and stable convergence [11]. In our example, mean square error (MSE) is converged to 2.52×10^{-16} in 12 iterations. The convergence of the mean square error (MSE) with 12 epochs during the training of an ANN is shown in Figure 6. The MSE, plotted on a logarithmic scale, shows a rapid decrease from an initial value to 2.52×10^{-16} at the 12th epoch. This MSE value proves that ANN is capable of matching Λh and Λg coefficient matrices to the corresponding mixed element network structure perfectly.

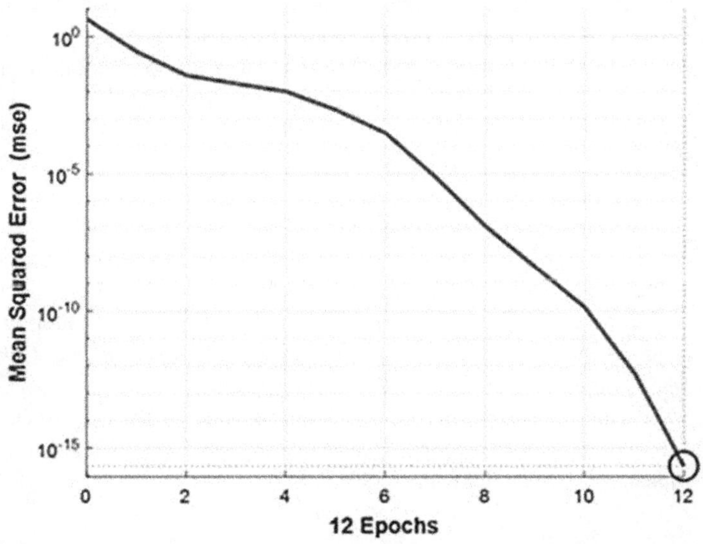

Fig. 6. MSE performance of the trained ANN.

3. Conclusions

In this work, an ANN based mixed element network structure determination approach is proposed. In literature, it is possible to determine only some of the possible network structures. But as shown here, the proposed approach can determine all of the possible mixed element structures with 100% accuracy.

A similar procedure can be followed for each type (low pass, high pass, band pass and band stop) and degree of the mixed element networks. Therefore, after designing a network with mixed elements and obtaining coefficient matrices, the user can identify the network structure via a trained ANN. As future work, based on the results obtained here, determination of circuit structure from Λ_h and Λ_g

coefficient matrices can be realized easily for different type of mixed element networks with higher degrees.

References

1. P. I. Richards, Resistor transmission-line circuits, *Proc. Inst. Radio Eng.* 36(2), 217 (1948).
2. L. A. Riederer and L. Weinberg, Synthesis of lumped-distributed networks: Lossless cascades, *IEEE Trans. Circuits Syst.* 27(10), 943 (1980).
3. S. Basu and A. Fettweis, On synthesizable multidimensional lossless two-ports, *IEEE Trans. CAS* 35(12), 1478 (1988).
4. M. Şengül, Design of practical broadband matching networks with mixed lumped and distributed elements, *IEEE Trans. CAS II, Exp. Briefs* 61(11), 875 (2014).
5. A. Aksen, Design of lossless two-port with mixed lumped and distributed elements for broadband matching, Ph.D. thesis, Ruhr Univ., Bochum, Germany (1994).
6. A. Sertbaş, Two-variable scattering formulas to describe some classes of lossless two-ports with mixed lumped elements and commensurate stubs, *Turk. J. Elect. Eng.* 13(2), 231 (2005).
7. M. Şengül, Construction of lossless ladder networks with simple lumped elements connected via commensurate transmission lines, *IEEE Trans. CAS II, Exp. Briefs* 56(1), 1 (2009).
8. S. Haykin, *Neural Networks, a Comprehensive Foundation.* 2nd Edition, Prentice-Hall Inc., New Jersey (1999).
9. M. Şengül, A. Özmen and M. Yılmaz, A new synthesis algorithm for mixed lumped-distributed low-pass ladder networks via artificial neural networks. in *International Conference on Electronics and Computer Engineering*, Bursa, Turkey (2003).
10. I. Goodfellow, Y. Bengio and A. Courville, A. *Deep Learning*, MIT Press (2017).
11. H. Yu and B. M. Wilamowski, *Intelligent Systems: Levenberg-Marquardt Training, Industrial Electronics Handbook*, 2nd edn. CRC Press (2011).

Chapter 25

Vibration Analysis of FGM Plate: A Hybrid Analytical and Machine Learning Approach

Emin Emre Özdilek*, Emircan Gündoğdu†, Murat Çelik*,‡ and Erol Demirkan*

*Istanbul Technical University, Department of Civil Engineering, Istanbul, Turkey
†Istanbul Technical University, Department of Computer Engineering, Istanbul, Turkey
‡celikmur15@itu.edu.tr

This study explores the free vibration characteristics of functionally graded material (FGM) plates using both analytical solutions and machine learning (ML) techniques. Initially, we derive equilibrium equations for FGM plates through Hamilton's principle to determine their natural frequencies under various conditions. We then develop an Artificial Neural Network (ANN) model, trained on a comprehensive dataset from prior research, to predict these vibrational frequencies [1]. The ANN's predictions are rigorously compared with our analytical results and validated against existing studies, demonstrating its high precision and computational efficiency. Our research highlights that the ANN model can significantly streamline the analysis process, effectively handling complex patterns that challenge traditional methods. This integration of analytical rigor with ML innovation presents a novel approach to enhancing the structural analysis and design of FGM plates, potentially transforming material science and engineering practices. The accuracy with which the ANN model predicts natural frequencies across diverse FGM plates underscores the power of data-driven approaches in engineering analysis. Integrating ML not only enhances the precision of traditional methods but also introduces adaptability and scalability previously unattainable. Our findings suggest that combining computational mechanics with artificial intelligence holds great promise for future material design and optimization, offering a more comprehensive understanding of the dynamic properties of FGM plates. This study also sets a precedent for applying ML to complex engineering problems, encouraging further exploration into hybrid methodologies that bridge theoretical analysis and practical engineering solutions [2]. Through this approach, we aim to advance FGM technology, paving the way for more resilient and efficient structural components.

1. Introduction

Functionally graded materials (FGMs) represent a significant advancement in material science, characterized by a gradual variation in composition and properties over a specific dimension. This innovative design approach has led to the development of functionally graded plates (FGPs), which are increasingly used in high-performance applications due to their superior mechanical properties and enhanced durability. Unlike traditional homogeneous materials, FGPs exhibit a continuous transition in properties such as thermal conductivity, elasticity, and strength, tailored to meet specific operational demands.

In this paper, we investigate the dynamic behavior of functionally graded plates using a simple shear deformation theory. This theory provides an efficient and accurate framework for capturing shear deformation effects, which are particularly significant in thick plates or those subjected to high transverse loads. Our primary objective is to determine the natural frequencies of these FG plates, a critical parameter in assessing their dynamic performance and structural integrity. The static behavior of functionally graded plates was studied by Reddy [3] using third-order shear deformation theory.

The deflections of a simply supported functionally graded polygonal plate, as described by both first-order and third-order shear deformation theories, have been correlated to those of an equivalent homogeneous Kirchhoff plate [4]. The free vibration analysis of functionally graded simply-supported square plates was conducted by [5] using a higher-order finite element formulation, as part of their broader research study.

As demonstrated in numerous past studies, the deformation of functionally graded plates becomes increasingly complex due to both their geometry and the external loads applied to them, making the analytical solution of these problems very challenging. Today, with the integration of artificial intelligence applications into the field of engineering, such problems can be addressed effectively through a machine learning perspective. In the present study, we analyzed the natural frequency values of functionally graded plates for the simply-supported beam in the first mode, using both analytical methods and the Artificial Neural Network (ANN) approach. To train the ANN system, we used the plate dimensions (*a, b, h*) and power law index value (*p*) as inputs. For the functionally graded system, we analyzed the natural frequencies for different (*a/h*) and (*a/b*) values of square and rectangular plates analytically, with the aid of Python and Mathematica programs [6]. The natural frequency values obtained for different geometries were then used to train the ANN system.

The most significant aspect of this study is that, through the machine learning approach, natural frequencies can be calculated with very high accuracy within seconds for any desired FG plate size and load configuration. This capability makes the study both unique and valuable in the field of structural analysis.

In the analytical solution of the system, the governing equations were derived using Hamilton's principle. Then, using the Navier approach, the natural frequencies of the functionally graded plate were determined through a simple mathematical transformation. To align with the literature, non-dimensional frequencies were obtained.

2. Mathematical Model of FG Plate

The mathematical model is derived based on the assumption that the mechanical properties of the plate vary through the thickness according to a predefined function, typically a power law or exponential function. Figure 1 illustrates a functionally graded (FG) plate, characterized by dimensions: Length a, width b, and thick-ness h. The composition gradient is established with a ceramic-rich upper surface at $z = +h/2$ and transitions progressively to a metal-rich lower surface at $z = -h/2$.

The variation in material properties, designated by P, across the plate's thickness adheres to the established rule of mixture, as described subsequently.

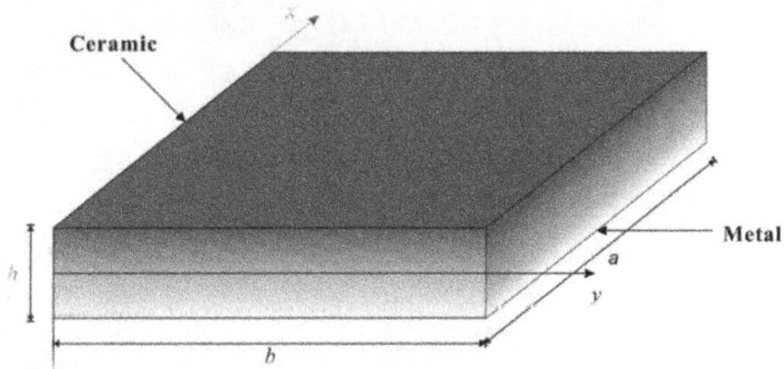

Fig. 1. FG Plate model.

$$P(z) = P_c V_c + P_m V_m \tag{1}$$

Within this framework, P_c and P_m articulate the material properties, whereas V_c and V_m define the volume fractions of the ceramic and metallic components, respectively. The total volume fraction of the components is delineated as follows:

$$V_c + V_m = 1 \tag{2}$$

The volume fraction of the ceramic constituent, conforming to a power-law distribution, is articulated as,

$$V_c = \left(\frac{z}{h} + \frac{1}{2}\right)^p \tag{3}$$

Where p represents a non-negative value, signifying the power law index. The material properties of the functionally graded plate may be characterized by the subsequent expressions:

$$E(z) = (E_c - E_m)V_c + E_m \tag{4}$$

$$\rho(z) = (\rho_c - \rho_m)V_c + \rho_m \tag{5}$$

In this model, E and ρ respectively signify the Young's modulus and the mass density of the functionally graded plate. The equations of the FG plate can be formulated in constitutive equation matrix form as below [7]:

$$\begin{Bmatrix} \sigma_x \\ \sigma_y \\ \tau_{xy} \\ \tau_{yz} \\ \tau_{xz} \end{Bmatrix} = \begin{bmatrix} Q_{11} & Q_{12} & 0 & 0 & 0 \\ Q_{12} & Q_{22} & 0 & 0 & 0 \\ 0 & 0 & Q_{44} & 0 & 0 \\ 0 & 0 & 0 & Q_{55} & 0 \\ 0 & 0 & 0 & 0 & Q_{66} \end{bmatrix} \begin{Bmatrix} \varepsilon_x \\ \varepsilon_y \\ \gamma_{xy} \\ \gamma_{yz} \\ \gamma_{xz} \end{Bmatrix} \tag{6}$$

Where σ_x, σ_y, τ_{xy}, τ_{yz}, τ_{xz}, and ε_x, ε_y, γ_{xy}, γ_{yz}, γ_{xz} are the components of stress and strain, respectively. The stiffness coefficients, represented by the known C_{ij} matrix can be characterized as,

$$Q_{11} = Q_{22} = \frac{E(z)}{1-\nu^2}$$

$$Q_{12} = \frac{\nu E(z)}{1-\nu^2}$$

$$Q_{44} = Q_{55} = Q_{66} = \frac{E(z)}{2(1+\nu)} \tag{7}$$

2.1. *Kinematics of the plate model*

Primarily, these assumptions are posited within the article: Citing the neutral surface of the FG plate, the cartesian coordinate system is established. Displacements remain small in comparison to the plate thickness, resulting in infinitesimal strains. The transverse shear stress is assumed by this theory, necessitating the inclusion of a shear correction factor to meet plate boundary conditions on both upper and lower surfaces.

The displacement field of the plate:

$$u(x,y,z,t) = u_0(x,y,t) - z\frac{\partial \phi}{\partial x} \tag{8a}$$

$$v(x,y,z,t) = v_0(x,y,t) - z\frac{\partial \phi}{\partial y} \tag{8b}$$

$$w(x,y,z,t) = w(x,y,t) \tag{8c}$$

In this context u, v and w denote displacements along the x, y and z axes, respectively, where u_0 and v_0 signify displacements at the neutral surface. The function is a function of the coordinates x and y as well as the temporal variable t.

The strains associated with the displacements of FG plates depending on the applied plate theory.

$$\varepsilon_x = \frac{\partial u_0}{\partial x} - z\frac{\partial^2 \phi}{\partial x^2} \tag{9}$$

$$\varepsilon_y = \frac{\partial v_0}{\partial x} - z\frac{\partial^2 \phi}{\partial y^2} \tag{10}$$

$$\gamma_{xy} = \frac{\partial u_0}{\partial y} + \frac{\partial v_0}{\partial x} - 2z\frac{\partial^2 \phi}{\partial x \partial y} \tag{11}$$

$$\gamma_{yz} = \frac{\partial w}{\partial y} - \frac{\partial \phi}{\partial y} \tag{12}$$

$$\gamma_{xz} = \frac{\partial w}{\partial x} - \frac{\partial \phi}{\partial x} \tag{13}$$

2.2. *Principal formulas and laws*

The equilibrium equations are derived using the Hamilton's principle. In this context, the principle of virtual work is applied to yield the following:

$$\int_0^T (\delta U + \delta V - \delta K)dt = 0 \tag{14}$$

Here, δU represents the variation in strain energy, δv denotes the variation in potential energy, and δK signifes the variation in kinetic energy. By incorporating Eqs. (8a)–(8c) and (9)–(13) into Eq. (14), the variation in strain energy can be determined as follows:

$$\int_V \left(\sigma_x \delta \varepsilon_x + \sigma_y \delta \varepsilon_y + \tau_{xy} \delta \gamma_{xy} + \tau_{yz} \delta \gamma_{yz} + \tau_{xz} \delta \gamma_{xz}\right) dAdz - \int_A (q \delta w)\, dA = 0 \tag{15}$$

Rearranging and consolidating Eqs. (8a)–(8c) and (9)–(13) into Eq. (14) yields the following:

$$\int_A \Bigg(N_x \frac{\partial \delta u_0}{\partial x} - M_x \frac{\partial^2 \delta \phi}{\partial x^2} + N_y \frac{\partial \delta u_0}{\partial y} - M_y \frac{\partial^2 \delta \phi}{\partial y^2} + N_{xy}\left(\frac{\partial \delta u_0}{\partial y} + \frac{\partial \delta u_0}{\partial x}\right) - 2M_{xy} \frac{\partial^2 \delta \phi}{\partial x \partial y} + Q_{yz}\left(\frac{\partial \delta w}{\partial y} - \frac{\partial \delta \phi}{\partial x}\right) + Q_{xz}\left(\frac{\partial \delta w}{\partial x} - \frac{\partial \delta \phi}{\partial x}\right)\Bigg) dA - \int_A (q \delta w)\, dA = 0 \tag{16}$$

Where the stress resultants are expressed as:

$$\{N_x, N_y, N_{xy}\} = \int_{-h/2}^{h/2} \{\sigma_x, \sigma_y, \tau_{xy}\}\, dz$$
$$\{M_x, M_y, M_{xy}\} = \int_{-h/2}^{h/2} \{\sigma_x, \sigma_y, \tau_{xy}\}\, zdz$$
$$\{Q_{xz}, Q_{yz}\} = \int_{-h/2}^{h/2} \{\tau_{xz}, \tau_{yz}\}\, dz \tag{17}$$

And the variational kinetic energy is,

$$\delta K = \int_V \rho(z)(\dot{u}\delta\dot{u} + \dot{v}\delta\dot{v} + \dot{w}\delta\dot{w})\, dAdz$$
$$= \int_A \Bigg[J_1(\dot{u}_0 \delta \dot{u}_0 + \dot{v}_0 \delta \dot{v}_0 + \dot{w}\delta\dot{w}) - J_2\left(\dot{u}_0 \frac{\partial \delta \dot{\phi}}{\partial x} + \frac{\partial \dot{\phi}}{\partial x} \delta \dot{u}_0 + \dot{v}_0 \frac{\partial \delta \dot{\phi}}{\partial y} + \frac{\partial \dot{\phi}}{\partial y} \delta \dot{v}_0\right) + J_3\left(\frac{\partial^2 \delta \dot{\phi}}{\partial x^2} + \frac{\partial^2 \delta \dot{\phi}}{\partial y^2}\right)\Bigg] dA \tag{18}$$

Denoting differentiation with respect to the time variable t by the superscript dot, the terms J_1, J_2, J_3 define the mass moments of inertia of the FG plate as follows:

$$\{J_1, J_2, J_3\} = \int_{-h/2}^{h/2} \rho(z)(1, z, z^2)\, dz \tag{19}$$

Incorporating Eqs. (16) and (18) into Eq. (14) and subsequently conducting rigorous computations and simplifications, we deduce the ensuing equations:

$$\delta u_0: \frac{\partial N_x}{\partial x} + \frac{\partial N_{xy}}{\partial y} = J_1 \ddot{u}_0 - J_2 \frac{\partial}{\partial x} \ddot{\phi} \tag{20}$$

$$\delta v_0: \frac{\partial N_{xy}}{\partial x} + \frac{\partial N_y}{\partial y} = J_1 \ddot{v}_0 - J_2 \frac{\partial}{\partial y} \ddot{\phi} \tag{21}$$

$$\delta \phi: \frac{\partial^2 M_x}{\partial x^2} + 2\frac{\partial^2 M_{xy}}{\partial x \partial y} + \frac{\partial^2 M_y}{\partial y^2} - \frac{\partial Q_{xz}}{\partial x} - \frac{\partial Q_{yz}}{\partial y} = J_2 \frac{\partial \ddot{u}_0}{\partial x} + J_2 \frac{\partial \ddot{v}_0}{\partial x} - J_3 \nabla^2 \ddot{\phi} \tag{22}$$

$$\delta w: \frac{\partial Q_{yz}}{\partial Y} + \frac{\partial Q_{xz}}{\partial x} + q = J_1 \ddot{w} \tag{23}$$

If Eqs. (9)–(13) are restated into Eqs. (20)–(23) with the necessary adjustments, following equations are obtained:

$$A_{11} \frac{\partial^2 u_0}{\partial x^2} + A_{66} \frac{\partial^2 u_0}{\partial y^2} + A_{12} \frac{\partial^2 v_0}{\partial x \partial y} + A_{66} \frac{\partial^2 v_0}{\partial x \partial y} = J_1 \frac{\partial \ddot{u}_0}{\partial t} - J_2 \frac{\partial^2 \phi}{\partial x \partial t} \tag{24a}$$

$$A_{11} \frac{\partial^2 v_0}{\partial y^2} + A_{66} \frac{\partial^2 v_0}{\partial x^2} + A_{12} \frac{\partial^2 u_0}{\partial x \partial y} + A_{66} \frac{\partial^2 u_0}{\partial x \partial y} = J_1 \frac{\partial \ddot{v}_0}{\partial t} - J_2 \frac{\partial^2 \phi}{\partial y \partial t}$$

$$-D_{66} \frac{\partial^4 \phi}{\partial x^4} - 2(D_{11} + 2D_{66}) \frac{\partial^4 \phi}{\partial x^2 \partial y^2} - D_{11} \frac{\partial^4 \phi}{\partial y^4} - A_{66} \left(\frac{\partial^2 w}{\partial x^2} - \frac{\partial^2 \phi}{\partial x^2} \right) \tag{24b}$$

$$-A_{66} \left(\frac{\partial^2 w}{\partial y^2} - \frac{\partial^2 \phi}{\partial y^2} \right) = J_2 \frac{\partial \ddot{u}_0}{\partial x} + J_2 \frac{\partial \ddot{v}_0}{\partial y} - J_2 \nabla \ddot{\phi} \tag{24c}$$

$$A_{66} \left(\frac{\partial^2 w}{\partial y^2} - \frac{\partial^2 \phi}{\partial y^2} \right) + A_{66} \left(\frac{\partial^2 w}{\partial x^2} - \frac{\partial^2 \phi}{\partial x^2} \right) + q = J_1 \ddot{w} \tag{24d}$$

In which A_{ij} and D_{ij} are the stiffness of the plate and defined as:

$$A_{11} = \int_{-h/2}^{h/2} Q_{11} dz,\ D_{11} = \int_{-h/2}^{h/2} Q_{11} z^2 dz$$

$$A_{12} = \int_{-h/2}^{h/2} Q_{12} dz,\ D_{12} = \int_{-h/2}^{h/2} Q_{12} z^2 dz$$

$$A_{66} = \int_{-h/2}^{h/2} Q_{66} dz,\ D_{66} = \int_{-h/2}^{h/2} Q_{66} z^2 dz \tag{25}$$

The stiffness coefficient Q_{ij} which is defined as Eq. (7) can be written as above.

3. The Solution to The Problem

3.1. *Analytical solution*

FG plates are commonly categorized according to the type of support employed. In this study, attention is given to the exact solution Eqs. (24a)–(24d) for a simply-supported FG plate. The following geometric and essential boundary conditions are imposed at the side edges:

$$u_0 = w = \phi = N_x = M_x = 0, at x = 0, a \tag{26a}$$

$$v_0 = w = \phi = N_y = M_y = 0, at y = 0, b \tag{26b}$$

Adhering to the Navier's solution procedure, we posit a solution form for u_0, v_0, ϕ and w that complies with the prescribed boundary conditions. The expression above illustrates this proposed solution, where each variable is expressed as a sum over m and n terms:

$$\begin{bmatrix} u_0 \\ v_0 \\ \phi \\ w \end{bmatrix} = \sum_{m=1}^{\infty}\sum_{n=1}^{\infty} \begin{Bmatrix} U_{mn} \cos(\lambda x) \sin(\mu y)\, e^{iwt} \\ V_{mn} \sin(\lambda x) \cos(\mu y)\, e^{iwt} \\ \Psi_{mn} \sin(\lambda x) \sin(\mu y)\, e^{iwt} \\ W_{mn} \sin(\lambda x) \cos(\mu y)\, e^{iwt} \end{Bmatrix} \tag{27}$$

Here, U_{mn}, V_{mn}, Ψ_{mn}, and W_{mn} represent arbitrary parameters to be determined. w denotes the eigenfrequency associated with the (m,n)-th eigenmode, where $\lambda = \frac{m\pi}{a}$ and $\mu = \frac{n\pi}{b}$. For the scenario involving a sinusoidally distributed load.

Upon substituting the expressions from Eq. (27) into Eqs. (24a)–(24d), the implicit solution of the problem emerges from the following expression. A pivotal aspect of our investigation lies in the dual exploration of both the bending and free vibration behavior of the FG plate. Notably, sections denoted as (*a* and *b*) delineate the solutions to the bending and free vibration problems, respectively.

$$([K] - \omega^2[M])[\Delta] = [F] \tag{28}$$

3.2. *Artificial neural network method (ANN)*

An artificial neural network (ANN) is a computational model inspired by the human brain's neural structure. It consists of interconnected neurons organized into layers. Each neuron in the ANN receives one or more input values $\{x_1, x_2,..., x_n\}$ from the previous layer, each associated with a weight $\{W_1, W_2,..., W_n\}$. The neuron then computes a weighted sum, adds a bias b, and applies an activation function $g(.)$. This activation function, such as sigmoid, tanh, ReLU, or leaky ReLU, introduces non-linearity into the model [8].

Through these interconnected layers and activation functions, ANNs process input data to produce meaningful output. In our study, a supervised learning approach is employed where known outputs correspond to given inputs. This architecture allows ANNs to effectively learn and map complex relationships between inputs and outputs, making them powerful tools for various computational tasks.

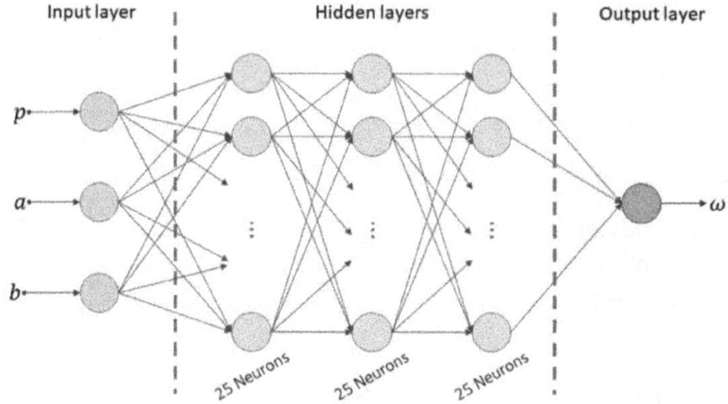

Fig. 2. ANN model architecture.

4. Results and Discussions

The numerical study focuses on a square FGM plate composed of ceramic (Alumina) and metal (Aluminum). The plate's composition transitions from Aluminum at the bottom to Alumina at the top, with non-homogeneous elastic properties varying through its thickness according to a power law. The dimensions of the FGM plate in the *x*, *y*, and *z* directions are denoted as *a*, *b*, and *h*, respectively. The Young's modulus of the metal (E_m) is 70 GPa, and that of the ceramic (E_c) is 380 GPa. A constant Poisson's ratio is assumed, with $v = 0.3$. All results obtained from both analytical methods and Artificial Neural Network (ANN) models are compared with existing literature and demonstrated using graphics.

In this section, the results of several numerical examples for different (*a/h*) ratios are presented. The specific material properties of the FG plate constructed of ceramic (Al_2O_3) and metal (aluminum) in this study are given as follows:

Alumina (Al_2O_3): $E_c = 380$ GPa, $\rho_c = 3960$ kg/m^3, $v = 0.3$ and Aluminum (Al); $E_m = 70$ GPa, $\rho_m = 2702$ kg/m^3, v = 0.3.

The dimensional frequency can be calculated using the following formula [9]:

$$\bar{\omega} = \omega h \sqrt{\frac{\rho_c}{E_c}} \tag{29}$$

Table 1. Dimensionless frequencies ($\overline{w}$) of FG plate for the first mode.

a/h	Theories/Approaches	Gradient index (p)			
		0	1	4	10
	Present	0.2112	0.1757	0.1562	0.1402
5	ANN	0.2116	0.1644	0.1571	0.14085
	Ref [9]	0.2112	0.1628	0.1375	0.1300
	Present	0.05769	0.04799	0.04336	0.03906
10	ANN	0.05755	0.04866	0.04150	0.03883
	Ref [9]	0.05769	0.04417	0.03804	0.03635
	Present	0.01480	0.01231	0.01118	0.01008
20	ANN	0.01416	0.01247	0.01123	0.01031
	Ref [9]	0.01480	0.01130	0.00980	0.00940

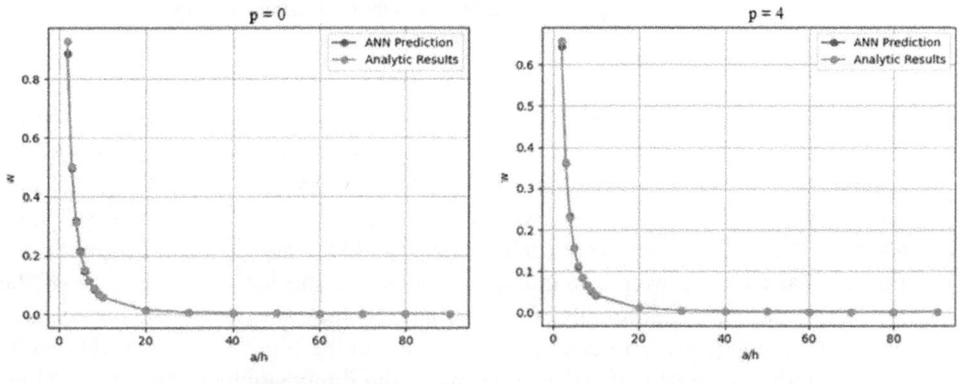

(*a*) (*b*)

Fig. 3. Non-dimensional frequency values obtained ANN and analytical solution for (*a*) p = 0, (*b*) p = 4.

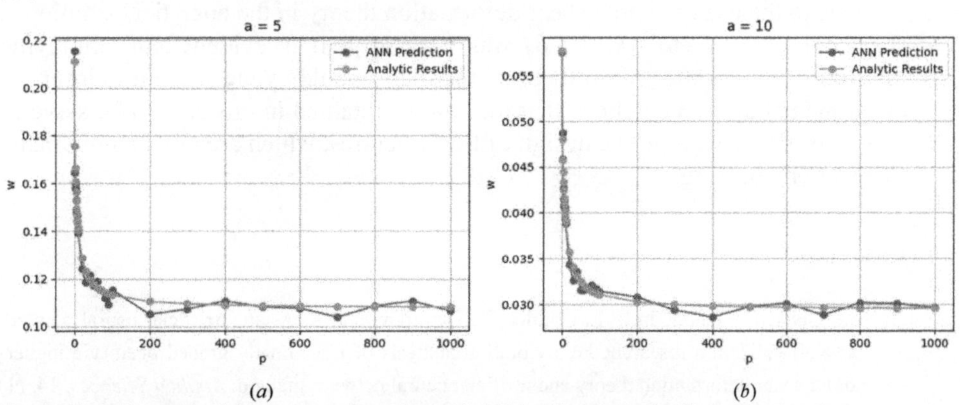

(*a*) (*b*)

Fig. 4. Non-dimensional frequency values obtained ANN and analytical solution for the plate dimension along the x direction for (*a*) a = 5, (*b*) a = 10.

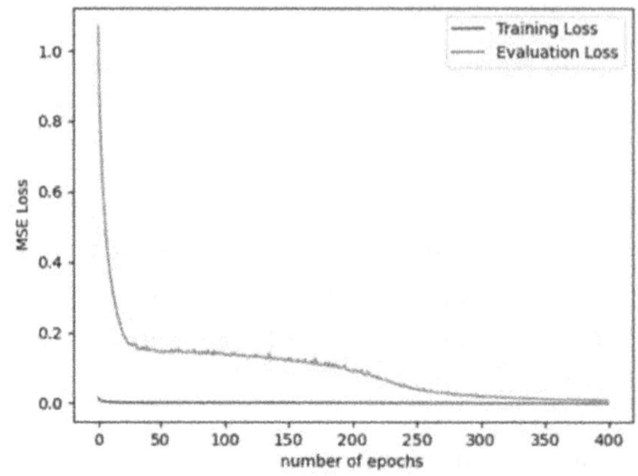

Fig. 5. Training and evaluation losses through training epochs.

5. Conclusions

In this study, the free vibration of a functionally graded (FG) plate is analyzed using simple shear deformation theory and an artificial neural network (ANN) approach. Dimensionless frequency values for various dimensions of square and rectangular supported FG plates were first solved analytically for the first mode, and then these parameters were used to train the ANN model. It can be seen in the figures that as the power law index (p) increases, the natural frequency of the system decreases. Additionally, it was observed that the natural frequency also decreases as the (a/b) ratio decreases as shown in the Figure 4. The ANN results show a high degree of convergence between the dimensionless frequency values obtained from the analytical solution and those estimated by the ANN, as well as with previous studies.

Due to the use of simple shear deformation theory in the analytical solution, a shear correction factor (k_s = 5/6) was necessary. It is evident that employing different higher-order deformation theories would yield results closer to experimental data. We believe that the results obtained in this study will serve as an important resource in the design of FG structures, which are prevalent in many fields of engineering.

References

1. M. Çelik, E. Gündoğdu, E.E. Özdilek, E. Demirkan and R. Artan, Artificial neural network (ANN) validation research: Free vibration analysis of functionally graded beam via higher-order shear deformation theory and artificial neural network method. *Applied Sciences*, 14, (1) 217 (2024).

2. Vaishali, P.K. Karsh, S. Kushari, R.R. Kumar, and S. Dey, Stochastic free vibration and impact responses of functionally graded plates: A support vector machine learning model approach. *Journal of Vibration Engineering & Technologies*, 11, 2927–2943 (2023).
3. J.N. Reddy, Analysis of functionally graded plates. *International Journal for Numerical Methods in Engineering*, *47*, 663–684 (2000).
4. Z.Q. Cheng and R.C. Batra, Deflection relationships between the homogeneous Kirchhoff plate theory and different functionally graded plate theories. *Archives of Mechanics*, 52 (1), 143–158 (2000).
5. S. Pradyumna and J.N. Bandyopadhyay, Free vibration analysis of functionally graded curved panels using a higher-order finite element formulation. *Journal of Sound and Vibration*, 318, 176–192 (2008).
6. Mathematica, (version 14) [Computer Software], Wolfram Research.
7. M. Talha and B.N. Singh, Static response and free vibration analysis of FGM plates using higher order shear deformation theory. *Applied Mathematical Modelling*, 34 (12) 3991–4011 (2010).
8. N. Srivastava, G. Hinton, A. Krizhevsky, I. Sutskever and R. Salakhutdinov, Dropout: A simple way to prevent neural networks from overfitting. *Journal of Machine Learning Research*, 15, 1929–1958 (2014).
9. A. Benachour, H.D. Tahar, H.A. Atmane, A. Tounsi and M.A. Ahmed, A four variable refined plate theory for free vibrations of functionally graded plates with arbitrary gradient. *Composites Part B: Engineering*, 42, (6) 1386–1394 (2011).

Chapter 26

Wind Energy Forecasting Using Various Machine Learning Techniques

Ayça Zenginoğlu[*,‡] and Sakhi Mohammad Hamidy[†,§]

[*]*Technology Transfer Office, Aksaray University*
Aksaray, 68100, Türkiye
[†]*Institute of Natural and Applied Sciences, Department of Industrial Engineering,*
Çukurova University, Adana, 01130, Türkiye
[‡]*ayca.zenginoglu@aksaray.edu.tr*
[§]*sakhimh@gmail.com*

The rapid increase in global energy demand is increasing the use of fossil fuels, leading to greenhouse gas emissions and thus climate change and environmental degradation. This situation increases the need to turn to environmentally friendly and renewable energy sources day by day. In this context, in this study, wind energy generation forecasting models covering the years 2019–2024 for Aksaray province and nine districts have been developed. The unpredictable nature of wind energy increases the importance of short-term forecasts and plays a vital role in making strategic decisions in energy markets. ANFIS (Adaptive Network Based Fuzzy Inference System) and Artificial Neural Networks (ANN), two machine learning techniques, are used in the study. Both models made predictions using the power outputs of different wind turbines and high accuracy rates were obtained. The findings show that Agacoren district has the highest wind energy potential and the Siemens-Gamesa turbine is the most efficient. These results provide important contributions to decision-making processes for more effective and efficient use of wind energy and suggest that factors such as cost, installation time, and production cost should also be considered in the future.

1. Introduction

Energy is one of the most important elements of economic and social development, determining the level of development of countries. Depending on technological progress, countries' need for electrical energy is constantly increasing. In recent years, it has gained great importance to produce energy with the most reliable, cheapest and highest-quality methods.

Waste of fossil energy resources and localized pollution are important factors that threaten public health. Therefore, reducing greenhouse gas emissions is as necessary as energy production in daily life. Governments should increase public awareness about the sharp rise in fuel-based electricity generation and implement effective measures to address primary sources of atmospheric pollution, such as refrigerant chemicals and aerosols. Renewable energy sources offer new opportunities to diversify energy supply and reduce air pollution. Wind energy is the most cost-effective and cleanest form of renewable energy [1].

Recently, the depletion of fossil fuels such as coal and oil and the overdependence of traditional industry on these fuels have led to serious environmental pollution problems and increased the demand for reducing carbon dioxide and other greenhouse gas emissions. Therefore, the use of new energy sources has started to be seen as a solution to the global energy crisis and climate change problems. Renewable energy sources such as solar, wind, and geothermal have attracted a great deal of attention in the 21st century. Especially wind energy, which is an abundant, renewable, widespread, and clean resource, stands out as an important option to support sustainable development worldwide [2].

Recent years have witnessed a notable increase in renewable energy generation both globally and in Türkiye. As illustrated in Figure 1, hydraulic energy has been the dominant source of modern renewable energy production worldwide for many years, while wind and solar energy have been rising rapidly. Wind energy has shown a significant increase in recent years, while solar energy has become the fastest-growing source. In Türkiye, the share of wind and solar energy has been increasing despite the fluctuating production of hydraulic energy. These data reveal that diversification in the field of renewable energy is accelerating both globally and regionally, and new sources are coming to the fore [3].

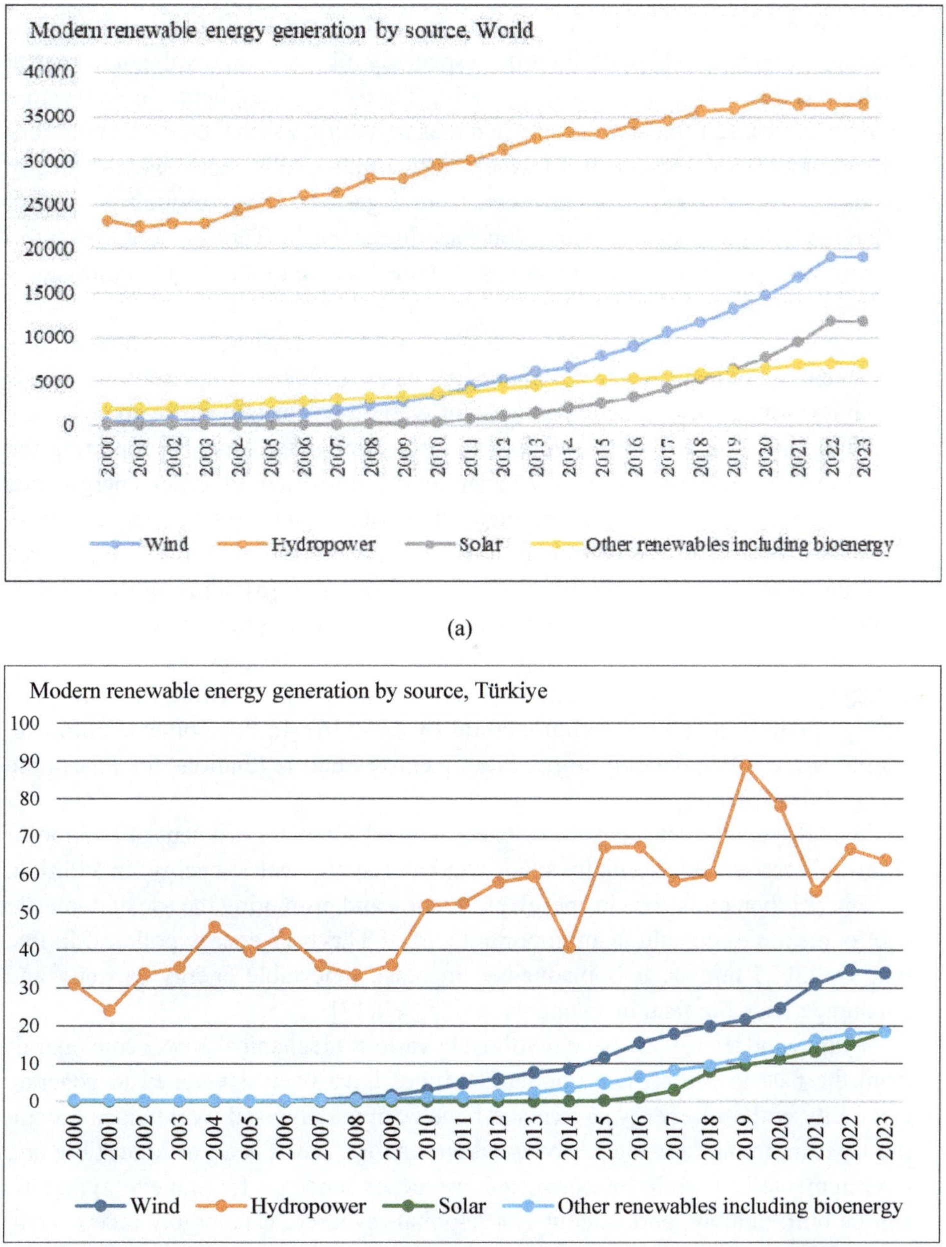

Fig.1. Comparisons of modern renewable energy production by source (a) Modern renewable energy production in the world (b) Modern renewable energy production in Türkiye [3]

The rise of renewable energy sources has influenced fossil fuel consumption. As seen in Figure 2, while fossil fuels—especially oil—remain dominant in energy production both in Türkiye and globally, Türkiye's investments in renewable energy have caused fluctuations and declines in natural gas and coal consumption. When comparing global and Türkiye's fossil fuel trends, both show a steady increase in oil consumption. However, natural gas and coal consumption shows differences. Natural gas consumption has fluctuated in Türkiye. But increased steadily worldwide, while coal use has remained stable globally but significantly declined in Türkiye [4].

In this regard, understanding the regulatory frameworks and international agreements guiding renewable energy policies becomes important to assess Türkiye's strategic direction and alignment with global sustainability goals.

Important regulatory frameworks and international agreements supporting the shift towards renewable energy accelerate the transition to clean energy. For example, the Paris Climate Agreement encourages countries to reduce carbon emissions and increase investments in renewable energy, aiming to keep global warming below 2 °C and limit it to 1.5 °C if possible [5]. This agreement has contributed to many countries prioritizing renewable energy projects.

Furthermore, the European Green Deal aims to increase the use of renewable energy sources and reduce fossil fuel consumption, while setting the European Union's goal of becoming carbon-neutral by 2050 [6]. In this context, European countries are implementing large-scale incentives and regulations for renewable energy infrastructure.

In Türkiye, concrete steps have been planned for more efficient utilization of renewable resources, especially solar, wind and hydroelectric energy. In addition, reducing carbon emissions in energy production and promoting the use of domestic energy resources constitute an important part of Türkiye's energy policies. In line with the 2035 targets, it is planned to increase renewable energy capacity and encourage technological investments in this field [7].

While wind energy has been involved in various mechanical power conversions from the past to the present, modern turbines have been developed to generate electricity with this energy source and have been widely used worldwide. Among the most important advantages of wind energy are that it is an inexhaustible and environmentally friendly resource, reduces dependence on foreign energy, plants can be built quickly, and since it is a regional resource, it is largely independent from global markets [1, 2].

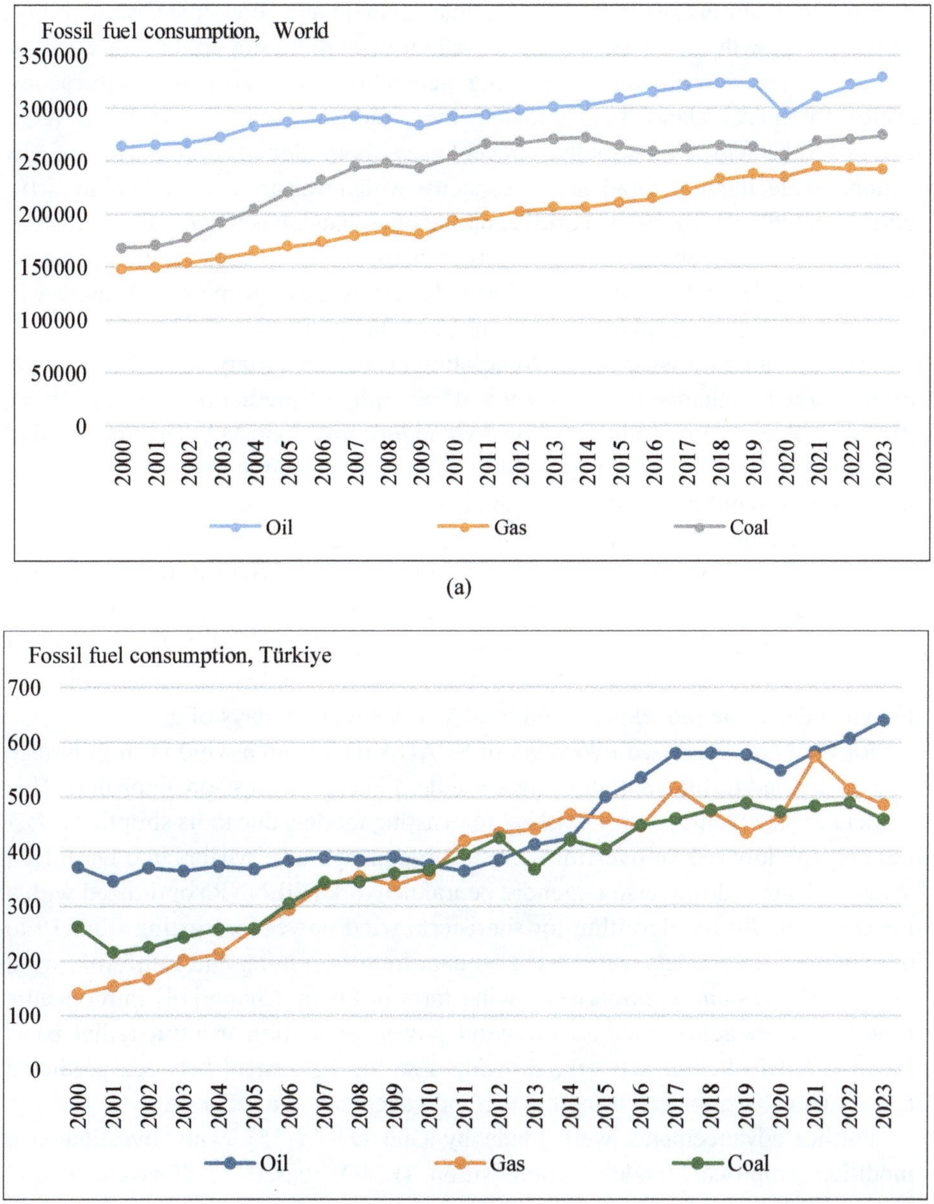

(b)

Fig. 2. Comparison of fossil fuel consumption by source (a) fossil fuel consumption in the world (b) fossil fuel consumption in Türkiye [4]

Wind energy has become an increasingly important area among renewable energy sources. In this context, forecasting wind energy generation is a critical research topic for energy planning, grid management and economic sustainability. This rapid growth in wind energy has made it even more important to accurately forecast energy production. Increasing generation capacity makes forecasting critical for energy planning, grid management and economic sustainability. Wind energy has been growing rapidly in recent years, both globally and in Türkiye. On a global scale, the total wind energy capacity, which was around 200 GW in 2010, will reach 900 GW by 2023. Türkiye, on the other hand, has taken important steps in the field of wind energy, especially after 2010, and has a capacity exceeding 12 GW by 2023 [8]. This growth has been shaped because of increased interest in renewable energy and environmental sustainability goals.

The literature on wind power forecasting presents a variety of methodologies, each aiming to enhance the accuracy and reliability of predictions. Qureshi *et al.* [9] used Python Programming for two algorithms, ARIMA (Autoregressive integrated moving average) and GRU (Gated recurrent unit), to analyze the prediction of wind power using some historical data obtained from wind turbine data. Delgado and Fahim [10] analyzed SCADA data using the Long Short-Term Memory (LSTM) model to predict wind power. LSTM provided the lowest error rate compared to Moving Average Techniques (MA) and Multilayer Perceptrons and best modeled the dynamics of wind power. Shabbir *et al.* [11] SVM-based machine learning is used for short-term forecasting of wind power generation with Estonian data. The model is trained in MATLAB with 30 days of data.

Rashid *et al.* [12] used two years of SCADA data from a wind farm in France to predict wind turbine output using a Random Forest Regression algorithm. This model has an advantage over existing forecasting models due to its simplicity, fast training and low risk of overfitting. In a subsequent study, Mishra and Dash [13] suggested a pseudo-inverse Legendre neural network (PILNNR) optimized with a metaheuristic firefly algorithm for short-term wind power forecasting from 10 to 60 minutes. The models were tested on data from Wyoming and California wind farms in the US and the Sotavento wind farm in Spain. Chang [14] introduced a time series prediction method for wind power generation using a radial basis function (RBF) neural network, showing a strong agreement between predicted and actual values, underscoring the method's accuracy and reliability.

Further advancements were made by Guo *et al.* [15], who investigated a modified empirical mode decomposition (EMD) based feedforward neural network (FNN) for wind forecasting. Their method outperformed basic FNN and unmodified EMD-based FNN in multi-step forecasts of monthly and daily wind speeds in Zhangye, China. Yang *et al.* [16] contributed to the field by proposing an ANFIS method for interpolating missing and invalid wind data. Tests using data from a wind farm in North China confirmed the effectiveness of this approach.

In addition, Zhou *et al.* [17] conducted a systematic study on fine-tuning least-squares support vector machines (LSSVM) for one step ahead of wind speed forecasting, demonstrating that LSSVM methods generally outperformed traditional persistence models. Shi *et al.* [18] explored hybrid models, ARIMA-ANN and ARIMA-SVM, for wind speed and power forecasting, highlighting the viability of these approaches for various forecasting time horizons, although they did not always provide superior performance.

Guo *et al.* [19] developed a hybrid wind speed forecasting method that integrated back propagation neural networks with seasonal exponential adjustment to eliminate seasonal effects, achieving better performance than standalone neural networks. Finally, Catalão *et al.* [20] proposed a hybrid approach combining artificial neural networks with wavelet transform for short-term wind power forecasting in Portugal, showing significant improvement in prediction accuracy through the decomposition of wind power series into more manageable components. Yıldırım and Yavuz [21] proposed the efficiency of wind energy investments using artificial neural networks (ANN). Trained with 36 months of wind speed data, the model made very accurate predictions, and the average prediction error was calculated as 0.48%.

All these above-mentioned studies highlighted the importance of forecasting algorithms for wind power generation.

The remainder of this paper is organized in three sections. The second two presents the dataset and forecasting methodology. Section three presents the findings from the proposed model. Finally, the fourth section presents the conclusions.

2. Material and Methods

2.1. *Material*

In this study, ANFIS and Artificial Neural Network (ANN) models are developed to estimate the energy that could be produced by using various wind turbines. The prediction models are developed in MATLAB software package. For this purpose of study, two data sets are used.

The daily wind data for nine districts of Aksaray provincial is obtained from the Directorate General of Meteorology [22]. Unlike other studies, the data set covers daily wind speed of the last five years (01.01.2019–31.03.2024). The data set includes the daily average wind speed for each of the nine districts of Aksaray in (m/s).

There are several types of turbines used in energy production. 10 different types of turbines which are widely preferred in wind energy production field are select alternatives for the purpose of the study [23]. These turbines are Nordex N117, Vestas V112, Enercon E82, GE Energy 2,75/120, Siemens-Gamesa G132, Sinovel SL 3000, Suzlon S88, Neg Micon NM92, Senvion 3.2 M112 and Alstom Power 112. The second dataset which includes the amount of power (kW) produced by a turbine within a specific wind speed (m/s) is taken from the official web sites of the turbines [24].

2.2. *Adaptive network based fuzzy inference system (ANFIS)*

ANFIS (Adaptive Neuro-Fuzzy Inference System) is an adaptive multilayer feed-forward network used in nonlinear forecasting, where historical data points are utilized to predict future values. ANFIS combines the learning capabilities of neural networks with the linguistic representation of fuzzy inference systems [25]. The structure of the ANFIS network consists of five layers, each containing nodes defined by specific node functions. It combines fuzzy theory and artificial neural networks, so it performs better than other neural networks in some computational cases [25]. However, it also has some obvious disadvantages like non-convergence and over-fitting phenomenon. The standard modeling steps of the ANFIS network can be found in detail in [16], [20], and [25].

2.2.1. *Modeling with ANFIS*

An ANFIS prediction model is built to predict the amount of energy produced by any type of turbine. The input for this model is the wind speed and the output is the amount of energy produced. For each type of turbine, models of ANFIS and ANN are built. To build an ANFIS model parameters such as Membership functions, Number of membership functions and Maximum number of training epochs should be determined.

To determine the best values for these parameters, combinations of various options are assigned and the combination with the best result is set as final parameters. Table 1 describes the options for each of the parameters. For each type of turbine 504 ($8 \times 21 \times 3$) ANFIS model has been built.

Table 1. Parameters of ANFIS models

Membership functions	(1) Trimf (2) Trapmf (3) Gbellmf (4) Gaussmf (5) Gauss2mf (6) Pimf (7) Dsigmf (8) Psigmf
Number of membership functions	(1) 5 (2) 6 (3) 7 (4) 8 …. (21) 25
Maximum number of training epochs	(1) 50 (2) 100 (3) 150

After determining the best parameters for these models are used to predict the total energy that could be produced in time arrival of 2019–2024 for each of nine districts of Aksaray.

2.3. *Artificial neural networks (ANN)*

ANNs have become an increasingly widely used field by researchers in all fields, including engineering and medicine. ANNs are useful in the design of adaptive and predictive systems and work on input-output interaction. By varying the synapses/weights of the hidden layer, a neural network tries to achieve targeted outputs based on what the network has learned from its previous history of input-output interaction [26, 27].

A neural network can be trained just like the human brain; it learns from previous interactions. It can be used to solve problems that are considered too complex to solve analytically. A neuron is the basic element of an ANN, and this structure consists of an input layer, a hidden layer and an output layer [28].

A single neural cell can be expressed by the following mathematical formula:

$$a = f(w \times P + b) \tag{1}$$

The weight and bias (w and b) of each neuron in a neural network are adjustable scalar parameters of the neurons. Neural network models can represent complex nonlinear relationships and extract dependencies between variables through the training process [27]. Studies have shown that three-layer feed-forward neural networks are particularly suitable for modeling unpredictable complexities, using sigmoid functions in the hidden layer and linear functions in the output layer. This structure is considered a universal mapper if there are enough units in the hidden layer [28].

2.3.1. *Modeling with ANN*

To build an ANN model parameter such as training function, number of hidden layers and performance measure (error type) should be determined.

To determine the best values for these parameters, combinations of various options are assigned and the combination with the best result is set as final parameters. Table 2 describes the options for each of the parameters.

Table 2. Parameters of ANN models

Training function	(1) Traingdx (2) Trainlm (3) Trainbr (4) Trainsc
Number of hidden layers	(1) 1 (2) 2 (3) 3 (4) 4 .... (10) 10
Performance measure	(1) MAE (2) MSE (3) SAE (4) SSE (5) Cross entropy

For each type of turbine 200 (4 × 10 × 5) ANN model has been built. Since the division dataset to test and training data contains randomness, each model has been run 10 times and the average result of all 10 replications of the model is assigned as the final result of the model. After determining the best parameters for these models are used to predict the total energy that could be produced in time arrival of 2019–2024 for each of nine districts of Aksaray.

3. Results

3.1. *ANFIS model results*

For each type of turbine separately all the related ANFIS models with different parameter sets are compared with each other and the one with the best adjusted regression value of test data and over all data is selected as the final predictor model. Table 3 shows the best models for each type of turbine.

By using the best prediction ANFIS models the total energy that could be produced by the turbines in nine districts of Aksaray is predicted and Table 4 shows the result of this prediction. As a result of this prediction the rank of the districts from the best to worst is as: Agacoren, Hasandagi, Eskil, Sultanhani, Gulagac, Ortakoy, Sariyahsi, Guzelyurt and Aksaray.

3.2. *ANN model results*

For each type of turbine separately all the related ANN models with different parameter sets are compared with each other and the one with the best adjusted regression value of test data and overall data is selected as the final predictor model. In addition, the training function used for all turbine models was 'Trainbr', which gave the best performance for each model. Table 5 shows the best models for each type of turbine.

Table 3. Best models for ANFIS

Turbine	Nordex	Vestas	Enercon	GE Energy	Si. Gamesa	Sinovel	Suzlon	Neg Micon	Senvion	Als. Power
Membership Fun	Gaussmf	Gbellmf	Gaussmf	Gbellmf	Gaussmf	Trimf	Gbellmf	Trimf	Trimf	Gaussmf
Number of MFs	9	25	8	8	7	12	9	24	13	9
Max.# of Epoch	50	150	50	100	100	150	100	150	100	50
Test dt, adj. Reg.	0.999982	0.999985	0.999890	0.999991	0.999960	0.999985	0.999991	0.999986	0.999964	0.999966
Overall adj. Reg.	0.999990	0.999994	0.999959	0.999996	0.999975	0.999988	0.999997	0.999988	0.999990	0.999983

Table 4. Power production predictions with ANFIS models (a) Power production predictions with ANFIS models for five turbines (b) Power production predictions with ANFIS models for other five turbines

(a)

	Nordex	Vestas	Enercon	GE Energy	Si. Gamesa
Agacoren	1459794.14	906130.61	790450.04	1421228.59	2880584.50
Aksaray	2217.57	3226.18	8230.97	9500.07	142562.05
Hasandagi	1079746.01	695365.16	594875.99	1022658.54	2149156.63
Ortakoy	238103.71	132825.22	145471.30	263289.18	1084742.44
Sultanhani	441094.12	255150.50	249278.79	472335.21	1479311.33
Eskil	671873.92	386969.85	369428.20	707209.88	2002265.12
Gulagac	284369.18	159162.58	166326.95	319952.25	1211823.69
Guzelyurt	30735.05	22058.94	29805.44	48715.90	347885.22
Sariyahsi	134644.45	74458.38	89823.00	160902.67	830104.43

Table 4. *Continued*

(b)

	Sinovel	Suzlon	Neg Micon	Senvion	Als. Power
Agacoren	1137308.30	707374.81	942539.63	1463203.47	1424442.09
Aksaray	7034.67	1384.91	37694.97	20757.41	10534.71
Hasandagi	854204.11	527115.36	740946.98	1058976.54	1022371.57
Ortakoy	193831.62	86986.19	217280.65	255483.12	269497.77
Sultanhani	348926.45	174604.93	325252.88	462532.51	479693.01
Eskil	524631.93	269608.33	455150.08	688490.05	715724.35
Gulagac	228628.49	96797.23	232210.50	307841.05	327368.82
Guzelyurt	35321.74	11702.98	73175.00	34844.51	51509.41
Sariyahsi	115788.48	42888.31	153803.55	153831.22	166767.83

Table 5. Best models for ANN

Turbine	Nordex	Vestas	Enercon	GE Energy	Si. Gamesa	Sinovel	Suzlon	Neg Micon	Senvion	Als. Power
Train function	Trainbr	Trainbr	Trainbr	Trainbr	Trainbr	Trainbr	Trainbr	Trainbr	Trainbr	Trainbr
# of train fun.	6	9	10	10	9	8	9	9	10	5
Perf, measure	MSE	SSE	MSE	MSE	MSE	MSE	SSE	MSE	MSE	SSE
Test dt, adj. Reg.	0.999986	0.999982	0.999969	0.999997	0.999966	0.999980	0.999997	0.999977	0.999983	0.999987
Overall adj. Reg.	0.999991	0.999967	0.999926	0.999995	0.999909	0.999978	0.999988	0.999956	0.999942	0.999975

By using the best prediction ANN models, the total energy that could be produced by the turbines in nine districts of Aksaray is predicted and Table 6 shows the result of this prediction. As a result of this prediction the rank of the districts from the best to worst is as: Agacoren, Hasandagi, Eskil, Sultanhani, Gulagac, Ortakoy, Sariyahsi, Guzelyurt and Aksaray.

Table 6. Power production predictions with ANN models (a) Power production predictions with ANN models for five turbines (b) Power production predictions with ANN models for other five turbines

(a)

	Nordex	Vestas	Enercon	GE Energy	Si. Gamesa
Agacoren	1459794.14	906130.61	790450.04	1421228.59	2880584.50
Aksaray	2217.57	3226.18	8230.97	9500.07	142562.05
Hasandagi	1079746.01	695365.16	594875.99	1022658.54	2149156.63
Ortakoy	238103.71	132825.22	145471.30	263289.18	1084742.44
Sultanhani	441094.12	255150.50	249278.79	472335.21	1479311.33
Eskil	671873.92	386969.85	369428.20	707209.88	2002265.12
Gulagac	284369.18	159162.58	166326.95	319952.25	1211823.69
Guzelyurt	30735.05	22058.94	29805.44	48715.90	347885.22
Sariyahsi	134644.45	74458.38	89823.00	160902.67	830104.43

(b)

	Sinovel	Suzlon	Neg Micon	Senvion	Als. Power
Agacoren	1137308.30	707374.81	942539.63	1463203.47	1424442.09
Aksaray	7034.67	1384.91	37694.97	20757.41	10534.71
Hasandagi	854204.11	527115.36	740946.98	1058976.54	1022371.57
Ortakoy	193831.62	86986.19	217280.65	255483.12	269497.77
Sultanhani	348926.45	174604.93	325252.88	462532.51	479693.01
Eskil	524631.93	269608.33	455150.08	688490.05	715724.35
Gulagac	228628.49	96797.23	232210.50	307841.05	327368.82
Guzelyurt	35321.74	11702.98	73175.00	34844.51	51509.41
Sariyahsi	115788.48	42888.31	153803.55	153831.22	166767.83

3.3. *Turbine selection*

Considering the total energy prediction tables (both ANFIS and ANN) the rank from the best to worst turbine is given in Table 7. According to this Table Siemens-Gamesa is the best option while Suzlon is the worst option to be used for the wind energy production in Aksaray city.

Table 7. Final ranking of turbines

Rank	Turbine
1	Siemens-Gamesa
2	Alstom Power
3	GE Energy
4	Senvion
5	Nordex
6	Neg Micon
7	Sinovel
8	Enercon
9	Vestas
10	Suzlon

4. Conclusion

The forecast for energy generation is of extreme importance as it can have a significant effect on cost effectiveness and better demand-side management. The main objective of this study is to develop prediction models for ten different wind turbines and apply forecasting methods to predict the electricity generation from wind energy in Aksaray province with the lowest margin of error. In this study, speed values provided by turbine manufacturers and turbine power data are used as datasets. The developed models are trained, and electrical energy production prediction was made. Examining the modeling results shows that the predictions provided by the ANFIS method are as good as the ANN method. The analysis shows that Agacoren is the best district of Aksaray province for wind production. Also, Siemens-Gamesa would produce much more energy than other turbines. For further studies in this case the other important factors such as installation cost, installation time and production cost which are also effective in selection of turbines can be considered.

References

1. I. Mert, C. Karakuş and F. Üneş, Estimating the energy production of the wind turbine using artificial neural network, *Neural Computing and Applications,* 27, 1231–1244 (2016).
2. Q. Hu, Y. Wang, Z. Xie, P. Zhu and D. Yu, On estimating uncertainty of wind energy with mixture of distributions, *Energy,* 112, 935–962 (2016).
3. Modern renewable energy generation by source, https://www.energyinst.org/statistical-review (Accessed 23 October 2024).
4. Fossil fuel consumption, https://www.energyinst.org/statistical-review (Accessed 23 October 2024).
5. Paris climate agreement, https://www.mfa.gov.tr/paris-anlasmasi.tr.mfa (Accessed 24 October 2024).
6. European green deal, https://www.abre.gov.tr/avrupa-yesil-mutabakati_53729.html (Accessed 24 October 2024).
7. Türkiye's 2035 Roadmap for Renewable Energy, https://enerji.gov.tr/duyuru-detay?id=30481 (Accessed 24 October 2024).

8. Wind energy, https://www.irena.org/Energy-Transition/Technology/Wind-energy (Accessed 24 October 2024).
9. S. Qureshi, F. Shaikh, L. Kumar, F. Ali, M. Awais and A. E. Gürel, Short-term forecasting of wind power generation using artificial intelligence, *Environmental Challenges,* 11, 100722 (2023).
10. I. Delgado and M. Fahim, Wind turbine data analysis and LSTM-based prediction in SCADA system, *Energies,* 14(1), 125 (2021).
11. N. Shabbir, R. Ahmadi Ahangar, L. Kütt, M. N. Iqbal and A. Rosin, Forecasting short term wind energy generation using machine learning. In *2019 IEEE 60th International Scientific Conference on Power and Electrical Engineering of Riga Technical University (RTUCON) IEEE*, 1–4 (2019).
12. H. Rashid, W. Haider and C. Batunlu, Forecasting of wind turbine output power using machine learning, In *2020 10th International Conference on Advanced Computer Information Technologies (ACIT) IEEE*, 396–399 (2020).
13. S. P. Mishra and P. K. Dash, Short-term prediction of wind power using a hybrid pseudo-inverse Legendre neural network and adaptive firefly algorithm, *Neural Computing and Applications,* 31(7), 2243–2268 (2019).
14. W. Y. Chang, Wind energy conversion system power forecasting using radial basis function neural network, *Applied Mechanics and Materials*, 284–287: 1067–1071 (2013). Doi: 10.4028/www.scientific.net/AMM.284-287.1067
15. Z. H. Guo, W. G. Zhao, H. Y. Lu and J. Z. Wang, Multi-step forecasting for wind speed using a modified emd-based artificial neural network model, *Renewable Energy,* 37, 241–249 (2012).
16. Z. L. Yang, Y.Q. Liu and C. R. Li, Interpolation of missing wind data based on ANFIS, Renewable Energy, 36, 993–998 (2011).
17. J. Y. Zhou, J. Shi and G. Li, Fine tuning support vector machines for short-term wind speed forecasting, *Energy Conversion and Management* 52, 1990–1998 (2011).
18. J. Shi, J. M. Guo and S. T. Zheng, Evaluation of hybrid forecasting approaches for wind speed and power generation time series, *Renewable and Sustainable Energy Reviews,* 16, 3471–3480 (2012).
19. Z. H. Guo, J. Wu, H. Y. Lu, and J. Z. Wang, A case study on a hybrid wind speed forecasting method using BP neural network, *Knowledge-Based Systems,* 24, 1048–1056 (2011).
20. J. P. S. Catalao, H. M. I. Pousinho and V. M. F. Mendes, Hybrid wavelet-PSO-ANFIS approach for short-term wind power forecasting in Portugal, *IEEE Transactions on Sustainable Energy,* 2(1), 50–59 (2011).
21. Yildirim, H. H., and Yavuz, M., Evaluation of wind energy investment with artificial neural networks. *An International Journal of Optimization and Control: Theories & Applications (IJOCTA)*, 9(2), 142–147 (2019).
22. Daily wind speed data, https://www.mgm.gov.tr (Accessed 18 April 2024).
23. Wind turbines database, https://www.enerjiatlasi.com/ruzgar-turbini, (Accessed 18 April 2024).
24. Wind turbines manufacturer, https://en.wind-turbine-models.com/turbines, (Accessed 18 April 2024).
25. J. S. R. Jang, ANFIS adaptive-network-based fuzzy inference system, *IEEE Transactions on Systems, Man, And Cybernetics*, 23(3), 665–685 (1993).

26. T. Hill, L. Marquez, M. O'Connor and W. Remus, Artificial neural network models for forecasting and decision making, *Int J Forecast,* 10, 5–15 (1994).
27. A. K. Jain and J. Mao, Artificial neural network: a tutorial, *Communications,* 29, 31–44 (1996).
28. M. Jamil and M. Zeeshan, A comparative analysis of ANN and chaotic approach-based wind speed prediction in India, *Neural Computing and Applications,* 31, 6807–6819 (2019).

Chapter 27

Transient Performance Improvement of Induction Motor Drive Systems Using PSO Optimised Interval Type-2 Fuzzy PID Controller

Fuat Kılıç

Department of Electrical and Electronics Engineering,
Balıkesir University, Balıkesir, Turkey
fuatkilic@balikesir.edu.tr

This study investigates interval type-2 fuzzy PID (IT2-FPID) method, whose parameters are optimized using the particle swarm optimization (PSO) metaheuristic algorithm, to improve the dynamic transient and steady state response in the speed control of a three-phase induction motor. The efficient implementation of variable speed motor applications is crucial nowadays, since they have become indispensable in the industrial sector. Induction motors are one of the prime candidates for such applications due to their simple structure, long life, high efficiency, low maintenance and relatively low cost. Induction motors, on the other hand, have a nonlinear, tightly coupled model with uncertainty. Indirect field-oriented vector control approach, which allows torque and flux components of the motor to be regulated independently, is utilized to take advantage of the properties of the induction motor and overcome the coupling problem. Fuzzy logic (FL) based controllers do not require knowledge of motor parameters and at the same time have a robust behavior in controlling against disturbances and uncertainties. Nevertheless, type 2 fuzzy logic controllers are recognized to provide better results in the presence of disturbances, uncertainties and oscillations compared with type-1 FL algorithm. In this study, the input and output parameters of IT2-FPID control algorithm are optimised to obtain the best control response in addition to these properties. Simulation of the speed control of induction motor is carried out under sudden changes of speed and load torque, which require fast and robust dynamic response. The acquired simulation results are compared to the performances of conventional PI control, FL control, and IT2-FPID with optimised parameters controller. The results reveal that IT2-FPID controller with optimised parameters has superior speed control performance, with a shorter rise time, reduced overshoot, and faster acceleration in speed response.

1. Introduction

The asynchronous AC motor, also known as an induction motor (IM), is particularly well-represented by the squirrel cage-type, which is widely used due to its cost-effectiveness. This type of motor has gained popularity over the past few decades, especially in sensor-less IM drive applications, thanks to its many advantages, including mechanical durability, low maintenance, and straightforward construction [1]. These motors are commonly employed in various applications, such as fans, pumps, machine tools, robotics, hybrid electric vehicles, household appliances, heating and cooling systems, rolling mills, wind power, and more [2]. With the advancement of vector control techniques, the deployment of induction motors in variable speed drive systems has expanded across numerous industries. The principle behind torque generation in an IM is similar to that of a DC machine [3].

To ensure high-performance drive systems that remain unaffected by load torque disturbance effects, model uncertainties and parameter variations, the motor's speed must closely follow a predefined reference trajectory [4]. Achieving this level of performance typically involves using vector control for the IM drive. However, the control algorithm plays a critical role in determining the system's overall efficiencies. Variations in motor parameters can significantly affect the separation characteristics of vector-controlled IMs [5].

Proportional–Integral–Derivative (PID) control is the most prevalent control method in industrial motion control applications [6]. However, traditional PID control encounters several significant practical issues, including integrator windup, high sensitivity to load disturbances, and parameter variations.

For more than three decades, the synchronous reference frame PI controller has been widely employed controller [7]. Nevertheless, it introduces the cross-coupling effect between the direct and the quadrature axis. The converters switching frequency is consistently restricted by losses, particularly in high-power drives resulting in a substantial the control system's delay [8–9]. Conventional PI controllers are incapable of eliminating the delay effect, which results in intense cross-coupling effects and a decrease in the overall system's performance.

To overcome these challenges, a range of advanced control techniques has been introduced to improve plant robustness. It is imperative to employ sophisticated nonlinear control techniques for regulating nonlinear systems, such as induction motors. Various methods such as fuzzy logic control, adaptive control, robust control, model predictive control (MPC), neural networks, and sliding mode control (SMC) have been created and studied in the literature [10].

Fuzzy Logic Controllers (FLCs) have been introduced for motor control applications. The foundational concepts of fuzzy sets, introduced by L. A., inspired further developments by researchers such as Mamdani [11], Takagi, Sugeno [12], and others in the fuzzy control modeling field. Fuzzy logic controllers have proven successful in managing a variety of complex and nonlinear processes, including water purification, automated train operation, container crane operation, nuclear reactor control, and in fuzzy hardware and memory devices [13]. An inherent benefit of fuzzy logic control, in comparison to conventional control approaches, is its independence from the need for mathematical modeling in controller design. Fuzzy logic has proven to be a successful approach in controlling systems that are either poorly comprehended or extremely intricate, where accurate modeling is difficult or unattainable. Fuzzy logic provides an alternative to conventional feedback control approaches in motion control systems. Research has demonstrated that implementing nonlinear speed control methods, such as fuzzy control, can improve the dynamic performance of electric drives and increase their resistance to changes in parameters [14]. FLCs rely largely on human logic models, necessitating the use of sense, skilled expertise, and technical decision in their design. FLCs are categorized based on their input and output factors. For example, when factors like error, variation in error, and error summing are utilized individually or in mixture, FLCs transform into Fuzzy Proportional, Fuzzy Integral, and other variations [15].

In this study, IT2-FPID is being designed as an alternative to the traditional proportional-integral (PI) controller and conventional FLC. Fuzzy logic control provides the benefit of being robust and having a relatively uncomplicated approach as it does not necessitate a comprehensive model of the system. The performance of IT2-FPID controller has been compared with that of the PI and FLC controllers. The subsequent sections of this study are structured as follows: the second section elucidates the model of the asynchronous motor and vector control, the third section examines fuzzy logic controllers and the IT2-FPID controller, and the fourth and ultimate section showcases the simulation results and ensuing discussions.

2. Induction Motor Model and Vector Control

The mathematical description of an induction motor can be expressed in three distinct reference frames: The stationary reference frame ($\alpha\beta$), or the rotating (dq) reference frame. The following formulas represent the mathematical model of asynchronous machines in the (dq) synchronous reference frame. The formulas for flux linkage are provided as:

$$v_{qs} = R_s i_{qs} + \frac{d\psi_{qs}}{dt} \tag{1}$$

$$v_{ds} = R_s i_{ds} + \frac{d\psi_{ds}}{dt} \tag{2}$$

$$0 = R_r i_{qr} + \frac{d\psi_{qr}}{dt} - \omega_r \psi_{dr} \tag{3}$$

$$0 = R_r i_{dr} + \frac{d\psi_{dr}}{dt} - \omega_r \psi_{qr} \tag{4}$$

The flux linkage expressions are written in terms of current and inductance.

$$\psi_{qs} = L_{ls} i_{qs} + L_{mm}(i_{qs} + i_{qr}) \tag{5}$$

$$\psi_{qr} = L_{lr} i_{qr} + L_{mm}(i_{qs} + i_{qr}) \tag{6}$$

$$\psi_{ds} = L_{ls} i_{ds} + L_{mm}(i_{ds} + i_{dr}) \tag{7}$$

$$\psi_{dr} = L_{lr} i_{dr} + L_{mm}(i_{ds} + i_{dr}) \tag{8}$$

The expressions for the electromagnetic torque, rotor speed, and rotor electrical speed are listed below:

$$Te = \frac{3}{2}(\frac{P}{2})(\psi_{ds} i_{qs} - \psi_{qs} i_{ds}) \tag{9}$$

$$T_e = J\frac{d\omega_r}{dt} - B\omega + T_m \tag{10}$$

$$\omega_e = \frac{P}{2}\omega_r \tag{11}$$

where *d, q* direct and quadrature axis, *s* stator, *r* rotor, *R* phase resistance, *L* inductance, L_{mm} magnetizing inductance, L_l leakage inductance, ψ flux linkage, *P* poles, ω_r rotor speed, ω_e electrical speed, *J* inertia, *B* friction coefficient, T_e electrical torque, T_m mechanical torque and λ flux is represented respectively [16]. In indirect field-oriented (vector) control, $\lambda_{qr} = d\lambda_{qr}/dt = 0$. Additionally, due to the nature of the machine, $V_{qr} = V_{dr} = 0$. In steady-state operation of the machine, $d\lambda_{dr}/dt = i_{dr} = 0$. Here, it is clear that the rotor voltages are zero, and since vector control only creates a flux along the rotor direct-axis, the flux on the rotor quadrature-axis and its derivative are zero. In steady-state operation, all derivatives in the equations are zero. When voltage equations are rearranged to account for this steady-state condition, the following ω_{sl} slip is obtained.

$$\omega_{sl} = -\frac{R_r i_{qr}}{\lambda_{dr}} \tag{12}$$

Rotor current equation is rearranged as:

$$i_{qr} = -\frac{L_m}{L_r} i_{qs} \tag{13}$$

Slip equations is expanded as follows:

$$\omega_{sl} = \frac{R_r}{L_r}\frac{L_m}{\lambda_{dr}} i_{qs} \tag{14}$$

The slip equality can be written in terms of $\tau_r = L_r / R_r$ the rotor time constant as follows:

$$\omega_{sl} = \frac{L_m}{\tau_r \lambda_{dr}} i_{qs} \tag{15}$$

According to these, the electrical position slip and rotor speed are obtained by integrating them.

$$\theta_e = \int \omega_e dt = \int\left(\frac{P}{2}\omega_r + \omega_{sl}\right)dt \tag{16}$$

Finally, rotor d-axis flux equation is written as:

$$\lambda_{dr} = \left[\frac{1}{1+s\tau_r}\right](L_m i_{ds}) \tag{17}$$

The transfer function above is a first-order low-pass filter. Rotor field-oriented control is implemented using the following equations. The effectiveness of this control system is greatly influenced by motor settings, particularly the rotor time constant [17].

Figure 1 depicts an indirect vector controlled induction motor drive employing three controllers. The outer part of the speed controller detects the difference between the desired and actual speeds of the rotor, and adjusts accordingly. The controller's output produces the reference current i_{qs}. The two inside loop current regulators evaluate the reference values of ids and i_{qs} with their measurements. Both the speed and current regulators can be implemented as either PI controllers or Type-2 Fuzzy-PID controllers.

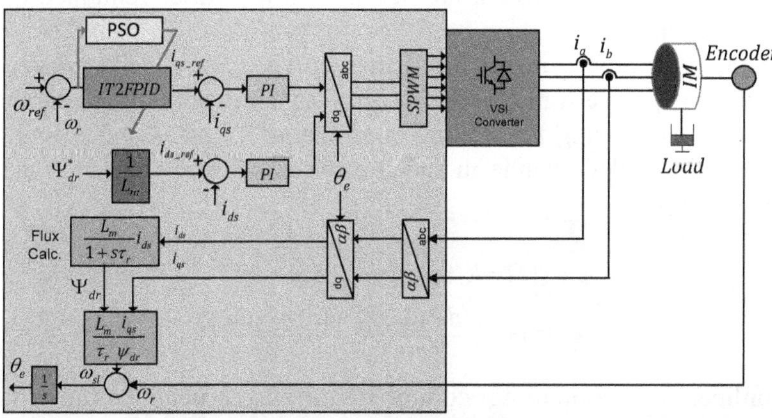

Fig. 1. Block diagram of field oriented control and IT2-FPID control scheme of induction motor control

3. Speed Controller with Optimized Parameters

3.1. *Interval type 2 fuzzy PID controller*

This section provides a comprehensive explanation of the design process for fuzzy rules in the speed regulation loop. This study employs a typical FLCs and IT2-FPID block diagram, as depicted in Figure 2. The controller is divided into three distinct stages: Prior processing, Fuzzy logic controller, and after processing. The FLC arrangement has two input variables to symbolize speed error and the rate of variation in speed error, and utilizes a single output parameter for regulating output.

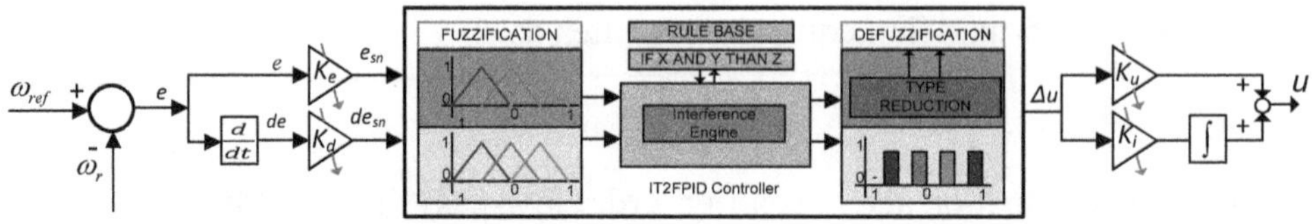

Fig. 2. Block diagram of FLC/IT2-FPID controller

Reference [18] indicate that the performance of the FLC is greatly influenced by three main variables: The Scaling Factor (SF), fuzzy regulations, and membership role. The FLC was originally constructed using computed scaling factors and chosen membership function sets. Figure 3 illustrates the utilization of triangular membership functions (MF) for both the input and output variables. Triangular roles were chosen based on their ease of use and computational effectiveness [19–20]. The MFs had a symmetrical placement with a 50% intersect between neighboring triangles. They were categorized as positive (P), zero (Z), negative (N). During the analysis, the SF and MF quantities remained consistent for both the regular FLC.

During the prior processing phase of the regulator, the speed error (e) and the rate of change of speed error (de) were utilized as the fuzzy input parameters. The inputs were adjusted using the scaling coefficients K_e and K_d. The scaled speed error, e_{sn}, and the scaled change in speed error, de_{sn}, were computed using Eqs. (18) and (19) [21].

$$e_{sn}(t) = K_e(e(t)) = K_e(\omega_{ref} - \omega_r) \tag{18}$$

$$de_{sn}(t) = K_d \frac{de(t)}{T_s} = K_d \frac{e(t) - e(t-1)}{T_s} \tag{19}$$

In a nonlinear induction motor control system, it is crucial for the controller to exhibit robust and stable dynamic performance, especially in situations such as sudden changes in speed reference, the existence of external disruptions such as load torque and variations in variables. Therefore, in this nonlinear system structure, an appropriate controller is needed to achieve a response that meets dynamic and control system criteria and enhances system performance. For this reason, an IT2-FPID controller has been chosen. Type-2 fuzzy sets have more advanced features than Type-1 fuzzy sets in modeling situations arising from uncertainties in linguistic variables. In order to mitigate the impact of unpredictability, Type-2 fuzzy sets have been proposed as an effective method and are widely used in many control systems. There is an interval type-2 fuzzy control structure that reduces the computational load of Type-2 fuzzy regulators. In this control structure, the Takagi-Sugeno-Kang (TSK) inference method is used. As in the traditional FLC controller; the inputs are scaled by the K_e and K_d coefficients and applied to the fuzzification block. The error and error change are given in Eq. (20).

$$e_{sn} = K_e e, \quad e_{sn} \in [-1,1]$$
$$de_{sn} = K_d \dot{e}, \quad de_{sn} \in [-1,1] \tag{20}$$

At the output of the defuzzifier, the control input, u to the system is formed by scaling again with the *Ku* and *Ki* coefficients. The magnitude of the controller output is given in Eq. (21).

$$u = K_u U + K_i \int U dt, \qquad U \in [-1,1] \tag{21}$$

where U represents output crisp value of controller and Δu is control output changes in accordance with sampling time. Table 1 presents the rule table used in this study. The concept of Type-2 fuzzy controllers (T2-FLC) is actually a generalized form of Type-1 fuzzy controllers (T1-FLC). T2 and T1-FLCs use similar mathematical operators. However, it is not as easy to mathematically define T2-FLCs as it is with T1-FLCs.

Table 1. Rule table

e/De	N	S	P
N	NB	NO	S
S	NO	S	PO
P	S	PO	PB

The rule table in Table 1 is established based on the regions above and below the error generated in response to the controller input, taking into account a constant reference speed value. The error is classified as positive, negative, positive error change, or negative error change as appropriate.

The T2 fuzzy set depicted in Figure 3 is established via the membership function specified below. The mathematical expression of the fuzzy set is given in Eqs. (22) and (23).

$$\tilde{A} = \left\{(x,u), \mu_{\tilde{A}}(x,u) | \forall x \in X, \forall u \in J_x \subseteq [0,1]\right\} \tag{22}$$

Here, $0 \le \mu_{\tilde{A}}(x,u) \le 1$ and $u \in J_x \subseteq [0,1]$ is a numerical value within the range of the primary membership value. In the continuous domain, the set $\tilde{A}$ can be represented as follows:

$$\tilde{A} = \int_{x \in X} \int_{u \in J_x} \mu_{\tilde{A}}(x,u) / (x,u), J_x \subseteq [0,1] \tag{23}$$

Here, $\iint$ denotes the common values of x and u throughout the domain. The term J_x refers to the main membership function of the variable x, $\mu\tilde{A}(x,u)$ the term secondary membership function refers to the additional membership function associated with variable x. In a T2 fuzzy set, $\bar{\mu}_{\tilde{A}}$, the uncertainty is expressed in a region of the primary membership function. In this case, this region is referred to as the footprint of uncertainty. In the type-2 fuzzy logic system, set $\tilde{A}$ is defined with the footprint of uncertainty in Eqs. (22) and (23) [22]. In this case, the type-2 controller provides a more accurate response when the controlled system is nonlinear. This is clearly observed by comparing Figures 4 and 5.

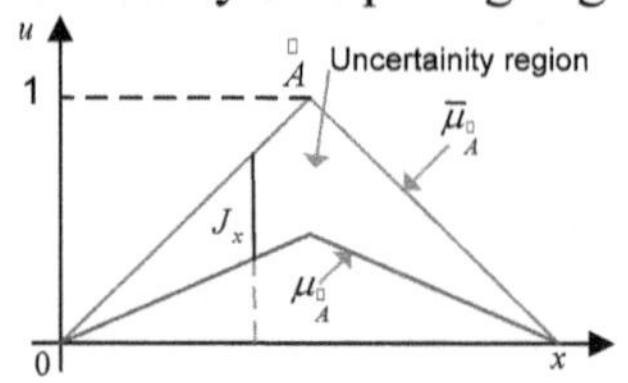

Fig. 3. Footprint of uncertainty region formed by IT2-FPID membership functions

Figures 4 and 5 presents the membership functions of conventional FLCs and type-2 fuzzy logic control. According to these membership functions, the change of interval-valued membership functions for the type-2 fuzzy logic regulator is defined according to the definition seen in Figure 3 [23].

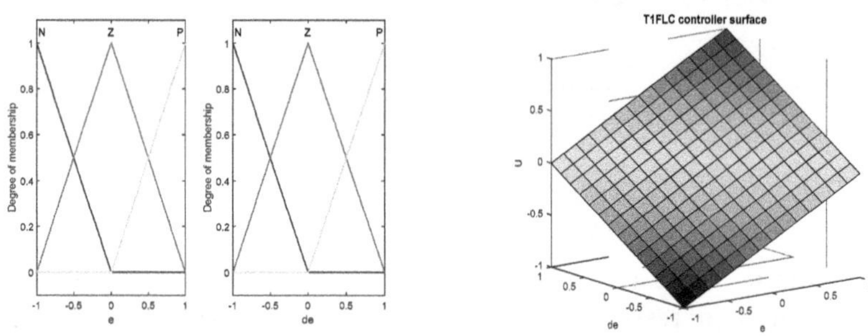

Fig. 4. T1FLC scaled membership functions, T1FLC controller surface respectively

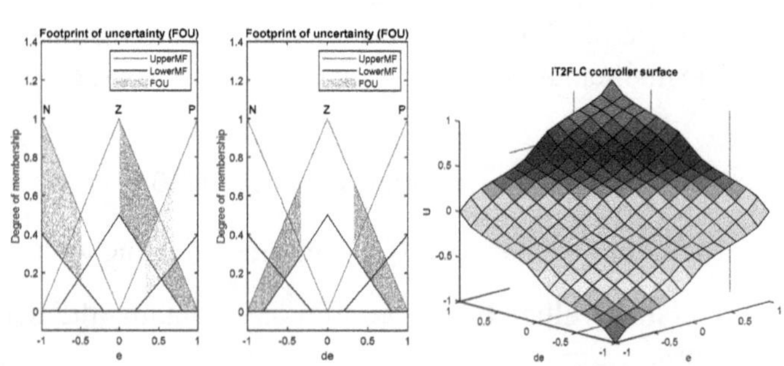

Fig. 5. IT2FLC scaled membership functions, T1FLC controller surface respectively

3.2. *Parameter optimization via particle swarm method*

Particle Swarm Optimization (PSO) is a meta-heuristic algorithm that relies on swarm intelligence. Every individual in the swarm possesses a distinct position, velocity, and a personal best (pbest). The location of a particle corresponds to a potential solution to the optimization issue, and its effectiveness is assessed using a fitness function. A particle's individual best refers to the position at which it has attained the most favorable outcomes and is in closest proximity to the ideal outcome. The term "global best" (gbest) denotes the particle inside the swarm that possesses the highest fitness value. During each iteration, the algorithm identifies both a global best and an individual best for every particle. In each iteration, the velocity of each particle is updated depending on its individual best and the global best, as outlined in Eq. (24). The locations of the particles are subsequently modified in each cycle using the newly calculated velocity, according to Eq. (25). The iterative process persists until the ideal solution meets the convergence requirements. The formula for updating velocity is as listed below:

$$v_i^{t+1} = \omega \times v_i^t + c_1 \times rand() \times (g_{best}^t - x_i^t) + c_2 \times rand() \times (p_{best,i}^t - x_i^t) \quad (24)$$

$$x_i^{t+1} = x_i^t + v_i^{t+1} \quad (25)$$

The velocity of the i_{th} particle is represented by v_i, the iteration number is denoted by t, the inertia factor is indicated by w which is multiplied by 0.99 after each iteration, c_1 and c_2 represent the global and individual learning factors respectively, and x_i represents the position of the i_{th} particle [24].

In the classical PID control algorithm, the PID parameters require adjustment and necessitating the determination of the control parameters. The PID parameters adjustment methodology is often categorized into three primary types:

First approach use theoretical analysis to determine PID control parameters derived from the model of the controlled system. This method is challenging to apply in complicated controlled objects and uncertain model systems, and it is computationally complicated and difficult to execute. Second method involves parameter setting as an engineering approach grounded in empirical formulation, which derives the essential dynamic characteristic parameters of the controlled object through dynamic experiments conducted on the actual system. The prevalent techniques are the extended critical ratio approach and the extended response curve method, which complicate the acquisition of optimal PID parameters or parameters that approximate optimality. The last one is a self-tuning method for PID parameters that employs an optimization algorithm to autonomously modify PID parameters. The particle swarm optimization algorithm is a method for intelligent self-optimization and parameter change [25].

To use the advantageous features of the IT2FPID algorithm, the input derivative and controller output proportional and integral parameters must be set to their ideal values for the controller to yield the most effective output.

Optimization approaches necessitate a particular indicator to demonstrate whether the system is being optimized with each iteration. Consequently, the objective function assigns a numerical number to the method to evaluate the proximity of its response. Various objective functions impose specific requirements for optimizing the algorithm's efficacy within the system. In the scenario of an algorithm enhancing instantaneous response, the algorithm would evaluate its performance based on the computed PID constants.

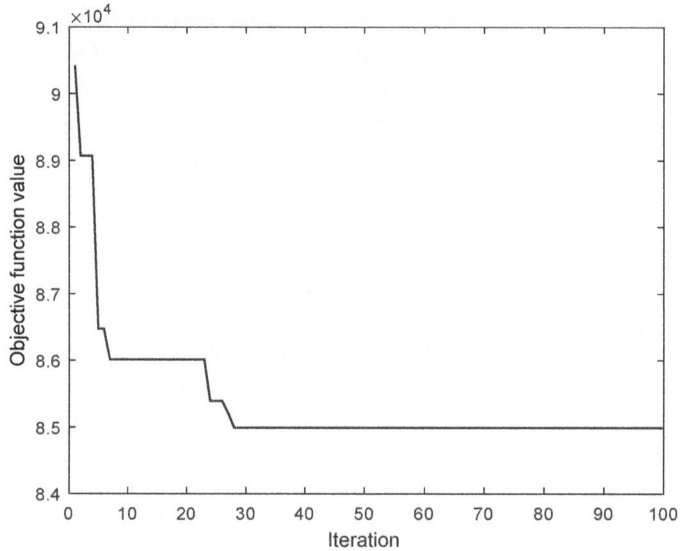

Fig. 6. Objective function values- iteration changes graph

Figure 6 shows selected objective function value in each iteration. These objective functions consider integral of time-weighted absolute error (ITAE) and integral of square of control output (ISCO). A lower value of *J* indicates a smaller error between the system response and the desired output response. Two design objectives are identified: (1) minimizing the error (e) between the reference signal and the output, and (2) minimizing the control (u). Decreasing the first target enhances reference tracking and improves disturbance rejection. The second term or ISCO is employed to mitigate excessive control output that may lead to saturation. The mathematical delineation of these objectives is provided below.

$$\left.\begin{aligned} Obj\_J_1 &= \int_0^{\infty} t|e(t)|dt \\ Obj\_J_2 &= \int_0^{\infty} u^2(t)dt \\ J &= \gamma_1 Obj\_J_1 + \gamma_2 Obj\_J_2 \end{aligned}\right\} \tag{26}$$

where e(t) represents the closed-loop error, i.e., the difference between the reference value and the actual value, and u(t) denotes the control output signal. The two weights, γ_1 and γ_2, are employed to equilibrate the objective function Eq. (26), facilitating the balance of the ITAE and ISCO signals. Weights are regarded as equal and set to one to assign equal significance to the minimizing of the error index and control output. Weights are selected as 0.5.

4. Simulation Results

In this study, speed control of a three-phase induction motor was performed using different types of controllers. The parameters of the motor and controllers are provided in Table 2. Simulation is executed in MATLAB/Simulink program. Simulation results comparing the speed performance of the controllers under different speed references and varying load torques are presented below. To control the induction motor, the current and speed loops are implemented using the conventional PI method. The current control parameters were determined according to the amplitude optimum method, while the speed control parameters were determined according to the symmetric optimum method. In Table 2, the magnitudes of the control parameters are provided. In the optimization of the IT2FPID algorithm parameters, the lower and upper boundary parameters were selected based on traditional PI parameters. While the lower limit was set to 0, the upper limit was set to 1000. The optimum ratio obtained from the PSO algorithm, with integral and derivative parameters $K_p = 36.04$, $K_i = 895.89$, and $K_d = 0.56$, has been achieved.

Table 2. Induction motor and speed controller parameters

Parameter	Value
Power Rating	37300 W
Stator resistance per phase	0.087 Ω
Stator inductance per phase	0.0001 H
Magnetizing inductance per phase	0.0347 H
Moment of inertia of rotating mass	0.662 kgm^2
Damping coefficient	0.1 N.m.s
No of Poles	4
Proportional part of speed PI controller	20
Integral part of speed PI controller	260
Proportional part of speed FLC controller	63
Integral part of speed FLC controller	1000
Derivative part of speed FLC controller	1/1000

In Figure 7, the responses of the PI, traditional fuzzy controller, and the proposed adaptive IT2-FPID controller are compared for reference speed values of 100, 500, 750, 900, −500, and −750 rpm.

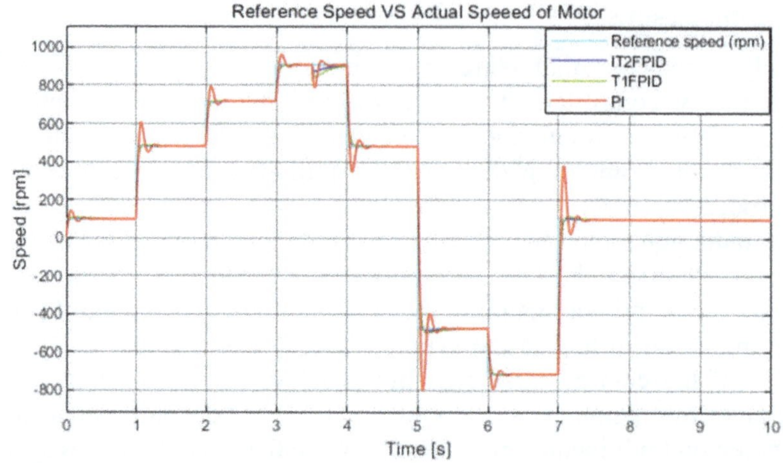

Fig. 7. PI, T1FPID, IT2FPID Speed controllers dynamic response graphs under various speed references

In the graph, the reference values and the controller responses are shown in blue, green, red, and turquoise colors, respectively. At first glance, it is evident from the overshoot and undershoot values that the PI controller exhibits the worst performance. To examine the controller behaviors in more detail, zoomed-in graphs are provided in Figures 8 and 9.

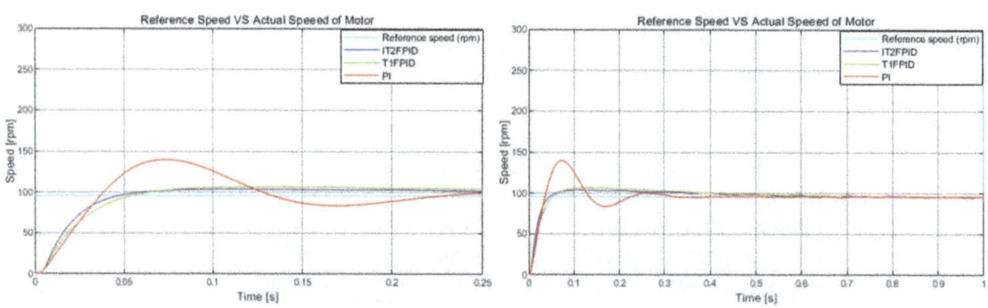

Fig. 8. PI, T1FPID, IT2FPID Speed controllers dynamic responses in 0–1 second between respectively

In Figure 8, when the graphs are evaluated in terms of rise time, settling time, overshoot, and oscillations, it is observed that the IT2-FPID controller has the best dynamic response. The order of the controllers based on the fastest rise time is IT2-FPID, T1-FPID, and PI controller. In terms of overshoot, both fuzzy logic controllers remain below the 5% overshoot value, which is considered an acceptable performance criterion.

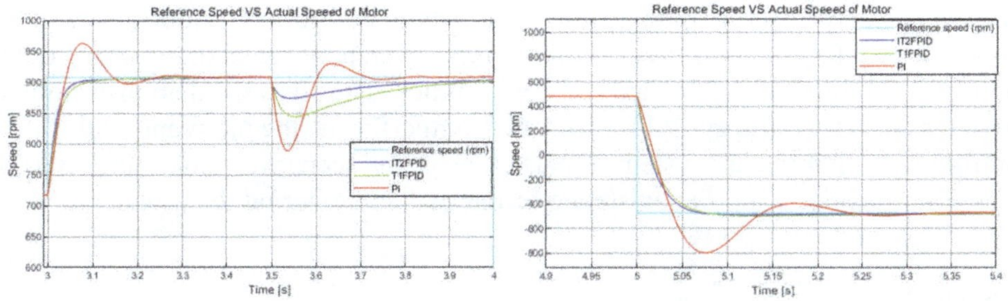

Fig. 9. PI, T1FPID, IT2FPID Speed controllers dynamic responses in 3–5 second between respectively

In terms of oscillations, the fuzzy logic controllers exhibit almost no oscillations, while the PI controller shows significantly more oscillations and has a much longer settling time compared to the other controllers. In Figure 9, graphs are analyzed within the 3–5 second time interval to evaluate the performance of load torque and speed reversal. It is clear that the IT2-FPID controller has the best recovery time and value in response to the instantaneous load torque. The same performance ranking is maintained in the speed reversal response. In Figure 10, the variation of the electromagnetic torque is presented with respect to the quadrature axis current in vector control. While similar responses are observed across all controllers, the instantaneous value of the PI controller is higher compared to the others.

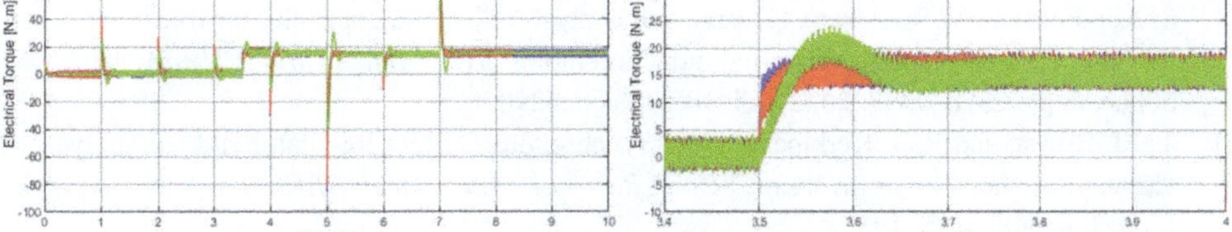

Fig. 10. PI, T1FPID, IT2FPID electromagnetic torque related to quadrature axis current

5. Conclusions

This research presents a novel approach where an Interval Type-2 Fuzzy Logic Controller (IT2FLC) is integrated with a traditional Proportional-Integral-Derivative (PID) controller to provide a robust fuzzy PID control system for controlling the indirect vector of an induction motor. The practical swarm optimization (PSO) technique was employed to get the most suitable parameters for the proposed controller. In order to showcase the efficacy of the proposed controller in comparison to traditional approaches, a sequence of simulations will be conducted using Matlab/Simulink. Through quantitative assessments with IT2-FPID, FLC, and PID controllers, it has been verified that the suggested control method has a quicker response time than other controllers. Additionally, the proposed IT2-FPID controller exhibits less overshoot/undershoot than other controllers in most of the carried out case studies. The findings demonstrate that the interval type-2 fuzzy PID controller outperforms the conventional fuzzy controller in nonlinear systems, particularly in scenarios involving abrupt reference transitions and the existence of internal and external disturbances.

References

[1] G. S. Buja and M. P. Kazmierkowski, "Direct torque control of PWM inverter-fed AC motors - A survey," *IEEE Trans. Ind. Electron.*, vol. 51, no. 4, pp. 744–757, 2004, doi: 10.1109/TIE.2004.831717.

[2] N. I. Nahin, S. P. Biswas, G. S. Member, and K. Hosain, "Predictive Torque Control with Reduced Voltage Vectors for Voltage Source Inverter Driven Induction Motor Drives," *IEEE Trans. Appl. Supercond.*, vol. 34, no.8, pp. 1–5, 2024, doi: 10.1109/TASC.2024.3446291.

[3] L. Wu *et al.*, "Design of Complex Vector Controller for High-Power Induction Machine Drive," *IEEE Trans. Transp. Electrif.*, vol. 10, no. 1, pp. 1816–1826, 2024, doi: 10.1109/TTE.2023.3265044.

[4] S. Nassiri, M. Labbadi, C. Chatri, and M. Cherkaoui, "Optimal Recursive Terminal Sliding-Mode Control Using Super-Twisting Algorithm for Improving High Efficiency and Reliability of Pump Systems," *IEEE Control Syst. Lett.*, vol. 7, pp. 3385–3390, 2023, doi: 10.1109/LCSYS.2023.3330748.

[5] N. I. Nahin, S. Prokash Biswas, S. Mondal, M. K. Hosain, M. Rabiul Islam, and S. S. Haque, "Improved Vector Control Scheme for Voltage Source Inverter Driven Induction Motor Drives," *2023 10th IEEE Int. Conf. Power Syst. ICPS 2023*, pp. 1–6, 2023, doi: 10.1109/ICPS60393.2023.10428760.

[6] S. Adigintla and M. V. Aware, "Design and analysis of a speed controller for fractional-order-modeled voltage-source-inverter-fed induction motor drive," *Int. J. Circuit Theory Appl.*, vol. 50, no. 7, pp. 2378–2397, 2022, doi: 10.1002/cta.3290.

[7] T. M. Rowan and R. J. Kerkman, "New Synchronous Current Regulator and an Analysis of Current Regulated Pwm Inverters.," *Conf. Rec. - IAS Annu. Meet. (IEEE Ind. Appl. Soc.)*, vol. I, no. 4, pp. 487–495, 1985.

[8] J. Holtz and X. Qi, "Optimal control of medium-voltage drives-an overview," *IEEE Trans. Ind. Electron.*, vol. 60, no. 12, pp. 5472–5481, 2013, doi: 10.1109/TIE.2012.2230594.

[9] K. Berringer, J. Marvin, and P. Perruchoud, "Semiconductor power losses in AC inverters," *Conf. Rec. - IAS Annu. Meet. (IEEE Ind. Appl. Soc.)*, vol. 1, pp. 882–888, 1995, doi: 10.1109/ias.1995.530391.

[10] E. Zerdali, M. Rivera, P. Zanchetta, P. Wheeler, and L. Ristic, "Encoderless Predictive Speed and Torque Control of an Induction Motor," *Proc. 22nd Int. Symp. Power Electron. Ee 2023*, 2023, doi: 10.1109/Ee59906.2023.10346148.

[11] E. H. Mamdani and S. Assilian, "Experiment in linguistic synthesis with a fuzzy logic controller," 1999. doi: 10.1006/ijhc.1973.0303.

[12] M. Sugeno and K. Murakami, "Fuzzy Parking Control of Model Car.," *Proc. IEEE Conf. Decis. Control*, pp. 902–903, 1984, doi: 10.1109/cdc.1984.272144.

[13] A. Mishra, "Design of Speed Controller for Squirrel-cage Induction Motor using Fuzzy logic based Techniques," *Int. J. Comput. Appl.*, vol. 58, no. 22, pp. 975–8887, 2012.

[14] S. Galichet and L. Foulloy, "Fuzzy Controllers: Synthesis and Equivalences," *IEEE Trans. Fuzzy Syst.*, vol. 3, no. 2, pp. 140–148, 1995, doi: 10.1109/91.388169.

[15] P. F. Controllers, R. K. Mudi, and N. R. Pal, "A Robust Self-Tuning Scheme for," *Ieee Trans. Fuzzy Syst.*, vol. 7, no. 1, pp. 2–16, 1999.

[16] T. Ohtani, "Vector control of induction motor without shaft encoder," *IEEE Trans. Ind. Appl.*, vol. 28, no. 1, pp. 157–164, 1992.

[17] A. Olajube and O. M. Anubi, "Model-based Loss Minimization Control of a Squirrel cage Induction Motor Drive with shorted Rotor under Indirect Field Orientation," *2023 IEEE Conf. Control Technol. Appl. CCTA 2023*, pp. 759–765, 2023, doi: 10.1109/CCTA54093.2023.10252896.

[18] Z. M. S. Elbarbary, "Experimental comparison between four and six switch inverters fed FLC based speed sensorless IM drives at very low speed," *Int. J. Power Electron. Drive Syst.*, vol. 8, no. 1, pp. 130–146, 2017, doi: 10.11591/ijpeds.v8i1.pp130-146.

[19] M. Nasir Uddin, Z. R. Huang, and A. B. M. Siddique Hossain, "Development and implementation of a simplified self-tuned neuro-fuzzy-based im drive," *IEEE Trans. Ind. Appl.*, vol. 50, no. 1, pp. 51–59, 2014, doi: 10.1109/TIA.2013.2269131.

[20] C. B. Butt, M. A. Hoque, and M. A. Rahman, "Simplified fuzzy-logic-based MTPA speed control of IPMSM drive," *IEEE Trans. Ind. Appl.*, vol. 40, no. 6, pp. 1529–1535, 2004, doi: 10.1109/TIA.2004.836312.

[21] M. H. N. Talib *et al.*, "An improved simplified rules Fuzzy Logic Speed Controller method applied for induction motor drive," *ISA Trans.*, vol. 105, pp. 230–239, 2020, doi: 10.1016/j.isatra.2020.05.040.

[22] T. Kumbasar, "Interval type-2 fuzzy PID controllers and an online self-tuning mechanism," *Pamukkale Univ. J. Eng. Sci.*, vol. 22, no. 8, pp. 643–649, 2016, doi: 10.5505/pajes.2016.49344.

[23] T. Kumbasar, C. Ozturk, E. Yesil, and H. Hagras, "Performance evaluation of interval type-2 and online rule weighing based Type-1 Fuzzy PID controllers on a pH process," *IEEE Int. Conf. Fuzzy Syst.*, pp. 1875–1882, 2014, doi: 10.1109/FUZZ-IEEE.2014.6891627.

[24] O. Chao and L. Weixing, "Comparison between PSO and GA for parameters optimization of PID controller," *2006 IEEE Int. Conf. Mechatronics Autom. ICMA 2006*, vol. 2006, pp. 2471–2475, 2006, doi: 10.1109/ICMA.2006.257739.

[25] M. Yang and S. Zhang, "PSO-based PID controller design for unwinding tension system of web press," *Proc. 13th IEEE Conf. Ind. Electron. Appl. ICIEA 2018*, pp. 1399–1403, 2018, doi: 10.1109/ICIEA.2018.8397928.

Chapter 28

A Strategic Approach for Sustainable Logistics: Environmental Performance Evaluation of Logistics Companies Using Grey Relational Analysis

Cansu Aksu
Business Administration Department, Zonguldak Bülent Ecevit University, Zonguldak, Turkey
cansuaksu@beun.edu.tr

This study presents an innovative methodology for measuring the environmental sustainability performance of companies operating within the logistics sector. Utilizing data from Global Reporting Initiative (GRI) sustainability reports, it evaluates the environmental performance of these companies through the application of Grey Relational Analysis (GRA). Grey Relational Analysis approach, one of the multi criteria decision-making methods serves as a significant tool for analyzing uncertain and multidimensional environmental data, objectively comparing and ranking companies based on various performance criteria. In this context, the analysis identifies leading companies in terms of environmental sustainability within the sector and provides valuable data for the development of sustainable operational strategies. Furthermore, this research is considered a new methodological approach to measuring and improving environmental sustainability performance in the logistics sector. Consequently, the study offers a comprehensive analysis and strategic approach to evaluating and enhancing the environmental sustainability performance of logistics companies. This approach emphasizes the importance of sustainable logistics management and aims to increase awareness and practices related to environmental sustainability within the industry.

1. Introduction

Environmental sustainability is a critical approach designed to preserve natural resources and minimize environmental degradation, ensuring the availability of these resources for future generations. In the logistics sector, this concept is particularly pertinent as it involves reducing energy consumption, managing waste effectively, and minimizing carbon emissions, which are key to mitigating the sector's environmental impact.

The logistics sector's activities, including transportation, storage, and distribution, have significant environmental implications. These processes consume large amounts of energy and contribute to emissions, directly affecting environmental sustainability. Consequently, the adoption of sustainable practices is crucial not only for reducing environmental impacts but also for providing long-term competitive advantages to companies within the sector [1].

This study employs the Grey Relational Analysis (GRA) method to evaluate the environmental performance of logistics companies, with data processed using Microsoft Excel. In multi-criteria decision-making (MCDM) problems, the weighting of criteria could be approached in two distinct ways: subjectively, based on expert knowledge and experience, or objectively, based on empirical measurement data. For this study, the selected environmental performance criteria were objectively weighted using the CRITIC (Criteria Importance Through Intercriteria Correlation) method, followed by the application of GRA to rank the companies according to their environmental performance [2, 3].

To calculate the Grey Relational Grade from the Grey Relational Coefficient matrix, two methods can be utilized: assigning equal weights to each criterion or weighting them according to their relative importance. In this study, the CRITIC method was chosen for its objective nature. This method weights criteria by evaluating the interrelationships among them and considering their standard deviations, thus providing a more robust and unbiased evaluation [2].

The primary aim of this study is to introduce an innovative methodology for assessing the environmental sustainability performance of companies operating in the logistics sector. By applying this methodology, the study seeks to contribute to the ongoing discourse on sustainable logistics management and provide actionable insights for enhancing environmental performance in the industry.

2. Literature Review

GRA is a versatile and robust technique widely used for analyzing and comparing complex systems, particularly in situations where data is scarce, incomplete, or uncertain. GRA is highly effective in scenarios where information may be ambiguous, offering a comprehensive framework to evaluate relationships among multiple factors. The strength of GRA lies in its ability to manage uncertain and multidimensional data, making it adaptable to a broad range of applications. This flexibility has led to its extensive use in fields such as engineering, economics, and environmental science, where decision-making often requires the evaluation of complex, interrelated criteria [4]. Literature review is listed in Table 1.

Table 1. Literature review.

Study	Method used	Application area	Key findings	Contribution of GRA
Chen and Tsuei-Yang (2011) [5]	Grey Markov Model	Financial Crisis Prediction	Accurate forecasting	Extends GRA's application in finance
Lin and Shun-Jyh (2011) [6]	GRA	Financial Crisis Prediction	Superior to conventional techniques	Highlights GRA's predictive 8capabilities
Xia and Wong (2014) [7]	GRA	Fashion Retailing	Effective in uncertain environments	Demonstrates GRA's robustness in handling uncertainty
Ayaydın *et al.* (2017) [8]	GRA	Turkish Logistics Companies	Objective performance measurement	Applies GRA in logistics performance evaluation
Li *et al.* (2020) [9]	GRA & ANP	Water Resource Security	Evaluated water resource security in Guizhou Province	Combines GRA with ANP for comprehensive water resource security assessment
Yıldırım *et al.* (2021) [10]	GRA	Financial Performance of Steel Companies	Analyzed financial performance of steel companies on BIST	Applies GRA for financial performance assessment
Silva *et al.* (2023) [11]	CRITIC & GRA	Investment Portfolio Selection	Optimized portfolio based on financial indicators	Combines GRA with CRITIC for financial decision-making
Ersoy and Taslak (2023) [12]	GRA, PIV, ROV, MARCOS	Corporate Sustainability in Energy Sector	Ranked energy companies by sustainability	Uses GRA in multi-criteria sustainability evaluation
Shil *et al.* (2023) [13]	GRA & Possibility Theory	Climate Change Impact on Hydropower	Identified most important factors affecting hydropower	Applies GRA in a possibility-based decision-making environment
Peng and Li (2024) [14]	GRA	Environmental Load Assessment	Evaluation of environmental loads in Hubei Province	Demonstrates GRA's effectiveness in environmental management
Ranjan and Saha (2024) [15]	GRA & PCA	3D Printing Optimization	Enhanced tensile strength and reduced surface roughness	Combines GRA with PCA for optimized 3D printing parameters
Priyalatha *et al.* (2024) [16]	GRA	Antenna Design	Optimized bandwidth and gain for X-band antennas	Demonstrates GRA's utility in multi-objective antenna optimization
Sandhu and Kaushal (2024) [17]	GRA & ANOVA	Anaerobic Digestion	Improved methane and biogas yields using manure	Applies GRA for optimizing biogas production parameters
Rajesh *et al.* (2024) [18]	GRA & DOE	EDM Process Optimization	Enhanced material removal rate with minimal tool wear	Optimizes electro discharge machining parameters using GRA

One key reason for GRA's widespread use is its ability to handle uncertain, multidimensional data, making it suitable for real-world problems. Chen and Tsuei-Yang [5] demonstrated this through a Grey Markov Model for financial crisis prediction, proving GRA's effectiveness in uncertain scenarios. Lin and Shun-Jyh [6] further validated GRA's superiority in predicting financial crises over conventional methods, particularly in managing market complexities. In retail, Xia and Wong [7] applied GRA to improve fashion sales forecasting, while

Ayaydın *et al.* [8] used it to evaluate Turkish logistics companies' operational efficiency. Yıldırım *et al.* [10] also used GRA to analyze the financial performance of steel companies on Borsa Istanbul (BIST).

GRA has seen applications in environmental sustainability, as Li *et al.* [9] combined it with ANP for water resource security in China, and Silva *et al.* [11] integrated it with CRITIC for sustainable investment portfolios. Ersoy and Taslak [12] applied GRA in evaluating corporate sustainability in the energy sector. In engineering, Ranjan and Saha [15] utilized GRA with PCA for optimizing 3D printing parameters, while Rajesh *et al.* [18] employed it for optimizing Electro Discharge Machining (EDM) processes. Priyalatha *et al.* [16] applied GRA in antenna design, enhancing bandwidth and gain.

However, GRA's application in evaluating environmental sustainability in logistics is limited. While some studies, such as Sandhu and Kaushal [17], have applied GRA for energy and emissions criteria, comprehensive multi-criteria approaches remain scarce. This gap underscores the need for more robust methodologies that integrate multiple criteria for a holistic evaluation of environmental sustainability in logistics.

3. Methodology

3.1. *Grey Relational Analysis (GRA)*

Grey Relational Analysis (GRA) was introduced as part of Grey System Theory by Deng in 1982 [3]. GRA is designed to evaluate the relationships between multiple factors to identify patterns and rank alternatives, making it a valuable tool in decision-making processes. This method is particularly effective in situations where information is incomplete or ambiguous and has been extensively utilized across various fields due to its robustness in handling uncertain and multidimensional data [4].

GRA employs grey relational grades to rank alternatives based on their similarity to a reference series. Higher grey relational grades indicate greater similarity to the reference series and are therefore preferred in decision-making contexts.

The steps involved in the application of GRA are as follows [20]:

(1) **Determination of the reference series:** The reference series is established with a length of "n" data points [4].

(2) **Normalization of data:** Data normalization is performed to standardize the values, ensuring that all criteria are comparable. The normalization process can be conducted as follows [20]:

For criteria where higher values are better (maximization):

$$X_i(k) = \frac{x_i^0(k) - \min x_i^0(k)}{\max x_i^0(k) - \min x_i^0(k)} \tag{1}$$

For criteria where lower values are better (minimization):

$$X_i(k) = \frac{\max x_i^0(k) - x_i^0(k)}{\max x_i^0(k) - \min x_i^0(k)} \tag{2}$$

For reference criteria:

$$X_i(k) = 1 - \left|\frac{x_i^0(k) - x^0}{\max x_i^0(k) - x^0}\right| \tag{3}$$

(3) Determination of the series to be compared: The difference between the reference series and each alternative series is calculated as [3, 4, 20]:

$$\Delta_{0i}(k) = |X_0^*(k) - X_i^*(k)| \tag{4}$$

(4) Determination of grey relational coefficients: The grey relational coefficients are computed to evaluate the degree of relation between the reference series and the alternative series [3, 20]:

$$\in \left(X_0(k), X_i(k)\right) = \frac{\Delta_{min} + \xi\Delta_{max}}{\Delta_{0i} + \xi\Delta_{max}} \tag{5}$$

where ξ is the distinguishing coefficient, typically between 0 and 1.

(5) Determination of grey relational grade: The grey relational grade is calculated by averaging the grey relational coefficients over all criteria [4]:

$$\gamma(X_0, X_i) = \frac{1}{n}\sum_{k=1}^{n} \varepsilon\left(\chi_0(k), \chi_i(k)\right) \tag{6}$$

Alternatively, weighted grey relational grades can be calculated by applying weights to each criterion:

$$\gamma(X_0, X_i) = \sum_{k=1}^{n} w_i(k) \times \varepsilon(\chi_0(k), \chi_i(k)) \tag{7}$$

GRA is particularly suitable for evaluating environmental performance as it provides objective comparisons and rankings based on multiple performance criteria. In this study, GRA will be used to analyze data from Global Reporting Initiative (GRI) sustainability reports, offering a robust framework for performance assessment. This method will help identify leading companies in environmental sustainability and pinpoint areas for improvement.

3.2. *CRITIC method*

CRITIC (Criteria Importance Through Intercriteria Correlation) is an objective criterion weighting method widely utilized in multicriteria decision-making (MCDM) problems. Introduced by Diakoulaki, Mavrotas, and Papayannakis in 1995, the CRITIC method provides a robust framework for objectively weighting evaluation criteria based on empirical data, while considering the degree of interaction between the criteria themselves [2]. The method leverages the standard deviations and correlations of criteria to determine their relative importance.

The steps involved in the application of the CRITIC method are as follows [2, 21–25]:

(1) Creation of the decision matrix: A decision matrix X is created, consisting of alternatives and criteria [21, 25]:

$$X = [X_{ij}]_{m*n} \begin{bmatrix} X_{11} & \cdots & X_{1n} \\ \vdots & \ddots & \vdots \\ X_{m1} & \cdots & X_{mn} \end{bmatrix} \quad (8)$$

This matrix includes the evaluation of each alternative concerning each criterion.

(2) Normalization: The values of the criteria are normalized between the maximum and minimum values of the relevant criterion to ensure comparability [21, 23–25]:

For benefit criteria:
$$r_{ij} = \frac{x_{ij} - x_j^{min}}{x_j^{max} - x_j^{min}} \quad (9)$$

For cost criteria:
$$r_{ij} = \frac{x_j^{max} - x_{ij}}{x_j^{max} - x_j^{min}} \quad (10)$$

The values of the criteria are normalized between the maximum and minimum values of the relevant criterion. This ensures the scaling of values in the decision matrix.

(3) Calculation of correlation coefficients: Correlation coefficients between criteria are calculated using the normalized values, which measure the degree of relationship between criteria [23–25]:

$$\rho_{jk} = \frac{\sum_{i=1}^{m}(r_{ij} - \overline{r_j})(r_{ik} - \overline{r_k})}{\sqrt{\sum_{i=1}^{m}(r_{ij} - \overline{r_j})^2 \sum_{i=1}^{m}(r_{ik} - \overline{r_k})^2}}, \quad j, k = 1, 2, \ldots n \quad (11)$$

(4) **Calculation of information amount:** The information amount for each criterion is calculated as [2, 21, 23–25]:

$$C_j = \sigma_j \sum_{k=1}^{n} 1 - \rho_{jk} \quad (12)$$

where σ_j is the standard deviation of the criterion j.

(5) **Determination of criteria weights:** The weight of each criterion is determined by dividing the information amount of the criterion by the total information amount of all criteria [2, 21, 23–25]:

$$W_j = C_j / \sum_{k=1}^{n} C_k \quad (13)$$

This process results in the objective determination of the importance levels of criteria, providing a data-driven basis for evaluating alternatives in MCDM problems.

4. Application

In this study, the environmental performance of logistics companies was assessed using the GRA method, with data processed in Microsoft Excel. The environmental performance indicators—energy consumption, water consumption, greenhouse gas emissions, hazardous waste, and non-hazardous waste—were derived from the publicly available sustainability reports of companies listed in the Fortune 500 Turkey for 2022, specifically within the "Storage, Transportation and Logistics Services" and "Travel and Transportation Services" categories.

In MCDM problems, criteria weights can be determined either subjectively, based on expert knowledge and experience, or objectively, using empirical measurement data. For the purpose of evaluating environmental performance, the selected criteria were objectively weighted using the CRITIC method. The CRITIC method is particularly effective as it accounts for the correlation between criteria and their variability, providing a robust and unbiased set of weights. After determining the criteria weights through the CRITIC method, GRA was applied to rank the companies according to their environmental performance.

The primary research question guiding this study is: "Which logistics companies are leading in terms of environmental sustainability based on GRA?"

4.1. *Data source*

The Global Reporting Initiative (GRI) provides standardized guidelines for sustainability reporting that are widely adopted by companies globally. These reports offer comprehensive data on various environmental performance indicators, such as energy consumption, waste management, and emissions, ensuring consistency and comparability across different organizations. Given its reliability and the breadth of data it offers, the GRI framework was selected as the primary source for environmental performance evaluations in this study.

For the sample selection, this study utilized the Fortune 500 Turkey list for 2022 [26], which ranks the largest companies in Turkey by annual net sales revenue. This list is widely recognized in the business community as a benchmark for identifying the most influential companies in Turkey across various industries.

The analysis focused on logistics companies listed under the "Storage, Transportation and Logistics Services" and "Travel and Transportation Services" categories in the 2022 Fortune 500 Turkey list [26]. Only those companies that publicly disclosed their GRI-compliant sustainability reports for 2022 were included in the study. This ensures that the evaluation is based on the most current and relevant environmental performance data, capturing a representative sample of the logistics sector in Turkey.

4.2. *Environmental performance indicators*

The environmental performance indicators utilized in this study were adapted from the framework established by Senir and Büyükkeklik [27]. These indicators provide a comprehensive evaluation of a company's environmental impact across key areas, such as resource efficiency, energy consumption, waste management, and emissions control [27]. By focusing on these specific aspects, the indicators enable an in-depth assessment of how well companies manage their environmental responsibilities and contribute to sustainability goals.

Energy Consumption: This indicator quantifies the total energy usage by the company, encompassing both renewable and non-renewable energy sources [27]. The consumption is measured in kilowatt-hours (kWh). Environmental goal is to minimize energy consumption to reduce the environmental impact and improve energy efficiency.

Water Consumption: This indicator measures the total volume of water utilized by the company, which includes water used in operational processes, cooling systems, and other industrial activities [27]. The volume is measured in cubic meters (m^3). Environmental goal is to minimize water consumption to ensure sustainable water use and reduce the company's overall environmental impact.

Greenhouse Gas Emissions: This indicator assesses the total emissions of greenhouse gases (GHGs) produced by the company, expressed in equivalent tonnes of carbon dioxide (t CO_2 eq.) [27]. Environmental goal is to minimize GHG emissions to mitigate the company's contribution to climate change.

Hazardous Waste: This indicator tracks the total amount of hazardous waste generated by the company. Hazardous waste includes materials that are toxic, flammable, corrosive, or otherwise pose significant risks to human health and the environment [27]. The waste volume is measured in tonnes. Environmental goal is to minimize the generation of hazardous waste to reduce potential environmental and health risks, and to comply with regulatory requirements.

Non-hazardous Waste: This indicator measures the total amount of non-hazardous waste generated by the company, including general waste, organic waste, and recyclable materials [27]. The volume is measured in tonnes. Environmental goal is to minimize non-hazardous waste to promote recycling, reduce landfill use, and enhance sustainable waste management practices.

4.3. *Data set*

The data set for this study comprises environmental performance indicators collected from the sustainability reports of logistics companies listed in the Fortune 500 Turkey for 2022 [26, 28–32]. The companies were selected from the "Storage, Transportation and Logistics Services" and "Travel and Transportation Services" categories. The indicators include energy consumption, water consumption, greenhouse gas emissions, hazardous waste, and non-hazardous waste (Table 2). These indicators provide a comprehensive view of each company's environmental impact.

Table 2. Data set.

Firm name	Fortune 500 ranking [18]	Energy consumption (kWh)	Water consumption (m^3)	Greenhouse gas emissions (t CO_2 eq.)	Hazardous waste (tonnes)	Non-hazardous waste (tonnes)
Turkish Airlines [28]	3	70,149,163,777	364,508	22,983,093	2,312	2,313
Ekol Logistics [29]	109	28,381,008	119,630	398,518	23.4	3761
Mars Logistics [30]	156	2,248,423	24,342	43,202	17.896	155.866
DSV Air and Sea Transport [31]	174	1,414,118,980	1,274,000	15,930,000	652	27,540
Noatum Logistics [32]	195	48,303,000	44,053	9,561	59.376	2,821.905

4.4. *Normalization matrix*

The normalization matrix is an essential component in MCDM, allowing for the comparison of diverse criteria on a standardized scale [33]. In this study, the environmental performance indicators—energy consumption, water consumption, greenhouse gas emissions, hazardous waste, and non-hazardous waste—were normalized to values between 0 and 1. This process ensures that each criterion, regardless of its original unit of measurement, can be directly compared to the others.
The normalized values for each logistics company are presented in Table 3.

Table 3. Data set.

Firm name	Energy consumption (kWh)	Water consumption (m^3)	Greenhouse gas emissions (t CO_2 eq.)	Hazardous waste (tonnes)	Non-hazardous waste (tonnes)
Turkish Airlines	0.0000	0.7278	0.0000	0.0000	0.9212
Ekol Logistics	0.9996	0.9237	0.9831	0.9976	0.8683
Mars Logistics	1.0000	1.0000	0.9985	1.0000	1.0000
DSV Air and Sea Transport	0.9799	0.0000	0.3070	0.7236	0.0000
Noatum Logistics	0.9993	0.9842	1.0000	0.9819	0.9026

4.5. *Distance matrix*

The distance matrix represents the deviations between each company's environmental performance and the ideal (or worst) performance based on the normalized values [34].

4.6. *Grey Relational Degree (GRD)*

The Grey Relational Degree (GRD) is a fundamental metric in the GRA method, utilized to evaluate the extent to which an alternative's performance aligns with an ideal reference across multiple criteria. This metric is particularly useful in decision-making processes involving complex systems characterized by limited or uncertain data [4]. The robustness of GRD in such contexts makes it a valuable tool in various fields, including engineering, economics, and environmental management.

Table 4. Distance matrix.

Firm name	Energy consumption (kWh)	Water consumption (m^3)	Greenhouse gas emissions (t CO_2 eq.)	Hazardous waste (tonnes)	Non-hazardous waste (tonnes)
Turkish Airlines	1.0000	0.2722	1.0000	1.0000	0.0788
Ekol Logistics	0.0004	0.0763	0.0169	0.0024	0.1317
Mars Logistics	0.0000	0.0000	0.0015	0.0000	0.0000
DSV Air and Sea Transport	0.0201	1.0000	0.6930	0.2764	1.0000
Noatum Logistics	0.0007	0.0158	0.0000	0.0181	0.0974

4.7. *Grey Relational Coefficient*

The Grey Relational Coefficient (GRC) is a pivotal element in the GRA method, used to quantify the relationship between each alternative and a reference series across multiple criteria. This coefficient effectively measures the closeness of each alternative to the ideal solution, facilitating a comprehensive ranking of alternatives [34].

In the context of this study, the Grey Relational Grade can be calculated using two approaches: (1) by assigning equal weights to each criterion or (2) by applying weights based on the relative importance of each criterion. Adopted the latter approach, utilizing the CRITIC method. The CRITIC method is an objective technique that assigns weights to criteria based on their standard deviations and the correlations between them. This ensures that the most significant criteria, those with the greatest variability and least redundancy, are given appropriate emphasis in the analysis.

Table 5 presents the Grey Relational Coefficient matrix for the selected logistics companies. The coefficients reflect the degree to which each company's performance aligns with the ideal reference, with higher coefficients indicating stronger alignment. The weights applied to each criterion were derived using the CRITIC method, ensuring a robust and data-driven evaluation.

Table 5. Grey Relational Coefficient.

Weights	**0.24**	**0.20**	**0.13**	**0.17**	**0.26**
Firm name	Energy consumption (kWh)	Water consumption (m^3)	Greenhouse gas emissions (t CO_2 eq.)	Hazardous waste (tonnage)	Non-hazardous waste (tonnage)
Turkish Airlines	0.3333	0.6475	0.3333	0.3333	0.8639
Ekol Logistics	0.9993	0.8677	0.9672	0.9952	0.7916
Mars Logistics	1.0000	1.0000	0.9971	1.0000	1.0000
DSV Air and Sea Transport	0.9613	0.3333	0.4191	0.6440	0.3333
Noatum Logistics	0.9987	0.9694	1.0000	0.9651	0.8370

4.8. *Grey Relational Grade*

The Grey Relational Grade (GRG) is a comprehensive measure derived from the Grey Relational Coefficients, reflecting the overall performance of each alternative relative to the ideal reference [35]. Table 6 presents the Grey Relational Grades and the corresponding ranks for the logistics companies analyzed in this study.

Table 6. Grey Relational Grade.

Firm name	Fortune 500 ranking	Grey Relational Grade	Rank
Mars Logistics	156	0.9996189	1
Noatum Logistics	195	0.9454399	2
Ekol Logistics	109	0.9147257	3
DSV Air and Sea Transport	174	0.5495584	4
Turkish Airlines	3	0.5325066	5

4.9. *Evaluation of findings*

The results of this study offer a comprehensive evaluation of the environmental performance of logistics companies. Through the application of the GRA method, we were able to systematically rank the companies based on their environmental sustainability indicators. The key findings of this analysis are as follows:

Mars Logistics achieved the highest GRG, signifying its superior environmental performance. This indicates that Mars Logistics is most closely aligned with the ideal benchmark across the evaluated criteria.

Noatum Logistics and Ekol Logistics also demonstrated strong environmental performance, securing the second and third ranks, respectively. These companies show a solid commitment to sustainability, reflected in their high GRGs.

Grey Relational Grades of DSV Air and Sea Transport and Turkish Airline suggest areas where these companies can improve to enhance their sustainability practices.

The findings underscore the effectiveness of the GRA method in managing uncertain and multidimensional data, providing a robust and reliable framework for assessing environmental performance. Companies that excel in environmental performance not only contribute significantly to global sustainability goals but also secure long-term competitive advantages and ensure compliance with increasingly stringent regulatory requirements.

5. Conclusions

This study utilized the GRA method to assess the environmental sustainability performance of logistics companies. The findings underscore the effectiveness of GRA in managing uncertain and multidimensional data, offering a robust and reliable framework for evaluating performance. The results suggest that companies with superior environmental practices not only secure long-term competitive advantages but also ensure compliance with regulatory requirements.

Future research could focus on refining the criteria further and expanding the dataset to include a larger number of companies as well as additional environmental indicators. Such advancements would enhance the robustness of the analysis and provide deeper insights into the environmental performance of the logistics sector.

References

1. A. McKinnon, *Decarbonizing Logistics: Distributing Goods in a Low Carbon World* (Kogan Page Publishers, 2018).
2. D. Diakoulaki, G. Mavrotas, and L. Papayannakis, Determining objective weights in multiple criteria problems: The CRITIC method, *Computers & Operations Research*, 22(7), 763–770 (1995). https://doi.org/10.1016/0305-0548(94)00059-H.
3. J. Deng, Control problems of grey systems, *Systems & Control Letters*, 1(5), 288–294 (1982). https://doi.org/10.1016/S0167-6911(82)80025-X.
4. S. Liu, and Y. Lin, *Grey Systems: Theory and Applications* (Springer, 2011). https://link.springer.com/book/10.1007/978-3-642-16158-2.
5. L.-H. Chen and T.-Y. Guo, Forecasting financial crises for an enterprise by using the Grey Markov forecasting model, *Quality & Quantity: International Journal of Methodology*, 45, 911–922 (2011). https://doi.org/10.1007/s11135-010-9403-z.
6. S.-L. Lin and S.-J. Wu, Is grey relational analysis superior to the conventional techniques in predicting financial crisis?, *Expert Systems with Applications*, 38, 5119–5124 (2011). https://doi.org/10.1016/j.eswa.2010.09.139.

7. H M. Xia and W. K. Wong, Knowledge-based systems: A seasonal discrete grey forecasting model for fashion retailing, *Knowledge-Based Systems,* 57, 119–126 (2014). https://doi.org/10.1016/j.knosys.2013.12.014.

8. H. Ayaydın, S. Durmuş, and F. Pala, Gri ilişkisel analiz yöntemiyle Türk lojistik firmalarında performans ölçümü, *Gümüşhane Üniversitesi Sosyal Bilimler Enstitüsü Elektronik Dergisi,* 8(21), 76–94 (2017).

9. B. Li, Q. Wu, and W. Zhang, Water resources security evaluation model based on grey relational analysis and analytic network process: A case study of Guizhou Province, *Journal of Water Process Engineer*ing, 37, 101429 (2020). https://doi.org/10.1016/j.jwpe.2020.101429.

10. M. Yıldırım, K. Bal, and M. Doğan, GRI ilişkisel analiz yöntemi ile finansal performans analizi: bist'te işlem gören demir çelik şirketleri üzerinde bir uygulama, *Muhasebe Bilim Dünyası Dergisi,* 23(1), 122–143 (2021). https://doi.org/10.31460/mbdd.788840.

11. N. F. Silva, M. Santos, and C. F. S. Gomes, An integrated CRITIC and grey relational analysis approach for investment portfolio selection, *Decision Analytics Journal*, 8, 100285 (2023). https://doi.org/10.1016/j.dajour.2023.100285.

12. N. Ersoy and S. Taslak, Comparative analysis of MCDM methods for corporate sustainability performance in the energy sector, *Ege Academic Review*, 23(3), 341–362 (2023). https://doi.org/10.21121/eab.986122.

13. B. Shil, P. Sinha, and S. Das, GRA-Based MADM strategy in possibility environment: Application to climate change and hydropower, *Process Integration and Optimization for Sustainability,* 7, 599–608 (2023). https://doi.org/10.1007/s41660-023-00319-w.

14. W. Peng and Y. Li, Application of grey relational analysis in evaluating the environmental loads in Hubei province, China, during 1995–2019, *Polish Journal of Environmental Studies,* 33(3), 2773–2786 (2024). https://doi.org/10.15244/pjoes/175262.

15. R. Ranjan and A. Saha, A novel hybrid multi-criteria optimization of 3D printing process using grey relational analysis (GRA) coupled with principal component analysis (PCA), *Engineering Research Express,* 6(015080) (2024). https://doi.org/10.1088/2631-8695/ad2320.

16. P. Priyalatha, L. Ramayee, and R. Kumari, Multi-objective optimization of a compact wideband antenna using grey relational analysis, *International Journal of Electronics and Communications,* 175, 155063 (2024). https://doi.org/10.1016/j.aeue.2023.155063.

17. S. Sandhu and R. Kaushal, Optimisation of anaerobic digestion of layer manure, breeding manure and cow dung using grey relational analysis, *Biomass Conversion and Biorefinery*, 14, 4715–4727 (2024). https://doi.org/10.1007/s13399-022-02677-w.

18. M. Rajesh, M. N. S. Sri, S. Jeyakrishnan, and P. Anusha, Optimization parameters for electro discharge machining on Nimonic 80A alloy using grey relational analysis, *International Journal on Interactive Design and Manufacturing,* 18, 1429–1442 (2024). https://doi.org/10.1007/s12008-023-01616-x.

19. Q. Zhu and J. Sarkis, Relationships between operational practices and performance among early adopters of green supply chain management practices in Chinese manufacturing enterprises,

Journal of Operations Management, 22(3), 265–289 (2004). https://doi.org/10.1016/j.jom.2004.01.005.

20. G. Uzun, Analysis of grey relational method of the effects on machinability performance on austempered vermicular graphite cast irons, *Measurement*, 142, 122–130 (2019). https://doi.org/10.1016/j.measurement.2019.04.059.

21. M. F. Bayramoğlu and G. Topaloğlu, Türkiye imalat sanayiindeki alt sektörlerin finansal risk derecelendirmesi, *Yönetim, Ekonomi, Edebiyat, İslami ve Politik Bilimler Dergisi*, 8(1), 1–22 (2023). https://doi.org/10.24013/jomelips.1259548.

22. H. Fidan, Critic ve mairca çok kriterli karar verme yöntemi ile uluslararası hedef pazar seçimi, *Karamanoğlu Mehmetbey Üniversitesi Sosyal ve Ekonomik Araştırmalar Dergisi*, 23(41), 291–309 (2021).

23. S. Çakır and S. Perçin, Çok kriterli karar verme teknikleriyle lojistik firmalarında performans ölçümü, *Ege Akademik Bakış*, 13(4), 449–459 (2013).

24. U. Ünlü, N. Yalçın, and İ. Yağlı, Kurumsal yönetim ve firma performansı: TOPSIS yöntemi ile BIST 30 firmaları üzerine bir uygulama, *Dokuz Eylül Üniversitesi Sosyal Bilimler Enstitüsü Dergisi*, 19(1), 63–81 (2017). https://doi.org/10.16953/deusbed.09673.

25. Y. Akgül, Borsa İstanbul'da işlem gören ticari bankaların finansal performansının bütünleşik CRITIC COCOSO modeliyle analizi, *Journal of Economics and Financial Researches*, 3(2), 71–90 (2021).

26. *Fortune 500 Turkey 2022 Rankings*, Retrieved from https://www.fortuneturkey.com/fortune500?yil=2022&tip=1.

27. G. Senir and A. Büyükkeklik, Sürdürülebilirlik raporlaması ve lojistik şirketler üzerine bir uygulama, *The International New Issues in Social Sciences*, 5(5), 119–138 (2017).

28. Turkish Airlines, *Turkish Airlines Sustainability Report 2022.* https://investor.turkishairlines.com/documents/sustainability/2022-sustainability-report.pdf.

29. Ekol Lojistik, *Sürdürülebilirlik Raporu 2022.* https://www.ekol.com/wp-content/uploads/2023/07/surdurulebilirlik_raporu_2022.pdf.

30. Mars Logistics, *Sürdürülebilirlik Raporu 2020-2022.* https://www.marslogistics.com/content/images/web/pdf/2022-surdurulebIlIrlIk-raporu-tr.pdf.

31. DSV, *Sustainability Report 2022.* https://docs.dsv.com/group/group-compliance/corporate-sustainability-report-2022/?page=1.

32. Noatum, *Sustainability Report 2022.* https://www.noatum.com/wp-content/uploads/2023/07/NOATUM_NFS_en_2022.pdf.

33. F. F. Gündüz and H. Ataş. *Gri Sistemler Teorisi Gri İlişkisel Analiz: Teori ve Uygulama* (Gazi Kitabevi, Ankara, 2022).

34. B. C. Giri, M. U. Molla and P. Biswas. Grey relational analysis method for SVTrNN based multi-attribute decision making with partially known or completely unknown weight information, *Granular Computing,* 5, 561–570 (2020). https://doi.org/10.1007/s41066-019-00174-6.

35. Y. Kuo, T. Yang and G. W. Huang. The use of grey relational analysis in solving multiple attribute decision-making problems. *Computers & Industrial Engineering,* 55(1), 80–93 (2008). https://doi.org/10.1016/j.cie.2007.12.002.

Chapter 29

Assessing Handwritten Examination Papers Using Artificial Intelligence Techniques

Nursena Akbuğa*, Yusuf Kuvvetli* and Adem Erik†,‡

*Department of Industrial Engineering, Çukurova University, Adana, Turkiye

† Project Office, Tarsus University, Mersin, 33400, Turkiye

‡ademerik@tarsus.edu.tr

Evaluating examination papers is a critical task for many educational institutions, demanding significant effort, time, and expertise to be completed within a constrained timeframe. This study aims to address the challenges of evaluating examination papers by developing a system to automatically read and assess handwritten Turkish exam papers. The proposed approach begins with extracting characters from the examination papers. The complexities involved in deciphering handwritten papers stem from factors such as a large number of characters, overlapping letters, and the inherent variability in individual handwriting, even within different samples from the same person. The system for assessing handwritten short answers utilizes Optical Character Recognition (OCR) techniques to capture characters from the papers. In this study, OCR is used to obtain handwritten Turkish characters, and its capabilities are enhanced with the OpenCV library to improve text detection rates. The second phase involves using natural language processing methods to extract keywords or sentence patterns from both student responses and correct answers. Finally, the similarity between the given and correct answers is calculated to grade the examination papers. This approach ensures an unambiguous assessment process for both academics and students and speeds up the grading process by enabling quick feedback.

1. Introduction

Handwriting Detection refers to the process of identifying and analyzing handwritten text in a document. This technology is similar to Optical Character Recognition (OCR) but focuses on interpreting and digitizing handwritten material. Handwritten examination papers remain a fundamental component of educational assessment across various levels of schooling. This method is valued for its simplicity, accessibility, and ability to reflect a student's raw thought processes. However, the growing digital transformation in education has led to an increasing interest in the application of advanced technologies like handwriting recognition and automated scoring systems for handwritten texts [1]. Recent research in handwriting recognition has focused on developing robust models to address the challenges posed by different languages and scripts. For instance, the MuLTReNets model was designed to recognize multiple languages and scripts simultaneously, reflecting the multilingual nature of many educational settings [2]. Similarly, the MSdocTr-Lite, a lightweight transformer model, was proposed for efficient recognition of handwritten texts in various scripts, enhancing the capability to handle complex, multilingual examination papers [3].

Moreover, advancements in deep learning techniques, such as convolutional neural networks (CNNs) and bidirectional long short-term memory (BiLSTM) networks, have demonstrated significant improvements in the accuracy and efficiency of handwriting recognition [4]. These models have been tested on widely used datasets, like the IAM dataset for English handwriting, and have been found to offer comprehensive solutions for accurately recognizing and scoring handwritten responses [5].

The shift towards automated scoring and recognition systems is not just limited to the English language. For example, specific research has been conducted to improve the recognition of Arabic [6], Kazakh [7], and Mongolian [8] handwriting, among others. This wide-ranging research emphasizes the growing need for inclusive technologies that accommodate diverse linguistic contexts, which is crucial in maintaining fairness and equity in educational assessments.

Despite these advancements, challenges remain, particularly in recognizing cursive and complex scripts, which are often used in handwritten examinations. Studies have highlighted the difficulties in recognizing such texts due to the natural variability in handwriting styles and the lack of standardization in handwriting practices [9]. Moreover, agglutinative languages like Turkish present unique challenges for handwriting recognition systems due to their complex morphological structures [10].

In this study, two main topics were focused on and investigated. The first one is that investigate Turkish handwriting recognition/detection for evaluating exams. The second one is to propose an autonomous method to evaluate answers consisting of defined characters. In this context, the objectives of the study are listed below:

- To ensure accurate recognition of Turkish characters.
- To distinguish between questions and answers on the examination paper.
- To achieve scores that closely align with those of manual exam evaluations.
- To perceive text while maintaining the integrity of sentences and words.

- To match answer keys with the corresponding answers for exam evaluation.

The problem addressed in this study pertains to the automatic and fair evaluation of examination papers, which remain one of the most widely used methods for assessing educational outcomes. In this context, the first step involves recognizing the handwriting on the examination papers and converting it into a digital format. The next step is to compare the digitized text with the answer keys to determine the student's exam score. At this stage, the type of questions plays a crucial role. The case study discussed in this work includes two types of questions: those requiring exact matches and those evaluated based on similarity scores. The problem statement addressed in this study is illustrated in Figure 1.

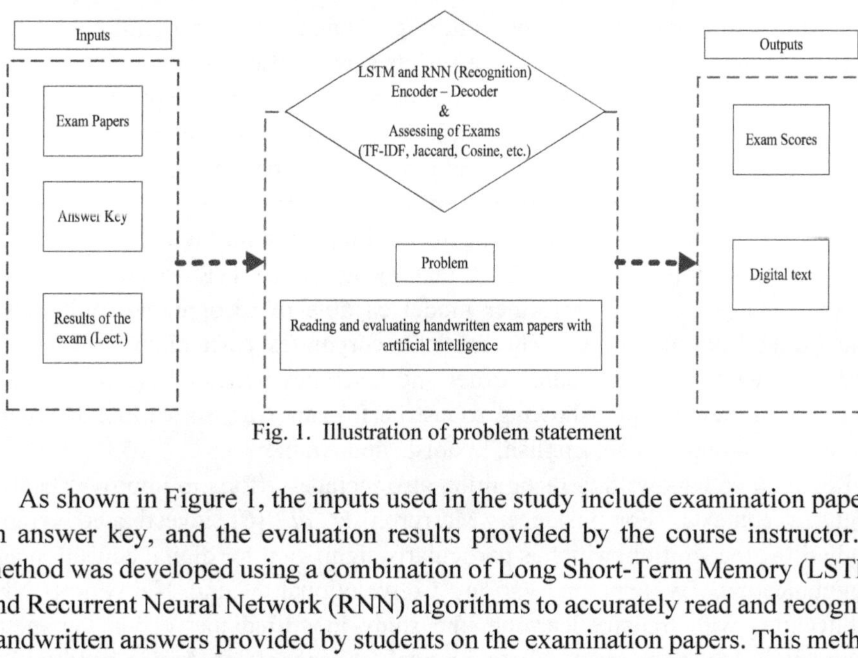

Fig. 1. Illustration of problem statement

As shown in Figure 1, the inputs used in the study include examination papers, an answer key, and the evaluation results provided by the course instructor. A method was developed using a combination of Long Short-Term Memory (LSTM) and Recurrent Neural Network (RNN) algorithms to accurately read and recognize handwritten answers provided by students on the examination papers. This method converts the handwriting into digital text. For comparing the students' answers obtained from the examination papers with the answer key, similarity models such as Jaccard, TF-IDF, and Cosine were employed. Finally, the exam scores generated by these models were compared with the scores determined by the course instructor to evaluate the model's performance.

2. Background

Handwriting recognition has been an area of active research, particularly for educational applications where handwritten examination papers are still commonly used. Recent advancements in deep learning have enabled the development of systems that integrate handwriting recognition with automated essay scoring. For example, Sharma *et al.* [1] introduced a system that utilizes deep learning models adapted to an unconstrained dataset from a national essay competition, which effectively scores essays and identifies key features that impact rubric-level scores. Despite some performance degradation due to transcription errors, this approach demonstrates the potential of combining hand-writing recognition with automated assessment methods.

Several studies have focused on multilingual handwriting recognition systems, which are crucial for handling diverse scripts across different languages. Chen *et al.* [2] proposed a framework that simultaneously performs script identification and handwriting recognition for multiple languages, including English, French, Kannada, Urdu, and Bangla. This framework incorporates Long Short-Term Memory (LSTM) and Connectionist Temporal Classification (CTC) methods to enhance the accuracy of handwriting recognition in multilingual contexts. To address data scarcity in handwriting text recognition (HTR), Dhiaf *et al.* [3] proposed a lightweight transformer model capable of recognizing multi-script handwriting from full pages. The model incorporates curriculum learning for understanding page-level reading order and leverages transfer learning for easy adaptation to new scripts, showing robust performance across various language datasets, including French, English, Spanish, and Arabic.

Research on handwriting recognition also includes efforts to improve specific language contexts. For instance, Masruroh *et al.* [6] investigated Arabic handwriting recognition, which is particularly significant for digital education and communication. By applying various Convolutional Neural Network (CNN) architectures with transfer learning, the study identified the Vision Geometry Group (VGG16) as the most effective model for recognizing Arabic handwriting on diverse datasets. In English handwriting recognition, Kizilirmak and Yanikoglu [4] developed a system combining CNN and Bidirectional Long Short-Term Memory (BiLSTM) models, achieving notable results on the IAM dataset. The system integrates a CTC layer, enhancing its performance through data augmentation techniques and thorough error analysis.

Other studies have explored less commonly researched languages, such as Arabic children's handwriting [5] and Kazakh handwriting [7]. Bin Durayhim *et al.* [5] proposed a deep learning model specifically for recognizing Arabic children's handwriting, outperforming other models in the literature. Yeleussinov *et al.* [7] introduced a generative adversarial network (GAN) approach to enhance Kazakh handwriting recognition, demonstrating significant improvements in accuracy over traditional methods. Fan *et al.* [8] developed a method for Mongolian handwriting recognition. The method integrates a pretrained language model into a sequence-to-sequence model with attention mechanisms. This approach significantly reduces character and word error rates, highlighting the benefits of using language models for script recognition. Ali and Abdulrazzaq [9] and Madane *et al.* [11] explored neural network-based systems for handwriting recognition, emphasizing the effectiveness of Convolutional Neural Networks (CNNs) in capturing the structural characteristics crucial for accurate character recognition. They evaluated several methodologies and datasets, demonstrating the robustness of CNNs in handwriting recognition tasks.

In the context of Turkish handwriting recognition, Yanikoglu and Kholmatov [10] addressed the challenges posed by Turkish's agglutinative morphology through a system that uses a prefix recognizer. Further studies by Çapar *et al.* [12] and Vural *et al.* [13] expanded on the recognition of isolated Turkish handwritten letters and the development of online handwriting recognition systems, respectively. Aydemir *et al.* [14] compared two systems for recognizing Ottoman and Turkish handwriting, demonstrating the superior performance of a Recurrent Neural Network (RNN)-based system over a Hidden Markov Model (HMM)-based approach. Bartos *et al.* [15] extended this line of research by evaluating different feature extraction techniques for Hungarian handwriting recognition, revealing that combining various techniques enhances classification accuracy. Bilgin and Yesilyurt [16] focused on large vocabulary online Turkish handwriting recognition using hidden Markov models, achieving promising results through preprocessing and bigram language modeling. Kuncan *et al.* [17] and Kabakus and Erdogmus [18] developed systems using artificial neural networks (ANNs) and novel convolutional neural network models, respectively, to improve recognition accuracy for Turkish handwritten letters. Lastly, Taşdemir [19] employed transfer learning in an online Turkish handwriting recognition system, demonstrating that fine-tuning a CNN-BLSTM model with a smaller Turkish dataset significantly improves recognition rates, addressing the challenge of data scarcity in Turkish handwriting recognition. Together, these studies illustrate the breadth of research on handwriting recognition across various languages and contexts, highlighting the importance of developing advanced models and techniques to improve accuracy and efficiency in digitizing handwritten content.

3. Material and Method

3.1. *Material*

The material used in the study consists of two data sets. The first of these is a data set in English consisting of 36000 words used for the training and validation of the encoder-decoder model [20]. The data are divided into two parts that are labels/file paths and images. A part of this data set consisting of 24000 words was used as training data, and the other part consisting of 12000 words was used as validation data. The second is a data set consisting of 19200 words obtained from the 43 exam papers of a course by students to test the model. The test data used are the handwritten answers given to the exam questions in Turkish.

3.2. *Method*

3.2.1. *Model for handwriting recognition*

The model consists of two main components the encoder and the decoder. The encoder part processes the input data (likely images in this context) using CNN layers followed by max-pooling layers and then maps it to a dense representation. The decoder part takes the dense representation from the encoder and generates the output sequence using Bidirectional LSTM (BiLSTM) layers. The primary role of the CNN encoder is to extract spatial features from the input images (handwritten text) and map these features into a dense representation, which can then be used by the decoder to generate an output sequence (recognized text). Figure 2 shows the encoder illustration in the model for recognition handwriting.

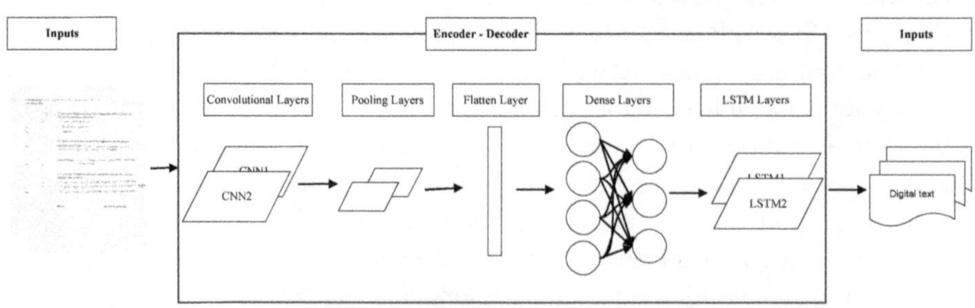

Fig. 2. Illustration of encoder-decoder model for handwriting recognition

There are convolutional, pooling, flatten and dense layers in the encoder section. Input data (exam paper images) is first processed by two convolutional layers. Convolutional layers are used to extract spatial features from the input. Eqs. (1) and (2) show the convolutional layers operations [4].

$$Conv1(X) = \sigma(W_1 * \mathrm{X} + b_1) \tag{1}$$

$$Conv2\big(Conv1(X)\big) = \sigma(W_2 * Conv1(X) + b_2) \tag{2}$$

The output of the first convolutional layer, $Conv1(X)$ is obtained by performing a convolution of the input X with the weight matrix W_1, adding the bias b_1, and then applying the activation function σ (ReLU). This operation detects specific patterns (edges, corners) in the input data and transforms it into an initial set of feature maps. The second convolutional layer takes the output from the first layer, $Conv1(X)$, and applies another convolution operation using a new set of filters, W_2, and bias b_2. The activation function σ is then applied to the resulting feature maps. This step further extracts more complex features by combining the simpler features learned in the first layer.

Each convolutional layer is followed by a max-pooling layer which reduces the spatial dimensions of the feature maps while retaining the most significant features. Eqs. (3) and (4) shows the pooling layers operations.

$$MaxPool1\,\big(Conv1(X)\big) = \max_{poolsize}\big(Conv1(X)\big) \tag{3}$$

$$\begin{aligned} &MaxPool2\Big(Conv2\Big(MaxPool1\,\big(Conv1(X)\big)\Big)\Big) \\ &\quad = \max_{poolsize}\Big(Conv2\Big(MaxPool1\,\big(Conv1(X)\big)\Big)\Big) \end{aligned} \tag{4}$$

The first max pooling layer (MaxPool1) takes the feature map output from the first convolutional layer, Conv1(X), and applies a max pooling operation over it. The pooling operation scans through the feature map with a fixed-size window, selects the maximum value within each window and constructs a new, down-sampled feature map. This operation reduces the size of the feature map while retaining the most prominent features, which makes the model more efficient and less prone to overfitting. The second max pooling layer (MaxPool2) takes the output of the second convolutional layer, Conv2(MaxPool1(Conv1(X))), and applies another max pooling operation. This further reduces the dimensionality of the feature map obtained from the second convolutional layer, focusing on the most salient features learned from the input data.

After processing the input data through the convolutional and pooling layers, the resulting feature maps are transformed into a dense, lower-dimensional representation to serve as the encoding space for the model. This process involves flattening the multi-dimensional feature maps into a single-dimensional vector and passing them through a fully connected (dense) layer, which learns a compressed representation of the input features. Equation (5) shows the flatten and dense layers operations.

$$Dense\ (X) = \sigma(W_d * flatten(MaxPool2) + b_d \tag{5}$$

Combining the flattening process and dense layer transformation, the model constructs a compact, meaningful encoding of the input data that serves as a foundation for further processing, such as sequence generation or classification in the decoder stage. This encoding space efficiently captures the essential patterns and structures within the input, supporting the model's overall performance and accuracy.

The decoder network is designed to generate an output sequence from the dense representation produced by the encoder. It consists of two BiLSTM layers. These layers are particularly well-suited for tasks such as sequence prediction or language generation, as they can learn complex temporal dependencies from the data. The decoder network consists of two BiLSTM layers. The first BiLSTM layer takes the dense representation h from the encoder as input and processes it in both directions. The Eq. (6) shows this operation [21].

$$BILSTM1(h) = \left[\overrightarrow{LSTM_1(h)}, \overleftarrow{LSTM_1(h)}\right] \tag{6}$$

The output of this layer, BILSTM1(h), is a concatenation of the forward and backward LSTM outputs, capturing dependencies from both directions. The output from the first BiLSTM layer is then fed into the second BiLSTM layer. The second BiLSTM operation is shown in Eq. (7).

$$BILSTM2(BILSTM1(h)) = \left[\overrightarrow{LSTM_2(BILSTM1(h))}, \overleftarrow{LSTM_2(BILSTM1(h))}\right] \tag{7}$$

The output of this layer, $BILSTM1(h)$, is a concatenation of the forward and backward LSTM outputs, capturing dependencies from both directions. The output from the first BiLSTM layer is then fed into the second BiLSTM layer.

3.2.2. *Models for Similarity*

In the study, four different models were used to compare the exam papers with the answer key and to find the similarity rate. The model that gave the best results was determined by comparing the exam scores determined by the academician teaching the course with the results given by the models. The first of these models is the Bag of Words model (BoW). The BoW model is a fundamental and widely used technique in Natural Language Processing (NLP) for text representation. This model simplifies the text by treating it as an unordered collection of words, disregarding grammar, syntax, and the original word order while retaining the multiplicity or frequency of the words.

The BoW model focuses solely on the occurrence of words to represent the content of the text. The first step involves constructing a vocabulary that contains all the unique words found in the entire corpus of text data. This vocabulary serves as a reference list against which each text document is evaluated. Each text document is then represented as a vector in which each dimension corresponds to a word in the vocabulary. The value assigned to each dimension is determined by the frequency of the corresponding word in the given text. The resulting vector is a numerical representation that captures the presence and frequency of words within the text, but not their order or context [22].

The second model is Jaccard Similarity Model. The Jaccard similarity model is a statistical measure used to evaluate the similarity and diversity between two finite sets. It quantifies how similar two sets are by comparing their common elements to the total number of unique elements present in both sets. Specifically, the Jaccard similarity coefficient is defined as the ratio of the size of the intersection of the sets to the size of their union. This measure is widely used in various domains, including information retrieval, data mining, and natural language processing, to compare and match text, documents, or other data structures. Jaccard similarity measures the similarity between finite sample sets and is defined as the size of the intersection divided by the size of the union of the sample sets [23]. The similarity rate of the Jaccard model can be determined with Eq. (8).

$$J(A,B) = \frac{(|A \cap B|)}{(|A \cup B|)} \tag{8}$$

Here, $|A \cap B|$ represents the cardinality (or size) of the intersection of sets A and B, indicating the number of elements that are common to both sets. $|A \cup B|$ represents the cardinality of the union of sets A and B, indicating the total number of unique elements present in either set.

The third model, which is Cosine similarity, is a metric used to measure the similarity between two non-zero vectors in a multi-dimensional space. It is widely used in various fields, particularly in text analysis and natural language processing, to quantify the similarity between two documents, sentences, or words. This measure considers the cosine of the angle between the vectors, rather than their magnitude, providing a normalized measure of similarity that is independent of the size of the vectors [24]. The similarity rate of the Cosine model can be determined with Eq. (9).

$$Cosine\ Similarity = \cos(\theta) = \frac{A.B}{||A||\ ||B||} \tag{9}$$

By dividing the dot product of the two vectors by the product of their magnitudes, cosine similarity produces a value ranging from -1 to 1. A cosine similarity of 1 indicates that the vectors are identical (i.e., they point in the same direction), while a similarity of 0 suggests that they are orthogonal (i.e., they share no similarity). A cosine similarity of -1 indicates that the vectors are diametrically opposed. Here A · B denotes the dot product of vectors A and B, ||A|| and ||B|| are the magnitudes (or Euclidean norms) of vectors A and B, respectively.

The fourth model for calculating similarity rate is TF-IDF, or Term Frequency-Inverse Document Frequency. The model is a statistical measure widely used in text mining and information retrieval to assess the importance of a word within a specific document relative to a collection or corpus of documents. It helps in identifying the most relevant words for each document, enhancing the effectiveness of tasks such as text classification, clustering, and information retrieval [25]. The TF-IDF value for a given term t in a document d is calculated as the product of two separate metrics: Term Frequency (TF) and Inverse Document Frequency (IDF). The calculation formula is shown in Eq. (10).

$$TF - IDF(t,d) = TF(t,d)\ x\ IDF(t) \tag{10}$$

Term Frequency measures the occurrence of a term t within a specific document d. It provides an indication of how frequently a word appears in a given document. The term frequency is calculated as Eq. (11).

$$TF(t,d) = \frac{\text{Number of times term t appears in document d}}{\text{Total number of terms in document d}} \tag{11}$$

This metric quantifies the relative frequency of a term within a document, normalizing the count of the term by the total number of words in that document. Higher values of TF indicate that the term is more prevalent in the document. Inverse Document Frequency measures the informativeness of a term by evaluating its rarity across the entire collection of documents. It assigns greater weight to less common terms, which are more likely to be significant for identifying the content of the document. The IDF is calculated as Eq. (12).

$$IDF(t) = \log\left(\frac{\text{Total number of documents}}{\text{Number of documents containing term t}}\right) \tag{12}$$

A higher IDF value corresponds to a term that appears in fewer documents, indicating that it is more distinctive and informative for differentiating the content of the documents.

The last model for calculating similarity rate is N-Gram model. The N-Gram model is a foundational concept in natural language processing (NLP) and computational linguistics. It represents a contiguous sequence of n items from a given sample of text or speech, where each item can either be a character or a word. N-grams are used to predict the next item in a sequence, analyze text patterns, and improve the performance of language models by capturing local context within text data [26]. There are several types of N-grams based on the value of n. They are listed below.

- Unigrams (1-grams): Unigrams represent individual words in a sequence, capturing the simplest form of context by treating each word as an independent unit. For example, given the sentence "The cat sat on the mat," the unigrams would be: ["The", "cat", "sat", "on", "the", "mat"] Unigrams are useful for basic frequency analysis and as features in simple text classification tasks.
- Bigrams (2-grams): Bigrams consist of pairs of consecutive words, capturing the immediate context between two adjacent words. This allows for the modeling of word pairs and their likelihood of occurring together. For the same sentence, the bigrams would be ["The cat", "cat sat", "sat on", "on the", "the mat"]. Bigrams are particularly useful for tasks such as sentiment analysis and part-of-speech tagging, where the relationship between pairs of words is significant.
- Trigrams (3-grams): Trigrams extend the context to sequences of three consecutive words, offering a more detailed representation of local dependencies within the text. For the sentence "The cat sat on the mat," the trigrams would be ["The cat sat", "cat sat on", "sat on the", "on the mat"]. Trigrams are beneficial for capturing more complex language structures, such as idiomatic expressions or common phrases, which are often missed by simpler models.

4. Findings

In this study, the initial task involved reading and digitizing the handwritten responses provided on exam papers. To achieve this, an encoder-decoder model was developed using Convolutional Neural Networks (CNN) and Bidirectional Long Short-Term Memory (BILSTM) algorithms. The performance of this model was evaluated using two standard metrics from the literature: Character Error Rate (CER) and Word Error Rate (WER). The Character Error Rate (CER) measures the number of incorrect characters in the model's output (substitutions), additional characters that are not present in the reference text (insertions), and characters that are present in the reference but missing from the hypothesis (deletions). Upon evaluation, the developed method achieved a CER of 2.77%, which is considered a reasonable rate. This performance is comparable to, or even better than, similar studies in literature. Similarly, the Word Error Rate (WER) assesses errors at the word level, including incorrect words (substitutions), extra words do not present in the reference text (insertions), and missing words that are present in the reference (deletions). The model's WER was found to be 5.10%, a value deemed acceptable according to established literature standards. Overall, the results suggest that the proposed model performs effectively in digitizing handwritten text, yielding competitive or superior error rates compared to previous studies.

Table 1 presents a comparative analysis of five different similarity models (Jaccard, TF-IDF, Cosine, BoW, and N-Gram) used for evaluating exam papers. The models were assessed based on four key statistical metrics: Mean, Standard deviation (Std. Dev.), Mean squared error (MSE), and Mean absolute percentage error (MAPE).

Table 1. The performance of similarity models for evaluating exam papers

	Jaccard	TF-IDF	Cosine	BoW	N-Gram
Mean	14.89	24.25	2.38	21.33	22.22
Std. Dev.	15.50	15.18	13.565	15.22	14.73
MSE	456.27	813.62	185.55	681.58	705.68
MAPE	0.25	0.37	0.198	0.33	0.33

The mean values indicate the average performance of each model in terms of similarity measurement, while the standard deviation reflects the variability or dispersion around these means. Among the models, the Cosine similarity method shows the lowest mean value (2.38) and the lowest standard deviation (13.57), suggesting that it not only provides the most consistent and stable results but also the most accurate average similarity score when compared to the other methods. The MSE metric evaluates the average squared difference between the estimated and actual values, serving as an indicator of model accuracy. A lower MSE value reflects a higher level of precision. Here, the Cosine similarity model has the lowest MSE (185.55), indicating superior accuracy in estimating the similarity between exam papers compared to all other models. MAPE measures the accuracy of predictions as a percentage, with lower values indicating better performance. The Cosine similarity method again exhibits the lowest MAPE (0.198), reinforcing its position as the most effective and reliable model for this specific application.

The analysis suggests that the Cosine similarity model outperforms the other models across all evaluated metrics, with the lowest means, standard deviation, MSE, and MAPE. This indicates that the Cosine similarity approach is the most effective for evaluating exam papers, providing consistent and accurate results. In contrast, the TF-IDF model performs the poorest, with the highest MSE and

MAPE, implying significant errors in its similarity assessments. The Jaccard, BoW, and N-Gram models show intermediate performance, with relatively higher variability and error rates compared to the Cosine model but still providing useful insights into similarity measurement. Figure 3 shows a line graph comparing the exam score values of different similarity or distance measures for 43 students.

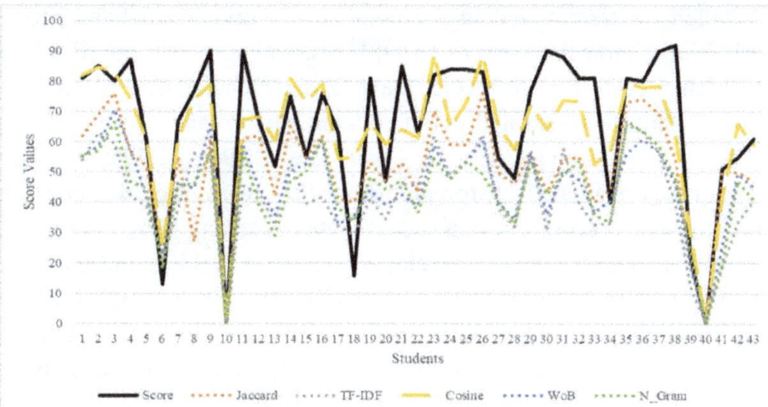

Fig. 3. Illustration of encoder-decoder model for handwriting recognition

The exam scores determined by the academician (represented by the black line) display considerable fluctuations across different students, indicating a wide range of performance levels. Some students achieved high scores, while others scored notably lower. The Cosine similarity measure (depiction by the yellow solid line) generally follows a pattern similar to the actual exam scores, albeit with some deviations. In contrast, the Jaccard and N-Gram measures (illustrated by the orange and green dotted lines, respectively) tend to maintain values within a mid-range level, showing lower fluctuation. The TF-IDF (gray dotted line) and Bag of Words (BoW, blue dotted line) exhibit distinct patterns that occasionally diverge significantly from the actual exam scores. Figure 3 provides a comparative analysis of how these different similarity measures align with the actual scores across a group of students. The figure reveals varying degrees of correlation between the similarity measures and the actual scores, with certain measures—such as Cosine similarity—demonstrating a stronger correlation. However, it also highlights instances where the measures deviate markedly from the actual scores, suggesting areas that may require further investigation or refinement in the application of these similarity measures for evaluating student performance.

5. Conclusion

This study successfully developed and tested an effective artificial intelligence algorithm for handwriting recognition, using real exam papers as test data. Among the five different methods employed to compare the exam responses with the answer key, cosine similarity emerged as the most effective approach. However, specific challenges were noted in the context of Turkish language processing, particularly with characters such as Ç, Ö, Ü, Ğ, and Ş, which posed a barrier to achieving higher performance levels. Despite these challenges, the handwriting recognition system demonstrated commendable performance with a Word Error Rate (WER) of 5.10% and a Character Error Rate (CER) of 2.77%. Furthermore, the student grades evaluated by the algorithm had an average deviation of 2.38 points from the actual scores, with a success rate exceeding 80%.

These results suggest that the developed algorithm holds significant potential for practical applications in automated grading systems, especially in the context of the Turkish language. However, several areas for improvement have been identified for future research. Firstly, implementing a dictionary-based correction mechanism for misread words could enhance overall accuracy. Secondly, refining the weight determination for the similarity methods used in comparison could further increase the success rate. Additionally, the integration of multi-criteria decision-making methods may enable more nuanced and precise scoring. Expanding the algorithm's capabilities to include the recognition and evaluation of mathematical expressions and numerical exams could broaden its applicability. Finally, enhancing the detection of identical expressions in open-ended questions would allow for the identification of correlations or potential instances of cheating between exam papers.

Acknowledgments

This study is derived from the project titled "Evaluation of Handwritten Examination Paper Using Image Processing and Multi-Criteria Decision-Making Methods" supported by TÜBİTAK 2209-A Program. Besides, Adem Erik is supported by TÜBİTAK 2211 program. The authors would like to thank the Scientific and Technological Research Council of Turkey (TÜBİTAK) for their support. The authors would like to thank Assist. Prof. Dr. Müfide Narlı for her valuable contributions to the study through exam evaluation.

References

1. Sharma, A., Katlaa, R., Kaur, G., and Jayagopi, D. B. Full-page handwriting recognition and automated essay scoring for in-the-wild essays. *Multimedia Tools and Applications*, 82(23), 35253–35276, (2023).
2. Chen, Z., Yin, F., Zhang, X. Y., Yang, Q., and Liu, C. L. MuLTReNets: Multilingual text recognition networks for simultaneous script identification and handwriting recognition. *Pattern Recognition*, 108, 107555, (2020).
3. Dhiaf, M., Rouhou, A. C., Kessentini, Y., and Salem, S. B. MSdocTr-Lite: A lite transformer for full page multi-script handwriting recognition. *Pattern Recognition Letters*, 169, 28–34, (2023).
4. Kizilirmak, F. and Yanikoglu, B. CNN-BiLSTM model for English Handwriting Recognition: Comprehensive Evaluation on the IAM Dataset. arXiv preprint arXiv:2307.00664, (2023).
5. Bin Durayhim, A., Al-Ajlan, A., Al-Turaiki, I., and Altwaijry, N. Towards accurate children's arabic handwriting recognition via deep learning. *Applied Sciences*, 13(3), 1692, (2023).
6. Masruroh, S. U., Syahid, M. F., Munthaha, F., Muharram, A. T., and Putri, R. A. Deep Convolutional Neural Networks Transfer Learning Comparison on Arabic Handwriting Recognition System. *JOIV: International Journal on Informatics Visualization*, 7(2), 330–337, (2023).
7. Yeleussinov, A., Amirgaliyev, Y., and Cherikbayeva, L. Improving OCR accuracy for Kazakh handwriting recognition using gan models. *Applied Sciences*, 13(9), 5677, (2023).
8. Fan, D., Sun, Y., Wang, Z., and Peng, Y. Online Mongolian Handwriting Recognition Based on Encoder–Decoder Structure with Language Model. *Electronics*, 12(20), 4194, (2023).
9. Ali, S. and Abdulrazzaq, M. A comprehensive overview of handwritten recognition techniques: A survey. *Journal of Computational Science*, 19, 569–587, (2023).
10. Yanikoglu, B. A., and Kholmatov, A. Turkish handwritten text recognition: A case of agglutinative languages. In *Document Recognition and Retrieval X*, 5010, 227–233. SPIE, (2003).
11. Madane, V., Ovhal, K., and Bhong, M. Handwriting recognition using artificial intelligence neural network and image processing, *International Research Journal of Modernization in Engineering Technology and Science*, 5(3), 1433–1447 (2020).
12. Çapar, A., Taşdemir, K., Kılıc, Ö., and Gökmen, M. A Turkish handprint character recognition system. In *Computer and Information Sciences-ISCIS 2003: 18th International Symposium*, Antalya, Turkey, Proceedings 18 (447–456). Springer Berlin Heidelberg, (Nov. 2003).
13. Vural, E., Erdogan, H., Oflazer, K., and Yanikoglu, B. An online handwriting recognition system for Turkish. In *Proceedings of the IEEE 12th Signal Processing and Communications Applications Conference*, 2004. 607–610. IEEE, (April, 2004).
14. Aydemir, M. S., Aydın, B., Kaya, H., Karlıağa, İ., and Demir, C. Tübitak Turkish-Ottoman handwritten recognition system. In *22nd Signal Processing and Communications Applications Conference (SIU)*, 1918–1921. IEEE, (April, 2014).
15. Bartos, G. E., Hajnal, É., and Hoşcan, Y. Comparison of Feature Extraction Techniques for Handwriting Recognition. In *IEEE 12th International Symposium on Applied Computational Intelligence and Informatics (SACI)*, 405–410. IEEE (May 2014).

16. Bilgin, E. F., and Yeşilyurt, A. B. Y. Large vocabulary recognition for online Turkish handwriting with sublexical units. *Turkish Journal of Electrical Engineering and Computer Sciences*, 26(5), 2218–2233, (2018).
17. Kuncan, M., Vardar, E., Kaplan, K., and Ertunç, H. M. Turkish handwriting recognition system using multi-layer perceptron. *Journal of Mechatronics and Artificial Intelligence in Engineering*, 1(2), 41–52, (2020).
18. Kabakus, A. T., and Erdogmus, P. A novel handwritten Turkish letter recognition model based on convolutional neural network. *Concurrency and Computation: Practice and Experience*, 33(21), e6429, (2021).
19. Taşdemir, E. F. B. Recognition of Online Turkish Handwriting using Transfer Learning. *Gazi University Journal of Science Part C: Design and Technology*, 11(3), 719–726, (2023).
20. Sarkar, S., Handwriting Recognition (OCR), Kaggle, (2020). https://www.kaggle.com/datasets/ssarkar445/handwriting-recognitionocr/data.
21. Lu, W., Li, J., Wang, J., and Qin, L. A CNN-BiLSTM-AM method for stock price prediction. Neural Computing and Applications, 33(10), 4741–4753, (2021).
22. Qader, W. A., Ameen, M. M., and Ahmed, B. I. An overview of bag of words; importance, implementation, applications, and challenges. In *2019 international engineering conference (IEC)*, 200-204. IEEE, (June 2019).
23. Bag, S., Kumar, S. K., and Tiwari, M. K. An efficient recommendation generation using relevant Jaccard similarity. *Information Sciences*, 483, 53–64, (2019).
24. Dehak, N., Dehak, R., Glass, J. R., Reynolds, D. A., and Kenny, P. Cosine similarity scoring without score normalization techniques. In *Odyssey*, 15, (June 2010).
25. Bafna, P., Pramod, D., and Vaidya, A. Document clustering: TF-IDF approach. In *2016 International Conference on Electrical, Electronics, and Optimization Techniques (ICEEOT)*, 61–66. IEEE, (March 2016).
26. Kondrak, G. (2005, November). N-gram similarity and distance. In *International symposium on string processing and information retrieval*, 115–126. Berlin, Heidelberg: Springer Berlin Heidelberg.

Chapter 30

Architecting and Evaluating a RAG-Based Question Answering System for SQuAD Dataset

Nimet Aksoy[*,‡], Murat Osman Ünalır[*,§] and Zekeriya Anıl Güven[†,¶]

[*]*Computer Engineering Department, Ege University, İzmir, Turkey*
[†]*Computer Engineering Department, İzmir Bakırçay University, İzmir, Turkey*
[‡]*nimet.aksoy@balikesir.edu.tr*
[§]*murat.osman.unalir@ege.edu.tr*
[¶]*zekeriyaanil.guven@bakircay.edu.tr*

Large language models have gained great popularity in recent years and have started to be used in various applications in the field of natural language processing [1, 2]. The main advantage of these models is that they have high-level language comprehension abilities [3]. However, since they are designed to understand general knowledge, they may have difficulty achieving the desired success in specialized subjects. RAG (Retriever-Augmented Generation) technique has been developed to close this gap. RAG focuses on improving the accuracy and reliability of large language models by using information and data from external sources [4, 5]. In this study, a question-answering system was designed using the RAG technique on the SQuAD 2.0 dataset. This dataset contains two different types of questions, the answer of which is included in the context and the answer is not included in the context. The aim is to determine the most appropriate parameter configuration to increase RAG performance with the questions whose answers are included in the content. For this purpose, a series of experiments were conducted with different independent variables. As a result of the experiments, it was determined that "chunk_size" and "search type" parameters were the independent variables that most affected RAG performance, depending on the characteristics of the dataset used. System performance was evaluated by comparing LLM-generated responses with reference responses included in the dataset. This evaluation was performed automatically using an LLM-supported evaluation model and used the answer correctness metric [4]. As a result of different experiments, accuracy rates between 78% and 82% were obtained. Questions with answers not included in the context were used to test the LLM's tendency to produce hallucinations. This shows that the model produces unrealistic or incorrect answers. Different "prompts" were identified in the research, and it was observed that LLM did not create hallucinations when a correct "prompt" was used. However, when the "prompt" was not used, LLM was observed to answer questions not included in the dataset and produce hallucinations.

1. Introduction

We use question answering systems in many areas like Chatbots, virtual assistants, educational apps, etc [6]. Especially after the emergence of large language models, we have more powerful tools. Large language models have tremendous natural language understanding. Because they were trained on large amounts of text data from the internet and other external sources. Despite the ability to understand language, large language models have been criticized for reliability. This issue is called hallucination and there are many publications on the subject. Hallucination means that the language model presents things that do not actually exist as real or changes existing facts [7–9]. Although large language models have general language understanding capabilities because they are trained on large and diverse text data, they do not have the same success with domain-specific data and tend to suffer from hallucination like problems. In addition, the cost of retraining these models is high [2]. The RAG (Retrieval-Augmented Generation) method, developed to overcome these weaknesses and minimize costs, uses information obtained from external sources by storing it in the vector database [2, 5, 10]. When a user question comes up, a similarity search is made, and the most similar information is selected and used by the large language model (LLM). This method allows general language models to be effective in overcoming deficiencies in specific subjects and contributes to the model gaining domain-specific expertise through information obtained from external sources. In this study, a question answering system is proposed on a domain-specific dataset (SQuAD 2.0) with the RAG approach. We tested the "LLM" performance with the RAG approach on SQuAD 2.0 and tested the "LLM" hallucination. In other sections, problem definition, the proposed method, results, and evaluation sections will be explained.

2. Problem Definition

Large language models were trained on huge amounts of text data, and they have the high ability to understand natural languages. Retrieval-augmented generation (RAG) is a technique to increase the accuracy and reliability of large language models with real/true information from external sources. Especially on domain-specific datasets, RAGs can improve gaining domain-specific expertise through information obtained from external sources [11]. In this study, we used the RAG (Retrieval-Augmented Generation) approach on the SQuAD 2.0 dataset developed by Stanford University. The SQuAD 2.0 dataset contains two types of data based on answer type. The first type consists of classic questionanswer pairs, in which the answers are in the corresponding paragraph. The other type includes questions that are not answered in the relevant paragraph, called unanswered questions.

Our study consisted of two main aims:

(1) We experimented with different parameters using the first dataset to identify the optimal parameter configuration for enhancing RAG performance.

(2) We evaluated the hallucination problem of large language models using a dataset containing unanswered questions.

3. Proposed Method

In this section, the road map, materials, and steps are defined. First, we have described our road map in Figure 1.

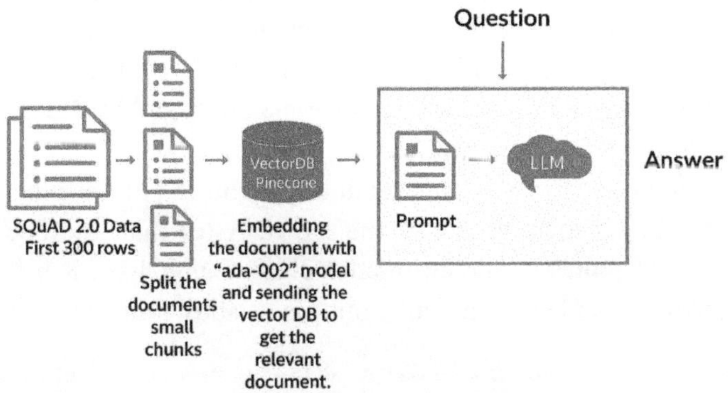

Fig. 1. Road Map of Study.

In this study, the SQuAD 2.0 dataset developed by Stanford University was used. The Stanford Question Answering Dataset (SQuAD) is a reading comprehension dataset consisting of questions asked by crowd workers on a set of Wikipedia articles. For this dataset, more than 100,000 question-answer pairs were obtained using 536 articles [12]. The dataset contains many paragraphs from a Wikipedia article. There are also many question-answer pairs obtained from each paragraph. An example content and the question-answer pairs of this content are shown in Figure 2.

Beyoncé Giselle Knowles-Carter (/biːˈjɒnseɪ/ bee-YON-say) (born September 4, 1981) is an American singer, songwriter, record producer and actress. Born and raised in Houston, Texas, she performed in various singing and dancing competitions as a child and rose to fame in the late 1990s as lead singer of R&B girl-group Destiny's Child. Managed by her father, Mathew Knowles, the group became one of the world's best-selling girl groups of all time. Their hiatus saw the release of Beyoncé's debut album, Dangerously in Love (2003), which established her as a solo artist worldwide, earned five Grammy Awards and featured the Billboard Hot 100 number-one singles "Crazy in Love" and "Baby Boy".	**When did Beyonce start becoming popular?** Ground Truth Answers: in the late 1990s.
	What areas did Beyonce compete in when she was growing up? Ground Truth Answers: singing and dancing.
	Who managed the Destiny's Child group? Ground Truth Answers: Mathew Knowles

Fig 2. SQuAD Data Example.

A total of 53775 unanswered question-answer pairs in the same documents were added to SQuAD 2.0 version. Thus, it was aimed to test the difficulty level of the QA systems [13]. Each row contains title, context, question and answer fields. Title is subject of the paragraph, context is paragraph, question is related question obtained in paragraph and answer is real answer to the question.
While working on SQuAD 2.0.

(1) We did many experiments with different independent variables on SQuAD 2.0 (the answer is in the relevant document) and evaluated the results.
(2) We did experiments (with the best parameter of the result of the first experiment) on unanswerable questions on SQuAD 2.0 and tested the hallucination of the large language model.

Answers were removed from the dataset at the beginning of the experiment to prevent them from being memorized by the LLM. After that, the documents were load-ed with LangChain's document loader class. LangChain is a framework for the RAG approach. Loaded document example as shown in Figure 3.

page_content='Beyoncé Giselle Knowles-Carter (/biːˈjɒnseɪ/ bee-YON-say) (born September 4, 1981) is an American singer, songwriter, record producer and actress. Born and raised in Houston, Texas, she performed in various singing and dancing competitions as a child and rose to fame in the late 1990s as lead singer of R&B girl-group Destiny\'s Child. Managed by her father, Mathew Knowles, the group became one of the worlds\'s best-selling girl groups of all time. Their hiatus saw the release of Beyoncé\'s debut album, Dangerously in Love (2003), which established her as a solo artist worldwide, earned five Grammy Awards and featured the Billboard Hot 100 number-one singles "Crazy in Love" and "Baby Boy".',

metadata= {'id': '56be85543aeaaa14008c9063', 'title': 'Beyoncé', 'question': 'When did Beyonce start becoming popular?'}

Fig 3. Loaded Document Example.

Based on the experiments we conducted, the impact of chunk size, search type, and temperature on the performance of RAG systems became clearly evident, as each of these independent variables significantly influenced the quality and accuracy of the results. We carefully examined how varying the values of these parameters affected the system's ability to retrieve relevant information and generate coherent answers. The possible values and definitions of these variables, along with their specific roles in shaping the outcomes of the experiments, are

detailed in Table 1, providing a comprehensive overview of their contributions to the overall performance of the RAG model.
Different independent variables in Table 1 will directly affect the result. One is chunk size parameter, which splits the uploaded documents into small pieces before calculating the embedding. The purpose of breaking data into smaller pieces is to divide information into multiple smaller pieces to store information more efficiently and meaningfully. At this stage, the dimensions of dividing the data into pieces are important in determining question-answering performance. These dimensions may vary depending on the characteristics of the data and the embedding model used [14]. Chunk size is important to establish the semantic relationship between texts. In the research, the minimum and average values of the paragraphs were calculated. These values were used for chunk size. Also, there are suitable chunk size values for usage embedding models, appropriate chunk size values for our embedding model are 256 and 512 [15–16] . We performed four experiments for different chunk sizes and compared the results, using the other independent variables the same while per-forming these experiments.

Table 1. Independent Variables and Possible Values.

Name of Independent Variable	Definition	Usage Values
Chunk size	The number of maximum char. in the chunk	min lenght of the document mean len of document according to the embedding model ideal values: 256 and 512
Search type	Type of the search when retrieving the relevant document from the vector database. Search type default value is similarity.	mmr similarity search_kwargs: "k" relevant documents, the k value must be an integer. similarity_score_threshold: threshold value, the threshold value must be between 0 and 1
Temperature	The temperature variable used by LLM models establishes a balance between randomness and determinism when creating the output. As the value of this variable approaches 1, randomness increases, increasing the likelihood of the model producing hallucinations. However, as this value approaches zero, the model gives more consistent responses, and the likelihood of hallucinations decreases [7–8].	Between 0 and 1

After splitting the document into small chunks, we calculated the usage embedding models and the embedding vectors of the small chunks. Embedding refers to the numerical representation of words, phrases, sentences or documents. The generated embedding vectors represent the content of document chunks. After the embedding vectors of all chunks were calculated, they were stored in the vector database (Figure 4).

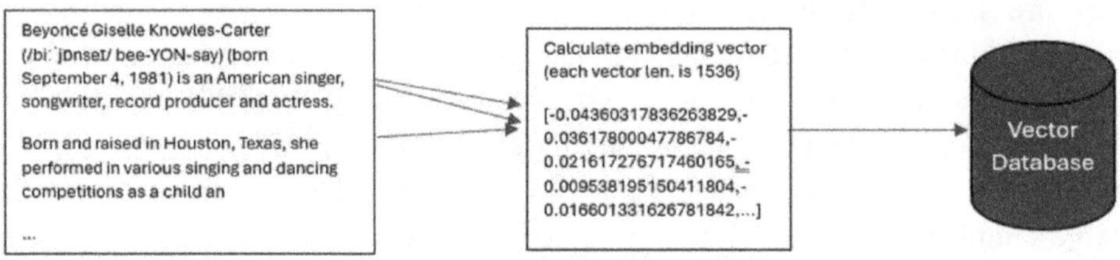

Fig 4. Split-Embedding-Store.

Vector databases are designed to efficiently store, manage and index large amounts of high-dimensional vector data. Unlike classical databases, vector databases often include special indexing algorithms and similarity search methods. These are similarity, threshold and MMR. By determining "k" values in the similarity search type, the most similar "k" documents are retrieved in the vector database. In the threshold search type, a threshold value is determined and documents whose similarity is greater than or equal to this threshold value are retrieved from the vector database. In the MMR method, it first selects the documents that are most similar to the query and then filters these documents by a certain diversity factor. In this way, it is aimed to present various documents to the user that are compatible with the query. In this study, we used cosine similarity to find the similarity of vectors. We also used the MMR search type to retrieve the relevant document in the vector database [5, 17, 18].

When a question is asked to the system, the relevant document of the question is taken from the vector database and sent to LLM. Therefore, LLM can know the scope of the answer. We can also use prompts to control scope. Prompts consists of special query expression and user's instructions to used to most accurate answer from LLM. This process is designed to increase the LLM's ability to obtain correct answers and extract the desired information more effectively [19].

4. Result and Discussion

In this study, we used gpt 3.5 turbo model, ada-002 embedding model, pinecone vector database and LangChain framework for RAG approach. We used different part of SQuAD 2.0 dataset; first part is first 300 rows are that question-answer pair the answer is in the related context. The second part is the last 100 rows in SQuAD 2.0 that question answer pair, the answer is not in the related context-no answer question-.

For the first part of the data, we did different experiments with different independent variables using the prompt. This prompt was used to determine the scope of the QA system to reduce hallucination. First, we did four experiments for different chunk size values, these values as shown in Table 1. We calculated min and mean length of document and also, we used suitable chunk size for ada-002 embedding model. We used RAGAs framework for evaluation, there are many different metrics for evaluation for context recall, context precision, answer relevancy, answer correctness and so on [4].

In our evaluation, we used the answer correctness metric to assess whether the model's responses were correct. Specifically, the answer correctness metric quantifies the accuracy of the model's answers by generating a value between 0 and 1, where 0 indicates an entirely incorrect answer and 1 represents a fully correct answer. The closer the score is to 1, the more accurate the answer is considered. To determine the threshold for what qualifies as a correct answer, we analyzed the distribution of answer correctness scores across various experiments. Based on this analysis, we observed that answers with a correctness score greater than 0.5 were generally considered valid, as this value represented a clear distinction between correct and incorrect responses. Consequently, we set the threshold for correctness at 0.5, which means that any answer with a correctness score above this value was categorized as correct, while answers below this threshold were classified as incorrect. This threshold provided a reliable way to evaluate the overall performance of the model across different experimental conditions and parameter settings. The correct answer numbers calculated with this method are shown in Figure 5.

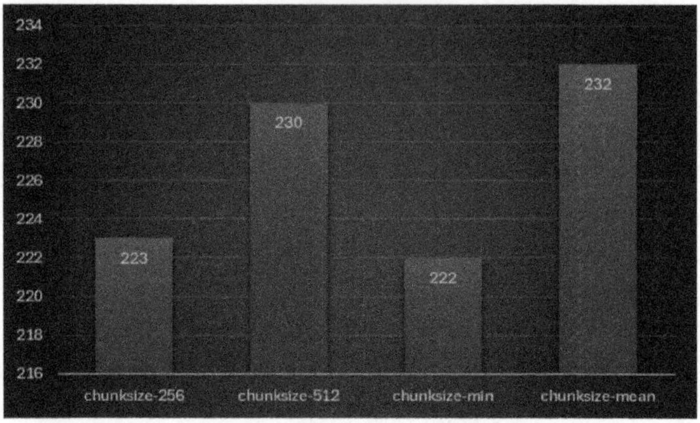

Fig 5. Number of LLM's correct answer in 300 questions according to chunk size.

Figures 6 and 7 have shown that answer correctness values of chunk sizes 256 and 900 values. These values are before the threshold calculation.

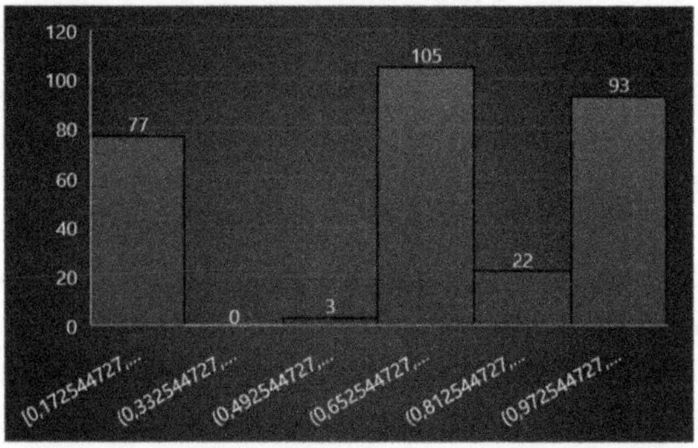

Fig 6. Answer Correctness Values for chunk size 256.

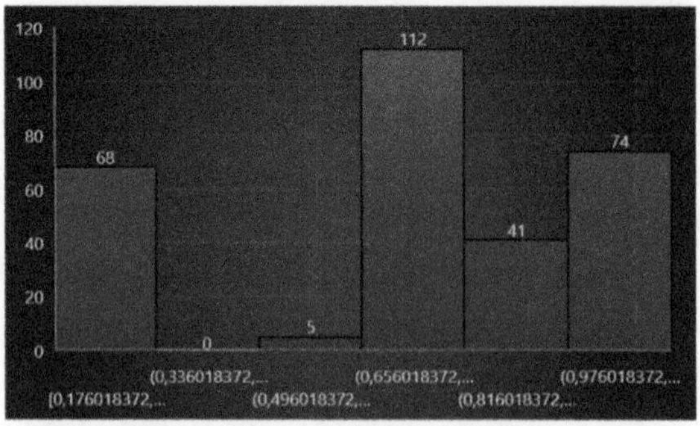

Fig 7. Answer correctness values for chunk size 900.

We employed the MMR (Maximal Marginal Relevance) search type to explore a diverse set of potential contexts for answering each question. Unlike traditional search types, which prioritize retrieving the most similar document based on a similarity metric, MMR aims to balance both relevance and diversity in the context selection process. By considering multiple potential answers, MMR helps avoid over-reliance on a single document, which can lead to biased or incomplete responses. In our experiments, MMR was applied to the best-performing results from the initial experiments, where we observed its effectiveness in producing more varied and relevant context for the answers.

To further assess the impact of different search types, we also experimented with a traditional similarity search on the same set of questions. In this case, the search was focused on retrieving the most similar document based purely on context similarity, without considering diversity. Interestingly, the results from this experiment showed a decrease in the number of correct answers compared to MMR. This suggests that while similarity-based search may prioritize more relevant documents, it lacks the ability to explore alternative context possibilities that could improve the answer's accuracy. The full results of this comparison, including the impact on answer correctness, are presented in Figure 8.

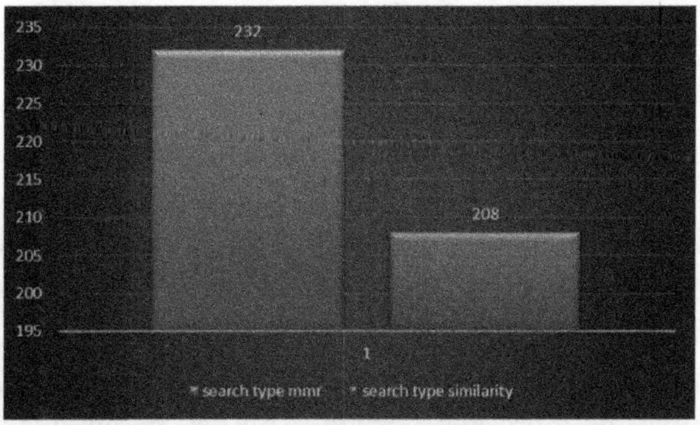

Fig 8. Answer Correctness Value with Different Search Type.

We did experiment without custom prompt on the best result of first experiments. In these situations, answer correctness ratio has increased.

In the second part of our study, we focused on the last 100 questions from the SQuAD 2.0 dataset, which are categorized as "unanswered" questions because the answers do not appear in the associated context. These questions were selected specifically to test the occurrence of hallucinations in the large language model (LLM). In this experiment, we initially applied the best-performing parameters from the previous experiments, without using any special prompt, in order to observe the model's natural tendency to generate hallucinations when no explicit instruction is given. The results revealed that the LLM consistently generated incorrect answers for all the unanswered questions. This behavior is characteristic

of hallucination, where the model fabricates information when it cannot find a suitable answer in the context.
To further evaluate the model's performance and ability to manage hallucinations, we then used a custom prompt in the same scenario. This prompt was specifically designed to encourage the model to refrain from fabricating an answer when the correct response is unavailable. When the prompt was applied, the LLM responded to each unanswered question with the phrase, "I don't know the answer." This response demonstrated the effectiveness of the custom prompt in preventing hallucination, as the model refrained from generating incorrect or fabricated answers, instead acknowledging its inability to provide a correct response. This experiment highlights the importance of prompt design in managing LLM hallucinations and improving the reliability of the system when faced with unanswered or ambiguous questions.

5. Conclusion

At the end of the study, we found that RAG is a vital approach for domain-specific data. However, we need to find the parameter that best fits the data. Particle size affected the results, but the particle size values were relevant to our data. When we examined the results for different chunk sizes, we saw that the average content length and values close to these values gave the best results. Because these values were valid for most of our content. The other important variable is search type. MMR fits our data because the contents of our data are related to each other. MMR took a chance on different documents that could be linked and used them in the results. However, if each content is not related to each other in different data sets, similarity search may be more appropriate. And key parameter is "prompt". The best way to prevent or reduce hallucination is "prompt". It may reduce the accuracy of the study but increase the reliability. We will test different embedding types and models for RAG in future studies. We will also examine in detail the questions that cannot be answered by the large language model and try to find solutions with knowledge graphs.

References

1. Yogi, K. (2024). Building a Question-Answering System Using LLM on 'Private Text'. Plain English AI, Mar. 15. [Online]. Available: https://ai.plainenglish.io/building-a-question-answering-system-using-llm-on-private-text-437125cd05ff.

2. Arefeen, M. A., Debnath, B. and Chakradhar, S. (2024). LeanContext: Cost-Efficient Domain-Specific Question Answering Using LLMs. *Natural Language Processing Journal*, 7, 100065. https://doi.org/10.1016/j.nlp.2024.100065.
3. Pan, S., Luo, L., Wang, Y., Chen, C., Wang, J. and Wu, X. (2024). Unifying Large Language Models and Knowledge Graphs: A Roadmap. arXiv preprint, arXiv:2306.08302.
4. Es, S., James, J., Espinosa-Anke, L. and Schockaert, S. (2023). RAGAS: Automated Evaluation of Retrieval Augmented Generation, https://doi.org/10.48550/arXiv.2309.15217.
5. Besbes, A. (2024). 3 Advanced Document Retrieval Techniques to Improve RAG Systems. Towards Data Science, Mar. 12. [Online]. Available: https://towardsdatascience.com/3-advanced-document-retrieval-techniques-toimprove-rag-systems-0703a2375e1c.
6. Zahra, A. (2023). Types of Questions in Question Answering Systems. Medium, Oct. 16. [Online]. Available: https://medium.com/mlearning-ai/types-ofquestions-in-questions-answering-systems-1d040bd42d08.
7. Neugebauer, F. (2023). Understanding LLM Hallucinations. Towards Data Science, Oct. 31. [Online]. Available: https://towardsdatascience.com/llm-hallucinations-ec831dcd7786.
8. Ji, Z., Lee, N., Frieske, R., Yu, T., Su, D., Xu, Y., Ishii, E., Bang, Y. J., Madotto, A. and Fung, P. (2023). Survey of Hallucination in Natural Language Generation. *ACM Computing Surveys*, 55(12), 1–38, Dec. https://doi.org/10.1145/3571730.
9. FHIRFLY (2024). Tuning the Temperature of GPT Models: A Guide to Reducing Hallucinations. Medium, Mar. 12. [Online]. Available: https://medium.com/@fhirfly/tuning-the-temperature-of-gpt-models-a-guide-toreducing-hallucinations-a92af019c534.
10. Xie, Q., Bishop, J. A., Tiwari, P. and Ananiadou, S. (2022). Pre-trained language models with domain knowledge for biomedical extractive summarization. *Knowledge-Based Systems*, 252, 109460, Sep. https://doi.org/10.1016/j.knosys.2022.109460.
11. Lewis, P., Perez, E., Piktus, A., Petroni, F., Karpukhin, V., Goyal, N., Küttler, H., Lewis, M., Yih, W. T., Rocktäschel, T., Riedel, S. and Kiela, D. (2020). Retrieval-augmented generation for knowledge-intensive NLP tasks. In *Advances in Neural Information Processing Systems (NeurIPS)*, 33, 9459–9474.
12. Rajpurkar, P., Zhang, J., Lopyrev, K. and Liang, P. (2016). SQuAD: 100,000+ questions for machine comprehension of text. In *Proceedings of the 2016 Conference on Empirical Methods in Natural Language Processing*, Austin, USA, 2383–2392. https://doi.org/10.18653/v1/d16-1264.
13. Rajpurkar, P., Jia, R. and Liang, P. (2018). Know what you don't know: Unanswerable questions for SQuAD. In *Proceedings of the 56th Annual Meeting of the Association for Computational Linguistics*, Melbourne, Australia, 784–789. https://doi.org/10.18653/v1/p18-2124.
14. Actalyst (2024). Optimize LLM Enterprise Applications through Embeddings and Chunking Strategy. Medium, Mar. 11. [Online]. Available: https://actalyst.medium.com/optimize-llm-enterprise-applications-throughembeddings-and-chunking-strategy-1bbdb03bedae.

15. Naarla, R. (2024). Prompt Engineering - Chunking Strategies. LinkedIn, Mar. 11. [Online]. https://www.linkedin.com/pulse/prompt-engineeringchunking-strategies-ravi-naarla.
16. Liu, N. F., Lin, K., Hewitt, J., Paranjape, A., Bevilacqua, M., Petroni, F. and Liang, P. (2023). Lost in the Middle: How Language Models Use Long Contexts. arXiv preprint, arXiv:2307.03172.
17. Han, Y., Liu, C. and Wang, P. (2023). A Comprehensive Survey on Vector Database: Storage and Retrieval Technique, Challenge, https://doi.org/10.48550/arXiv.2310.11703.
18. Carbonell, J. and Goldstein, J. (2017). The Use of MMR, Diversity-Based Reranking for Reordering Documents and Producing Summaries. *ACM SIGIR Forum*, 51(2), 209–210. https://doi.org/10.1145/3130348.3130369.
19. Coursera (2024). What Is Prompt Engineering? Definition and Examples. Mar. 12. [Online]. Available: https://www.coursera.org/articles/what-isprompt-engineering.

Chapter 31

The Circular Grid on Microstructure Stainless Steel Image Using Correspondence Analysis with K-Means Clustering

Yuli Sri Afrianti[*,†,|], Udjianna Sekteria Pasaribu[*,**], Husaini Ardy[‡,††], Fadhil Hanif Sulaiman[§,‡‡] and Grace Angelia[¶,§§]

[*]*Statistics Research Division, Faculty of Mathematics and Natural Science, Bandung Institute of Technology, Ganesa Street No. 10 Bandung, 40132, Indonesia*

[†]*Mathematics Doctoral program, Faculty of Mathematics and Natural Science, Bandung Institute of Technology, Ganesa Street No. 10 Bandung, 40132, Indonesia*

[‡]*Material Engineering and Science Research Division, Faculty of Mechanical and Aerospace Engineering Bandung Institute of Technology, Ganesa Street No. 10 Bandung, 40132, Indonesia*

[§]*Graduate Program in Computational Science, Faculty of Mathematics and Natural Science, Bandung Institute of Technology, Ganesha Street No 10, Bandung, 40132, Indonesia*

[¶]*Undergraduate Program in Mathematics, Faculty of Mathematics and Natural Science, Bandung Institute of Technology, Ganesha Street No 10, Bandung, 40132, Indonesia*

[|]*yuli.afrianti@itb.ac.id*
[**]*udjianna@itb.ac.id*
[††]*husaini@itb.ac.id*
[‡‡]*10117053@mahasiswa.itb.ac.id*
[§§]*10120010@mahasiswa.itb.ac.id*

Stainless steel (SS) is a material extensively used in daily life; however, exposure to high temperatures can lead to the formation of the sigma phase, reducing corrosion resistance. This phase appears in five colors, ranging from light brown (LBr), dark brown (DBr), bluish brown (BBr), light blue (LBl), to dark blue (DBl), indicating the gradual emergence of the sigma phase. In this study, microimages were obtained through laboratory processes, and three observation locations – A, B, and C – were selected based on temperature differences. The independence between the phase color and three locations will be tested using a Correspondence Analysis (CA), facilitated by color classification through KMeans Clustering and image processing. The result is that location A is dependent of DBl, B with BBr, and C with LBr. This indicates that location A, which was affected by higher temperatures, has high risk of corossion compared to other locations. This result research can become a basis for processing images then used to determine the condition of stainless steel specimen. Future studies should explore similar associations in other stainless steel types to enhance the understanding of corrosion mechanisms an inform industrial practices.

1. Introduction

Since Stainless Steel (SS) is the primary supporting material in lower-side superheater (LSSH) convection pipes for natural gas or petroleum, corrosion on SS is a basic issue that jeopardizes the sustainability of industry and infrastructure [1]. Because corrosion can lead to structural flaws and lower the quality of a product, more research and analysis is needed to determine whether using stainless steel as the primary pipe support material is feasible, particularly in terms of corrosion resistance. Elements dispersed in SS include carbon, silicon, manganese, iron, chromium, nickel, and manganese [2]. The protective oxide layer of chromium (Cr) acts as a self-healing barrier to stop corrosion [3]. Fe-Cr intermetallic precipitates (sigma phase) can be formed by Cr and Fe alloys when SS is heated for an extended period at temperatures ranging from 560 to 980°C [4–5]. These deposits are intermetallic materials. The color change on the surface of the sigma phase provides insight into its morphological features. Over time, color gradually becomes apparent. The surface of brown delta ferrite will turn blue after being subjected to high temperatures for an extended period; this is an indication that the sigma phase has formed. Aditya (2020) provided an explanation of how δ-ferrite changes into the sigma phase [5]. The degree of corrosion risk at each observation site is shown by distinct colors. In addition to determining the risk level of corrosion, this also enables the identification of regions, such as the center or edge, that are more vulnerable to corrosion. Thus, the investigation of SS sigma phase patterns according to color and location is the main goal of this study. First, Light Brown (LBr), followed by Dark Brown (DBr), Brownish Blue (BrBl), Light Blue (LBl), and Dark Blue (DBl), are the five color groups that will represent the corrosion process. This study will be conducted at three different locations: A, B, and C.

2. Methodology

The methodology of this study involves processing microimages of the specimens, discretizing them using circular girds, clustering the colors with K-Means, validating the models, and detecting associations between the location and color variables through correspondence analysis. This process produces a correspondence map, which is used to analyze the results. The research flowchart is presented in Figure 1.

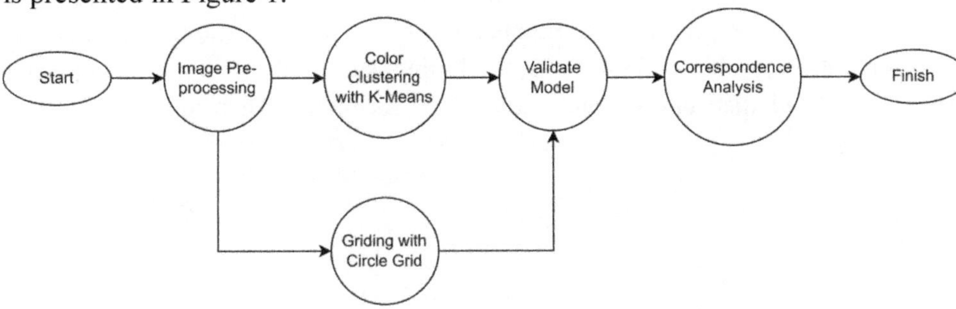

Fig. 1. Flowchart research.

2.1 *Data collection*

Several stages of procedures were carried out in the Material Engineering laboratory to process SS specimens into microimages. Starting with cutting, etching resin, and observing them on a microcomputer. After the images were obtained, three observation locations were selected: A, B, and C. It is suspected that these three locations will provide different analysis results because of their different positions. Microimages A, B, and C arc presented in Figure 2.

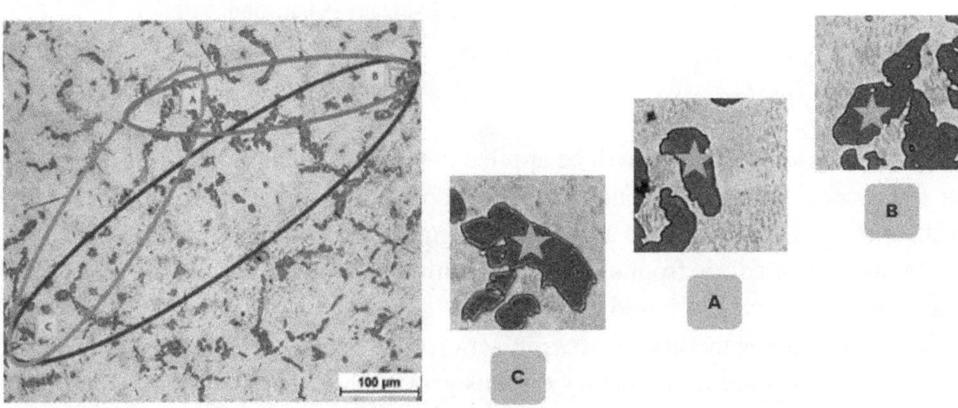

Fig. 2. Observation data.

2.2 *K-Means clustering*

Based on the sigma phase, the observation data were aggregated and categorized into five color groups using K-Means clustering. As shown in Figure 3, K-Means is: (i) a multivariate analysis technique that divides objects into clusters by minimizing the distance to the nearest means, or centroid [6–7], (ii) a method that partitions n data points into k clusters, where k initial centroids are selected and iteratively updated as the mean vector of each cluster [6], (iii) widely recognized as one of the most popular data mining techniques in the scientific community [8], and (iv) known to have limitations, including susceptibility to local optima, unevem centroid distribution, and reduced performance when applied to large datasets [9].

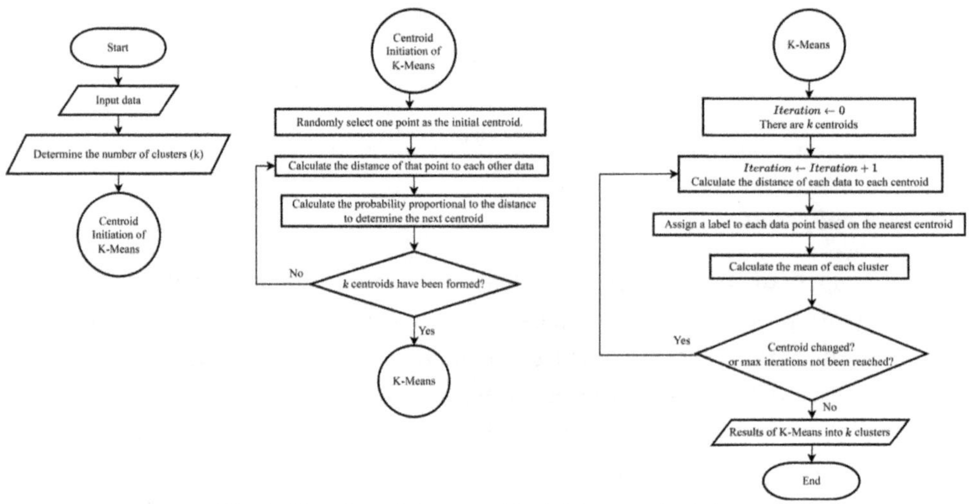

Fig. 3. Flowchart K-Means with K-Means++ initiation.

2.3 *Circular griding*

A weighted grid (i.e., circle) will be applied to this investigation to provide context for the observation images. The spatial point process concept is used to create the griding process as a stochastic approach to sigma phase problems [10]. The method of obtaining new colors from a circle grid with a diameter pixel. Each color in the grid is assigned a weight based on the area of the circle. A 12×12 pixel image in Figure 4a illustrates the grid process in action. The integers represent each pixel's weight and $d = 4$ are used to carry out this grid method (Figure 4b).

2.4 *Correspondence analysis*

A graphical method for determining the relationship between two categorical variables in a contingency table $N_{I \times J}$ is called Correspondence Analysis (CA). It is customary to use $k = 3$ (which includes a so-called trivial dimension) as the k dimensional approximation where a contingency table's rows and columns are optimally represented when employing correspondence analysis (CA). Since CA is a dimension reduction approach, Willemsen *et al.* [11] may assume that the k-dimensional approximation is not unique, meaning that the same k-dimensional approximation could exist in many contingency tables. The CA mechanism will be explained using the frequency of the sigma phase color in Table 1. Two categorical variables' frequencies are shown in Table 1 as a contingency table (n_{ij}), along with total rows ($n_{i.}$), total columns ($n_{.j}$), and total data (n). The relative frequency of the entire data and the marginal relative frequency of rows $\left(p_{i.} = \frac{n_{i.}}{n}\right)$ (and columns $\left(p_{.j} = \frac{n_{.j}}{n}\right)$) are represented by the values in brackets in the correspondence matrix $\left(p_{ij} = \frac{n_{ij}}{n}\right)$.

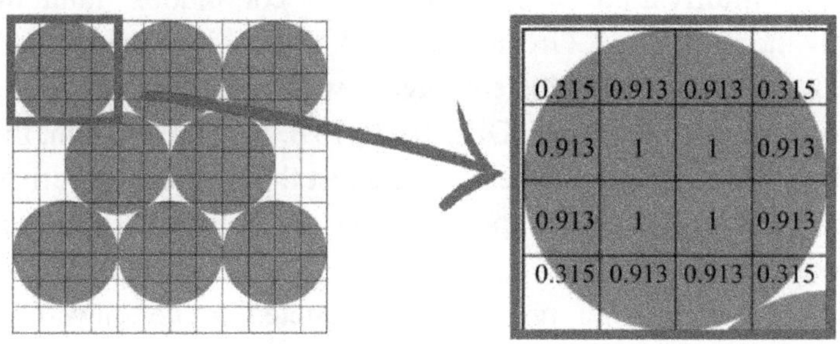

(a) **(b)**

Fig. 4. Example process of circular griding in the 12 × 12 pixel image with $d = 4$. Blue circles are new colors that represent each grid and (**b**) Weight of each pixel calculated by area of blue circle divided by total area of blue circle.

The association that CA is looking for is independence between row and column categories. Assumption of independence between a row and a column occurs when $p_{ij} = p_{i.}\, p_{.j}$ which is equivalent to $b_j = k$ that is when the entire column profile is the same as the centroid, or it is called the profile homogeneity assumption. To measure the significance of independence between rows and columns, the Chi Square statistic test can be used. To depict the profile in lower dimensions, orthogonal projection is carried out with the criterion of maximizing the accuracy of the display or the inertia $I = \chi^2/n$ [12]. This projection is carried out by weighting each vector $\vec{w}_j$ with $\sqrt{c_j}$ (purple colored points in Figure 5D) which is usually notated as a standard residual matrix (S with $s_{ij} = (n \times n_{ij} - n_{i.}n_{.j})/n\sqrt{n_{i.}n_{.j}}$). This procedure makes use of SVD, or $S = UDV^T$, where matrix D^2 represents the display's precision (the λ) and matrix U represents the projection axis. With D_c and D_k representing the diagonal matrices of the column mass and centroid, respectively, this equation also provides the coordinates of the profile projection $G = D_c^{-1/2}VD$ and the vertex projection $\Phi = D_r^{-1/2}U$. Lastly, by transposing the contingency table, symmetry is achieved in the column and row profile analysis. A scaling factor of D can be used to determine the coordinates of this dual analysis from other coordinates. Using the weighted grid, the researchers finally contrasted the CA results with the reference model.

3. Results

During preprocessing, images are cleaned. Masking is done by manually drawing boundaries, while background cleaning is performed using K-Means clustering with $k = 11$. This selection of k is based on preliminary analyses that suggest this number effectively captures the necessary background. Next, the cropping stage is done to focus on the image, facilitating the gridding process. The preprocessing stage produces multivariate data, with three continuous random variables represented by the RGB (Red, Green, Blue) color intensity at each pixel within the interval $[0, 1]$. This data then clustered again with K-Means clustering with $k = 20$ to get 5 color categories: LBr, DBr, BrBl, LBl, and DBl. Again, this selection of k is based on preliminary analyses that suggest this number effectively captures the necessary variations in color representation. This clustered data will be called reference model.

Next, the circular griding process is conducted to get new color (image) representation. As example an image of the observed location with $d = 4$ is shown in Figure 5a. The new color in circular grid then aligns with the clusters in the reference model (Figure 5b), resulting in 5 clusters (Figure 5c).

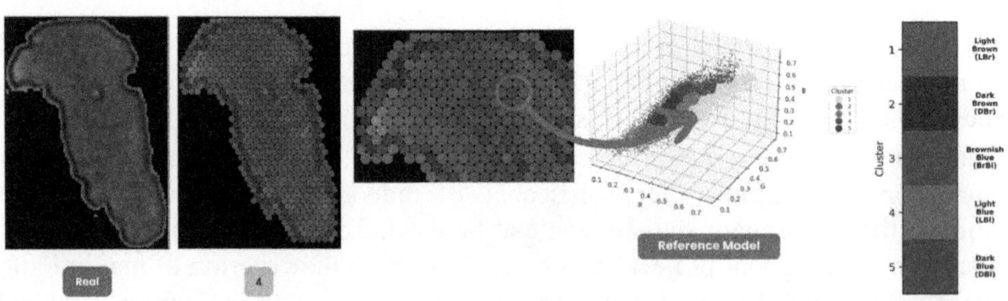

(a) (b) (c)

Fig. 5. Example of color classification process for griding with $d = 4$. (a) An observed location image and its corresponding circular grid with $d = 4$, (b) color classification is adjusted to the clusters in the reference model, and (**c**) the five clusters LBr, DBr, BrBl, LBl, and DBl represented by their controid color.

After the colors are classified into five categories, the frequency of each color at each location is tabulated in a contingency table. Table 1 displays the contingency table for a circular grid with a diameter of 4. It shows that the frequency of LBr is highest at location C, DBr is most frequent at location A, BrBl is also highest at location C, LBl appears most at location B, and DBl is abundant at location A. This contingency table serves as the input for Correspondence Analysis, which is used to identify associations between location and color variables. These associations are visualized in a correspondence map as the output of Correspondence Analysis. A one-dimensional correspondence map (asymmetric) for a circular grid with a diameter of 4 is shown in Figure 6.

Table 1. Contingency table when the diameter of the circular grid is 4.

$d = 4$	LBr	DBr	BrBl	LBl	DBl
A	23	37	260	69	236
B	375	27	384	21	153
C	687	17	524	47	108

Based on the correspondence map for different diameters, it is observed that location C is closely associated with LBr, location B with BrBl, and location A with DBl and DBr. This indicates that C has a strong relationship with LBr, B with BrBl, and A with DBl and DBr. Consequently, it can be concluded that location A tends to be blue, as the sigma phase, indicated by blue, is frequently found there. This is consistent with the position of A at the center, where it absorbs the most heat, leading to the highest risk of corrosion. In contrast, location C, which appears lighter brown, has a lower presence of the sigma phase and therefore the lowest risk of corrosion. The risk level at location B is moderate, falling between A and C.

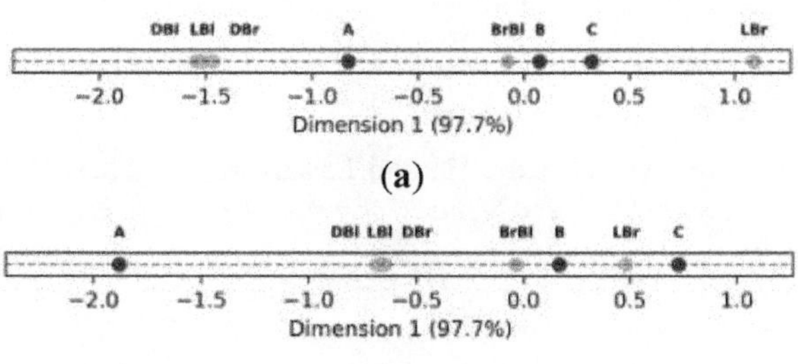

Fig. 6. Correspondence map 1D for the circular grid with $d = 4$. (a) Row analysis and (b) column analysis.

4. Conclusions

The modeling of the sigma phase as an indicator of the level of corrosion risk has been successfully conducted. A circular grid was applied, with observations made when $d = 4$. It was observed that location A tends to appear blue, as the sigma phase is represented by blue, indicating that the sigma phase is frequently found at this location. This result aligns with the fact that location A is in the center, where it absorbs the most heat, leading to the highest risk of corrosion. In contrast, location C tends to show lighter brown, indicating a lower presence of the sigma phase and thus the lowest corrosion risk. Location B, as expected, exhibits a moderate risk level, falling between A and C.

While these findings provide valuable insights into the relationship between the sigma phase and corrosion risk in this specific type of stainless steel, it is important to note that they may not be applicable to all stainless steel grades or other materials. Nonetheless, understanding how the sigma phase influences corrosion risk can inform future research and practices within the context of this particular material.

Future studies should focus on exploring similar relationships in other types of stainless steel and investigating the impact of different microstructural phases on corrosion behavior. By examining these variations, researchers can develop a more comprehensive understanding of corrosion mechanisms and enhance the application of this knowledge in relevant industrial contexts.

Acknowledgments

This research was funded by LPPM ITB through the "Riset Dosen Magang ITB 2024". The author would like to express gratitude for the support and funds provided for this research to be completed.

References

1. Afrianti, Y. S., Pasaribu, U. S., Ardy, H., and Lestari, K. E. (2024). *Elliptic Confidence on Correspondence Analysis Stainless Steels Corrosion Problem.* Engineering Letters, 32(7): 1332–1339
2. Gao, Q., Liu, Z., Sun, L., Ma, Q., Zhang, H., Bai, J., Lin, X., Yu, L., and Li, H. (2023). *Review on precipitates and high-temperature properties of alumina-forming austenitic stainless steel.* Journal of Materials Research and Technology, 25: 5372–5393. https://doi.org/10.1016/j.jmrt.2023.07.030.
3. Hsieh, C. and Wu, W. (2012). *Overview of intermetallic sigma (σ) phase precipitation in stainless steels.* ISRN Metallurgy (Online), 2012: 1–16. https://doi.org/10.5402/2012/732471.
4. Afrianti, Y. S., Ardy, H., Pasaribu, U. S., and Latief, F. D. E. (2021). *Identification of Inhomogeneous Temperature on Stainless Steel using Statistical Analysis.* Journal of Physics: Conference Series, 2084(1): 012008. https://doi.org/10.1088/1742-6596/2084/1/012008.
5. D. M. Aditya (2020). *Degradasi temperature tinggi material ASTM A 297 grade HH pada support pipas konveksi LSSH setelah 24 tahun operasi*, Bachelor Thesis, Institut Teknologi Bandung Indonesia.
6. Rencher, A. C. (2012). *Methods of multivariate analysis (2nd ed.).* John Wiley & Sons Ltd. Distribution, USA. https://doi.org/10.1002/9781118391686.

7. Charu, C.A. and Chandan, K.R. (2014). *Data Clustering Algorithms and Applications.* CRC Press, Boca Raton.
8. Miao, Dejun, Wenhao Wang, Yueying Lv, Lu Liu, Kaixin Yao, and Xiuhua Sui (2023). *Research on the Classification and Control of Human Factor Characteristics of Coal Mine Accidents Based on K-Means Clustering Analysis.* International Journal of Industrial Ergonomics 97: 103481. https://doi.org/10.1016/j.ergon.2023.103481.
9. Hu, Haize, Jianxun Liu, Xiangping Zhang, and Mengge Fang. *An Effective and Adaptable K-Means Algorithm for Big Data Cluster Analysis*. Pattern Recognition 139 (2023): 109404. https://doi.org/10.1016/j.patcog.2023.109404.
10. H. Taylor and S. Karlin, *An Introduction to Stochastic Modelling*, Academic Press. 2023.
11. Willemsen, Rick S.H., Michel van de Velden, and Wilco van den Heuvel. *On the Uniqueness of Correspondence Analysis Solutions*. Linear Algebra and its Applications 690 (2024): 162–85. https://doi.org/10.1016/j.laa.2024.03.014.
12. Greenacre, M. J. (1984). Theory and Applications of Correspondence Analysis. Academic Press, Great Britain.

Chapter 32

A Numerical Comparison of Heat Transfer in a Geological Medium Utilizing Analogy Between Solute and Heat Transfer Equations

Ayşe Özdoğan Dölçek

Geological Engineering, Balıkesir University, Balıkesir, 10145, Türkiye
aodolcek@balikesir.edu.tr

Numerical modeling plays a crucial role in understanding solute and heat transport phenomena within geological media for various applications such as groundwater management, contaminant remediation, and geothermal energy utilization. This study presents a comparative analysis of two prominent numerical modeling approaches: MODFLOW MT3DMS, widely used in hydrogeological studies, and Computational Fluid Dynamics (CFD), traditionally applied in fluid mechanics and heat transfer analyses. Both approaches rely on the advection-diffusion equation to describe the movement of solutes or heat through geological formations, with additional terms accounting for dispersion, reactions, and thermal gradients. Also, numerical discretization techniques utilized by MODFLOW MT3DMS and CFD to solve the governing equations, highlighting the similarities in finite difference, finite volume, or finite element schemes employed to approximate spatial and temporal derivatives. It discusses the challenges associated with accurately capturing complex transport phenomena, including heterogeneity, anisotropy, and non-linear reactions, and the strategies employed by both methodologies to address these challenges. In this study, problems taken from MT3DMS documentation and user guide [1] is benchmarked to get the familiarity of MODFLOW and MT3DMS uses. For the aim of this study, results of CFD model and MT3DMS model are compared to validate the CFD groundwater-coupled heat transport process through porous media, which has not been shown by previous studies. Hence, this study offers a comprehensive analysis of the analogies between MODFLOW MT3DMS and CFD in the context of solute and heat transport modeling in geological media, emphasizing their complementary roles and potential for advancing predictive capabilities in geoscience and engineering applications.

1. Introduction

Comprehensive knowledge of the mechanisms by which heat and solutes are transported in geological media is essential for several applications including the management of groundwater, remediation of contaminants, and usage of geothermal energy [2, 4]. The complexity of these processes arises from the interactions inherent in porous media, which are influenced by heterogeneity, anisotropy, chemical reactions, and temperature gradients [5, 6]. Quantitative modeling has become an essential technique for capturing these interactions, enabling the prediction of subsurface behavior under different circumstances.

In the field of groundwater flow and solute transport simulation, conventional models like MODFLOW and MT3DMS have been widely employed. MODFLOW utilizes a three-dimensional finite-difference approach to effectively model the movement of groundwater in confined, unconfined, or mixed aquifers, considering both steady-state and transient scenarios [7]. MT3DMS refers to a modular three-dimensional multispecies transport model that is used to simulate the processes of advection, dispersion, and chemical reactions of pollutants in groundwater system [1]. This code can also be used to simulate heat transport in porous media by using the analogy between solute and heat transport processes. Recent research [5, 6] offer evidence to support this claim. A study [2] investigated the effectiveness of MT3DMS for shallow geothermal systems. The MODFLOW and MT3DMS systems are interconnected, with MT3DMS utilising the flow regime estimation obtained from MODFLOW. Hence, the impact of temperature on water viscosity, density, and hydraulic conductivity is not considered. This is the primary constraint of MT3DMS. Nevertheless, the variability in temperature within the shallow subsurface is often minimal and deemed acceptable. Existing research has demonstrated that the total temperature differences across the model domain are below 5 K. Furthermore, there is no notable disparity between heat transport predictions produced using MT3DMS and heat transport predictions that consider variable density and variable viscosity flow [2, 4].

The aforementioned methods possess significant value in the modeling of complex transport phenomena. However, their constrained computing efficiency and lack of adaptability in managing unstructured grids pose difficulties for specific applications. On the other hand, cutting-edge developments in Computational Fluid Dynamics (CFD) have enabled the use of more adaptable modeling methods, especially for capturing complex geometry and boundary conditions [3]. CFD models, utilizing finite-volume techniques, offer enhanced adaptability in managing unstructured meshes found in fractured geological formations or drilling settings [8]. A comparison between CFD and conventional hydrogeological models such as MODFLOW–MT3DMS presents new possibilities to enhance the precision and effectiveness of transport simulations, especially in the context of integrating heat and groundwater movement.

This comparative analysis is based on the analogy between solute and heat transport equations. Both phenomena may be characterized by comparable advection-diffusion equations, with included terms for dispersion, reaction, and thermal gradients [1]. Drawing on these comparisons, recent studies have modified solute transport simulation software such as MT3DMS to simulate heat transfer in porous materials, demonstrating encouraging outcomes in both controlled laboratory environments and real-world scenarios [2, 4]. The objective of this study is to assess the efficacy of several numerical modeling methods, such as MODFLOW-MT3DMS and CFD, in accurate simulation of heat and solute movement in geological materials. The objective is to offer a thorough comprehension of how these techniques can be leveraged in conjunction to improve predictive skills in geoscience and engineering applications, especially in situations where computing efficiency and precision are of significance. This work compares the results of the CFD model and MT3DMS model to validate the CFD groundwater-coupled heat transport mechanism across porous media, a phenomenon not previously demonstrated in existing research.

2. Mathematical modeling

The general governing equation for solute transport solved by MT3DMS for our system is as follows in Eq. (1) [1]:

$$R\frac{\partial(nC)}{\partial t} = \nabla.(n\boldsymbol{D}.\nabla C) - \nabla.(\boldsymbol{q}C) + q_s C_s, \tag{1}$$

Where
C = mass concentration [kg m^{-3}]
t = time [d]
q = specific discharge vector (Darcy's velocity) [md^{-1}]
q_s = volumetric flow rate per unit volume of the aquifer (fluid sources) [m^{3}s^{-1}m^{-3}]
C_s = concentration of the source and sink flux [kg m^{-3}]
R = retardation factor [dimensionless]

$$R = \left(1 + \frac{\rho_b}{n} K_d\right),$$

where n is the porosity of the subsurface medium [dimensionless], ρ_b is the bulk density of the rock matrix, ρ_b = (1-n) ρ_s where ρ_s is the solid density, K_d is distribution coefficient [m^3 kg^{-1}]

D = hydrodynamic dispersion coefficient tensor [$m^2 s^{-1}$]

$$D = D_{mech} + D_m,$$

where D_{mech} is the mechanical dispersion coefficient tensor and D_m is the molecular diffusion coefficient.

Assuming the thermal equilibrium is maintained between the pore fluid and solid phase, the analogous heat transfer equation can be shown in Eq. (2) [8]:

$$\left(1 + \frac{1-n}{n}\frac{\rho_S}{\rho_W}\frac{c_S}{c_W}\right)\frac{\partial(nT)}{\partial t} = \nabla.\left(n\left(\frac{\lambda_{eff}}{n\rho_W c_W} + D_{mech}\right).\nabla T\right) - \nabla.(qT) + q_S T_S, \quad (2)$$

where

T = temperature [K]

T_s = source temperature [K]

c_s = specific heat capacity of the solid [J kg^{-1} K^{-1}]

c_w = specific heat capacity of the fluid [J kg^{-1} K^{-1}]

ρ_w = density of the fluid [kg m^{-3}]

q_s = volumetric heat source and sink [W m^{-3}]

λ_{eff} = effective thermal conductivity of the rock matrix [W m^{-1} K^{-1}]

$$\lambda_{eff} = n\lambda_w + (1 - n)\,\lambda_s,$$

where λw and λs are the thermal conductivities of the water and solid medium respectively.

The analogy between solute transport and heat transport equations is obvious from a comparison of Eq. (1) with Eq. (2). The first term on the right in solute transport Eq. (1) is diffusion and dispersion term. The diffusion term is the pure molecular diffusion D_m that denotes a process driven only by concentration gradient and equivalent to the thermal diffusivity driven by the temperature gradient in heat transfer equation as follows:

$$D_m = \frac{\lambda_{eff}}{n\rho_W c_W},$$

The dispersion term in Eq. (1) is the hydrodynamic dispersion that denotes a process driven by velocity gradient at pole scale:

$$D_{mech} = \alpha v,$$

where α is the dispersivity term [m], v is the seepage velocity [m/d]. The dispersion term is equivalent to natural convection in heat transport process, but it is neglected because MT3DMS does not account for density and velocity change due to the temperature gradient that leads to natural convection.

The second term in Eq. (1), which represents advection term, is analogous to transport of heat by groundwater flow Eq. (2). The third term in Eq. (1) represents fluid solute sinks/sources and is analogous to thermal sinks/sources in Eq. (2).

Comparison of Eq. (2) to Eq. (1) indicates that variables for the MT3DMS solute transport model can be re-interpreted for heat transport. The solute concentration becomes the temperature. The distribution coefficient (K_d) for solute transport is replaced by the thermal distribution factor which can be calculated for temperature using the following Eq. (3):

$$K_d = \frac{c_s}{\rho_w \, c_w}, \tag{3}$$

Retardation factor (R) and distribution coefficient (K_d) are represented as solute sorption in solute equation Eq. (1), is analogous to heat exchange between the solid and water in heat transfer equation Eq. (2). Details on using solute transport simulation for heat transport modeling can be found in [9].

3. Benchmarking: Problem of Two-Dimensional Transport in a Uniform Flow Field

This section presents the analogy between mathematical equations for solute and heat transport. The efficiency of MODFLOW and MT3DMS is assessed by benchmarking problems extracted from the MT3DMS documentation and user guide [1].

3.1 Model set-up

An analytical solution for two-dimensional transport of solute injected continuously from a point source in a steady-state uniform flow field is given by [10]. The analytical solution is applicable under these assumptions:

(a) the aquifer is infinite in areal extent and relatively thin in vertical extent, so that instantaneous vertical mixing can be assumed,

(b) the injection rate is insignificant compared with the ambient uniform flow.

A numerical model consisting of 46 columns, 31 rows and 1 layer is constructed for comparison with the analytical solution of [10]. The model parameters used in the simulation are listed in Table 1.

Table 1. Model parameters used in 2D transport of a uniform flow model.

Cell width along rows (Δx) = 10 m
Cell width along columns (Δy) = 10 m
Layer thickness (Δz) = 10 m
Groundwater seepage velocity (v) = 1/3 m/day
Porosity (θ) = 0.3
Longitudinal dispersivity = 10 m
Ratio of transverse to longitudinal dispersivity = 0.3
Volumetric injection rate = 1 m^3/day
Concentration of the injected water = 1000 ppm
Simulation time (t) = 365 days
West head = 46 m
East head = 1 m
Hydraulic conductivity (K) = 1 m/day

The flow model is surrounded by constant-head boundaries on the east and west borders and no-flow boundaries on the north and south borders. The hydraulic head values at the constant-head boundaries are arbitrarily chosen to establish the required hydraulic gradient. The simulation period is chosen so that the plume developed from the point source does not reach the boundaries.

3.2 Results

To evaluate the accuracy of flow conditions in MODFLOW, mass balance between inflow and outflow to/from model domain is performed. The results show less than 0.006% error that an appropriate mass balance is achieved (see Figure 1). The simulated results from MT3DMS are shown in Figure 2. It is clear that the solute is mostly transported by advection in the flow direction, some transport by hydrodynamic dispersion is also seen in transverse direction. To ensure the accuracy and reliability of these models, they need to be validated or verified against known solutions done by [1]. Thus, this numerical simulation is in good agreement with analytical solutions as seen in Figure 2.

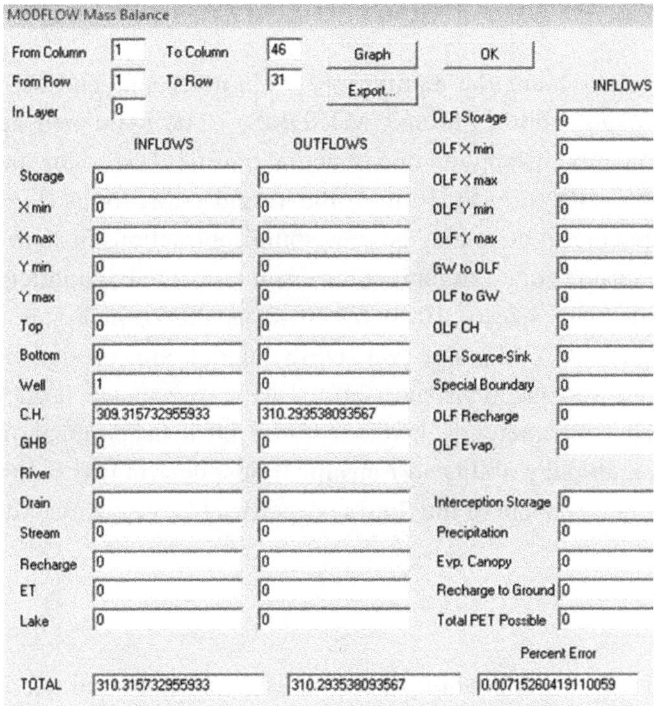

Fig. 1. Mass balance of 2D MODFLOW model.

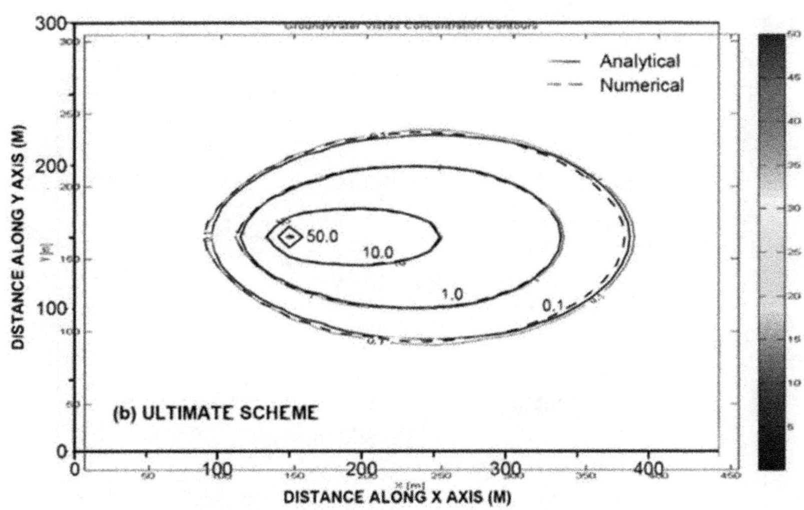

Fig. 2. Overlapping of concentrations contours from analytical solution and numerical solution (MT3DMS).

4. Comparison of CFD and MODFLOW-MT3DMS

Two-dimensional rectangular aquifer (490 m in horizontal and 310 m in vertical) is simulated using both CFD and MT3DMS. The simulated aquifer domain represents same dimension with one of actual borefield size. The model domain is discretized into cells 10 m by 10 m in the x- and y-directions (Figure 3). Due to the unstructured mesh, there may be seen some contractions in some places that do not affect the result. Solute/or temperature being injected to aquifer is represented as a red square with size of 10 m by 10 m as shown in Figure 3. The model parameters used in CFD and MT3DMS model are presented in Table 2. Additionally, a viscous resistance [$1/m^2$] is a parameter used only in CFD simulation. This parameter is inverse of the intrinsic permeability that is the measure of a material's ability to transmit fluids, determined by its internal pore structure, and independent of the fluid's properties like viscosity or density and can be defined using

$$\kappa = K\,\mu/\rho g$$

where κ is intrinsic permeability [m^2], K is hydraulic conductivity [m/s], μ is dynamic viscosity [kg/m·s], ρ is density [kg/m] and g is the acceleration due to gravity [m/s^2].

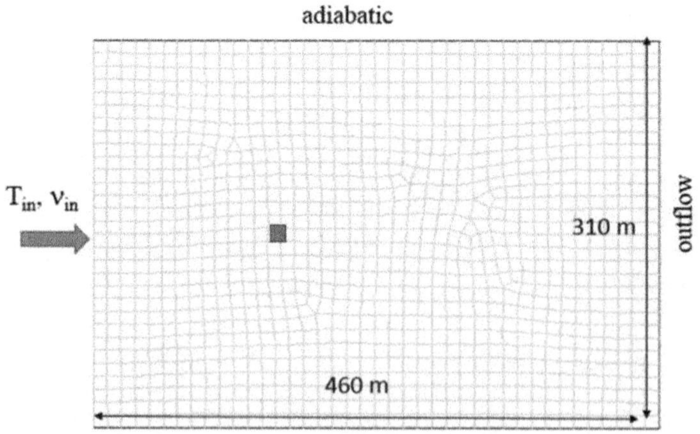

Fig. 3. Model geometry and boundary conditions used in CFD simulation.

Table 2. Model parameters used in CFD and MT3DMS simulations.

Geologic Unit	**Description**	**CFD**	**MT3DMS**
n [--]	Porosity	0.3	0.3
ρ_b [kg/m^{-3}]	Bulk density of geologic medium	[-]	157
ρ_s [kg/m^{-3}]	Density of solids in the geologic medium	224.3	[--]
c_s [Jkg^{-1}K^{-1}]	Specific heat capacity of the solid	7350	[--]
λ_s [Wm^{-1}K^{-1}]	Thermal conductivity of the solid	3	[--]
K [ms^{-1}]	Hydraulic conductivity of geologic medium	1.15E-05	1.15E-05
Fluid			
ρ_w [kgm^{-3}]	Density of water	998.2	[--]
c_w [Jkg^{-1}K^{-1}]	Specific heat capacity of the water	4182	[--]
λ_w [Wm^{-1}K^{-1}]	Thermal conductivity of the water	0.6	[--]
μ [kgm^{-1}s^{-1}]	Dynamic viscosity	0.001003	[--]
Boundary Conditions			
q [ms^{-1}]	Inlet velocity (Darcy's)	1.04E-06	[--]
v [ms^{-1}]	Seepage velocity	[--]	3.47E-06
T_i/C_i [K]	Initial ground temperature/or concentration	0	0
T_{in}/C_{in} [K]	Inlet fluid temperature/or concentration	1000	1000

5. Results

Figure 4 shows the temperature and concentration profile along the horizontal distance from the model simulated by CFD and MT3DMS, respectively. Heat transport by conduction was simulated and showed in Figure 4a. Heat transport by advection was also simulated by introducing groundwater flow using a seepage velocity of 3.47e-6 m/s in MT3DMS and Darcy's velocity ($q = v \times n$) of 1.04e-6 m/s in CFD as shown in Figure 4b. Both simulations are in strong agreement. Heat taken by ground water flow is transferring about a 50 m away from the source (Figure 4b). Contours for the temperature/or concentration around the source were also presented in Figure 5. MATLAB was used for post-processing the results for two-dimensional spatial distribution of concentration and temperature around the source that were derived from MT3DMS and CFD, respectively. Results shows that the concentration and temperature contours are in a good agreement. Some temperature disorder from the CFD results (Figure 5) can be seen due to discretization size used for surface meshing in CFD.

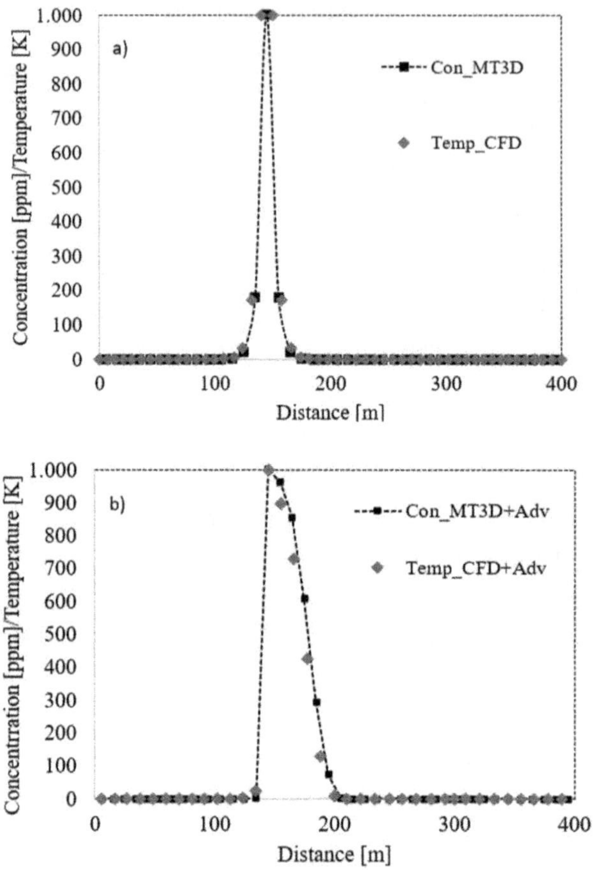

Fig. 4. CFD validation with MT3DMS, (a) heat conduction/ diffusion, (b) heat advection by groundwater flow.

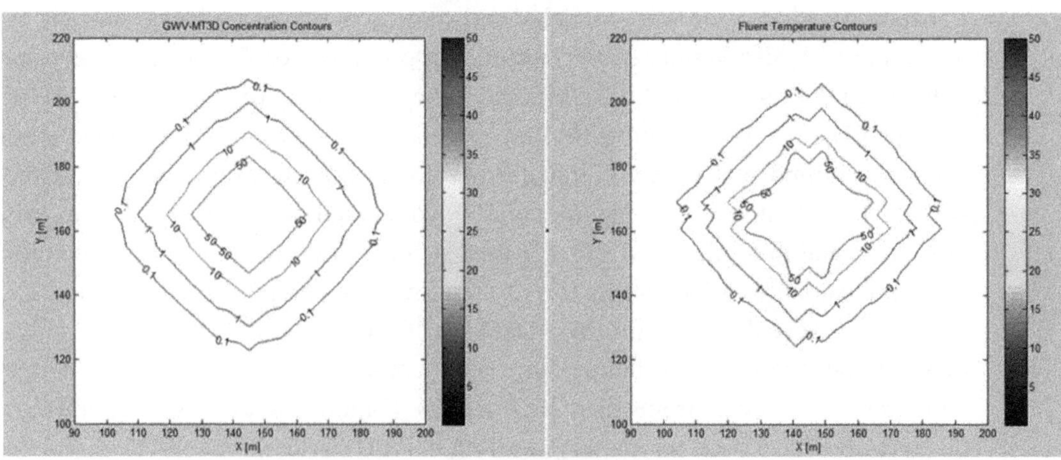

Fig. 5. Temperature contours from CFD (on the right) and concentration contours from MT3DMS (on the left) around the borehole.

The results of this benchmarking verify that both CFD and MT3DMS are suitable for solving the heat transfer problem after required coefficient and unit conversion are carefully defined. Model configuration and post-processing differ as MT3DMS is designed for the mass and solute transport equation particularly in underground conditions, whereas CFD is primarily used for the momentum and energy equation in complex fluid conditions. Applying the analogy between mass transport and heat transport equation MT3DMS may be more efficient in terms of computational time when dealing with a three-dimensional transient large-scale model. However, MT3DMS employs a finite-difference approach, which restricts its applicability to structured mesh. In contrast, CFD is based on finite volumes, allowing for greater flexibility in modeling both structured and unstructured mesh, such as circular boreholes, pipes, and U-pipes. In conclusion, this study demonstrates that the comparison between the CFD model and the MT3DMS model successfully validates the CFD approach for simulating groundwater-coupled heat transport in porous media, addressing a gap in existing research by confirming the applicability of CFD in contexts where such coupling effects have not been thoroughly explored previously.

Acknowledgements

Acknowledgements may be made to Prof. Dr. Dave Hart from Wisconsin Geological Survey and Prof. Christopher Choi from University of Wisconsin-Madison for their valuable guidance and support.

References

1. Zheng, C. and Wang, P.P. (1999) MT3DMS: A Modular Three-Dimensional Multispecies Transport Model for Simulation of Advection, Dispersion and Chemical Reactions of Contaminants in Groundwater Systems. *Documentation and User's Guide.*
2. Hecht-Mendez, J., Molina-Giraldo, N., Blum, P. and Bayer, P. (2010). Evaluating MT3DMS for heat transport simulation of closed shallow geothermal systems. *Ground Water*, 48(5), 741–756.
3. Ozdogan-Dolcek, A. (2015). Numerical Modeling of Heat Transport for Ground-Coupled Heat Pump (GCHP) Systems and Associated Life Cycle Assessments. *The University of Wisconsin-Madison.*
4. Ma, R. and Zheng, C. (2010). An evaluation of the impact of three-dimensional anisotropy on groundwater flow and heat transport modeling. *Advances in Water Resources*, 33(5), 748–760.

5. Martin, N. A. and Cathomen, N. (2001). Transport modeling in heterogeneous porous media: theory and experiment. *Water Resources Research*, 37(4), 1099–1108.
6. Cathomen, N. (2002). Warmetransport im Grundwasser—Auswirkung von Warmepumpenanlagen auf die Grund-wassertemperatur am Beispiel der Gemeinde Altach imVorarlberger Rheintal, *Diploma Thesis*, ETH Zurich, Switzerland.
7. Harbaugh, A. W. (2005). MODFLOW-2005, the U.S. Geological Survey Modular Ground-Water Model—the Ground-Water Flow Process. *U.S. Geological Survey Techniques and Methods*.
8. de Marsily, G. (1986). *Quantitative Hydrogeology: Groundwater Hydrology for Engineers*. Academic Press.
9. Thorne, D., Langevin, C. D., and Sukop, M. C. (2006). Addition of simultaneous heat and solute transport and variable fluid viscosity to SEAWAT. *Computers & geosciences*, 32(10), 1758–1768.
10. Wilson, J. L. and Miller, P. J. (1978). Two-dimensional plume in uniform ground-water flow. *Journal of the Hydraulics Division*, 104(4), 503–514. https://doi.org/10.1061/JYCEAJ.0004975.

Chapter 33

Preliminary Investigation of Energy Management Strategies in University Buildings: A Case Study

Halil Sinoplugil*,‡ and Gülşen Yaman†,§

*Institute of Science and Technology, Mechanical Engineering Department, Balikesir University, Turkiye

†Engineering Faculty, Mechanical Engineering Department, Balikesir University Turkiye

‡halil.sinoplugil@balikesir.edu.tr

§gyaman@balikesir.edu.tr

This study investigates energy management strategies in a university building to address increasing energy demands and environmental concerns. Focusing on a case study of the Faculty of Medicine Morphology Building at Balikesir University (BAUN), the study presents a preliminary work to model energy consumption in classrooms, workshops, and laboratories based on class schedules. Key objectives include optimizing energy efficiency and reducing environmental impact by implementing targeted energy-saving measures tailored to the building's unique demands. The findings highlight potential savings through real-time energy management and tariff analysis, underscoring the benefits of structured energy management systems in educational buildings. This study contributes to the ongoing research on sustainable energy management practices and offers practical implications for similar public institutions.

1. Introduction

Global CO2 emissions and energy consumption are projected to rise by 40% by 2040. In the United States, buildings contribute to 40% of the nation's CO2 emissions and energy usage, with 75% of that energy supplied by fossil fuels. According to the Intergovernmental Panel on Climate Change (IPCC), buildings were responsible for 32% of total global energy consumption and 19% of energy-related greenhouse gas (GHG) emissions (associated with electricity) in 2010. In Europe, buildings were responsible for approximately 40% of total energy use and 36% of greenhouse gas emissions. Mitigating this impact requires both improved building energy efficiency and increased use of renewable energy [1, 2].

Productivity, in short, is the ratio of raw materials to materials used in production. It can also be described as a tangible indicator of performance. The efficiency formula mentioned in the literature is shown below:

$$Efficiency = Effectiveness/Effective\ Activity \quad (1)$$

In other words, it is the ratio of the extent to which the goals are achieved (effectiveness) and the extent to which resources are used effectively [3]. The necessity of installing energy management systems (EMS) in our country's buildings, especially in buildings with high energy consumption, becomes even more prominent considering that these buildings have 2–3 times higher energy consumption compared to developed countries.

Although measures such as insulation are often considered important in new buildings, there is a significant energy saving potential, especially in the building sector. Public buildings, especially university buildings, have an important place in this potential [4]. This study serves as a preliminary investigation that lays the groundwork for future energy management modeling in university buildings. While specific modeling has not yet been conducted, the study offers a structured approach to guide subsequent modeling efforts.

2. Literature Review of Energy Management

Energy management is an interdisciplinary field structured to reduce unnecessary energy consumption, lower costs, protect the environment, and use energy efficiently without sacrificing product and service quality, safety, or environmental sensitivity. Energy management represents a systematic approach to energy with the aim of continuously improving an organization's energy performance and making the improvements sustainable [5]. Extensive research explores energy management across various sectors, yet practical, cost-effective EMS applications tailored to university buildings—particularly in settings with limited real-time monitoring capabilities—remain under-examined.

Effective energy management in buildings involves minimizing energy usage and expenditures while maintaining or enhancing comfort, service, and efficiency requirements. The goal of successful energy management in buildings is to keep energy usage and expenses as low as feasible while maintaining or enhancing comfort, service, and efficiency standards. This can only be achieved with the presence of a person (energy manager) or a unit responsible for energy efficiency. Additionally, the firm decision and support of top management are indispensable

[6]. Energy management steps for energy efficiency can be generally summarized as follows:

- Reviewing Historical Energy Use
- Conducting Energy Reviews
- Identifying Energy Management Possibilities
- Making Changes to Save Energy
- Monitoring the Energy Orientation Program, Setting Targets, Reviewing Developments

There are many different flow charts for establishing energy management systems in buildings. However, as shown above, in practice the five stages appear as a general method. When planning the energy management system, the organization should consider the conditions specified in the context and relevant parties and processes affecting energy performance and identify current and possible risks and opportunities. For this, methods such as SWOT analysis may be preferred [7].

When planning the energy management system, the organization should consider the specified conditions related to the context and relevant parties, as well as the processes affecting energy performance. It should identify existing and potential risks and opportunities. Energy objectives should be set [8]. An important aspect of energy management systems is energy efficiency awareness. This can also be considered one of the first steps of the EMS.

2.1. *Advances in energy monitoring and management*

A study proposes a framework aimed at increasing energy efficiency by monitoring and managing energy use in real-time. This framework includes standards for collecting, analyzing, and visualizing energy data. By giving real-time feedback on energy usage data, the study seeks to improve energy awareness. It highlights how crucial this is to attaining energy savings across a range of procedures. The research analyzes the current state of energy management systems using action research and case study methods and conducts a practical energy survey on an automotive manufacturer. Real-time analysis of energy data is emphasized as an important tool for improving energy performance and optimizing energy efficiency [9]. Although real-time monitoring is beneficial, many institutions, including universities, face limitations in implementing these technologies due to high costs. This study addresses these constraints by proposing strategies based on historical data and simulation, demonstrating achievable savings without immediate real-time infrastructure.

As mentioned above, EMS covers many different disciplines and various variables that need to be controlled. A study proposes an advanced energy management strategy for a hybrid microgrid that includes renewable resources, storage, backup power grids, and AC/DC loads. The mentioned study provides energy management and data analysis with a real-time monitoring interface on the Python platform using a model implemented in Matlab Simulink and applies an efficiency controller for optimal control of operations [10].

As seen in many applications, real-time consumption monitoring provides an important database and control capability for EMS. A study on real-time control demonstrates that using a wireless sensor network in an EMS optimization system, which consists of sensors, controllers, and actuators, enables real-time control of heating loads, thereby reducing energy consumption while maintaining thermal comfort [11].

2.2. *Control systems and optimization techniques*

A review article on control systems engineering for EMS addresses advanced control systems for energy savings and comfort management in buildings. The article focuses on multi-agent control systems that consider user preferences. It compares the advantages and disadvantages of traditional control systems with the applications of intelligent systems such as artificial intelligence and fuzzy logic in this field. As a result, simulations conducted using TRNSYS/MATLAB have shown that multi-agent control systems are successful in managing thermal and lighting comfort, indoor air quality, and energy savings [12].

Numerous techniques, including mathematical optimization, model predictive control, and heuristic control, have been introduced in the literature to help create effective operating schedules and make wise judgments on production and consumption within the context of EMS. Solution-oriented studies that address the comparative analysis of microgrid communication technologies and decision-making strategies and solution methods for energy management systems also offer a similar perspective. [13, 14]. Microgrids are distributed generation systems connected with AC, DC, or hybrid loads and energy storage systems. University campus microgrids, typically include distributed generation sources, energy storage, and electric vehicles. An advanced energy management system ensures uninterrupted energy flow to microgrids. EMS aims to provide sustainable, economical, and reliable energy for microgrids. Microgrids have different system types such as flexible loads, distributed generations, and energy storage systems. These systems include various controllers to manage loads and increase energy efficiency [15].

A study examining the mathematical models of energy technologies developed for environmentally friendly integrated energy management presents the mathematical models of building physics and popular building energy technologies (renewable energy systems, heating and cooling systems, energy distribution systems) and demonstrates how these models can be integrated. Increasing the energy performance of buildings, enhancing the use of clean and renewable energy systems to save resources, and combating climate change are highlighted as promising directions. Two common EMS strategies, rule-based control (RBC) and model predictive control (MPC), are emphasized [16].

It can be seen in the literature that smart energy management systems, addressed in various studies such as methods demonstrated with mathematical models, include optimizations based on weather conditions and applications of smart homes and buildings, IoT devices, and artificial intelligence. As previously mentioned, smart EMS applications save energy by monitoring and managing energy consumption in real-time. These systems can be divided into two main categories: direct control systems and indirect control systems. The studies give a novel framework that utilizes a variety of weather variable forecasts to improve commercial building systems through the application of a number of physical and statistical algorithms. Recommendations to assist users in making more effective use of energy are also provided. [17, 18].

In the presented study, even if real-time monitoring for energy management is not possible, detailed energy audits can still provide a sufficient database. Detailed energy audits are essential for creating successful energy management systems and are legally required to be undertaken on a regular basis. Efficiency-increasing projects should be prepared according to the energy audits to be carried out [19]. Unlike many studies focused on commercial or residential buildings, this study applies EMS principles to a university setting, taking into account the distinct energy demands of educational institutions and the potential for savings through schedule optimization and targeted interventions.

2.3. *Energy audits and energy management in buildings*

Energy audits include conducting measurements on energy-using equipment, measuring energy losses in facilities or equipment, analyzing flue gases, determining the suitability of the equipment used for heating, cooling, ventilation, or lighting needs, assessing insulation properties, and evaluating the maintenance and service life of the equipment [20].

The planned study aims to optimally schedule classrooms, workshops and similar uses in a university education building in terms of energy use, and to decide the amount of heating and cooling needs according to the location, architecture and usage conditions of the building. According to this need, it is aimed to analyze the current system and ultimately present projects with system simulation and optimization methods to increase energy efficiency and create energy modeling based on energy demand.

Universities are faced with managing a variety of facilities that no other organization encounters. Campuses may contain buildings of different types and for different purposes. Energy and safety requirements vary greatly by type of building. In some cases, systems operate as isolated units with different infrastructures. This is costly and poses a growth challenge. In university buildings, many energy-consuming elements such as heating, cooling, ventilation, computers, water and lighting create a high bill in terms of both cost and carbon emissions.

Energy consumption varies with location and climate. In some buildings, it is easier to reduce emissions due to small occupancy and limited service area; But large universities and medical schools and hospitals with laboratories present a more difficult situation. Automation technologies facilitate monitoring and controls regarding energy management. Automation technologies such as the Internet of Things (IoT), artificial intelligence, and edge computing can help reduce the environmental impact of energy-intensive equipment by using real-time data [11, 21].

As expected, electrical energy is used in all of the facilities. Natural gas, on the other hand, is a type of energy used for combustion and heating processes in some facilities. In addition to these two types of energy, liquefied petroleum gas, coal, diesel, hot oil, and sawdust may be used. Some facilities use solar energy and biomass-based energy, which are clean energy types. For facilities with combustion processes such as natural gas, coal, and sawdust, it is possible to achieve energy savings by developing waste heat recovery projects. With the analyzes to be made according to the densities of the consumption points of the electrical energy used, it will be possible to monitor the energy amounts of the machinery and equipment that consume in the operating processes.

Energy management strategies and policies to be developed in light of relevant international standards can be linked to procedures for purchasing energy-efficient products. These policies will foster a culture of determining the true purchase cost by considering the energy consumption values, purchase costs, operating costs, maintenance, and disposal costs of the products to be procured throughout their life cycles [22].

As mentioned above, energy audits are important for developing energy management strategies. However, energy audits can be expensive and time-consuming. In a study, an electrical energy audit application developed using the MATLAB® R2020a platform to demonstrate energy consumption in the selected building of a Public University in the Philippines used the traditional walk method in energy auditing as a data collection basis for model training parameters. This paper introduces the use of ANN (Artificial neural network) in lighting design and load balancing for an energy-efficient building. In the case of voltage and current balance, further studies on the use of ANN are proposed. This study aims to improve the building energy audit process and reduce the time and cost requirements for conducting a full physical audit [23].

The usage schedule and frequency of the areas within the building are also important for energy management and determining the energy load. A study conducted by researchers from Turkey and China revealed that when the course schedules of universities are arranged according to the course conditions, different occupancy hours may occur in the classrooms, which can lead to intermittent heating conditions. Heat transfer between adjacent walls contributes to the consumption of heat energy. It is stated that a new course should be programmed in which the consumption of intermittent heating energy should be reduced by reorganizing the heat transfer of the adjacent room. This study demonstrated that significant improvements in efficiency and carbon emissions were achieved simply by adjusting class schedules to reduce heat loss [24].

Various studies have shown that in the decision-making stages of energy management, the use of digital twin applications, future consumption predictions with artificial intelligence, and automation systems positively impact energy efficiency and the effectiveness of energy management [21, 25–26].

Energy-saving devices such as high efficiency motors (HEMs), variable speed drives (VSDs), economizers, leak avoidance, and pressure drop reduction have all been investigated in the literature. Based on the results of energy-saving technologies, it has been discovered that energy management systems can save a significant amount of electricity, emissions, and utility costs [19, 27].

These studies underscore the importance of tailored energy management strategies, especially within complex environments like university campuses where diverse energy demands must be managed efficiently. This study builds on these principles by proposing a structured approach to EMS in a university setting, with a focus on cost-effective, practical solutions that leverage historical data and targeted interventions to enhance energy efficiency.

3. BAUN Faculty of Medicine Morphology Building

Within the scope of the study, Balikesir University Faculty of Medicine Morphology Building was chosen. The building is located in Çağış Campus, which is located within the borders of Altieylül District of Balikesir. The building has a construction area of 19,330 m^2 and the construction permit date is 2014. The building was put into use in 2018. In the campus, which has been used since 1993, the fact that the building is relatively new as of the date of construction and accordingly the availability of construction projects in the digital environment, the observance of insulation and earthquake standards in the construction of the building, and the assumption that the occupancy rate will be more stable with student attendance has been preferred.

In terms of energy management, although there are more important energy consumers in the campus, demonstrated in Table 1, where the building is located, the selected building is in an important position by being in the top three in terms of energy consumption.

Table 1. The impact of campus buildings on natural gas and electricity consumption in the year 2023.

Natural Gas	Ratio (%)	Electricity	Ratio (%)
Hospital	35.68	Hospital	51.52
Faculty of Science And Literature	10.64	Rectorate (Including Eng.and Arch. Faculty)	12.38
Faculty of Medicine, Morphology	7.33	Faculty of Medicine (Morphology)	6.22
Rectorate	6.90	Dining Hall	4.27
Eng. & Tech. Faculty	6.55	Faculty of Science And Literature	4.19
Engineer Fac. (B Block)	5.01	Faculty of Veterinary Science	2.91
Dining Hall	4.55	Balıkesir Voc. School (Indoor Swimming Pool)	2.89
Library CCC	4.19	Experimental Animals Application and Research Center	2.74
Faculty of Economics anc Administrative Sciences	3.48	Library	2.71
Faculty of Veterinary Medicine	2.86	Balıkesir Voc. School (Existing Building)	2.34
Sports Science Faculty	2.77	Faculty of Economics and Administrative Sciences	2.23
Balıkesir Voc. School	2.72	Faculty of Theology / Law	1.54

Table 1. *(Continued)*

Natural Gas	Ratio (%)	Electricity	Ratio (%)
Faculty of Tourism	2.51	Faculty of Sport Sciences	1.13
Faculty of Fine Arts	1.96	Scientific Technical App. and Research Center Laboratory	0.98
Gym	1.55	Faculty of Tourism	0.97
Indoor Swimming Pool	1.12	Faculty of Fine Arts	0.67
Outdoor Sports Facility	0.19	Indoor Sports Hall	0.31

One of the initial actions that can be taken in energy management is tariff analysis. Tariff changes that can be applied to electricity tariffs are one of the methods to increase economic efficiency without requiring costs for electrical energy.

Table 2. Tariff analysis within the scope of the Faculty of Medicine Morphology Building 2023.

Faculty of Medicine Morphology Building January-December 2023 (Excluding Taxes) TL	
Current Tariff Amount	2,336,829.40 TL
Dual Term Tariff Amount	2,235,603.31 TL
Difference	101,226.09 TL
Difference %	4.33%

According to Table 2, it can be said that tariff earnings of around 100,000 TL, excluding annual taxes, can be obtained without requiring any expense with the tariff change. Equations 2 and 3 were used for the calculation given in Table 2.

$$Amount\ of\ Savings = \sum Kwhper \times EF - \sum Annual\ Modified\ traiff\ Amount \quad (2)$$

$$\sum Annual\ Modified\ traiff\ Amount = \sum(kWh \times DT) + (NP \times PC) \quad (3)$$

where *Kwhper* is Kwh consumed per month, *EF* is Unit Price, *NP* is Negotiate power (kw), *PC* is Power cost (₺) and *DT* is Dual Tariff Unit Price.

The power cost is determined by a monthly agreement and can be determined according to the highest demand value in the invoices or by energy analysis measurements to be made in transformers. If the specified power price is exceeded, the excess amount is doubled. Here, the agreement power price should not be chosen too high and paid extra every month, nor should you choose too low and pay extra every month.

One of the important steps in building energy management is the analysis of historical data. Since 2017, the historical energy consumption data of the entire

Faculty of Medicine Morphology Building have been determined monthly from the invoices of the building subscriptions. In order to predict the energy consumption of the building for the future and to analyze the efficiency increase, it is important to determine the consumption of elements such as lighting, heating, cooling, and plug loads separately. Currently, the general consumption data are presented in Figure 1 and Table 3.

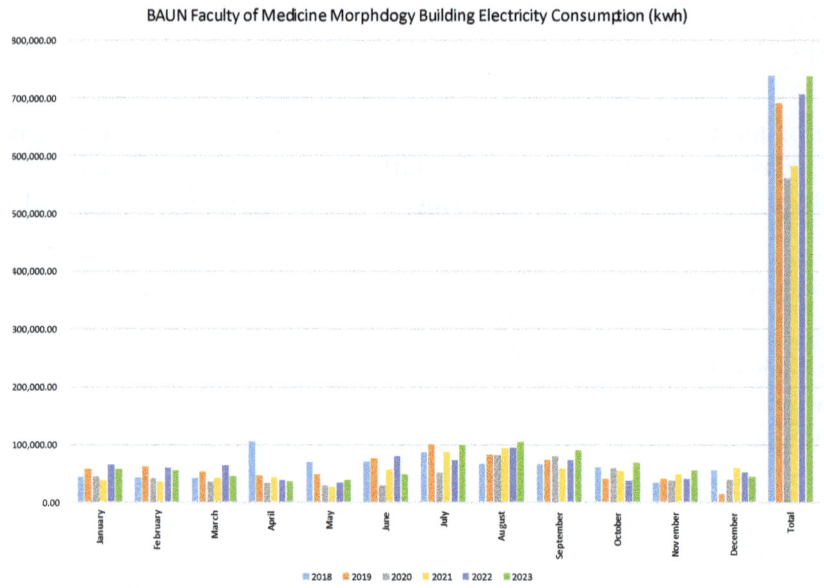

Fig. 1. BAUN Faculty of Medicine Morphology Building Electricity Consumption (kwh).

Table 3. Total electricity consumption for 2018–2023 (months).

Months	Kwh
August	520,673.31
July	495,793.98
September	385,006.035
June	358,517.79
October	318,109.32
January	303,430.95
April	300,617.82
February	291,725.1
March	278,977.68
December	260,118.27
November	254,990.12
May	243,716.004

When analyzing total electricity consumption, it is evident that the highest values occur during summer when cooling is necessary. High consumption is also observed in October and January after the summer months. Considering that the need for lighting decreases in the summer, it can be inferred that cooling consumption is the most significant energy consumption factor. Increasing the efficiency of cooling systems, ensuring proper insulation, and adjusting cooling levels for comfort and energy savings are valuable interventions for improving energy efficiency. Natural gas is used for heating in the building under consideration. There is also hot water production with natural gas. There is a separate dining hall belonging to the building, so there is no use of natural gas in food production. A graph of natural gas consumption is given in Figure 2.

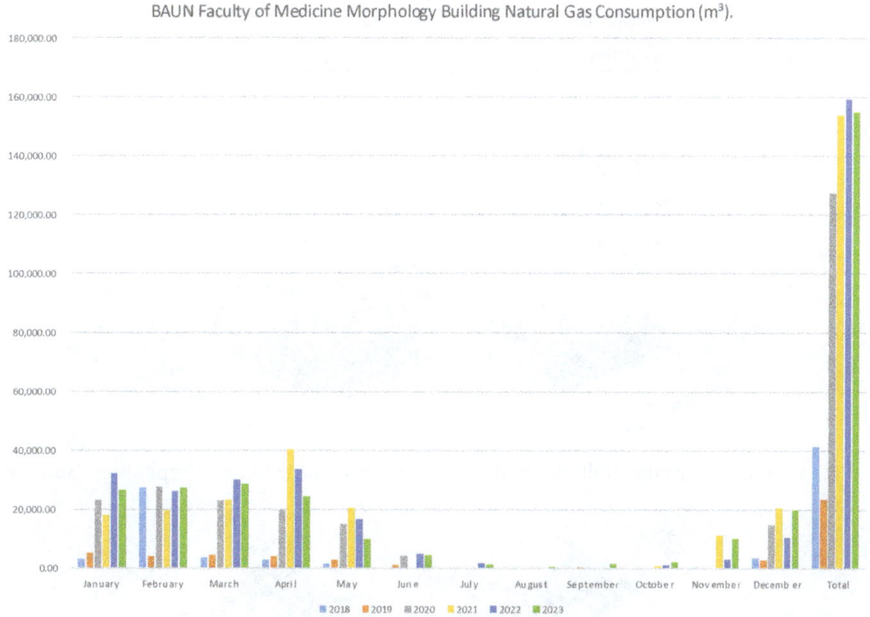

Fig. 2. BAUN Faculty of Medicine Morphology Building Natural Gas Consumption (m^3).

Total consumption of natural gas of the building is given in Table 4.

A three-dimensional drawing is made on Design Builder is shown in Figure 3 and Figure 4, and the operating hours of natural gas boilers during heating periods are given below.

Boilers are open between 06.00–17.00. Boilers are closed between 17.00 and 06.00 except for a single boiler between 01.00 and 03.00.

Table 4. Total natural gas consumption for months during 2018–2023 (m^3).

Months	Total Consumption (**m³**)
February	131,730
April	124,260
March	112,250
January	107,366
December	71,644
May	65,599
November	24,399
June	13,999
October	3,596
July	2,677
September	1,532
August	547

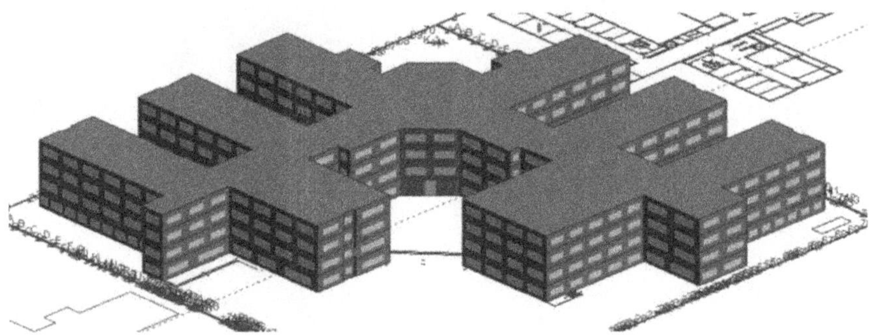

Fig. 3. Three-dimensional drawing of BAUN Faculty of Medicine Morphology Building.

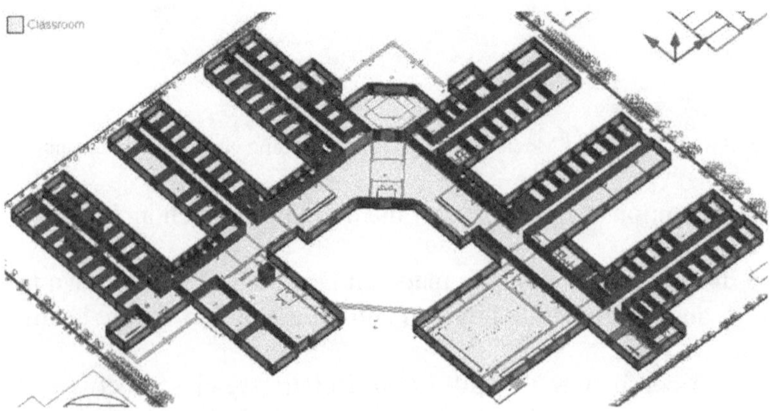

Figure 4. BAUN Faculty of Medicine Morphology Building 1st floor plan sample three-dimensional representation.

4. Steps to Sustainable Energy Management

As this study functions as a preliminary exploration, the steps outlined in this section represent a suggested framework for future modeling of the planned study.

4.1. *Energy consumption analysis*

Analysis of the past energy consumption of the building through invoices: Behavioral interventions, identifying areas for energy savings, and making energy consumption forecasts require the analysis of past data. Many studies in the literature have utilized retrospective analyses for deep learning, uncertainty analysis, and intelligent systems [28–30].

4.2. *Tariff analysis*

Tariff analysis of electricity bills: As in many parts of the world, different electricity tariffs based on consumption amount and time are applied in Turkey, determined by the Energy Market Regulatory Authority. If the tariff analysis is conducted according to the electricity consumption situation of the examined building, the lowest bill can be achieved for the same consumption [31]. The tariff analysis is expected to yield potential cost savings by aligning energy use with optimal tariffs. Preliminary calculations suggest that this adjustment alone could lead to annual savings of approximately 100,000 TL, demonstrating the financial benefits of tailored tariff strategies in university settings.

4.3. *Identification of key energy consumption points*

Identification of important energy consumption points within the building: Identifying key energy users is important to determine where to focus energy efficiency efforts and where the most gains can be achieved. Although methods such as machine learning, simulations and prediction models are presented in the literature to analyze important points in energy use; In this study, the data will be discussed based on past consumption and energy audit [32].

4.4. *Building modeling and simulation*

Designing the architectural structure of the building and the HVAC system in 3D in a program such as Design Builder: An application like DesignBuilder is beneficial for digitally modeling the building's architecture and simulating various energy consumption and efficiency scenarios. This method allows for the digital output of theoretical modifications [33, 34].

Measuring building energy consumption: Measuring the energy-producing and energy-consuming elements of the building is essential to reveal the potential for losses and the methods for increasing efficiency that can be applied [23]. The measurements will be carried out within the following scope:

- Boiler insulation detection (thermal imager) and flue gas analysis
- Detection of the presence of thermal bridges (thermal imager)
- Determination of the optimum insulation thickness (Wall sections and U-value measuring device)
- Determination of tightness and thermal permeability of windows
- Analysis of the electricity consumption of the building, heating and cooling systems with an energy analyzer
- Subtracting heating and cooling costs from monthly consumptions
- Determination of whether transformers have heating problems with thermal cameras
- Determination of compensation status

4.5 *Schedule optimization*

Analysis of the course schedule of the Faculty of Medicine and its optimization in terms of energy efficiency: Planning the course schedule in accordance with energy efficiency goals is important for making the most of natural light and optimizing the use of the air conditioning system. Factors such as holding classes at short intervals and having heated classrooms adjacent to each other can affect energy efficiency [24]. By optimizing the schedule based on peak usage and natural lighting, this method is expected to reduce the need for artificial lighting and heating. This adjustment aims to lower energy consumption during peak hours, potentially resulting in substantial energy savings and reduced operational costs, especially in the cooler months when heating demands are highest.

4.6 *Energy needs assessment*

Determination of energy needs by considering the use of buildings and classrooms according to the optimized course schedule: Determining the energy load based on an optimized course schedule according to the indoor and outdoor temperatures on specific dates will theoretically demonstrate the potential energy savings provided by the optimized course schedule. For this, the following studies are conducted:

- Determination of warm-up and cool-down times by measuring comfort in classrooms
- Determination of heating and cooling loads according to the requirements of indoor air temperatures
- Comparing the value of the heating and cooling loads, which is determined according to the course schedule adapted to energy efficiency and the improved theoretical characteristics of the building, with the current energy consumption and revealing the energy efficiency potential

The optimized energy load calculation based on this schedule adjustment demonstrates potential reductions in heating and cooling requirements, aligning with seasonal temperature changes. These preliminary findings indicate that targeted schedule and load adjustments could reduce total energy demand, emphasizing the effectiveness of adaptable management strategies in institutional buildings.

Mathematical modeling in which the improved course schedule and building conditions can be adapted according to changing conditions and the contribution of each improvement step to efficiency can be shown: According to the conditions brought by the provided education, there may be changes in student numbers and course schedules. Creating a mathematical model that can adjust to changes such as different course schedules to energy efficiency targets would improve the work's long-term viability. Additionally, mathematically modeling activities such as insulation, device efficiency, and lighting efficiency improvements that can be made in the building will allow the theoretical results of energy efficiency studies to be seen immediately. However, this methodology relies on historical data, which may not account for future variations in energy demand. While modeling provides valuable insights into potential savings, real-time adjustments and ongoing monitoring would be essential for full optimization.

5. Conclusions

This preliminary study provides a foundational approach for energy management modeling in university buildings, presenting a modeling framework and suggested interventions to support future empirical applications. This paper explored the energy-saving potential of a university building by examining historical consumption data, modeling architectural and operational energy demands, and proposing targeted energy management strategies. Using Balikesir University's Faculty of Medicine Morphology Building as a case study, this research demonstrated practical methods to enhance energy efficiency in educational settings, revealing potential energy savings and cost reductions through optimized usage schedules, targeted energy audits, and tariff adjustments. These findings illustrate the substantial economic benefits achievable through structured energy management in public institutions.

The proposed mathematical modeling approach provides a flexible tool for sustaining energy efficiency improvements, accommodating variable demands and environmental conditions over time. Future research could further investigate integrating real-time monitoring technologies for dynamic usage schedule refinement, as well as explore the applicability of these methods across other types of public buildings. This study offers a sustainable energy management framework that supports broader institutional goals of environmental stewardship and contributes to the field's efforts to reduce energy consumption across the public sector.

References

1. C. Poolla. *A Prediction and Decision Framework for Energy Management in Smart Buildings*. PhD. Thesis, Electrical and Computer Engineering, Carnegie Mellon University, (2016).
2. M. Koçyiğit. *Üniversite Binalarında Enerji Tasarruf Potansiyelinin Araştırılması: Balıkesir Üniversitesi Çağış Kampüsü Örneği*. Master's Degree Thesis, University of Balikesir, (2019).
3. M. K. Akyüz. *Havalimanlarında Sürdürülebilir Enerji Yönetim Modeli*, PhD. Thesis, University of Anadolu, (2018).
4. G. Demirdöğen, *A Framework For Integrating Building Information Modeling And Big Data Analytics For Sustainable Energy Management In Buildings*. PhD. Thesis, Department of Civil Engineering, Yıldız Technical University, (2022).
5. P. Özbakır, *Enerji Yönetimi*. Master's Degree Thesis, Yıldız Technical University, (2006).
6. Ministry of Energy and Natural Resources, Department of Energy Efficiency and Environment, *Enerji Yönetimi 1*, (Ankara, 2022).
7. Ministry of Energy and Natural Resources, Department of Energy Efficiency and Environment, *Enerji Yönetimi 3*, (Ankara, 2022).
8. Ministry of Energy and Natural Resources, Department of Energy Efficiency and Environment, *Enerji Yönetimi 2*, (Ankara, 2022).
9. K. Vikhorev, R. Greenough, and N. Brown, An advanced energy management framework to promote energy awareness, *Journal of Cleaner Production* 43, 103–112, (2013).
10. Zia Ullah, Shoarong Wang, Guoan Wu, Mengmeng Xiao, Jinmu Lai, and Mohamed R. Elkadeem, Advanced energy management strategy for microgrid using real-time monitoring interface, *Journal of Energy Storage*, 52, 104814, (2022).
11. N. H. Nguyen, Q. T. Tran, J. M. Leger and T. -P. Vuong, A real-time control using wireless sensor network for intelligent energy management system in buildings, *2010 IEEE Workshop on Environmental Energy and Structural Monitoring Systems*, Taranto, Italy, 87–92, (2010).
12. A.I. Dounis and C. Caraiscos, Advanced control systems engineering for energy and comfort management in a building environment—A review, *Renewable and Sustainable Energy Reviews* 13, 1246–1261, (2009).
13. M. Beaudin and H. Zareipour, Home energy management systems: A review of modelling and complexity, *Renewable and Sustainable Energy Reviews* 45, 318–335, (2015).
14. M. F. Zia, E. Elbouchikhi, and M. Benbouzida, Microgrids energy management systems: A critical review on methods, solutions, and prospects, *Applied Energy*, 222, 1033–1055, (2018).
15. H. A. Muqeet, H. Javed, M. N. Akhter, M. Shahzad, H. M. Munir, M. U. Nadeem, S. S. H. Bukhari, and M. Huba, Sustainable Solutions for Advanced Energy Management System of Campus Microgrids: Model Opportunities and Future Challenges, *Sensors*, 22, 2345, (2022).
16. Y. Zhang, B. Vand, and S. Baldi, A Review of Mathematical Models of Building Physics and Energy Technologies for Environmentally Friendly Integrated Energy Management Systems, *Buildings*, 12, 238, (2022).
17. S. Mischos, E. Dalagdi, D. Vrakas, Intelligent energy management systems: A review, *Artificial Intelligence Review*, 56, 11635–11674, (2023).

18. D. Lazos, A. B. Sproul, M. Kay, Optimisation of energy management in commercial buildings with weather forecasting inputs: A review, *Renewable and Sustainable Energy Reviews* 39, 587–603, (2014).
19. E.A. Abdelaziz , R. Saidur, and S. Mekhilef, A review on energy saving strategies in industrial sector, *Renewable and Sustainable Energy Reviews,* 15, 150, (2011).
20. T. M. Sooriyaarachchi, I. Tsai, S. El Khatib, and A. M. Farid, Job creation potentials and skill requirements in PV, CSP, wind, water-to-energy and energy efficiency value chains, *Renewable and Sustainable Energy Reviews,* 52, 653, (2015).
21. S. Bates, Creating an Efficient Campus, *Schneider Electric White Paper*, 1–15, (2011).
22. W. Khan, *Uncertainty analysis and management in building energy data mining: A bottom-up approach considering the temporal and spatial aspect of data,* PhD. Thesis, Eindhoven University of Technology, (2023).
23. F. L.Marcos and B. Plangklang, A high accurate user-friendly energy audit platform of a university building using ANN Bayesian regularization and Levenberg-Marquardt algorithm, *Energy Reports,* 11, 2220, (2024).
24. Y. Li, W. Li, X. Liu, Y. Jian, M. Arıcı, L. Zhang, T. Zhou, and Y. Cao, Adjacent heat transfer characteristics and energy efficiency optimization in university classrooms under partial-space heating by a numerical model based on field test, *Sustainable Energy Technologies and Assessments*, 63, 103666, (2024).
25. S. Pierce, F. Pallonetto, L. Donatis, and M. De Rosa, District energy modelling for decarbonisation strategies development-The case of a University campus, *Energy Reports,* 11, 1256, (2024).
26. Y. Yu, C. Zhang, G. Wang, X. Wang, and L. Yang, Short-term energy consumption prediction for university dormitory buildings based on energy consumption grouping by considering meteorological factors and teaching schedules, *Energy & Buildings,* 312, 114219, (2024).
27. L. P. Lombard, J. Ortis, and C. Pout, A review on buildings energy consumption information, *Renewable and Sustainable Energy Reviews,* 52, 653, (2015).
28. R. Mathumitha, P. Rathika, and K. Manimala, Intelligent deep learning techniques for energy consumption forecasting in smart buildings: A review, *Artificial Intelligence Review*, 57, 35, (2024).
29. X. Han and C. Wei, Household energy consumption: State of the art, research gaps, and future prospects, *Environment, Development and Sustainability*, 23, 12479–12504 (2021).
30. R. V. Klyuev, I. D. Morgoev, A. D. Morgoeva, O. A. Gavrina, N. V. Martyushev, E.A. Efremenkov, and Q. Mengxu, Methods of Forecasting Electric Energy Consumption: A Literature Review, *Energies*, 15, 8919, (2022).
31. D. A. Zaki and M. Hamdy, A Review of Electricity Tariffs and Enabling Solutions for Optimal Energy Management, *Energies*, 15, 8527, (2022).
32. M. Alnahhal, O. Antar, A. Sakhrieh, and M.A. Hazza, Analyzing Energy Consumption in Universities: A Literature Review, *International Journal of Energy Economics and Policy*, 14(3), 18–27 (2024).
33. https://web.mit.edu/sustainabledesignlab/projects/TeachingResources/EnergyModellingI_GettingStarted.pdf.

34. A. Fathalian and H. Kargarsharifabad, Actual validation of energy simulation and investigation of energy management strategies (Case Study: An office building in Semnan, Iran), *Case Studies in Thermal Engineering*, 12, 510–516, (2018).

Chapter 34

Classification of Wells in Bandung Regency by Mineral Element Using K-Means, K-Medians, and K-Medoids Clustering

Yuli Sri Afrianti[*,†,|], Utriweni Mukhaiyar[*,**], Grace Angelia[‡,††], Fadhil Hanif Sulaiman[§,‡‡], Katharina Oginawati[¶,§§], Indah Rachmatiah Siti Salami[¶,¶¶], and Nurul Fahimah[¶,||]

[*]*Statistics Research Division, Faculty of Mathematics and Natural Science, Bandung Institute of Technology, Ganesa Street No. 10 Bandung, 40132, Indonesia*

[†]*Mathematics Doctoral program, Faculty of Mathematics and Natural Science, Bandung Institute of Technology, Ganesa Street No. 10 Bandung, 40132, Indonesia*

[‡]*Undergraduate Program in Mathematics, Faculty of Mathematics and Natural Science, Bandung Institute of Technology, Ganesha Street No 10, Bandung, 40132, Indonesia*

[§]*Graduate Program in Computational Science, Faculty of Mathematics and Natural Science, Bandung Institute of Technology, Ganesha Street No 10, Bandung, 40132, Indonesia*

[¶]*Environmental Management Technology Research Group, Faculty of Civil and Environmental Engineering, Bandung Institute of Technology, Ganesha Street No 10, Bandung, 40132, Indonesia*

[|]*yuli.afrianti@itb.ac.id*
[**]*utriweni.mukhaiyar@itb.ac.id*
[††]*10120010@mahasiswa.itb.ac.id*
[‡‡]*10117053@mahasiswa.itb.ac.id*
[§§]*katharina.oginawati@gmail.com*
[¶¶]*indahrss28@gmail.com*
[||]*nurul.fahimah@itb.ac.id*

Wells provide potable water. However, their quality varies depending on the mineral element in them, which requires assessment of quality standards. This study evaluated the quality of well water in 160 locations in Bandung Regency, Indonesia, by analyzing six mineral elements (i.e., As, Cd, Fe, Mn, Pb, and Zn). The method used was Partitioning Clustering (i.e., K-Means, K-Medians, and K-Medoids), with Euclid and Manhattan distance metrics. This study aims to evaluate the distribution of well water quality in Bandung Regency with the best Partitioning Clustering method based on each distance. The number of optimal clusters was determined by the Elbow method, and then the best method from each distance was selected using three selection methods (i.e., Silhouette Coefficient, Calinski-Harabasz Index, and Xie-Beni Index). This study showed that K-Means Euclid and K-Medians Manhattan were the most effective methods. Evaluation of each cluster is conducted by reviewing drinking water quality standards, with quality levels ranging from 1 (very poor) to 4 (very good). These findings are important for managing water resources safely according to the existence of the region.

1. Introduction

Bandung Regency, with its varied topography from lowlands to hills and high rainfall and abundant river flow, is rich in groundwater resources, including well water [1]. However, due to this variation, the quality of well water as one of the main sources of drinking water for the community is often unstable and can be affected by mineral element and local environmental conditions. For example, Fahimah *et al.* [2] research examined Arsenic element, while Anggraeni *et al.* [3] focused on Lead and Zinc elements. In addition, human activities such as agriculture using fertilizers and pesticides also have an impact on groundwater quality.

Previous studies generally assessed well water quality based on observable physical characteristics, such as color, odor, and water clarity [4]. However, this approach is less effective when applied to large areas, because it requires a lot of time and effort. Therefore, this study uses the Partitioning clustering method with Euclid and Manhattan distances to group well water quality in seven areas in Bandung Regency, so that the analysis becomes more efficient and lightweight. The Partitioning clustering methods used include K-Means, K-Medians, and K-Medoids, with the aim of determining the best partitioning clustering method based on the effectiveness of each distance in data partitioning. This study focuses on six elements in well water, namely Arsenic (As), Cadmium (Cd), Iron (Fe), Manganese (Mn), Lead (Pb), and Zinc (Zinc). The results of this study are expected to be the basis for managing well water that is safer and in accordance with drinking water quality standards, especially in identifying areas that have the potential to have water quality that is hazardous to health.

2. Cluster Analysis Theoretic

Cluster analysis is a technique used to group multivariate observations into a cluster, with the main goal of identifying the unique characteristics of each cluster formed. One method that is often used to group observation vectors is partitioning clustering. The flowchart of this study can be seen in Figure 1.

2.1. *Partitioning clustering*

Partition clustering divides observations into several clusters, starting by randomly assigning initial partitions or cluster centers, and then reallocating observations based on an optimality criterion. The clustering of a data set is determined by the minimum distance to its cluster center [5].

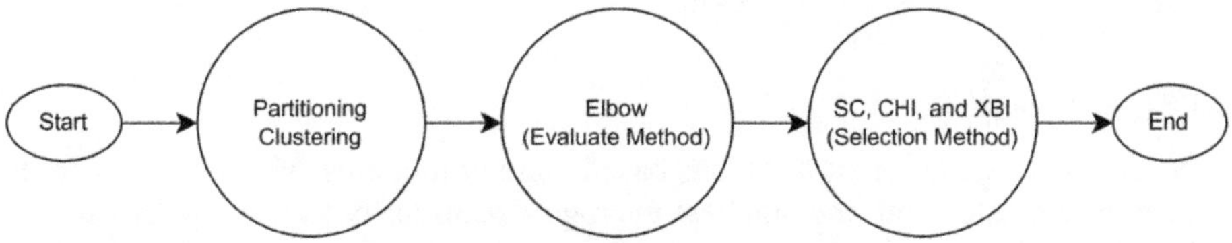

Fig. 1. Flowchart research.

2.1.1. *K-Means*

K-Means is a partitional clustering algorithm that groups n data that has p variables ($\vec{x}_i = [x_{i1}, x_{i2}, \dots, x_{ip}]^T$ with $i = 1, 2, \dots, n$) into k clusters $(C_j, j = 1, \dots, k)$ based on their means. Each cluster formed is determined by a cluster center or centroid ($\vec{c}_j = [c_{1j}, c_{2j}, \dots, c_{pj}]^T$ with $j = 1, 2, \dots, k$), which is the mean value of the cluster, that is:

$$\vec{c}_j = \frac{1}{|C_j|} \sum_{\vec{x}_q \in C_j} \vec{x}_q \tag{1}$$

The K-Means aims to choose centroids that minimize the inertia, or within-cluster Sum-of-Squares Error (SSE):

$$SSE = \sum_{j=1}^{k} \sum_{\vec{x}_q \in C_j} \|\vec{x}_q - \vec{c}_j\|^2 \tag{2}$$

2.1.2. *K-Medians*

K-Medians is a variation of K-Means based on their medians. Each cluster formed is determined by a centroid ($\vec{c}_j, j = 1, \dots, k$), which is the median value of the cluster, that is:

$$\vec{c}_j = median(\{\vec{x}_q : \forall \vec{x}_q \in C_j\}) \tag{3}$$

The goal of the K-Medians clustering is to determine a subset of median points that minimizes the cost of assignment of the data to the nearest median. The criterion function for K-Medians is defined as follows:

$$S_1 = \sum_{j=1}^{k} \sum_{\vec{x}_q \in C_j} |\vec{x}_q - \vec{c}_j| \tag{4}$$

2.1.3. *K-Medoids*

K-Medoids is a variation of K-Means based on their medoids. Medoids are central objects in a cluster with the smallest average dissimilarity to other points in the group. Each cluster formed is determined by a centroid ($\vec{c}_j, j = 1, \ldots, k$), which is the medoid value of the cluster, that is:

$$\vec{c}_j = arg\,\min_{\vec{x}_r \in C_j} \sum_{\vec{x}_s \in C_j} \|\vec{x}_r - \vec{x}_s\| \tag{5}$$

The goal of the K-Medoids clustering is to determine a subset of medoid points that minimizes the cost of assignment of the data to the nearest medoid. The criterion function for K-Medoids is defined as follows:

$$S_2 = \sum_{j=1}^{k} \sum_{\vec{x}_q \in C_j} |\vec{x}_q - \vec{c}_j| \tag{6}$$

2.2. ***Selection method***

Based on the results of cluster analysis obtained from each method, a selection was carried out using three methods to determine the best method for each distance metric. The best cluster analysis method was selected based on the method that provided the most optimal value [6].

2.2.1. *Silhouette coefficient (SC)*

The Silhouette Coefficient is an evaluation metric used to assess the quality of clusters in clustering by comparing the distances between and within clusters [7]. This coefficient measures how well a data point is clustered within its own cluster compared to its neighboring clusters. Let a_i represent the average distance of $\vec{x}_i$ to all points within the same cluster, and b_i be the average distance between $\vec{x}_i$ and the nearest cluster that does not include $\vec{x}_i$ (i.e., all points in the closest neighboring cluster) [8]. The Silhouette Coefficient is calculated using the following formula:

$$SC = \frac{\sum_{i=1}^{n} \frac{b_i - a_i}{\max(a_i, b_i)}}{n} \tag{7}$$

This coefficient ranges from -1 to 1, where values near 1 suggest that the element fits well within its cluster, and values near -1 indicate that the element may be incorrectly assigned to its cluster.

2.2.2. *Calinski-Harabasz index (CHI)*

Calinski-Harabasz Index (CHI) is an evaluation metric used to assess cluster quality by measuring the ratio between inter-cluster variability and intra-cluster variability [9]. This metric measures how well the clusters formed can separate different data groups and how uniform the data is within the cluster itself. In the equation for calculating CHI, a clustering that produces k clusters is given, with $B(k)$ denoting the inter-cluster sum of squares (the variability between the center points of a cluster and the center points of other clusters) and $W(k)$ denoting the intra-cluster sum of squares (the variability of points within a cluster). The equation for CHI can be expressed as follows.

$$CHI = \frac{\frac{B(k)}{(k-1)}}{\frac{W(k)}{n-k}} \tag{8}$$

CHI values range from 0 to ∞, with higher values indicating better clustering [10]. CHI assesses dense (small dispersion within clusters) and separated (large dispersion between clusters) clusters, and evaluates the overall quality of clustering.

2.2.3. *Xie-Beni index (XBI)*

Xie-Beni Index (XBI) is an evaluation metric used to measure cluster density and separation between clusters. XBI measures the extent to which data within a cluster are close together (compact) and how separated the clusters are from each other [11]. In this case, cluster density is measured by calculating how close the data points within a cluster are to their cluster centers. While cluster separation is measured by comparing the distance between cluster centers. The equation for XBI can be expressed as follows.

$$XBI = \frac{\sum_{j=1}^{k} \sum_{\vec{x}_p \in C_j} \left\| \vec{x}_p - c_j \right\|^2}{n * \min_{j \neq l} \left\| c_j - c_l \right\|^2} \tag{9}$$

XBI values ranges from 0 to ∞, with lower values indicating better cluster quality. Lower values indicate that the data point is closer to its cluster center than to the centers of other clusters.

2.3. *Drinking water quality standards (DWQS)*

Each mineral element in well water is compared to the relevant drinking water quality standards. This study uses the most recent standards from Government Regulation of the Republic of Indonesia Number 22 of 2021 and Regulation of the Minister of Health of the Republic of Indonesia Number 2 of 2021. The stricter standard from these sources was chosen, as shown in Table 1. This approach aims to evaluate the distribution of well water quality based on mineral elements and adherence to these standards, defining what is referred to as the well water quality level.

Table 1. Drinking water quality standards.

Mineral elements	DWQS (mg/L)	Mineral elements	DWQS (mg/L)
Arsenic (As)	0.01	Iron (Fe)	0.2
Cadmium (Cd)	0.003	Mercury (Hg)	0.001
Cobalt (Co)	0.2	Mangan (Mn)	0.1
Chromium (Cr)	0.05	Lead (Pb)	0.01
Copper (Cu)	0.02	Zinc (Zn)	0.05

3. Results

The Elbow method is used to identify the optimal number of clusters. This technique is used by plotting inertia against the number of clusters k. As shown in Figure 2, the graph shows a sharp decrease in inertia as the number of clusters increases, but after a certain point, the decrease becomes marginal, forming an elbow shape. This point is considered optimal because it represents a balance between minimizing intra-cluster fit and avoiding overfitting by having too many clusters. This method has been applied in various studies. This technique is crucial for balancing model complexity with accuracy [12]. Likas *et al.* (2003) further demonstrated its effectiveness in a variety of clustering applications, highlighting its widespread use in data science and machine learning [13].

After obtaining the number of optimal clusters from each cluster analysis method, the researcher used the selection method to obtain the best cluster analysis method from each distance. The results can be seen in Table 2, the best method for Euclid is K-Means, while for Manhattan is K-Medians.

Based on the results of cluster analysis from the best method that has been selected, a centroid or cluster center point is obtained. The centroid represents the condition of a cluster. The centroid value will be compared with the drinking water quality standards, which can be seen in Table 3. Because each mineral element consists of two criteria: Satisfying and not satisfying quality standards, the mineral element in the centroid is also made into two criteria. This aims to have a centroid that represents the feasibility or otherwise of a cluster.

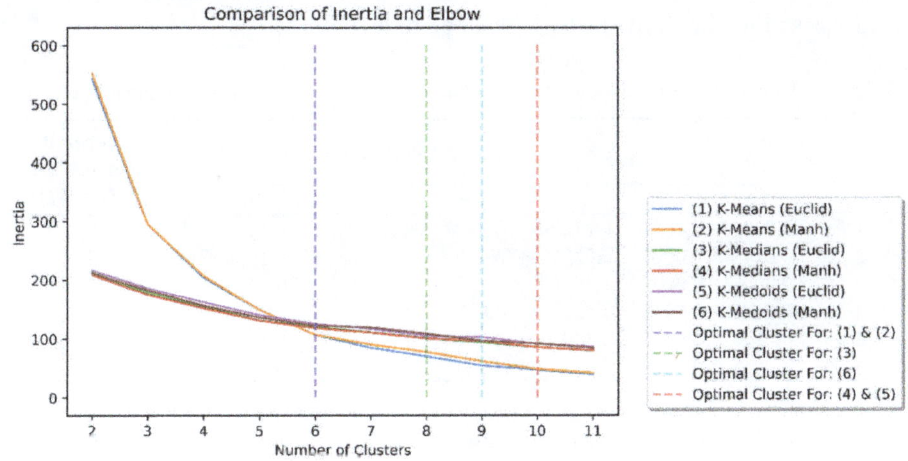

Fig. 2. Elbow method.

Table 2. Partitioning selection methods.

Cluster Analysis		Selection Methods		
		Silhouette coefficient (SC)	Calinski-Harabasz index (CHI)	Xie-Beni index (XBI)
Euclid	K-Means	0.6084	265.0843	2.8594
	K-Medians	0.5079	248.8966	32.2220
	K-Medoids	0.5238	355.5979	64.1508
Manhattan	K-Means	0.5853	265.0843	287.5425
	K-Medians	0.4938	346.4419	123.6926
	K-Medoids	0.4916	210.3653	301.4275

The best Euclid based on the selection method (red shading)
The best Manhattan based on the selection method (blue shading)

For example, the first, third, and sixth clusters in the As column are satisfy the quality standard. While the second, fourth, and fifth clusters don't satisfy the quality standard. However, these are not the case in Cd, Fe, and Mn which only contain 1 criterion. Therefore, we increase the number of clusters (k) until each element contains 2 criteria. Thus, for every element there is a centroid that represents satisfying the quality standard and not satisfying the quality standard. This process is fulfilled at $k = 19$.

Next, to review the distribution of well water quality, the centroid is mapped into the number of elements that satisfy drinking water quality standards, which are called well water quality levels. For example, in the first cluster, there are two elements that satisfy quality standards so that the first cluster is mapped to level 2. Another example, in the third cluster, there are three elements that satisfy quality standards so that the third cluster is mapped to level three.

Table 3. Mapping of well water quality levels based on drinking water quality standards $k = 6$

Cluster number	Centroid of mineral element (g/mL)						Number of elements that satisfy DWQS
	As	Cd	Fe	Mn	Pb	Zn	
1	0.0099	0.0009	0.27690	0.2308	0.0147	0.1881	2
2	0.0140	0.0009	0.72270	2.2551	0.0082	0.1983	2
3	0.0001	0.0005	14.9385	0.5337	0.0110	0.0404	3
4	0.0153	0.0018	7.52110	0.8792	0.0229	0.2368	1
5	0.0153	0.0012	2.45070	0.5710	0.0188	0.1725	1
6	0.0045	0.0025	3.19360	10.4203	0.0396	0.0409	3

Satisfy drinking water quality standards
Not satisfy drinking water quality standards

The same procedure is done on KMedians Manhattan (Table 4). At the optimal k, which is $k = 10$, each mineral element has 2 criteria, so no k will be added. However, in the third and fourth clusters, having the same criteria As, Cd, Fe, and Pb are both satisfied, and Mn and Zn are not satisfied. Therefore, a reduction in k is carried out to review whether the two separate clusters are the same cluster. This is obtained at $k = 9$. Then, these results are mapped to the well water quality level (Figure 3).

Table 4. Mapping of well water quality levels based on drinking water quality standards $k = 10$.

Cluster number	Centroid of mineral elements (g/mL)						Number of elements that satisfy DWQS
	As	Cd	Fe	Mn	Pb	Zn	
1	0.0175	0.00020	0.71460	2.69760	0.00160	0.06680	2
2	0.0051	0.00030	6.54940	12.2943	0.00010	0.04140	4
3	0.0091	0.00055	0.18620	1.21540	0.00780	0.08920	4
4	0.008	0.00010	0.16015	0.10195	0.00080	0.05595	4
5	0.0155	0.00110	7.67410	0.87850	0.01420	0.13700	1
	...	...	...	...	...	...	...
10	0.0042	0.00355	1.51575	9.48325	0.05935	0.04060	2

Satisfy drinking water quality standards
Not satisfy drinking water quality standards

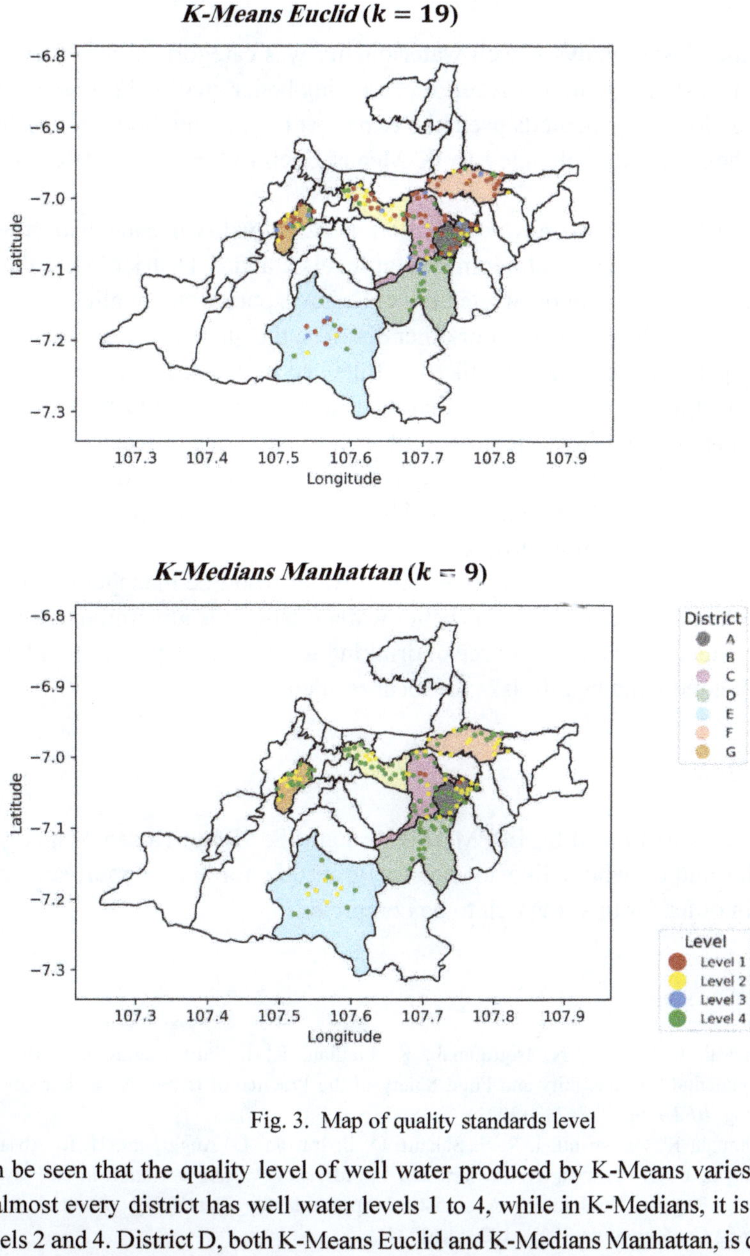

Fig. 3. Map of quality standards level

It can be seen that the quality level of well water produced by K-Means varies, meaning that almost every district has well water levels 1 to 4, while in K-Medians, it is dominant at levels 2 and 4. District D, both K-Means Euclid and K-Medians Manhattan, is dominated by the best water quality level. This is in contrast to district F, in K-Means Euclid it is dominated by level 1, while in K-Medians Manhattan it is dominated by level 2. This shows that in district F, the quality of well water is quite poor.

4. Conclusions

Based on the cluster analysis, well water quality was categorized into four levels within each cluster, with higher levels representing better quality. From the various partitioning clustering methods used, the two most optimal methods were selected based on their distance, obtained by K-Means Euclid ($k = 19$) and K-Medians Manhattan ($k = 9$).

K-Means Euclid shows a wide variation in water quality in each district, while K-Medians Manhattan tends to dominate at levels 2 and 4. District D has the best water quality in both methods, while district F shows poor water quality, especially with K-Means Euclid at level 1. For other districts, the quality of well water varies and is not specific, which is most likely influenced by the specific characteristics of each part of the district. Factors such as residential areas, industrial areas, and agricultural areas can affect the quality of well water. Further research is needed to identify the influence of the characteristics of each part of the district on water quality so that it can be better understood how various human activities play a role in changing the quality of well water.

This research is useful for more efficient water resource management in each district. By knowing an area that has better water quality, the government or related institutions can determine the source of drinking water from the district with better quality while reducing health risks for local residents.

Acknowledgments

This research was funded by LPPM ITB through the "Riset Dosen Magang ITB 2024". The author would like to express gratitude for the support and funds provided in order for this research to be completed.

References

1. S. Sunardi, I. Ghulam, N. Isqtiqomah, K. Fadilah, K. I. Safitri, and O.S. Abdoellah, Environmental Sustainability and Food Safety of the Practice of Urban Agriculture in Great Bandung, *IIETA*, pp. 737–743 (2023).
2. N. Fahimah, K. Oginawati, I. R. S. Salami, D. E. Irawan, D. Anggraeni, H. R. Absari, U. Mukhaiyar, U. S. Pasaribu, K. N. Sari, and L. Adiyani., Arsenic variation in soil layers in Bandung Regency, Indonesia: A statistical approach for identifying potential sources, *Emerging Contaminants*, 10(3), 100311 (2024).
3. D. Anggraeni, K. Oginawati, N. Fahimah, I. R. S. Salami, H. R. Absari, U. Mukhaiyar, U. S. Pasaribu, K. N. Sari, and L. Adiyani, Analysis of heavy metals (Pb and Cd) in soil layers of Indonesia: Spatial distribution, potential source, and groundwater effect, *Case Studies in Chemical and Environmental Engineering*, 9, 100652 (2024).

4. H. Lantapon, O. R. Pinontoan, and R. H. Akili, Analisis Kualitas Air Sumur Berdasarkan Parameter Fisik dan Derajat Keasaman (Ph) di Desa Moyongkota Kabupaten Bolaang Mongondow Timur, *Jurnal KESMAS*, 8(7), 161–166 (2019).
5. C. A. Charu, and K. R. Chandan, *Data Clustering Algorithms and Applications,* (Chapman & Hall Book, 2014), pp. 101–103.
6. A. C. Rencher, *Methods of Multivariate Analysis,* John Wiley & Sons, USA, pp. 455–465 (2007).
7. P. J. Rousseeuw, Shilhouettes: a graphical aid to the interpretation and validation of cluster analysis, *Journal of Computational and Applied Mathematics*, 20, pp. 53–65 (1987).
8. K. R. Shahapure and C. Nicholas, Cluster Quality Analysis Using Silhouette Score, *7th International Conference on Data Science and Advanced Analytics*, Sydney, NSW, Australia, (2020).
9. Ö. Akgüller, M. A. Balcı, and G. Cioca, Clustering Molecules at a Large Scale: Integrating Spectral Geometry with Deep Learning, *Molecules*, 29(16), 3902 (2020).
10. S. P. Lima and M. D. Cruz, A genetic algorithm using Calinski-Harabasz index for automatic clustering problem, *Revista Brasileira de Computação Aplicada*, 12(3), 97–106 (2020).
11. M. F. Lathief, I. Soesanti, and A. E. Permanasari, Combination of Fuzzy C-Means, Xie-Beni Index, and Backpropagation Neural Network for Better Forecasting Result, In *Proceedings of the International Conference on Creative Economics, Tourism and Information Management (ICCETIM 2019)- Creativity and Innovation Developments for Global Competitiveness and Sustainability, pp.* 72–77 (2019).
12. T. M. Kodinariya and P. R. Makwana, Review on determining number of Cluster in K-Means Clustering, *International Journal of Advance Research in Computer Science and Management Studies* 1, 90–95 (2013).
13. A. Likas, N. Vlassis, and J. J. Verbeek, The global k-means clustering algorithm, *Pattern Recognition*, 36(2), 451–461 (2003).

Chapter 35

Optimization of Ball Milling Parameters of AZ31/SiC Composite Materials with PSO

Harun Mert Ilbeyli*, Turker Turkoglu and Sare Celik

Department of Mechanical Engineering, Balikesir University, Balikesir, Turkiye
**mert.ilbeyli@balikesir.edu.tr*

This study investigates the production and optimization of AZ31 magnesium matrix composites reinforced with silicon carbide (SiC) to address the inherent limitations in the strength and ductility of magnesium and its alloys. The composite materials were produced using a high-energy ball milling process, followed by hot pressing. A central composite design was employed to systematically vary key production parameters, including ball milling speed, SiC content, and milling time. A mathematical model was developed based on experimental data and further optimized using Particle Swarm Optimization (PSO). The optimization process identified the optimal production conditions as 400 rpm ball milling speed, 4 wt.% SiC content, and 7.51 hours of milling time, yielding a maximum compressive strength of 353.19 MPa. Additionally, the robustness of the mathematical model is evidenced by its high R-sq value of 94.66%, indicating strong agreement with the experimental data. This study demonstrates that careful control and optimization of production parameters can effectively enhance the mechanical properties of magnesium matrix composites, making them suitable for advanced engineering applications.

1. Introduction

Magnesium and its alloys are foundational materials in modern technology, increasingly regarded as the "metal of the future." Ranking as the third most widely used metal in industry after iron and aluminum [1], magnesium is valued for its ultra-lightweight, weldability, high specific strength, and recyclability, making it suitable for applications across aerospace, automotive, and biomedical sectors [2]. However, due to its hexagonal close-packed (HCP) crystal structure, magnesium and its alloys exhibit limitations in production and formability, resulting in lower strength and ductility than aluminum and steel. To address these limitations, magnesium matrix composites have been developed by incorporating micro-sized ceramic particles into the matrix, effectively enhancing the mechanical properties of magnesium and its alloys [3].

Metal matrix composites (MMCs) are produced through various methods, including powder metallurgy, liquid phase sintering, and stir casting [4]. In the stir casting method, reinforcements are dispersed within the molten matrix. However, distributing reinforcements in molten magnesium poses significant challenges due to the high flammability of magnesium compared to other metals, making this approach hazardous [5]. Additionally, a notable limitation of stir casting is the variance in melting points among alloying elements; as temperature rises, elements with lower melting points begin to evaporate, compromising material homogeneity [6]. Powder metallurgy (PM) addresses these limitations by providing more consistent reinforcement distribution and enabling near-net-shape production, which enhances the mechanical properties of MMCs [7, 8]. Moreover, PM yields a superior surface quality for finished components compared to other methods. By incorporating ceramic reinforcements into the magnesium matrix, PM also enhances mechanical properties through precipitation and solid solution strengthening [5].

Mechanical alloying (MA) is a high-energy milling technique in which powder particles undergo repeated fracturing and cold welding until a stable composition is achieved. This process transfers mechanical energy to the powders, facilitating particles and grain size refinement while embedding reinforcements into the matrix to form the composite [9, 10]. MA typically involves placing alloy powders in a stainless-steel container with milling balls, preparing the mixture for the grinding process. Critical parameters influencing the MA process include milling speed, duration, and ball-to-powder ratio, which collectively control the balance between particle fracture and cold welding, thereby shaping the microstructure and, consequently, the mechanical properties of the final composite [11, 12]. Liao *et al.* [13] used high-energy ball milling (HEBM) to prepare reinforcement powders for a magnesium matrix composite produced via friction stir processing. Mg-Gd-BN powders (18.8 wt% Gd, 2 wt% BN) were processed with HEBM at 250 rpm for 2 hours, followed by 500 rpm for 12 hours. Similarly, Jayasathyakawin *et al.* [14] fabricated an SiC-reinforced composite with an Mg/3Al matrix using powder metallurgy. In their study, HEBM was employed at 120 rpm for 3 hours with a 10:1 ball-to-powder ratio, achieving homogeneous distribution of reinforcement within the matrix.

Nature-inspired computing is one of the most focused areas of applied mathematics in recent times. The aim here is to solve complex engineering problems with the help of appropriate algorithms [15, 16]. Particle swarm optimization (PSO) is one of the most effective optimization algorithms for solving complex engineering problems. PSO represents swarms of particles converging to the desired solution. Each particle in the PSO algorithm searches for the best position within the region of the problem at a certain speed. Each particle memorizes its previous best position. The current position of the particles is divided into two as personal best and global best position. In this way, particles balance global and local exploration within the search space, allowing especially multi-dimensional complex problems to be solved. Compared to Differential Evolution and Genetic Algorithms, PSO offers the option of convergence in a shorter time [17, 18]. Mi *et al.* [19] produced the Mg-Mn alloy with desired composition using a combination of active learning with regression and Bayesian optimization (BO). Adaan-Nyiak *et al.* [6] in their study, they made Taguchi experimental design according to the parameters they determined (PGA type, PGA amount, ball milling speed, ball milling time) and made multi-objective optimization using the mathematical model created according to the experimental results and genetic algorithm (GA). As a result of the optimization, they produced high entropy alloy (AlCoCrFe) according to the optimum values they wanted to get.

Sarhani *et al.* [15] stated in their study that the performance of classical approach algorithms deteriorates for problems with large search spaccs. They mentioned that the performance of the algorithms inspired by nature or biology, such as particle swarm optimization (PSO) and genetic algorithm (GA), increases and problems can be solved with more than one decision variable.

This study presents a pioneering approach to the optimization of production parameters for Mg matrix composites reinforced with SiC particles. By integrating mechanical alloying, central composite design, and Particle Swarm Optimization (PSO), the research fills a significant gap in literature, where studies often rely on empirical methodologies or limited single-variable optimizations. The proposed methodology combines a predictive mathematical model with advanced optimization techniques to navigate the multi-dimensional parameter space efficiently. This approach not only reduces the number of required experiments but also achieves notable enhancements in compressive strength. The study highlights the critical influence of SiC content, ball milling speed, and milling duration on composite performance, offering a comprehensive framework for future research. By providing a systematic and replicable method, this work establishes a new standard for tailoring the mechanical properties of magnesium composites to meet the demands of advanced engineering applications. The selection of the Particle Swarm Optimization (PSO) algorithm is rooted in its effectiveness for multi-variable optimization problems with non-linear characteristics. PSO demonstrates superior convergence efficiency and simplicity compared to traditional optimization methods such as Genetic Algorithms (GA) and Differential Evolution (DE). Its unique capability to balance global exploration and local exploitation allows for robust performance in optimizing interdependent parameters such as ball milling speed, SiC content, and milling duration. The algorithm's adaptability is further enhanced by its tunable parameters, including inertial weight and acceleration constants, which were calibrated to align with the specific requirements of this study. The computational efficiency of PSO is particularly advantageous given the time-intensive nature of the experimental processes, enabling precise and reliable optimization outcomes. This study underscores the suitability of PSO in addressing the complex challenges of material science research, setting a precedent for its application in optimizing composite material properties.

2. Methodology and experimental procedure

2.1. *Materials and processes*

In this study, AZ31 powder was selected as the matrix, while SiC was chosen as the reinforcement phase. AZ31 is a magnesium alloy commonly used in applications such as automotive and aerospace components, electronic enclosures, structural parts, sporting goods, medical devices, defense equipment, lightweight structures, heat sinks, and portable electronics. SiC, or silicon carbide, is characterized by its exceptional hardness, high thermal conductivity, low thermal expansion, excellent chemical stability, high strength, resistance to wear and abrasion, ability to withstand high temperatures, and its semiconductor properties. Figure 1 illustrates the SEM images of composite powders subjected to ball milling at 300, 350, and 400 rpm. These composite powders contain 8 wt.% SiC and have undergone mechanical alloying for 6 h.

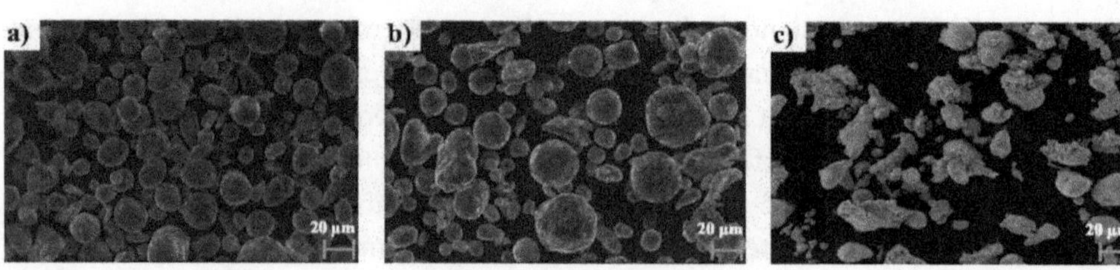

Fig. 1. (a) 300 rpm (b) 350 rpm (c) 400 rpm

2.2. *Experimental design and ball milling*

Central composite design was performed with the Minitab software using three variable parameters in Table 1. AZ31/SiC composite samples were subjected to ball milling process with different parameters (ball milling speed, wt% SiC, and ball milling time) according to the experimental design.

Table 1. Ball milling parameters.

No	Variables		Levels	
1	Ball milling speed (rpm)	300	350	400
2	SiC (wt.%)	4	8	12
3	Ball milling time (h)	4	6	8

The central composite experimental design based on the determined parameters is shown in Table 2.

Table 2. Central composite design created for the preparation of AZ31/SiC composite samples.

Sample No.	Ball milling speed (rpm)	SiC (wt.%)	Ball milling time (h)
1	400	4	8
2	400	4	4
3	300	12	4
4	400	12	4
5	300	4	8
6	300	4	4
7	300	12	8
8	400	12	8
9	300	8	6
10	400	8	6
11	350	4	6
12	350	12	6
13	350	8	4
14	350	8	8
15	350	8	6

The ball milling process was carried out on Nanjing Chishun QM-3SP2 planetary ball mill in Figure 2a. Each sample was prepared with a total composition weight of 30 g. Stearic acid process control agent (PCA) was added to the mixture to minimize cold welding during the process. The AZ31/SiC composite powder formed at the end of the ball milling process according to the experimental design is shown in Figure 2b.

2.3. *Sintering process*

The produced composite powders were taken into the mold produced from hot work steel with an inner diameter of 15 mm. Then, they were sintered at 600 °C, 50 MPa pressure and for 30 minutes in the MSE Technologies hot press machine shown in Figure 2c. Argon gas was used to prevent oxidation during sintering.

At the end of the hot pressing process, sintered AZ31/SiC composite samples with a diameter of 15 mm were obtained as shown in Figure 2d.

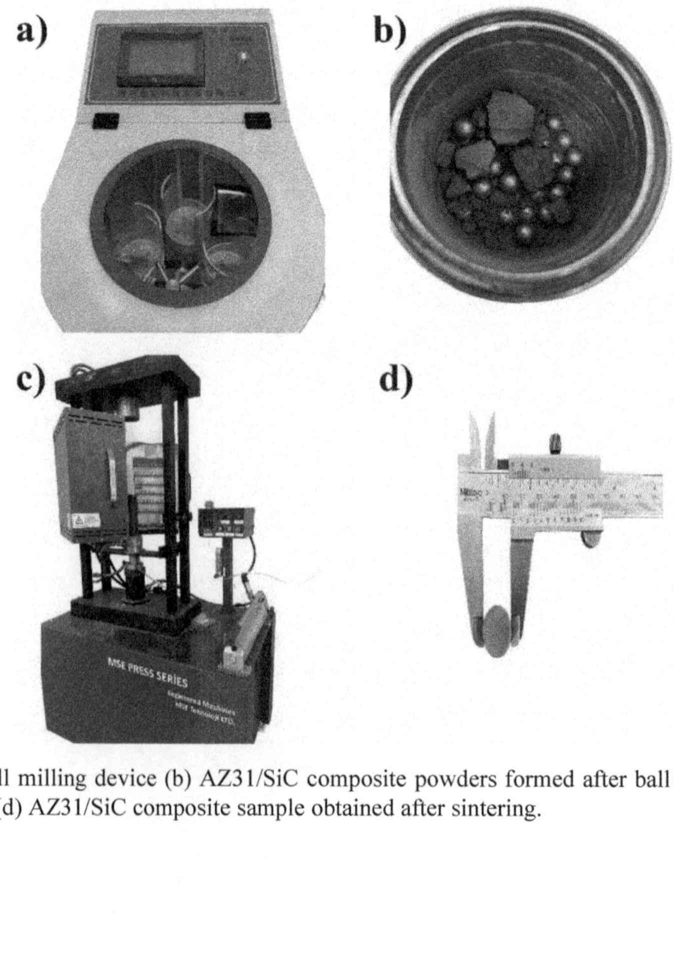

Fig. 2. (a) Ball milling device (b) AZ31/SiC composite powders formed after ball milling process (c) Hot press (d) AZ31/SiC composite sample obtained after sintering.

2.4. *Mathematical model and PSO application*

Compression test was applied to the samples created according to the experimental design in the Zwick/Roell device. Then, using the obtained compression strengths, the mathematical model symbolizing the experiment was obtained via Minitab software.

In the mathematical model created, the desired output value is the compressive strength (*CS*). The parameters affecting the output are ball milling speed (x_1), wt.% SiC (x_2) and ball milling time (x_3).

$$CS = 1141.92 - 3.34x_1 - 22.75x_2 - 93.36x_3 - 0.85x_2^2 + 3.23x_3^2 + 0.08x_1x_2 + 0.1x_1x_3 + 1.2x_2x_3. \quad (1)$$

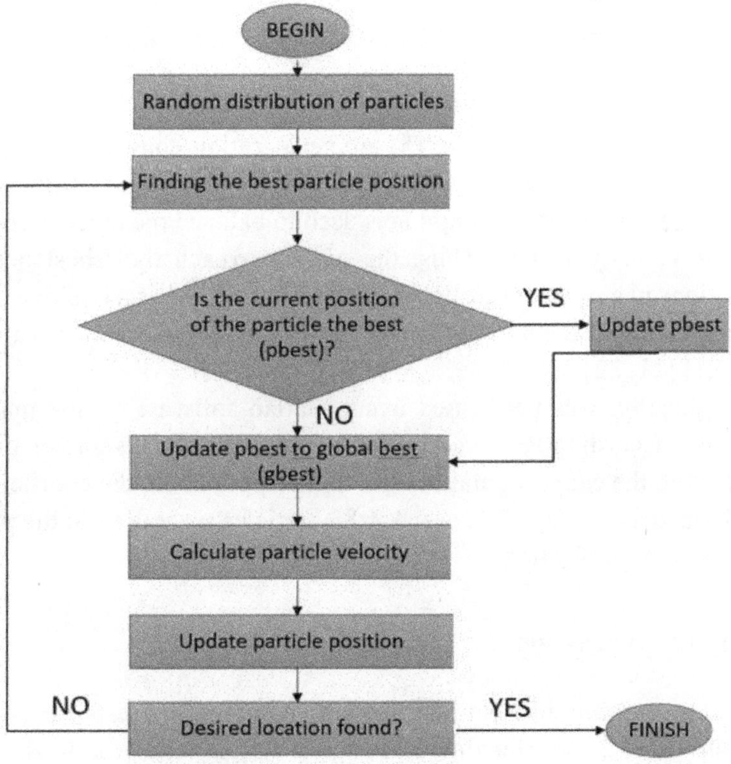

Fig. 3. PSO algorithm tree.

In PSO, each particle in the swarm flies over the objective function, searching for the desired solution. Particles constantly update their position in the objective function and the speed at which they should travel, based on their own flight experience and that of their neighbors [20]. The general operation of the PSO algorithm is shown in Figure 3.

For a function of N dimension, the position of the "k" th particle is expressed as:

$$X_k = (x_{k1}, x_{k2}, x_{k3}, \dots, x_{kN}). \quad (2)$$

Each particle memorizes its current personal best position is expressed as:

$$P_k = (p_{k1}, p_{k2}, p_{k3}, \dots, p_{kN}). \quad (3)$$

The global best position between particles is expressed as:

$$P_g = (p_{g1}, p_{g2}, p_{g3}, \dots, p_{gN}). \quad (4)$$

The velocity and position values used for the PSO algorithm are as follows:

$$v_{(k+1)N} = \omega v_{kN} + c_1 r_1 (p_{kN} - x_{kN}) + c_2 r_2 (p_{gN} - x_{kN}). \quad (5)$$

$$x_{(k+1)N} = x_{kN} + v_{(k+1)N}. \quad (6)$$

The values c_1 and c_2 in the Eq. (5) are acceleration constants. They represent the acceleration terms that bring each particle closer to their best personal and global positions. The inertial weight ω is used to balance the current velocity with the previous velocity values. Thus, the particles reach their best personal and global positions in a more controlled manner. It is defined as a positive integer and usually decreases over the iterations. The values r_1 and r_2 are random values chosen between [0,1] [21, 22].

PSO application was performed using Matlab software on the mathematical model created according to the compressive strength of the samples produced in accordance with the experimental design. In the application, the coefficients in Eq. (5) were defined as $c_1 = c_2 = 1.5$, $\omega = 0.7$. 80 particles were used in the process and 80 iterations were performed.

3. Results and Discussion

The compressive strengths obtained from the compression tests for each sample, as shown in Table 3, are illustrated in Figure 4. The results indicate significant variations in compressive strength depending on the experimental parameters, including ball milling speed, SiC content, and milling time. The highest compressive strength was achieved in experiment number 2, with a value of 312 MPa. This result highlights the critical influence of lower SiC content and optimized milling parameters on the mechanical performance of the composite. Lower compressive strength in other samples can be attributed to factors such as excessive SiC content leading to clustering or insufficient milling time, which affects the uniform dispersion of reinforcements. For instance, higher SiC contents (e.g., 12 wt.%) were observed to degrade mechanical performance, potentially due to agglomeration and reduced load transfer efficiency. Similarly, shorter milling times may result in incomplete bonding and mixing of the reinforcement phase within the matrix.

To better understand the relationship between the parameters and the compressive strength, regression analysis was conducted based on the experimental results. The regression model, as summarized in Table 4, yielded an R-sq value of 94.66% and an adjusted R-sq value of 89.86%, indicating that the model reliably represents experimental data. The probability plot (Figure 5) further supports the adequacy of the model by showing a close alignment between observed and predicted values. These statistical indicators confirm the robustness of the mathematical model for optimization purposes, which is essential for understanding and predicting composite performance under varying production conditions.

Table 3. Experimental tests.

	Composite Powders		
	Ball milling speed (rpm)	**SiC (wt.%)**	**Ball milling time (h)**
1	0 (Pure AZ31)	0 (Pure AZ31)	0 (Pure AZ31)
2	400	4	8
3	400	4	4
4	300	12	4
5	400	12	4
6	300	4	8
7	300	4	4
8	300	12	8
9	400	12	8
10	300	8	6
11	400	8	6
12	350	4	6
13	350	12	6
14	350	8	4
15	350	8	8
16	350	8	6

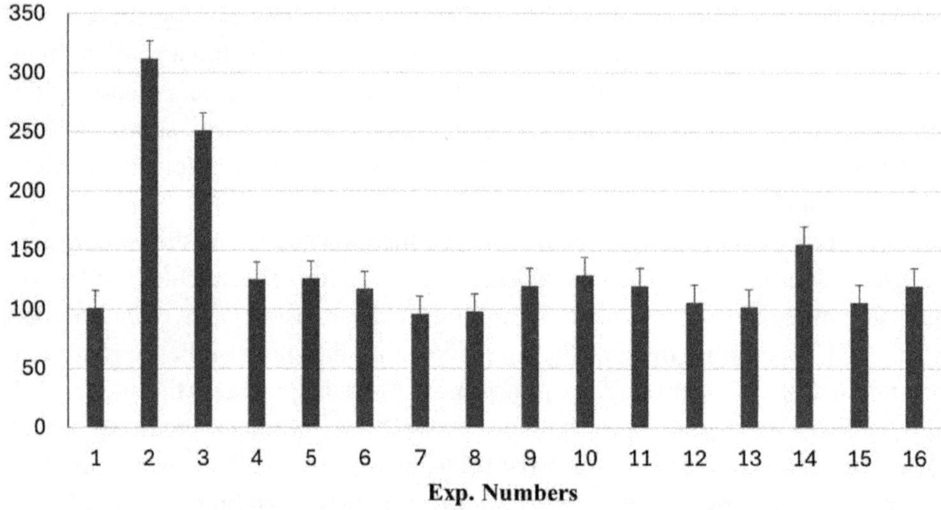

Fig. 4. Compressive strength (MPa) for each sample.

Table 4. Regression analysis.

Model summary		
S	R-sq	R-sq(adj)
6.92103	94.66%	89.86%

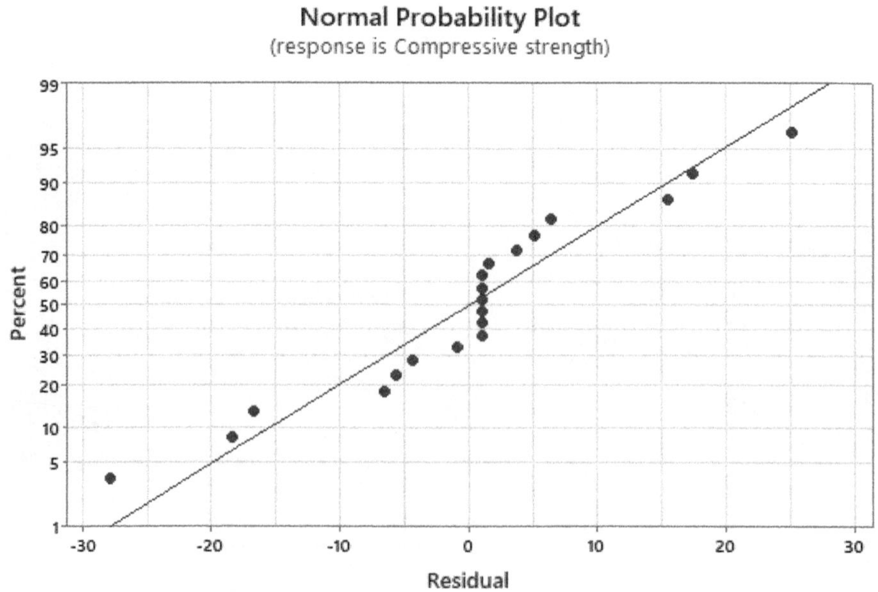

Fig. 5. Probability plot.

The regression analysis validates the use of the mathematical model for optimization purposes. The Particle Swarm Optimization (PSO) algorithm was then applied to determine the maximum compressive strength and the optimal production parameters. As shown in Figure 6, the maximum compressive strength of 353.19 MPa was achieved at a ball milling speed of 400 rpm, SiC content of 4 wt.%, and a milling time of 7.51 hours. The iterative optimization process revealed a convergence of the objective function, demonstrating the efficiency and robustness of the PSO algorithm.

The findings of this study underscore the critical role of optimization in achieving enhanced mechanical properties in magnesium matrix composites. The PSO optimization results revealed that the compressive strength of 353.19 MPa represents a substantial improvement compared to the highest experimental value of 312 MPa. This demonstrates the effectiveness of PSO in identifying optimal parameter combinations beyond conventional experimental boundaries.

The results highlight the interdependence of production parameters. For instance, a ball milling speed of 400 rpm appears to provide sufficient energy for effective particle fracturing and mixing, while a lower SiC content (4 wt.%) prevents agglomeration, maintaining homogeneous reinforcement distribution. The milling time of 7.51 hours balances particle refinement and bonding, ensuring optimal mechanical performance. These parameters collectively contribute to enhanced load transfer mechanisms and microstructural stability within the composite.

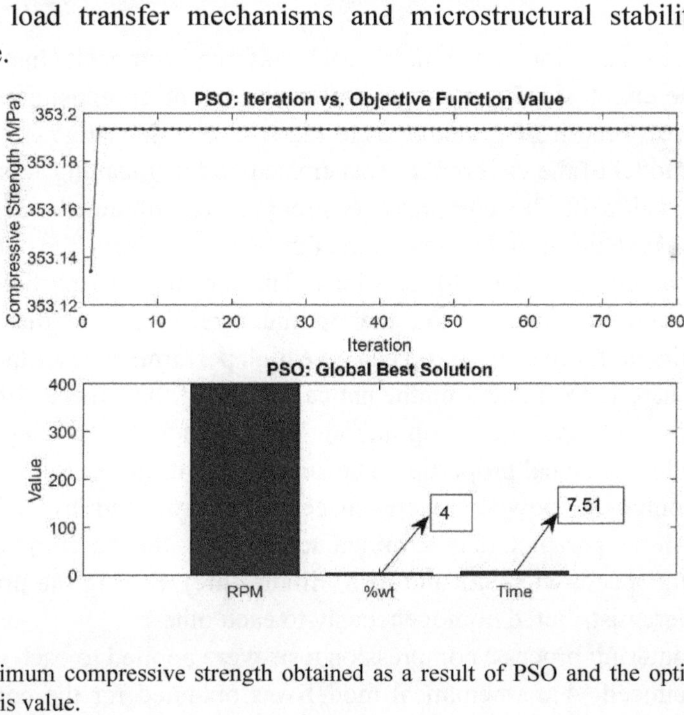

Fig. 6. Maximum compressive strength obtained as a result of PSO and the optimum parameters providing this value.

Comparison with similar studies in literature shows that the methodology employed in this study, particularly the integration of regression analysis with PSO, offers a more efficient and systematic approach to parameter optimization. Traditional methods often rely on trial-and-error experimentation, which is time-consuming and resource-intensive. By contrast, this study provides a replicable framework for optimizing multi-variable systems in material science.

The implications of these findings extend beyond this specific composite system. The methodology can be adapted to optimize other material systems where multiple interdependent parameters affect performance. Additionally, the insights gained into the relationship between ball milling parameters and mechanical properties can inform the design of lightweight, high-strength materials for applications in aerospace, automotive, and other advanced engineering fields.

In conclusion, the integration of experimental results with optimization techniques, supported by a robust mathematical model, represents a significant advancement in the field of composite material development. The PSO algorithm has proven to be a powerful tool for achieving superior material properties with minimal experimental effort, highlighting its potential for broader applications in material optimization.

4. Conclusions

In this study, the production of Mg matrix SiC reinforced composite materials was carried out. The effect of the production parameters of the composite material on the compressive strength of the composite material was investigated. After the mathematical model of the experiment was created and its meaning was checked, the maximum value of the compressive strength was obtained via the PSO algorithm. Thanks to the particle population of the PSO algorithm, the optimization process could be completed in 80 iterations. The fact that the particles can be populated at appropriate speeds on the optimum result shows that the PSO algorithm is suitable for this study and can give high performance results in a short time. The aim here is to create a mathematical model of the experiment with the minimum number of experiments, optimizing the model and producing materials with the desired mechanical properties. The general results are as follows.

(1) Mg/SiC composite powders were successfully produced by ball milling process with the parameters determined according to the experimental design (ball milling speed, wt.%SiC and ball milling time). During the process, the powders were distributed homogeneously to each other.
(2) After the sintering process, compression tests were applied to each composite sample produced. A mathematical model was obtained for the compressive

strength values obtained. Regression analysis was performed for the created model. Since the "R-sq(adj)" value was above 80%, it was determined that the model could represent the experiment and therefore was suitable for optimization.

(3) The maximum value of the objective function (compressive strength) generated as a result of the PSO process was determined as 353.19 MPa. The optimum experimental parameters providing the maximum output value were determined as 400 rpm, 4 wt.%SiC and 7.51 h.

References

1. Wang, T., Yang, L., Tang, Z., Wu, L., Yan, H., Liu, C., Ma, Y., Liu, W., and Yu, Z.: Microstructure, mechanical properties and deformation mechanism of powder metallurgy AZ31 magnesium alloy during rolling. Materials Science and Engineering: A. 844, 143042 (2022).
2. Sübütay, H. and Şavklıyıldız, İ.: The relationship between structural evolution and high energy ball milling duration in tin reinforced Mg alloys. Materials Today Communications. 35, 105868 (2023).
3. Ponhan, K., Tassenberg, K., Weston, D., Nicholls, K.G.M., and Thornton, R.: Effect of SiC nanoparticle content and milling time on the microstructural characteristics and properties of Mg-SiC nanocomposites synthesized with powder metallurgy incorporating high-energy ball milling. Ceramics International. 46, 26956–26969 (2020).
4. Purohit, R., Dewang, Y., Rana, R.S., Koli, D., and Dwivedi, S.: Fabrication of magnesium matrix composites using powder metallurgy process and testing of properties. Materials Today: Proceedings. 5, 6009–6017 (2018).
5. Nie, K.B., Wang, X.J., Deng, K.K., Hu, X.S., and Wu, K.: Magnesium matrix composite reinforced by nanoparticles – A review. Journal of Magnesium and Alloys. 9, 57–77 (2021).
6. Adaan-Nyiak, M.A., Alam, I., and Tiamiyu, A.A.: Ball milling process variables optimization for high-entropy alloy development using design of experiment and genetic algorithm. Powder Technology. 427, 118766 (2023).
7. Zegeye, D.B., Sinha, D.K., Venkatesh, C., Kumar, A., and Jha, P.: Mathematical modeling and mechanical characterization through process parameters optimization of AlMg-matrix SiCp reinforced composites produced by powder metallurgy route. Materials Today: Proceedings. 47, 3811–3818 (2021).
8. Ersoz, O.F., Turkoglu, T., and Celik, S.: Evaluation of reinforcement distribution of Al6061/TiC composites by computed tomography. Materials Today: Proceedings. 115, 1–4 (2023).
9. Canakci, A., Varol, T., and Nazik, C.: Effects of amount of methanol on characteristics of mechanically alloyed Al-Al 2O 3 composite powders. Materials Technology. 27, 320–327 (2012).

10. Verma, P.K., Singh, A., and Kumar, A.: Microstructure characterization and phase evolution of equiatomic AlCoMoFeNi high entropy alloy synthesized by mechanical alloying. Materials Chemistry and Physics. 318, 129325 (2024).
11. Ma, G., Xiao, H., Ye, J., and He, Y.: Research status and development of magnesium matrix composites. Materials Science and Technology. 36, 645–653 (2020).
12. Fan, L., Zhou, M., Zhang, Y., Dieringa, H., Qian, X., Zeng, Y., Lu, X., Huang, Y., and Quan, G.: Achieving high strength and ductility in a heterogeneous bimodal grain structured TiC/AZ61 magnesium nanocomposites via powder metallurgy. Materials Science and Engineering: A. 867, 144344 (2023).
13. Liao, H., Chen, J., Peng, L., Han, J., Yi, H., Zheng, F., Wu, Y., and Ding, W.: Fabrication and characterization of magnesium matrix composite processed by combination of friction stir processing and high-energy ball milling. Materials Science and Engineering: A. 683, 207–214 (2017).
14. Jayasathyakawin, S. and Ravichandran, M.: Fabrication and wear behaviour of Mg-3wt.%Al-x wt. % SiC composites. Heliyon. 9, 13679 (2023).
15. Sarhani, M. and Voß, S.: Chunking and cooperation in particle swarm optimization for feature selection. Annals of Mathematics and Artificial Intelligence. 90, 893–913 (2022).
16. Turkoglu, T. and Celik, S.: Process optimization for enhanced tribological properties of Al/MWCNT composites produced by powder metallurgy using artificial neural networks. Surface Topography: Metrology and Properties. 9, 045032 (2021).
17. Thangaraj, R., Pant, M., Abraham, A. and Bouvry, P.: Particle swarm optimization: Hybridization perspectives and experimental illustrations. Applied Mathematics and Computation. 217, 5208–5226 (2011).
18. Turkoglu, T. and Celik, S.: Synergistic effects of TiC/GNP strengthening on the mechanical and tribological properties of Al6061 matrix composites coupled with process optimization by artificial neural network. Proceedings of the Institution of Mechanical Engineers, Part E. 239, 181–193 (2023).
19. Mi, X., Dai, L., Jing, X., She, J., Holmedal, B., Tang, A., and Pan, F.: Accelerated design of high-performance Mg-Mn-based magnesium alloys based on novel bayesian optimization. Journal of Magnesium and Alloys. 12, 750–766 (2024).
20. Zhao, F., Ji, F., Xu, T., Zhu, N., and Jonrinaldi: Hierarchical parallel search with automatic parameter configuration for particle swarm optimization. Applied Soft Computing. 151, 111126 (2024).
21. Priya, G.V. and Ganguly, S.: Multi-swarm surrogate model assisted PSO algorithm to minimize distribution network energy losses. Applied Soft Computing. 159, 111616 (2024).
22. Han, G., Li, H., Wang, G., Liu, Y., Zhang, J., Hu, Z., You, X., and Xie, Y.: Prediction and control of profile for silicon steel strip in the whole tandem cold rolling based on PSO-BP algorithm. Journal of Manufacturing Processes. 120, 250–259 (2024).

Chapter 36

An Artificial Intelligence Approach to Parameter Estimation for an Addiction Model

Derya Avcı* and Sanem Sakarya

Department of Mathematics, Balıkesir University
Balıkesir, Çağış Campus, 10145, Turkey
**dkaradeniz@balikesir.edu.tr*

In this study, we introduce a system of differential equations adapted from an existing technology addiction model to simulate social media addiction. The model categorizes individuals into four primary levels of social media addiction: Susceptible, light/moderate users, heavy users, and those experiencing adverse mental effects. The transitions between these categories are modeled using dynamic rates, with the parameters of these transitions optimized through artificial intelligence techniques. The dynamic systems approach is adopted to analyze and understand the evolution of addiction levels over time. The primary goal of this study is to gain a deep understanding of the dynamics of social media addiction through mathematical modeling and machine learning methods and to develop effective strategies for managing these processes.

1. Introduction

Social media has evolved from an innovative communication method to an indispensable tool in modern life. The origins of social media can be traced back to the late 1990s and early 2000s with the emergence of platforms like Six Degrees, MySpace, and later, Facebook [1].

These platforms fundamentally changed the way people connect, share, and consume information, leading to the rapid expansion of social networks globally. As these platforms gained popularity, their impact on daily life grew exponentially [1, 2]. By the mid-2010s, social media had become deeply embedded in the fabric of society, with billions of users interacting daily on platforms like Facebook, Instagram, and Twitter [1, 3].

However, the widespread adoption of social media has also led to significant psychological and behavioral issues [2, 4]. Research indicates that excessive use of social media can lead to addiction-like symptoms, including obsessive checking of notifications, inability to reduce usage despite negative consequences, and the pervasive fear of missing out (FOMO) [1, 3]. Known as social media addiction, this phenomenon has been linked to various mental health issues, including anxiety, depression, and low self-esteem [3]. The ubiquity of smartphones and constant access to social media platforms have exacerbated these issues, making it increasingly difficult for individuals to disconnect [3, 5].

The rapid rise in social media addiction has prompted researchers to examine the underlying mechanisms and impacts of this phenomenon. There is an increasing need to understand how social media addiction develops and progresses over time. In response to this need, our study utilizes a differential equation system adapted from an existing technology addiction model [1, 3, 6]. This model categorizes individuals into four main levels of addiction susceptible, light/moderate user, heavy user, and those experiencing adverse mental effects and tracks the transitions between these categories through dynamic rates [3].

The model's key parameters, which represent the transition rates between different levels of addiction, have been optimized using artificial intelligence techniques [7]. Data was collected through surveys and processed through steps such as data cleaning, missing data imputation, and normalization. Supervised machine learning algorithms like Logistic Regression and Support Vector Machines (SVM) were employed to optimize these parameters. Logistic Regression was used to model the probabilities of different addiction levels, while SVM served as an effective classifier to distinguish between these levels in the dataset. To enhance the model's accuracy, techniques such as cross-validation and hyperparameter optimization were applied [8, 9, 10].

This study aims to comprehensively understand the dynamics of social media addiction through mathematical modeling and artificial intelligence techniques and to develop effective strategies to mitigate the negative impacts of this modern phenomenon [3, 6].

2. The Model

2.1. Differential equations system

In this study, social media addiction levels are modeled using a system of differential equations that categorizes individuals into four primary stages based on their level of addiction [1, 6]:

S (Susceptible): This stage represents individuals before they start using social media or when their usage is minimal, without showing signs of addiction.

L (Light/Moderate): At this stage, individuals' social media usage begins to increase, but it has not yet reached an addictive level.

H (Heavy): This stage signifies a significant increase in social media usage, where signs of addiction start to manifest.

B (Bad Mental Effects): At this stage, individuals experience severe negative mental effects due to excessive social media use, indicating full-blown addiction.

The dynamic system is formulated through a set of differential equations that define the transitions between these addiction levels. The equations are as follows: In these equations:

$$\frac{dS(t)}{dt} = -\alpha S(t) - kS(t) - qS(t)H(t) \quad (1)$$

$$\frac{dL(t)}{dt} = \alpha S(t) + kS(t)L(t) + qS(t)H(t) - \beta L(t) - cL(t)H(t) + \gamma H(t) + \epsilon B(t) \quad (2)$$

$$\frac{dH(t)}{dt} = \beta L(t) + cL(t)H(t) - \gamma H(t) - \delta H(t) - pH(t)2 \quad (3)$$

$$\frac{dB(t)}{dt} = \delta H(t) + pH(t)2 - \epsilon B(t) \quad (4)$$

where *S*(*t*), *L*(*t*), *H*(*t*) and *B*(*t*) are the state variables representing the population in each addiction stage at time t, and the parameters α, β, γ, δ, ϵ, k, q, c, and q are positive constants that represent the rates of transition between these states. These parameters can be interpreted as follows:

The parameters in this model represent the rates at which individuals transition between different addiction stages:

Table 1. This table summarizes the parameter definition and transition rates for the addiction model.

Parameter	
α	Rate of transition from *S* to *L*
β	Rate of transition from *L* to *H*
γ	Rate of transition from *H* to *L*
δ	Rate of transition from *H* to *B*
ϵ	Rate of transition from *B* to *L*
k	Rate constant for the transition of *S* individuals to *L* individuals due to interaction with *L* individuals.
q	Rate constant for the transition of *S* individuals to *L* individuals due to interaction with *H* individuals.
c	Rate constant for the transition of *L* individuals to *H* individuals, representing the influence of living or working with *H* individuals.
p	Rate constant for the transition of *H* individuals to *B* individuals, indicating the rate at which heavy users contribute to the population of problematic users (*B* group).

The system of differential equations provides a framework for understanding how the population transitions between different levels of addiction over time [3].

2.2 The role of artificial intelligence techniques in parameter estimation

The parameters within the differential equations (1)–(4) are critical for accurately simulating the addiction dynamics described by the model. Estimating these parameters requires sophisticated techniques that can handle the complexity of the data and the relationships between variables. In this study, we employed advanced artificial intelligence (AI) techniques, particularly supervised machine learning algorithms, to achieve precise parameter estimation [11, 12, 14, 15].

2.2.1 Logistic regression

Logistic regression is a generalized form of the linear regression model and is commonly used to model binary dependent variables (e.g., $Y = 0$ or $Y = 1$). In this study, logistic regression was applied to model the probabilities associated with transitions between addiction levels [24]. The logistic regression model is expressed as,

$$P(Y = 1 \mid X) = \frac{e^{\beta_0 X}}{1+e^{\beta_0 X}} \tag{5}$$

where $\beta_0 X$ is expanded as:

$$\beta_0 X = \beta_0 + \beta_1 x_1 + \beta_2 x_2 + \cdots + \beta_n x_n \tag{6}$$

In logistic regression analysis, unlike traditional multivariate regression that minimizes residuals using ordinary least squares, Maximum Likelihood Estimation (MLE) is employed to obtain the optimal estimation of independent variable parameters by maximizing the observed probability of the dependent variable. MLE is preferable as it can be applied to both linear models and more complex non-linear estimations, making it well-suited for the non-linear nature of logistic regression [10].

Each observed value is treated as a Bernoulli trial, where the random variable Y_i follows a Bernoulli probability distribution [10]. The expectation value is $E\{Y_i\} = p_i$, and the probability distribution is expressed as:

$$P(Y_i = y_i) = p_i^{y_i} \times (1 - p_i)^{1-y_i}, \qquad y_i = 0{,}1 \tag{7}$$

For independent random variables, the likelihood function, which needs to be maximized, is obtained as follows (Hauck & Donner, 1977):

$$L = P(Y_1 = y_1) \times P(Y_2 = y_2) \times \cdots \times P(Y_n = y_n) = \prod^{n} p_i^{y_i} \times (1 - p_i)^{1-y_i} \tag{8}$$

To facilitate calculation, this likelihood function is often transformed into a log-likelihood function:

$$log_e\, L(\beta) = \sum_{i=1}^{n}[y_i(\beta_0 X_i) - log_e(1 + e^{\beta_0 X_i})] \tag{9}$$

In logistic regression, the MLE of the parameter β is calculated by finding the maximum of this log-likelihood function. The derivative is given by:

$$S(\beta) = \frac{\partial log_e L(\beta)}{\partial \beta} = \sum_{i=1}^{n}\left[\frac{e^{\beta_0 X_i}}{i + e^{\beta_0 X_i}}\right] + \sum_{i=1}^{n} y_i X_i' = 0 \tag{10}$$

Since *S(β)* is a non-linear function of β, there is no closed-form solution, meaning a simple computational equation does not exist. Therefore, iterative computational methods are typically used to find the solution until the maximum value of the likelihood function for β converges.

Logistic regression is a critical method for accurately modeling the transition probabilities between addiction levels, enabling more realistic parameter estimates within the model, and thereby facilitating a better simulation of the addiction processes [10].

2.2.2 Support vector machines (SVM)

SVM was utilized to refine the boundaries between different addiction levels, thereby improving the estimation of parameters such as c, p, α, β, γ, δ, ϵ, k, and q, which govern interaction effects and transition rates as represented in Eq. (1) through (4).

The SVM approach works by identifying the optimal hyperplane that separates data points into different addiction levels [15, 17]. Mathematically, this is formulated as the following optimization problem:

$$min\frac{1}{2}||w||^2 + C\sum_{i=1}^{n}\xi_i \tag{11}$$

Subject to:

$$y_i(w.x_i + b) \geq 1 - \xi_i, \qquad \xi_i \geq 0 \tag{12}$$

Here, we represent the weight vector, x_i the input vectors, y_i the class labels, ξ_i the slack variables allowing some misclassifications, and C is the regularization parameter controlling the trade-off between maximizing the margin and minimizing classification errors [8].

For non-linear classification tasks, SVM employs kernel functions like the Radial Basis Function (RBF) to map the input space into a higher-dimensional space. This allows for the separation of non-linear addiction levels. The kernel function is expressed as [17]:

$$K(x_i, x_j) = \exp(-\mu\, ||x_i - x_j||^2) \tag{13}$$

where μ determines the influence of each training example. By using SVM, the model achieves more precise and accurate classification of different addiction levels, thereby improving the estimation of all parameters that define the model's dynamics [17]. This optimization process is further enhanced by cross-validation and hyperparameter tuning techniques, leading to increased predictive accuracy of the model.

3. Model Validation and Performance Analysis

In this section, we delve into the validation of the proposed social media addiction model and analyze its performance. The validation process is critical to ensure that the model accurately represents the dynamics of social media addiction and provides reliable insights for understanding and mitigating this phenomenon [8, 24].

3.1 Data collection and preprocessing

3.1.1 Survey design and data collection

To validate the model, a comprehensive survey was designed to capture a wide array of behavioral and psychological data related to social media usage. The survey included questions on frequency, duration, and emotional responses to social media use, as well as demographic information. The survey was distributed across various platforms to a diverse sample, ensuring a broad range of responses that could represent different levels of social media engagement.

The collected data was compiled into a structured dataset, where each row represented an individual respondent, and each column corresponded to a specific question or measured variable. This dataset formed the foundation for the subsequent data processing and model validation efforts [5].

3.1.2 Data cleaning and normalization

Before feeding the data into the model, extensive preprocessing was conducted to enhance the quality and usability of the dataset. The steps included [5, 7]:

3.1.2.1 Data cleaning

This involved identifying and handling missing, inconsistent, or outlier data points. Techniques such as mean or median imputation were applied to replace missing values, and outliers were carefully analyzed to determine whether they should be retained or removed.

3.1.2.2 Normalization

The dataset was normalized to ensure that all variables were on a comparable scale, which is crucial for accurate model performance. This was done by rescaling the data to a standard range, typically between 0 and 1, using min-max scaling or z-score normalization methods.

3.1.2.3 Feature engineering

New features were derived from the existing data to capture more nuanced aspects of social media usage. For example, interaction effects between different social media platforms or the cumulative time spent online per week were calculated and added as new variables.

3.1.2.4 Coding and tools

All data processing tasks were implemented using Python, specifically with the Pandas and NumPy libraries for data manipulation, and Scikit-learn for feature scaling and engineering. Below is a snippet of the code used for data normalization:

```
from sklearn.preprocessing import MinMaxScaler

# Initializing the scaler
scaler = MinMaxScaler()

# Selecting the features to normalize
features_to_scale = ['time_spent', 'emotional_response', 'frequency_of_use']

# Applying normalization
normalized_data = scaler.fit_transform(data[features_to_scale])

# Replacing the original features with normalized ones
data[features_to_scale] = normalized_data
```

Fig. 1. Normalization of selected features using MinMaxScaler

3.2 Model calibration

After preprocessing the data, the next step was to calibrate the model to ensure it accurately reflects real-world dynamics. This involved fine-tuning the model parameters to align with the observed data [16].

3.2.1 Parameter tuning and optimization

The differential equation system used in this model contains several parameters that represent the transition rates between different levels of addiction. These parameters α, β, γ, δ, ϵ, k, q, c and p optimized using advanced machine learning techniques [8, 14].

3.2.1.1 Logistic regression

This technique was employed to estimate the transition probabilities between different addiction levels. The logistic regression model was trained using the preprocessed survey data, and parameters were optimized using Maximum Likelihood Estimation (MLE) to ensure the best fit. The Python implementation relied on the Stats models library, and a sample code snippet is provided below [10, 18]:

```
import statsmodels.api as sm

# Defining the dependent and independent variables
X = data[['time_spent', 'emotional_response', 'frequency_of_use']]
y = data['addiction_level']

# Adding a constant to the model
X = sm.add_constant(X)

# Fitting the logistic regression model
logit_model = sm.Logit(y, X)
result = logit_model.fit()

# Outputting the results
print(result.summary())
```

Fig. 2. Implementation of the Logistic Regression Model

3.2.1.2 Support vector machines (SVM)

SVM was used to refine the boundaries between different addiction levels. By determining the optimal hyperplane separating different classes, SVM helped improve the accuracy of parameter estimates for transition rates such as c and p. The model was implemented using Scikit-learn, with kernel functions like Radial Basis Function (RBF) for non-linear classifications. Here is a sample of the SVM implementation [16, 19]:

```
from sklearn.svm import SVC
from sklearn.model_selection import GridSearchCV

# Defining the SVM model with RBF kernel
svm_model = SVC(kernel='rbf')

# Setting up a grid search for hyperparameter tuning
param_grid = {'C': [0.1, 1, 10], 'gamma': [1, 0.1, 0.01]}
grid = GridSearchCV(svm_model, param_grid, refit=True, verbose=2)
grid.fit(X, y)

# Outputting the best parameters
print(grid.best_params_)
```

Fig. 3. Implementation of the Support Vector Machine (SVM) Model

3.3 Model validation and performance evaluation

In this section, the validation process and performance evaluation of the proposed social media addiction model are detailed. The performance metrics used in this evaluation are crucial for determining the model's accuracy and reliability in simulating the dynamics of social media addiction. The evaluation consists of three main parts: Predictive Accuracy, Sensitivity Analysis, and Validation Against External Data [21, 23].

3.3.1 Predictive accuracy

The predictive accuracy of the model is one of the most critical aspects of its performance evaluation. After the model was calibrated, its predictive power was assessed using a holdout dataset that was not used during the training phase. The metrics used to quantify this accuracy included Mean Absolute Error (MAE) and the Coefficient of Determination (R^2). The R^2 value, in particular, provided insights into how well the model explained the variance in addiction levels observed in the data. The predictive accuracy was assessed using the following mathematical metrics [13, 16]:

3.3.1.1 Mean absolute error (MAE):

Mean Absolute Error (MAE) measures the average magnitude of errors between predicted values and actual values. It is calculated as:

$$MAE = \frac{1}{n}\sum_{i=1}^{n} |y_i - \hat{y}_i| \tag{14}$$

where y_i represents the actual values, $\hat{y}_i$ represents the predicted values, and n is the total number of samples in the dataset.

3.3.1.2 Coefficient of determination (R^2)

The Coefficient of Determination (R^2) indicates how well the model explains the variability in addiction levels. The R^2 value is calculated as:

$$R^2 = 1 - \sum_{i=1}^{n} \frac{(y_i - \hat{y}_i)^2}{(y_i - \bar{y}_i)^2} \tag{15}$$

where $\bar{y}_i$ represents the mean of the actual values. Below is the Python code used to calculate MAE and R^2:

```
from sklearn.metrics import mean_absolute_error, r2_score

# Calculate MAE and R²
mae = mean_absolute_error(y_test, y_pred)
r2 = r2_score(y_test, y_pred)

print(f"Mean Absolute Error (MAE): {mae}")
print(f"Coefficient of Determination (R²): {r2}")
```

Fig. 4. Calculation of MAE and R^2 Metrics

The results show that the MAE provides insight into the model's prediction error, while the R^2 value indicates the explanatory power of the model. These metrics are crucial for evaluating the model's overall performance.

3.3.2 Sensitivity analysis

Sensitivity analysis was conducted to explore how changes in model parameters affect its predictions. In this analysis, each parameter was systematically varied within a plausible range, and the impact on model outputs was assessed. Mathematically, sensitivity analysis can be expressed as [20]:

$$\text{Sensitivity} = \frac{\partial \hat{y}_i}{\partial \theta} \tag{16}$$

where $\hat{y}_i$ represents the predicted values, and θ represents the parameter under analysis.

The following Python code was used to conduct sensitivity analysis:

```
import numpy as np

# Sensitivity analysis for parameters
parameters = ['alpha', 'beta', 'gamma', 'delta', 'epsilon', 'k', 'q', 'c', 'p']
sensitivity_results = {}

for param in parameters:
    original_value = locals()[param]
    perturbation = original_value * 0.01  # Change parameter by 1%
    locals()[param] += perturbation

    # Update model predictions
    perturbed_solution = odeint(ode_technological, initialConditions, tspan, args=(alpha, beta, gamma, delta, epsilon, p))
    perturbed_output = perturbed_solution[:, 3]  # B level

    # Calculate sensitivity
    sensitivity = (perturbed_output - B) / perturbation
    sensitivity_results[param] = np.mean(np.abs(sensitivity))

    # Restore original parameter value
    locals()[param] = original_value

# Display results
print("Sensitivity Analysis Results:")
for param, sensitivity in sensitivity_results.items():
    print(f"{param}: {sensitivity}")
```

Fig. 5. Sensitivity Analysis of Model Parameters

This analysis identified the parameters that have the most significant impact on addiction dynamics. Specifically, parameters such as δ and p, which govern transitions to severe addiction levels, were found to play a critical role in the model.

3.3.3 Validation against external data

To further validate the model, its predictions were compared against external datasets from other studies on social media addiction. This external validation step ensured that the model's findings are generalizable across different populations and contexts. The following Python code was used for external validation [9, 21]:

```
# External dataset validation
external_data = pd.read_csv('external_dataset.csv')
external_predictions = logreg_model.predict(external_data)

# Compare model predictions with external data
external_accuracy = accuracy_score(external_data['true_labels'], external_predictions)
print(f"External Validation Accuracy: {external_accuracy}")
```

Fig. 6. External Validation of the Model

This comparison with external datasets reinforced the robustness of the model and confirmed the generalizability of its findings.

4. Results and Analysis

In this section, we present the outcomes of the numeric simulations, and the performance of the classification models applied to predict the different stages of social media addiction. The analysis includes population dynamics over time, label distribution, classification performance, and clustering results [3, 19, 22].

4.1 Population dynamics over time

Figure 7 illustrates the simulated population dynamics across the four addiction states. Susceptible (S), Light Addiction (L), Heavy Addiction (H), and Burnout (B), over a 40-day period. The simulation demonstrates a rapid initial transition, where the population in the Susceptible state (S) decreases significantly, leading to a corresponding rise in the more severe states of addiction. By the end of the simulation, the state B (Burnout) stabilizes around 0.45, indicating that nearly half of the population may end up in the most severe addiction state. States L and H also stabilize, with L around 0.35 and H around 0.15, while the Susceptible state stabilizes at a low level around 0.05.

The results indicate a significant shift towards more severe addiction states over time, with state B (Burnout) emerging as the most prevalent in the long term. The rapid decline in the Susceptible state suggests that individuals are quickly progressing into more severe states of addiction, emphasizing the need for early intervention. The stabilization of states L and H highlights the difficulty in reversing the addiction once individuals transition into these stages.

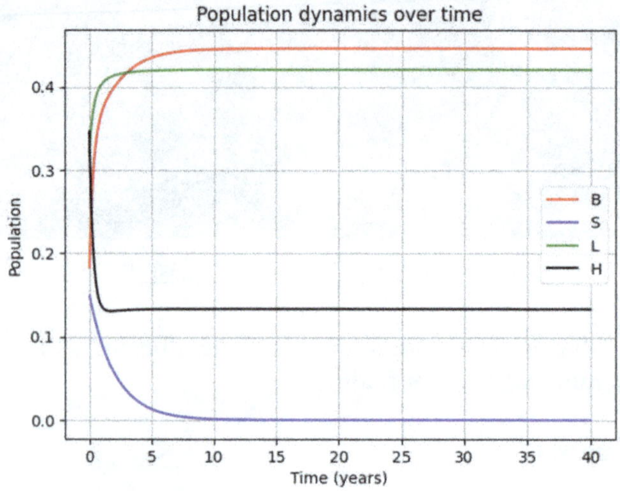

Fig. 7. Population Dynamics Over Time

4.2 Distribution of labels

The majority of the population falls within the Heavy Addiction (H) and Light Addiction (L) categories, with the Burnout (B) category also being significantly represented. The Susception (S) state, which indicates the lowest level of addiction, is the least represented.

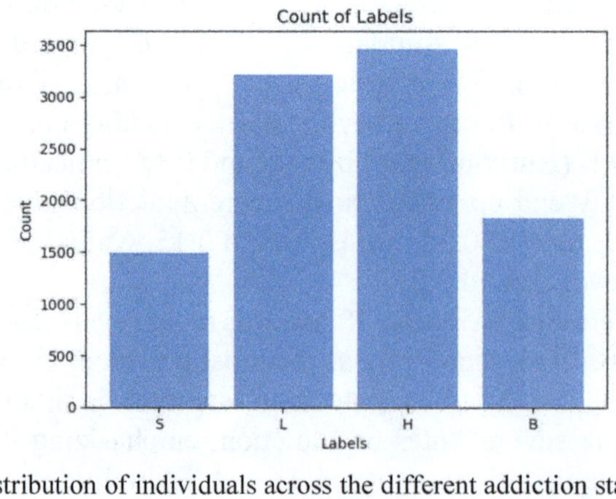

Fig. 8. Shows the distribution of individuals across the different addiction states (S, L, H, B) within the dataset.

This distribution reflects a worrying trend where most individuals have already progressed into moderate to severe levels of addiction. The low representation in the susceptible state indicates that the initial stages of addiction are short-lived, and individuals quickly move into higher addiction categories. The significant presence of individuals in the burnout state suggests that prolonged social media use may lead to severe psychological outcomes, further emphasizing the need for targeted intervention strategies.

4.3 ROC curve analysis

Figures 9 and 10 display the ROC curves for the Logistic Regression and SVM models, respectively. Both models exhibit strong classification performance, with AUC (Area Under the Curve) values ranging from 0.94 to 0.99. The Logistic Regression model slightly outperforms the SVM model, with a macro-average ROC AUC of 0.9732 compared to the SVM's 0.9515.

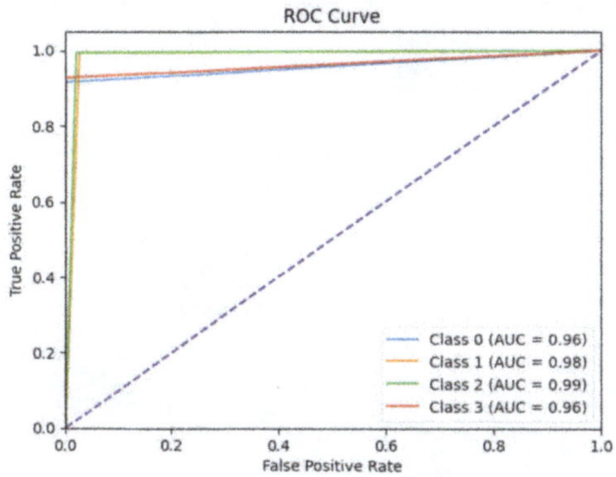

Fig. 9. ROC Curve Analysis for Logistic Regression

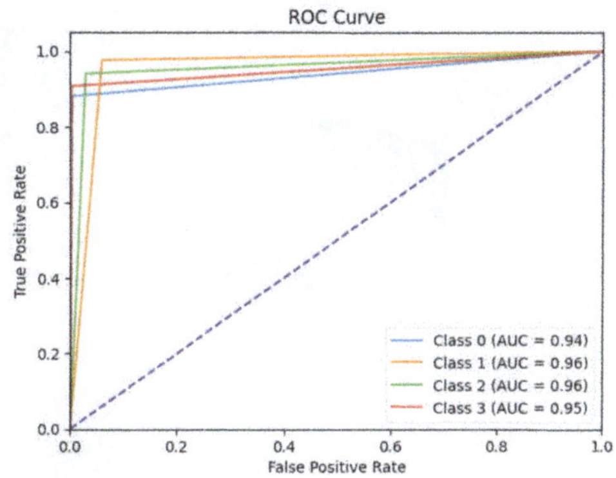

Fig. 10. ROC Curve Analysis for SVM

The high AUC values indicate that both models are highly effective in predicting the addiction state of individuals. The Logistic Regression model's macro-average ROC AUC of 0.9732 demonstrates its reliability in monitoring and diagnosing social media addiction, making it a robust choice for practical applications. The SVM model, with a macro-average ROC AUC of 0.9515, also performs well, though slightly less effectively than the Logistic Regression model. Overall, both models prove to be strong tools for classifying addiction levels, with the Logistic Regression model having a slight edge in performance.

4.4 Confusion matrix analysis

Figures 11 and 12 – These figures present the confusion matrices for the Logistic Regression and SVM models, respectively. The confusion matrices provide a detailed overview of the classification performance for both the Logistic Regression and SVM models. The Logistic Regression model achieved a high accuracy of 0.97, with a weighted precision of 0.9711. The majority of predictions correctly identified the addiction state, with minimal misclassifications primarily occurring between the Light Addiction (L) and Heavy Addiction (H) states. The SVM model also demonstrated strong performance, with an accuracy of 0.937 and a weighted precision of 0.9397. However, the SVM model showed slightly more misclassifications, particularly in distinguishing between the Heavy Addiction (H) and Light Addiction (L) states.

Both models are highly effective in predicting addiction states, as indicated by their confusion matrices. The Logistic Regression model slightly outperforms the SVM model in terms of accuracy and ROC AUC, suggesting it may be better suited for this application. The strong performance across both models underscores their potential utility in real-world scenarios for early detection and management of social media addiction.

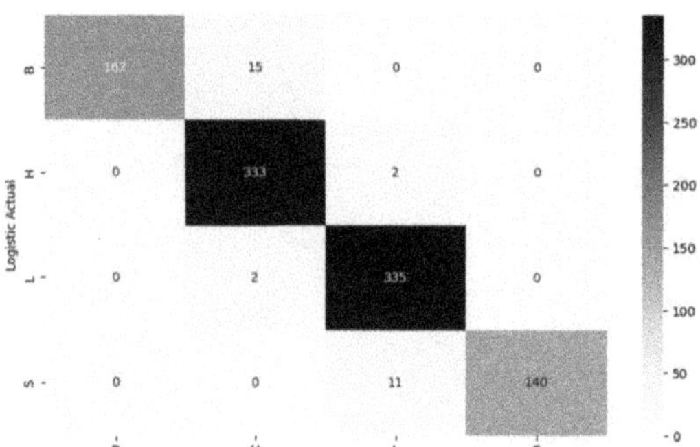

Fig. 11. Confusion Matrix Analysis for Logistic Regression

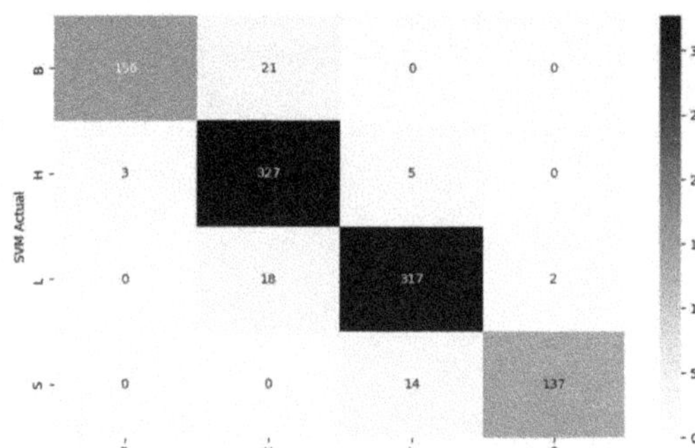

Fig. 12. Confusion Matrix Analysis for SVM

4.5 Addiction distribution

Figure 13 shows the percentage distribution of individuals in each addiction state within the population. The Heavy Addiction (H) category is the most prevalent, followed by Light Addiction (L), with Burnout (B) and Susceptible (S) being less common.

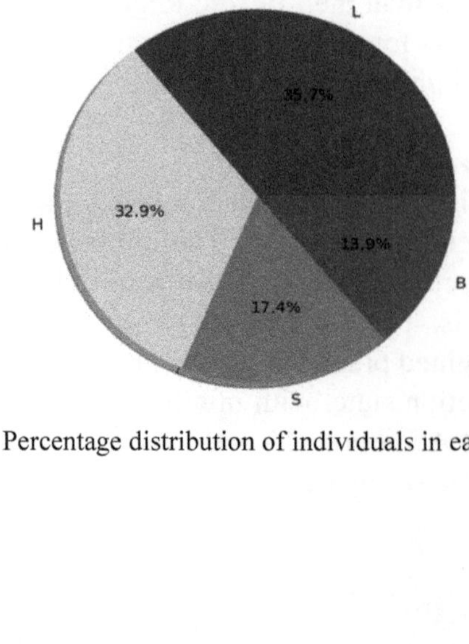

Fig. 13. Percentage distribution of individuals in each addiction state.

The addiction distribution highlights the severity of social media addiction within the population. The high prevalence of individuals in the Heavy Addiction State (H) suggests that a significant portion of the population is at high risk, requiring targeted interventions. The relatively high percentage of individuals in the Burnout state (B) is particularly concerning, as it indicates that a substantial number of users have progressed to a stage where the psychological impact of addiction may be profound. These findings underscore the need for comprehensive strategies to prevent and treat social media addiction.

5. Discussion and Conclusions

This study aimed to model and predict different stages of social media addiction using artificial intelligence (AI) techniques to identify individuals' addiction levels. The K-Means clustering analysis, followed by classification processes, successfully identified and categorized individuals' addiction states. The findings reveal that social media addiction has significant impacts on society, and these various stages of addiction can be distinctly separated.

AI played a crucial role in this study, particularly in parameter estimation. Analyzing large datasets and studying systems defined by dynamic differential equations can be time-consuming and complex using traditional methods. AI-based techniques overcome these challenges by accelerating the parameter estimation process and improving the accuracy of the results. The AI models used in this study enabled precise estimation of the parameters needed to model the underlying dynamic processes of social media addiction. This process was instrumental in accurately identifying and modeling addiction levels.

The ability to perform parameter estimation both quickly and accurately provides a significant advantage, especially when working with large datasets on dynamic systems. AI's effectiveness in this process allows for a better understanding and management of complex social issues like social media addiction. The K-Means clustering algorithm divided individuals into four main groups based on their social media addiction levels, proving to be an effective tool in determining addiction severity. The clustering results closely aligned with addiction states, strengthening the validity of this classification approach.

The subsequent classification processes were conducted using Logistic Regression and SVM models, both of which demonstrated high accuracy in predicting social media addiction. The Logistic Regression model achieved an impressive performance with a 97% accuracy rate and a macro-average ROC AUC value of 0.9732. The SVM model also showed strong performance with a 93.7% accuracy rate and a macro-average ROC AUC of 0.9515. These concrete results illustrate the power of AI-based approaches in analyzing complex social issues such as social media addiction.

The use of AI not only facilitated the identification of addiction levels but also provided opportunities to develop optimal control strategies for managing this addiction. These strategies help identify the optimal intervention points to manage social media use effectively and prevent the progression of addiction. Analyses conducted on dynamic systems demonstrate that the speed and accuracy of AI are crucial in developing such strategies.

The distribution of addiction states indicates that social media addiction is widespread across the population, with particularly high rates in the Heavy Addiction (H) and Burnout (B) categories. These findings highlight the potential for deepening psychological and social impacts of social media addiction, underscoring the need for intervention. Optimal control strategies could play a critical role in managing such addictions by helping to develop interventions that optimize timing and intensity to minimize addiction levels.

In conclusion, this study demonstrates the power of AI in addressing modern issues like social media addiction, providing practical solutions for identifying, managing, and developing optimal control strategies for this problem. AI has become an indispensable tool for researchers working on large datasets and dynamic systems, especially in terms of the speed and accuracy it provides in parameter estimation. The findings emphasize the importance of intervening in the early stages of social media addiction and suggest that preventive programs should be developed. Such programs, particularly tailored for younger individuals and those more prone to social media use, could prevent the progression to more severe stages of addiction.

In this context, the combined use of AI and optimal control strategies emerges as a powerful tool for monitoring, understanding, and managing social media addiction.

The high accuracy rates and ROC AUC values achieved by our study clearly demonstrate the effectiveness of these approaches in real-world applications. These strategies are crucial for reducing the negative impacts of social media use and helping individuals establish a healthier balance in the digital environment.

References

1. Simsek, A., Elciyar, K., and Kizilhan, T.: A comparative study on social media addiction of high school and university students. *Contemporary Educational Technology*, 10(2), 106–119 (2019).
2. Ali, A.S., Javeed, S., Faiz, Z., and Baleanu, D.: Mathematical modelling, analysis, and numerical simulation of social media addiction and depression. *PLoS One*, 19(3), e0293807 (2024).
3. Zara, M.C. and Monteiro, L.H.: The negative impact of technological advancements on mental health: An epidemiological approach. *Applied Mathematics and Computation*, 396, 125905 (2021).
4. Heo, J.: Epidemiological prediction using deep learning. *MS thesis*, Graduate School of UNIST (2020).

5. Van Velthoven, M.H., Powell, J., and Powell, G.: Problematic smartphone use: Digital approaches to an emerging public health problem. *Digital Health*, 4, 2055207618759167 (2018).
6. Sorhun, E.: *Machine Learning with Python*. Abakus Yayincilik, Istanbul (2021).
7. Kaur, I., Sandhu, A.K., and Kumar, Y.: Artificial intelligence techniques for predictive modeling of vector-borne diseases and its pathogens: A systematic review. *Archives of Computational Methods in Engineering*, 29(6), 3741–3771 (2022).
8. Karal, E. and Turan, M.: Hekime tanı koymada yardımcı, yapay zeka destekli hastalık tespit uzmanı. *Avrupa Bilim ve Teknoloji Dergisi*, 26, 100–116 (2021).
9. Pani, S.K., Dash, S., dos Santos, W.P., Bukhari, and S.A.C., Flammini, F. (Eds.): Assessing COVID-19 and other pandemics and epidemics using computational modelling and data analysis. Springer International Publishing (2022).
10. Chang, C.L. and Hsu, M.Y.: The study that applies artificial intelligence and logistic regression for assistance in differential diagnostic of pancreatic cancer. *Expert Systems with Applications*, 36(7), 10663–10672 (2009).
11. Nesa, M., Shaha, T.R., and Yoon, Y.: Prediction of juvenile crime in Bangladesh due to drug addiction using machine learning and explainable AI techniques. *Journal of Computational Social Science*, 5(2), 1467–1487 (2022).
12. Clement, J.C., Ponnusamy, V., Sriharipriya, K.C., and Nandakumar, R.: A survey on mathematical, machine learning, and deep learning models for COVID-19 transmission and diagnosis. *IEEE Reviews in Biomedical Engineering*, 15, 325–340 (2021).
13. Tessmer, H.L., Ito, K., and Omori, R.: Can machines learn respiratory virus epidemiology? A comparative study of likelihood-free methods for the estimation of epidemiological dynamics. *Frontiers in Microbiology*, 9, 343 (2018).
14. Schmidt, C.: Prediction of the influenza virus propagation by using different epidemiological and machine learning models. *PhD dissertation*, Hochschule Heilbronn (2019).
15. Rizvi, S.M.H.: Time series deep learning for robust steady-state load parameter estimation using 1D-CNN. *Arabian Journal for Science and Engineering*, 47(3), 2731–2744 (2022).
16. Wei, M., Wang, X., Fu, Y., Wang, D., Li, Y.: A review of LSSVM mathematical models and applications. In: *2024 5th International Conference on Mechatronics Technology and Intelligent Manufacturing (ICMTIM)*, 766–772. IEEE (2024).
17. Šušteršič, T., Blagojević, A., Cvetković, D., Cvetković, A., Lorencin, I., Šegota, S.B., Filipović, N.: Epidemiological predictive modeling of COVID-19 infection: Development, testing, and implementation on the population of the Benelux union. *Frontiers in Public Health*, 9, 727274 (2021).
18. Giraldo-Jiménez, C.F., Gaviria-Chavarro, J., Sarria-Paja, M., Bermeo Varón, L.A., Villarejo-Mayor, J.J., and Rodacki, A.L.F.: Smartphones dependency risk analysis using machine-learning predictive models. *Scientific Reports*, 12(1), 22649 (2022).
19. Afradi, A. and Ebrahimabadi, A.: Comparison of artificial neural networks (ANN), support vector machine (SVM), and gene expression programming (GEP) approaches for predicting TBM penetration rate. *SN Applied Sciences*, 2, 1–16 (2020).

20. Yang, S., Chen, H.C., Wu, C.H., Wu, M.N., and Yang, C.H.: Forecasting of the prevalence of dementia using the LSTM neural network in Taiwan. *Mathematics*, 9, 488 (2021).
21. Hassan, J., Saeed, S.M., Deka, L., Uddin, M.J., and Das, D.B.: Applications of machine learning (ML) and mathematical modeling (MM) in healthcare with special focus on cancer prognosis and anticancer therapy: Current status and challenges. *Pharmaceutics*, 16(2), 260 (2024).
22. Burkhardt, G.: Using machine learning to predict individual severity estimates of alcohol withdrawal syndrome in patients with alcohol dependence. *PhD dissertation*, LMU (2021).
23. Saleem, F., Al-Ghamdi, A.S.A.M., Alassafi, M.O., and AlGhamdi, S.A.: Machine learning, deep learning, and mathematical models to analyze forecasting and epidemiology of COVID-19: A systematic literature review. *International Journal of Environmental Research and Public Health*, 19(9), 5099 (2022).
24. Dhillon, A., Singh, A., Vohra, H., Ellis, C., Varghese, B., and Gill, S.S.: IoTPulse: Machine learning-based enterprise health information system to predict alcohol addiction in Punjab (India) using IoT and fog computing. *Enterprise Information Systems*, 16(7), 1820583 (2022).

Index

www.ingramcontent.com/pod-product-compliance
Lightning Source LLC
Chambersburg PA
CBHW080747281225
37357CB00006B/563
* 9 7 8 9 8 1 9 8 1 5 4 2 5 *